Parasitology Research Monographs

Volume 3

Series Editor:
Heinz Mehlhorn
Department of Zoomorphology
Cell Biology and Parasitology
Heinrich Heine University
Universitätsstrasse 1
40225 Düsseldorf
Germany

For further volumes:
http://www.springer.com/series/8816

Heinz Mehlhorn
Editor

Arthropods as Vectors of Emerging Diseases

Springer

Editor
Heinz Mehlhorn
Department of Zoomorphology
Cell Biology and Parasitology
Heinrich Heine University
Universitätsstrasse 1
40225 Düsseldorf
Germany

ISSN 2192-3671 ISSN 2192-368X (electronic)
ISBN 978-3-642-28841-8 ISBN 978-3-642-28842-5 (eBook)
DOI 10.1007/978-3-642-28842-5
Springer Heidelberg New York Dordrecht London

Library of Congress Control Number: 2012942852

Printed on acid-free paper

Springer is part of Springer Science+Business Media (www.springer.com)

Preface

As a result of their repeated exodus "out of Africa," the members of the species *Homo sapiens* settled finally all over the globe—even in very unpleasant biotopes. They used their growing brains—gifted with a droplet of divine illumination—to create a nearly unstoppable propagation, which now makes problems for seven billions of contemporaries in times of *globalization* and *global warming*. While the first process is surely man-made, the second hits the "human apprentices of magic" as victims, since only presumptuous individuals really may believe that humans as dwarfs in nature could stop climate processes that had thousand fold been repeated during the last billion of years in earth history.

Nevertheless the seven billions of humans exist and need food. Therefore a giant globalization process was initiated, transporting goods, animals, plants, and humans from one end of the earth to the other. This occurs now within one day, while in times of "good old pandemics" of cholera and plague it took weeks or months to travel from the "heart of darkness" into the supposed enlightened Europe or into "God's own country" of the Americas. Therefore daily confrontations occur at both ends of the world, when newcomers (plants, animals, bacteria, viruses) arrive in new regions, especially potential vectors of diseases, within which agents of disease may travel as stowaways and dangers during globalization. These vectors are the true masters of the world, where they live since more than 350 millions of years—unchanged and successfully swimming, flying, or walking by the help of name giving segmented legs (*Greek*: arthros = portion, segment).

These arthropods—spiders, mites, ticks, insects, and crustaceans—represent the animal phylum with the greatest number of described species (more than one million) as well as contain the greatest number of individuals when compared with any other group of animals including humans. These arthropods fully cover many levels in the pyramid of the food chain on earth. The disappearance in this peculiar "pyramid of life on earth" and their eventual lack of activity as destruents and decompositors of dead animals of any kind would have disastrous consequences.

However, on the other hand, insects, ticks, and mites may be involved in the transmission of microorganisms and parasites that threaten health and life of humans and animals. This occurs especially in those countries, where such diseases are freshly introduced as a consequence of the recent exorbitantly exaggerating globalization which is accompanied by the daily trips of millions of humans and daily transports of millions of containers with food and goods inside ships or planes reaching from one end of the world to the other. While in former times of rare contacts between continents, agents of diseases mostly remained limited to certain regions (where humans and animals had made "peace" with their agents of diseases or had died) and worldwide processing pandemics only seldom occurred. Nowadays outbreaks of diseases are explosive and run within days around the world. In countries, where these diseases had not yet been before, they were described as "emerging diseases," since their distribution starts at low levels, but—if the needed conditions of propagation are given—their expansion may speed up immediately, as e.g., it was the case of the so-called West Nile Fever virus that occurred—mosquito-transmitted—in 1999 and spread within 4 years all over the USA, and it expands its distribution sites even today.

Since bloodsucking insects, mites and ticks, and/or fecally dwelling insects come obligatorily into contact with agents of diseases of humans and animals and since they may have interchanging touchdowns on humans and on animals and/or their feces, they may transmit these agents mechanically by their feces not only by simple body contacts but also by bloodsucking. If agents of disease come from far away, they may meet potential local vectors or such relevant vectors are even imported bringing their "diseasing freight" with them. In both cases new endemic regions are opened for a propagation of diseases.

The present book reviews some of these diseases (*emerging or reemerging ones*) which are transmitted in various ways by arthropod vectors such as mosquitoes, flies, fleas, lice, bugs, and crustaceans and proposes means of control and prophylaxis. The future will show at what levels these diseases may emerge in the new countries. In any way the constant look at arthropods as potential vectors and their upmost control by old and new measurements are urgently needed.

Düsseldorf, January 2012 Heinz Mehlhorn

Acknowledgments

The quick and attractive publication of so many data is not possible without the help of many persons. Each group of authors has to thank heartily a broad spectrum of "helping ghosts" in their institutes. Mrs. Isabelle Mehlhorn and Mrs. Susanne Walter helped to arrange high-quality micrographs. We are indebted to Mrs. Dr. Koenig and Mrs. Lindqvist at Springer Publisher for their expertise to produce this book in an appreciable appearance.

Contents

Contributors

Jabbar Ahmed Veterinary Infection-Biology and Immunology, Research Center Borstel, Leibniz Center for Medicine and Biosciences, Borstel, Germany, jahmed@fz-borstel.de

Mohammed A. Bakheit Veterinary Infection-Biology and Immunology, Research Center Borstel, Leibniz Center for Medicine and Biosciences, Borstel, Germany

Carsten Balczun Group Zoology/Parasitology, Ruhr-Universität Bochum, Bochum, Germany, carsten.balczun@rub.de

T. Baldet Cirad, UMR Contrôle des Maladies, Montpellier, France

T. Balenghien Cirad, UMR Contrôle des Maladies, Montpellier, France

Burkhard Bauer Institute for Parasitology and Tropical Veterinary Medicine, Freie Universität Berlin, Berlin, Germany

Norbert Becker German Mosquito Control Association (KABS), Waldsee, Germany; University of Heidelberg, Heidelberg, Germany, norbertfbecker@web.de

Martin Beer Institute of Diagnostic Virology, Friedrich-Loeffler-Institut, Greifswald-Insel Riems, Germany

Markus Wilhelm Busch Medical Biodiversity and Parasitology, Biodiversity and Climate Research Centre (BiK-F), Frankfurt am Main, Germany; Senckenberg Gesellschaft für Naturforschung (SGN), Frankfurt am Main, Germany; Institute for Ecology, Evolution and Diversity, Goethe-University (GO), Frankfurt am Main, Germany, markus.busch@senckenberg.de

C. Cêtre-Sossah Cirad, UMR Contrôle des Maladies, Montpellier, France

Peter-Henning Clausen Institute for Parasitology and Tropical Veterinary Medicine, Freie Universität Berlin, Berlin, Germany, clausen.ph@vetmed.fu-berlin.de

Franz J. Conraths Institute of Epidemiology, Friedrich-Loeffler-Institut, Wusterhausen, Germany, franz.conraths@fli.bund.de

J.-C. Delécolle UdS, IPPTS, Faculté de Médecine, Strasbourg, France

Gerhard Dobler Department of Virology and Rickettsiology, Bundeswehr Institute of Microbiology, Munich, Germany, gerharddobler@bundeswehr.org

Michael Eschbaumer Institute of Diagnostic Virology, Friedrich-Loeffler-Institut, Greifswald-Insel Riems, Germany

Maike Förster Department of Zoology and Parasitology, Heinrich Heine University, Düsseldorf, Germany

Conrad Freuling Institute of Epidemiology, Friedrich-Loeffler-Institut, Wusterhausen, Germany

C. Garros Cirad, UMR Contrôle des Maladies, Montpellier, France

Falk Gestmann Department of Zoology and Parasitology, Heinrich Heine University, Düsseldorf, Germany

Jörn Gethmann Institute of Epidemiology, Friedrich-Loeffler-Institut, Wusterhausen, Germany

H. Guis Cirad, UMR Contrôle des Maladies, Montpellier, France

Bernd Hoffmann Institute of Diagnostic Virology, Friedrich-Loeffler-Institut, Greifswald-Insel Riems, Germany

Achim Kaiser German Mosquito Control Association (KABS), Waldsee, Germany

Helge Kampen Friedrich-Loeffler-Institut, Federal Research Institute for Animal Health, Greifswald-Insel Riems, Germany, helge.kampen@fli.bund.de

Sven Klimpel Medical Biodiversity and Parasitology, Biodiversity and Climate Research Centre (BiK-F), Frankfurt am Main, Germany; Senckenberg Gesellschaft für Naturforschung (SGN), Frankfurt am Main, Germany; Institute for Ecology, Evolution and Diversity, Goethe-University (GO), Frankfurt am Main, Germany, Sven.Klimpel@senckenberg.de

Astrid H. Kollien Group Zoology/Parasitology, Ruhr-Universität Bochum, Bochum, Germany

Matthias Kramer Institute of Epidemiology, Friedrich-Loeffler-Institut, Wusterhausen, Germany

Mandy Kronefeld Friedrich-Loeffler-Institut, Federal Research Institute for Animal Health, Greifswald-Insel Riems, Germany, mandy.kronefeld@fli.bund.de

Thomas Kuhn Medical Biodiversity and Parasitology, Biodiversity and Climate Research Centre (BiK-F), Frankfurt am Main, Germany; Senckenberg Gesellschaft für Naturforschung (SGN), Frankfurt am Main, Germany; Institute for Ecology, Evolution and Diversity, Goethe-University (GO), Frankfurt am Main, Germany

J. Languille Direction générale de l'alimentation, Ministère de l'Agriculture et de la Pêche, Paris, France

Abdalla A. Latif Parasites, Vectors and Vector-borne Diseases, Onderstepoort Veterinary Institute, University of Pretoria, Onderstepoort, South Africa

Julian Münster Medical Biodiversity and Parasitology, Biodiversity and Climate Research Centre (BiK-F), Frankfurt am Main, Germany; Senckenberg Gesellschaft für Naturforschung (SGN), Frankfurt am Main, Germany; Institute for Ecology, Evolution and Diversity, Goethe-University (GO), Frankfurt am Main, Germany

D. Martinez Cirad, UMR Contrôle des Maladies, Montpellier, France

B. Mathieu EID Méditerranée, Montpellier, France; UdS, IPPTS, Faculté de Médecine, Strasbourg, France

Heinz Mehlhorn Department of Zoology and Parasitology, Heinrich Heine University, Düsseldorf, Germany, mehlhorn@uni-duesseldorf.de

Dieter Mehlitz Institute for Parasitology and Tropical Veterinary Medicine, Freie Universität Berlin, Berlin, Germany

Christian K. Meiser Group Zoology/Parasitology, Ruhr-Universität Bochum, Bochum, Germany

Sabine Messler Medical Microbiology and Hospital Hygiene, Heinrich Heine University, Düsseldorf, Germany

S. Muders Department of Ecology and Parasitology, Zoology Institute, Karlsruhe Institute of Technology, Karlsruhe, Germany

Nicole Neuhausen Medical Microbiology and Hospital Hygiene, Heinrich Heine University, Düsseldorf, Germany

R. Oehme Landesgesundheitsamt Baden-Württemberg, Ref. 93, Allgemeine Hygiene und Infektionsschutz, Stuttgart, Germany

Sabine Petersdorf Medical Microbiology and Hospital Hygiene, Heinrich Heine University, Düsseldorf, Germany

T.N. Petney Department of Ecology and Parasitology, Zoology Institute, Karlsruhe Institute of Technology, Karlsruhe, Germany, petney@kit.edu

M. Pfäffle Department of Ecology and Parasitology, Zoology Institute, Karlsruhe Institute of Technology, Karlsruhe, Germany

Martin Pfeffer Institute of Animal Hygiene and Veterinary Public Health, University of Leipzig, Leipzig, Germany

Klaus Pfeffer Medical Microbiology and Hospital Hygiene, Heinrich Heine University, Düsseldorf, Germany, klaus.pfeffer@hhu.de

Björn Pluskota German Mosquito Control Association (KABS), Waldsee, Germany

Reiner Pospischil Bergheim Erft, Germany, reiner.pospischil@t-online.de

Carolina Probst Institute of Epidemiology, Friedrich-Loeffler-Institut, Wusterhausen, Germany

Francis Schaffner Vector Entomology Unit, Institute of Parasitology, University of Zürich, Zürich, Switzerland; Avia-GIS Agriculture and Veterinary Information and Analysis, Zoersel, Belgium

Günter A. Schaub Group Zoology/Parasitology, Ruhr-Universität Bochum, Bochum, Germany, guenter.schaub@rub.de

Ulrike Seitzer Veterinary Infection-Biology and Immunology, Research Center Borstel, Leibniz Center for Medicine and Biosciences, Borstel, Germany

M.-L. Setier-Rio EID Méditerranée, Montpellier, France

Kai Sievert Novartis Animal Health Inc., Basel, Switzerland, kai.sievert@novartis.com

J. Skuballa Department of Ecology and Parasitology, Zoology Institute, Karlsruhe Institute of Technology, Karlsruhe, Germany

Christoph Staubach Institute of Epidemiology, Friedrich-Loeffler-Institut, Wusterhausen, Germany

A. Tran Cirad, UMR AGIRS, Montpellier, France

Zati Vatansever Department of Parasitology, Kafkas University, Kars, Turkey, zativet@gmail.com

R. Venail EID Méditerranée, Montpellier, France, rvenail@eid-med.org

Doreen Werner Leibniz-Centre for Agricultural Landscape Research, Müncheberg, Germany, doreen.werner@zalf.de

C. Zetlmeisl Department of Ecology and Parasitology, Zoology Institute, Karlsruhe Institute of Technology, Karlsruhe, Germany

Chapter 1
Culicid Mosquitoes as Vectors of Disease Agents in Europe

Helge Kampen, Mandy Kronefeld, and Doreen Werner

Abstract The chikungunya fever epidemic in northern Italy in 2007 and the recent demonstration of the introduction of various exotic mosquito species alerted the European nations to a growing risk of invasion by vector mosquitoes and outbreaks of mosquito-borne infectious diseases. The importation of both mosquitoes and pathogens they are able to transmit is facilitated by increasing international trade and travel. Environmental and climatic changes do not only set the ground for the establishment of invasive mosquitoes away from their natural distribution areas but often also have beneficial effects on indigenous potential vector species, such as support of spread, growth in population density and extension of seasonal activity, thus increasing the probability of these biting a person or a reservoir animal infected with a mosquito-borne pathogen and transmitting it. While there is a considerable body of literature on invasive mosquito species and imported pathogens due to their relevance in their natural distribution areas, data on endemic mosquito species and mosquito-borne pathogens circulating in Europe are relatively scarce. With a few exceptions, these have in fact for several decades been of minor importance with respect to public health impact. Both the role of mosquitoes as vectors of disease agents and the scientific and political attention to mosquito-borne diseases, however, appear to be growing in Europe with ongoing globalization. We here discuss indigenous mosquito species that have historically been involved in pathogen transmission in Europe or have been demonstrated to be vectors elsewhere and

H. Kampen (✉) • M. Kronefeld
Friedrich-Loeffler-Institut, Federal Research Institute for Animal Health, Südufer 10, 17493 Greifswald-Insel Riems, Germany
e-mail: helge.kampen@fli.bund.de; mandy.kronefeld@fli.bund.de

D. Werner
Leibniz-Centre for Agricultural Landscape Research, Eberswalder Str. 84, 15374 Müncheberg, Germany
e-mail: doreen.werner@zalf.de

H. Mehlhorn (ed.), *Arthropods as Vectors of Emerging Diseases*,
Parasitology Research Monographs 3, DOI 10.1007/978-3-642-28842-5_1,

that may contribute to future disease outbreaks in Europe. Likewise, we present mosquito-borne pathogens that have been circulating in Europe or are considered probable to be introduced and established in the future.

Keywords Europe • Filariasis • Globalization • Malaria • Mosquito-borne pathogens • Mosquitoes • Viral diseases

1.1 Introduction

A widespread phenomenon of our modern times is the unprecedentedly quick and massive transportation of people, animals and goods all around the world (Tatem et al. 2006a). Not comparable in extent to historical times, the recent past has shown that some plants and animals haphazardly follow the global transportation routes and are increasingly imported into regions of the world that do not belong to their natural distribution areas. Provided that they find adequate ecological conditions, such as climate and feeding sources, they may succeed in establishing and spreading. Among these organisms are haematophagous arthropods including potential vectors of disease agents as well as the disease agents themselves (Tatem et al. 2006a, b). They are able to travel within shortest times to most distant places of the world, be it within a reservoir host, a subclinically infected vertebrate or a vector arthropod. At their place of destination, they may be transmitted, either by imported or by indigenous vector arthropods, and cause cases or outbreaks of disease. The chikungunya fever epidemic in Italy in 2007 (Angelini et al. 2008) is an example of transmission of imported pathogens by imported vectors, and the bluetongue disease outbreak in central Europe in 2006 (Kampen and Werner 2010) is an example of transmission of imported pathogens by indigenous vector-competent arthropods. Depending on vector–pathogen adaptation, imported vectors may of course also boost the circulation of pathogens already present in Europe which may not be spread very efficiently by endemic haematophagous arthropods.

Among the vector-borne diseases, mosquito-borne diseases have an outstanding global public health relevance. Primarily in the tropics, each year millions of people suffer and even die from malaria, dengue fever, yellow fever, Japanese encephalitis and lymphatic filariasis (Tolle 2009), just to mention the most important diseases caused by pathogens transmitted by mosquitoes. Historically, malaria, dengue fever and yellow fever did also occur in Europe; dengue and yellow fevers in subtropical Mediterranean regions (Hubálek 2008) and malaria much more widespread both in subtropical and in temperate regions as far north as Scandinavia (Bruce-Chwatt and de Zulueta 1980; Huldén et al. 2005). As numbers of persons returning from overseas travels to Europe with infections by mosquito-borne disease agents rise steadily (e.g. Odolini et al. 2011), concern increases that novel diseases emerge or diseases once eradicated resurge. This concern is additionally nourished by the ongoing import, establishment and spread of exotic mosquito species that are very efficient vectors and of pathogens that may encounter vector-competent indigenous mosquitoes. However, little up-to-date knowledge is available on the vector potency of indigenous European mosquito species which have made an appearance as vectors only exceptionally, both in space and time, during the last decades. Also, despite the occasional direct or indirect demonstration of mosquito-borne pathogens both in vertebrates and in the vectors, there is little data on the present circulation of these in Europe. This contribution will give an overview for Europe on important endemic mosquito vector species, on mosquito-borne pathogens already demonstrated to circulate or considered as risks for the near future and on recent outbreaks or cases of diseases caused by mosquito-borne pathogens.

1.2 Endemic Potential Mosquito Vector Species

Approximately 100 species of the dipteran family Culicidae occur in Europe (Snow and Ramsdale 1999). The females of most of them depend on feeding on blood for producing eggs, although in autogenous species not necessarily right from the first reproduction cycle, and thus might principally be potential vectors of disease agents, just like any other haematophagous arthropod. Whether a mosquito species indeed becomes a vector or not depends both on its innate vector competence for a certain pathogen and on various exogenous factors (Kampen 2009). However, there is arguably no mosquito species not able to transmit a pathogen, although probably with varying efficiency. Thus, comparable epidemiological impact with regard to pathogen transmission may be achieved by efficient vectors occurring in relatively low or moderate densities and less efficient vectors occurring in high densities. Generally, data are scarce, and lots of species have not been examined at all with regard to their theoretical vector competence and practical vector role.

European mosquito species involved in pathogen transmission belong to the culicid genera *Aedes*, *Anopheles*, *Culex*, *Culiseta*, *Ochlerotatus* and *Coquillettidia*, with the most important vectors being *Aedes*, *Anopheles*, *Culex* and *Ochlerotatus* species. According to their preferred breeding sites, these may be categorized ecologically into floodwater mosquitoes, woodland mosquitoes, salt-marsh mosquitoes and house mosquitoes. A few examples of important species of each of these groups will subsequently be discussed, however, with the malaria mosquito species of the genus *Anopheles* as a separate group.

1.2.1 Anopheles *Species*

The terms "malaria mosquito" and "fever mosquito" are used for mosquito species of the genus *Anopheles* and originate from some of them functioning as vectors of malaria parasites. Because the anopheline mosquitoes were in the focus of European research until the middle of the twentieth century when malaria was still endemic, the data available on the indigenous species are probably more numerous than those on most other indigenous mosquito species, although not necessarily up to date. *Anopheles* species have also been recognized in Europe as vectors of filariae and viruses.

Table 1.1 *Anopheles* species in Europe and their vector competence/efficiency for human pathogenic plasmodia

Species	Geographic distribution	Vector competence/ efficiency
An. atroparvus	Central and southern Europe	+++
An. labranchiae	Italy, Corsica, Balkans	+++
An. sacharovi	Balkans, Greece	+++
An. maculipennis	Central and southern Europe	++
An. claviger	Central and southern Europe	++
An. superpictus	Southeastern Europe	++
An. messeae	Whole Europe	++
An. algeriensis	Southern Europe	+
An. plumbeus	Central and southern Europe	+
An. sergentii	Pantelleria (island south of Sicily)	+
An. beklemishevi	Scandinavia, northern Russia	(+)
An. melanoon	Italy, southern France, Iberian Peninsula	(+)
An. subalpinus	Southern Europe	(+)
An. cinereus	Iberian Peninsula	(+)
An. hyrcanus	Southern Europe	(+)
An. marteri	Southern Europe	(+)
An. multicolour	Southern Spain	?
An. petragnani	Italy, southern France, Iberian Peninsula	–

+++: highly efficient, ++: moderately efficient, +: inefficient, (+): only exceptionally, ?: questionable, –: vector incompetent

Of the 18 *Anopheles* species occurring in Europe (Ramsdale and Snow 2000), at least three were most efficient vectors of human pathogenic plasmodia (Table 1.1): *An. atroparvus*, *An. labranchiae* and *An. sacharovi*. The latter two occurred, and still occur, in southern Europe, whereas the distribution area of *An. atroparvus* covers both southern and central Europe. Of mediate efficiency in transmitting the malarial agents were *An. maculipennis* and *An. claviger* in southern and central Europe, *An. superpictus* in southeastern Europe and *An. messeae* (and/or *An. daciae*, as a newly recognized cryptic member of the Maculipennis complex; Nicolescu et al. 2004; Linton et al. 2005) in the whole of Europe. Further *Anopheles* species with less transmission efficiency, no vector competence or unknown vector status are *An. algeriensis*, *An. plumbeus*, *An. sergentii*, *An. beklemishevi*, *An. melanoon*, *An. subalpinus*, *An. cinereus*, *An. hyrcanus*, *An. marteri* and *An. multicolor* (Table 1.1).

Despite its described limited vector efficiency, *An. plumbeus* has recently been shown to be able to develop *P. falciparum* oocysts and infectious sporozoites, respectively (Marchant et al. 1998; Eling et al. 2003). This species was also made responsible for two autochthonous malaria cases in Germany in 1997 (Krüger et al. 2001). Being originally a tree-hole breeder (Blacklock and Carter 1920), *An. plumbeus* appears to have recently adapted to other types of breeding sites and increasingly often colonizes cesspits and other kinds of underground water containers with high organic loads which allow mass development (Becker 2008).

Maculipennis complex mosquitoes which were not further identified to species were also found to be infected with West Nile virus in Portugal (Filipe 1972) and appear to be involved in *Dirofilaria* transmission in Italy (Cancrini et al. 2006). *An. claviger* is suspected to be able to transmit Ťahyňa virus (Pchelkina and Seledtsov 1978).

A challenge in *Anopheles* research is species differentiation between complex or sibling species, such as those of the Maculipennis and the Claviger complexes which are represented by eight and two members in Europe, respectively. These are very closely related species that are isomorphic or hardly distinguishable by morphological features in the larval, pupal and female adult stages. Yet, they are true species with characteristic biological traits, including vector competence. While species identification by egg structure and colouration was a more or less reliable standard technique for decades, with cytogenetics and zymotaxonomics as additional methods of certain relevance (Kampen 2004), DNA sequence-based PCR assays have meanwhile been introduced and become widely accepted (Proft et al. 1999; Kampen et al. 2003a, b; Kampen 2005).

1.2.2 Floodwater Species

Vector competence provided, floodwater species often play important roles in the epidemiology of mosquito-borne viral diseases. They breed in temporary water pools along rivers and other water bodies with fluctuating surface levels. Most of them are *Aedes* and *Ochlerotatus* species which oviposit on moist substrates shortly above the water surface. When the water levels rise high enough for the larvae to hatch, they tend to produce mass populations, thus allowing an intense circulation of pathogens. Moreover, their breeding sites are identical with the resting places of migratory or resident birds that are in many cases the reservoirs of viruses. Most of the floodwater species are aggressive biters and, although preferentially ornithophilic, readily feed on humans and other vertebrates, thereby serving as bridge vectors. As they are good flyers, they may cover considerable distances and contribute to pathogen dispersal.

Widely spread European floodwater species are *Ae. vexans*, *Ae. cinereus* and *Oc. sticticus*, all of which have been found infected with one or the other pathogenic virus or have even been demonstrated to be able to transmit these viruses. Thus, *Ae. vexans* may be involved in Ťahyňa and Rift Valley fever virus transmission (Simková et al. 1960; Turell et al. 2008), *Ae. cinereus* in Sindbis and Ťahyňa virus transmission (Danielová et al. 1976; Turell et al. 1990) and *Oc. sticticus* in Ťahyňa and West Nile virus transmission (Danielová and Holubová 1977; Andreadis et al. 2004).

1.2.3 Salt-Marsh Mosquitoes

Salt-marsh mosquitoes are a particular kind of floodwater mosquitoes and are similar in ecology and behaviour. They are tolerant, but not restricted, to brackish water. Floodplains in coastal areas and saltwater marshes often produce masses of individuals of these species. *Ochlerotatus caspius* may regionally be an annoying salt-marsh species which has been found infected with West Nile and Ťahyňa virus in the field (Hannoun et al. 1966; Pilaski 1987).

1.2.4 Woodland Mosquitoes

Breeding sites of woodland mosquito species include shallow ponds and swamp pools that develop after snowmelt and heavy rainfalls, as well as stagnant ditches, rock pools and tree holes. Usually these species have only one to two generations per year, and they are moderate flyers. Nevertheless, they appear to participate in the transmission of various pathogens. Frequent woodland species are, for instance, *Oc. cantans* which has been implicated in West Nile virus transmission (Labuda et al. 1974), *Oc. communis* which has been implicated in Sindbis, Batai and Inkoo virus transmission (Brummer-Korvenkontio et al. 1973; Lvov et al. 1984; Francy et al. 1989) and *Cx. modestus* which has been implicated in West Nile, Sindbis, Ťahyňa and Lednice virus transmission (Hannoun et al. 1964; Chippaux et al. 1970; Malková et al. 1974; Lvov et al. 1979).

1.2.5 House Mosquitoes

The banded house mosquito *Cs. annulata* is quite common and readily attacks humans. Although described as a vector of Ťahyňa virus (Danielová 1972), it is not comparable in public health relevance to the members of the Pipiens complex which play crucial roles in keeping natural viral transmission cycles among avian hosts going and as bridge vectors of viruses from birds to humans and other mammals.

The Pipiens complex in Europe is believed to be represented by *Cx. pipiens pipiens* including the two forms *Cx. pipiens pipiens* biotype *pipiens* and *Cx. p. pipiens* biotype *molestus*, as well as hybrids of these forms, and *Cx. torrentium* (Harbach et al. 1985; Becker et al. 2010; Reusken et al. 2010). The taxonomic affiliation of the latter is controversial (Farajollahi et al. 2011), but this species will be referred to as another member of the Pipiens complex here because of similar morphology and behaviour (Smith and Fonseca 2004).

Worldwide, mosquitoes of the Pipiens complex play key roles as vectors of many human and animal disease agents, including viruses, filarial worms and avian malaria parasites (Bailey et al. 1978; Balenghien et al. 2008; Smith and Fonseca 2004). In Europe, they are the principal vectors of human zoonotic pathogens with avian reservoir hosts: West Nile virus, Sindbis virus and Usutu virus (Francy et al. 1989; Savage et al. 1999; Weissenböck et al. 2007).

Members of the Pipiens complex are of special interest because there are no reliable morphological characteristics for discrimination, with the exception of details of the male genitalia (e.g. Onyeka 1982), but notable differences in biology (e.g. Byrne and Nichols 1999). For example, *C. p. pipiens* biotype *pipiens* requires a blood meal to produce the first egg raft (anautogeny), needs extensive space for mating (eurygamy), has a reproductive diapause in winter (heterodynamy) and has the affinity to feed on birds (ornithophily), while females of *Cx. p. pipiens* biotype *molestus* lay the first egg rafts without taking a blood meal (autogeny), are able to mate within a confined space without swarming (stenogamy), do not diapause (homodynamy) and preferably feed on mammals (mammophily).

According to blood meal investigations, it is hypothesized that North American hybrids of *Cx. p. pipiens* biotype *pipiens* and *Cx. p. pipiens* biotype *molestus* are indiscriminant biters without pronounced feeding preferences which act as bridge vectors of West Nile virus and other viruses from infective birds to humans and other mammals (Fonseca et al. 2004; Kilpatrick et al. 2007). An emerging occurrence of hybrids has therefore the potential to shift the dynamics of disease outcomes, especially for West Nile fever (WNF; Fonseca et al. 2004). In Europe, hybridization has so far only been observed in Portugal, southern France and the Netherlands (Fonseca et al. 2004; Gomes et al. 2009; Reusken et al. 2010).

Culex torrentium occurs in the same habitats and has a similar biology as *Cx. pipiens*. It is a proven vector of Sindbis virus in Sweden (Francy et al. 1989). Although *Cx. torrentium* has been reported from several European countries, only scattered information is available on its precise distribution, abundance and role as vector of viruses in Europe since it is often not discriminated from or confused with *Cx. pipiens*.

The ecological and behavioural variations between the sibling species and biotypes within the Pipiens complex may have an impact on their epidemiological significances and are important for the understanding of disease outcomes (Farajollahi et al. 2011). Thus, accurate species identification is of paramount relevance. To facilitate this, various methodologies including species-specific PCR assays targeting mitochondrial and nuclear genomes have been developed (e.g. Bahnck and Fonseca 2006; Hesson et al. 2010).

1.2.6 Invasive Mosquito Species

For several years, detections of invasive mosquito species have become increasingly frequent in Europe. Due to globalization and environmental changes, mosquitoes are accidentally imported by international trade and travel and sometimes succeed in establishing and spreading at their place of importation. These "newcomer mosquitoes," such as *Aedes albopictus*, *Ae. japonicus*,

Ae. aegypti, *Ae. atropalpus* and *Ae. koreicus* with efficient vectors of disease agents among them, are dealt with in a separate contribution to this volume.

1.3 Mosquito-Borne Pathogens and Associated Disease

Culicid mosquitoes collected in Europe have been found infected with several groups of disease agents including filariae, protozoa and viruses (Table 1.2). In addition, various mosquito-borne pathogens have been demonstrated to circulate in Europe by serological evidence or by clinical disease. The pathogens found circulating in Europe as well as Rift Valley fever virus, which is feared to be introduced into Europe, will be subsequently discussed.

The only known bacterial disease agent, *Francisella tularensis*, regarded by some authors (e.g. Olin 1942; Hanke et al. 2009; Triebenbach et al. 2010; Lundström et al. 2011) to be occasionally transmitted by mosquitoes is excluded because of negligible relevance of this pathogen as compared to other infection paths.

1.3.1 Dirofilaria

Dirofilariasis is predominantly a veterinary problem. Two species of dirofilariae, *Dirofilaria immitis* and *D. repens*, are widely distributed in southern Europe where they cause heartworm disease and tissue nodules, respectively, in dogs, cats and other carnivores. Humans are inadvertent and unsuitable hosts where the parasites usually do not develop to the mature and reproductive adult stage. Instead, in the case of human infection, the worms are generally found encapsulated in subcutaneous, pulmonary and intraorbital nodules induced by the immune defence (McCall et al. 2008; Simón et al. 2009).

Various mosquito species of the genera *Aedes*, *Anopheles*, *Culex* and *Ochlerotatus* take part in the transmission of dirofilariae in Europe (Pampiglione et al. 1995). In endemic areas in Italy, the most important vectors are probably *Oc. caspius* and *Ae. vexans* (Gratz 2004), but the newly introduced species *Ae. albopictus* also appears to play an important role in the epidemiology of dirofilariasis (Cancrini et al. 2003).

The highest prevalences of human dirofilariasis are to be found in Italy (66%), France including Corsica (22%), Greece (8%) and Spain (4%), but many cases are probably not diagnosed or notified (Pampiglione et al. 1995; Raccurt 1999). Notified human cases of *D. repens* infections have substantially increased in number during the past two decades or so in the Mediterranean (Pampiglione et al. 1995; Muro et al. 1999; Pampiglione and Rivasi 2000; Simón et al. 2005). Increasing travel seems to favour its emergence outside known endemic areas, and it is possible that both *D. immitis* and *D. repens* have spread to the south of Switzerland (Bucklar et al. 1998). Even more northerly, in the Netherlands and Germany, single autochthonous cases of *D. repens* infections in dogs were recently reported (Hermosilla et al. 2006; Overgaauw and Van Dijk 2009).

Table 1.2 Mosquito-borne viruses of human and veterinary importance that have been circulating in Europe or must be expected to be imported to Europe in the future (according to Lundström 1999, Hubálek 2008, and various other references)

Virus	Established/indigenous proven or suspected vector species	Human relevance/ symptoms	Veterinary relevance	Occurrence in Europe
(Probably) circulating in Europe				
West Nile virus (WNV)	*Cx. pipiens* *Cx. modestus* *Cq. richardii* *Oc. cantans* *Ae. albopictus* *An. maculipennis* s.l.	Fatal cases possible	Fatalities among horses, birds (esp. corvids, raptors)	Intermittently since 1962
Sindbis (Ockelbo) virus (SINV)	*Cx. pipiens* *Cx. torrentium* *Cx. modestus* *Cs. morsitans* *Cq. richardii* *Oc. communis* *Oc. excrucians* *Ae. cinereus* *Ae. albopictus* *An. hyrcanus* *An. maculipennis* s.l.	Influenza-like symptoms	Sporadic illness in birds, irregular deaths in old chickens	Intermittently since 1971
Ťahyňa virus (TAHV)	*Cx. pipiens* *Cx. modestus* *Cq. richardii* *Cs. annulata* *Oc. cantans* *Oc. caspius* *Oc. communis* *Oc. excrucians* *Oc. flavescens* *Oc. punctor* *Oc. sticticus* *Ae. cinereus* *Ae. hyrcanus* *An. maculipennis* s.l. *An. vexans*	Influenza-like symptoms	Unknown	Intermittently since 1958
Batai (Čalovo) virus (BATV)	*Cq. richardii* *Oc. communis* *Oc. punctor* *An. claviger* *An. maculipennis* s.l.	Influenza-like symptoms	Mild illness among sheep and goats	Intermittently since 1960
Inkoo virus (INKV)	*Oc. communis* *Oc. sticticus* *Oc. punctor* *Oc. hexodontus*	Sporadic, influenza-like symptoms	Unknown	Northern Europe since 1964
Lednice virus (LEDV)	*Cx. modestus*	Unknown	Unknown	Czech Republic 1972

(continued)

Table 1.2 (continued)

Virus	Established/indigenous proven or suspected vector species	Human relevance/ symptoms	Veterinary relevance	Occurrence in Europe
Usutu virus (USUV)	*Cx. pipiens* *Cx. hortensis* *Cx. territans* *Cs. annulata* *Ae. albopictus* *Ae. rossicus* *Ae. vexans*	Under discussion	High mortality among passerine birds and raptors	Various countries since 2001
At risk of being introduced				
Chikungunya virus (CHIKV)	*Ae.albopictus* *Ae.aegypti*	Fever, arthralgia	Unknown	Outbreak in Italy 2007 Autochthonous cases in France 2010
Dengue virus (DENV)	*Ae. aegypti* *Ae. albopictus*	Fatal cases	Unknown	Epidemic in Greece 1927–1928 Autochthonous cases in France and Croatia 2010
Yellow fever virus (YFV)	*Ae. aegypti*	Fatal cases	Unknown	Southern Europe in eighteenth and nineteenth centuries
Rift Valley fever virus (RVFV)	*Cx. perexiguus* *Cx. pipiens* *Cx. theileri* *Cx. tritaeniorhynchus* *Ae. albopictus* *Ae. caspius* *Ae. detritus* *Ae. vexans*	Fatal cases	High mortality among ruminants	Not yet emerged (historical distribution: Africa, Middle East)

The reasons for the observed geographic expansion of *Dirofilaria* worms are not clear, but climate change, the spread of vectors and increased uncontrolled movement of dogs and cats through Europe are supposed to be important driving forces (Genchi et al. 2005, 2009).

1.3.2 Malaria Parasites

Malaria was endemic in Europe until about the mid-twentieth century (Bruce-Chwatt and de Zulueta 1980). It did not only occur in the subtropical Mediterranean regions but as far north as Scandinavia (Huldén et al. 2005). *Plasmodium vivax*, *P. malariae* and, less commonly, *P. falciparum* were the malarial agents circulating in Europe. After centuries of human and parasitic coexistence (Reiter 2000), malaria was on the decline since the nineteenth century when the intensification of agriculture and livestock farming resulted in the large-scale drainage of swamps, marshes and other wetlands, thus substantially reducing *Anopheles* breeding sites. A general improvement in living and hygienic conditions as well as the spatial separation of human dwellings and livestock keeping facilities contributed to a further reduction of humans being exposed to mosquito bites. On a pan-European scale, however, malaria disappeared only as a consequence of the additional administration of efficient drugs (chloroquine) and insecticides (DDT) developed in the early twentieth century decades (Alten et al. 2007).

While the WHO declared malaria eradicated from Europe in the 1970s, isolated locally acquired cases continued to occur. Indigenous *Plasmodium* strains that had evolved in parallel to their European vector mosquitoes, however, appeared to have become extinct. Subsequent autochthonous malaria cases, plenty of which were recorded in continental Europe in the past 20 years or so (Sartori et al. 1989; Nikolaeva 1996; Baldari et al. 1998; Krüger et al. 2001; Cuadros et al. 2002; Kampen et al. 2002; Zoller et al. 2009; Santa-Ollala Peralta et al. 2010; Danis et al. 2011a), were most likely due to transmission of parasites imported with their human hosts from overseas endemic areas. Presumably due to missing coadaptation, experimental infection of European *Anopheles* species with tropical *Plasmodium* strains usually failed to produce infective parasite stages (Shute 1940; Ramsdale and Coluzzi 1975; Daškova and Rasnicyn 1982). In more recent studies, *An. plumbeus* developed *P. falciparum* oocysts and sporozoites, respectively, though (Marchant et al. 1998; Eling et al. 2003).

Given the still increasing travel activities to and from tropical countries with endemic malaria and a global warming scenario that allows both a higher reproduction rate of the mosquitoes, resulting in higher population densities, and a shorter developmental time of the malaria parasites in the mosquito (extrinsic incubation period) with rising temperatures, an increase in autochthonous malaria cases in Europe must be expected for the future. However, due to the high health standards, the lack of a zoonotic reservoir of the human plasmodia and their relatively short incubation period that makes disease cases quickly apparent, it is very unlikely that malaria ever becomes endemic again in central Europe.

1.3.3 West Nile Virus

Although West Nile virus (WNV; Flaviviridae, *Flavivirus*, Japanese encephalitis antigenic complex) was reported from the French Rhone delta already in the early 1960s (Hannoun et al. 1964) and Europe has henceforth faced occasional WNF epidemics, WNF is commonly considered emergent. This is probably due to the fact that, since the middle of the 1990s, outbreaks tend to occur increasingly often and more regularly present with severe neurological manifestations (Gubler 2007). While the majority of the majority remains asymptomatic, 20–40% of infected persons may develop clinical illness with about 1% displaying encephalitis, meningitis and flaccid paralysis (Kramer et al. 2003). Of the neuroinvasive cases, a substantial percentage will suffer from long-term disability or even die (Klee et al. 2004).

In addition to humans, equines and corvid birds are particularly susceptible to the virus. Numerous fatal cases of disease and several outbreaks among humans and equines with high morbidities and mortalities were observed during the past 15 years or so in various countries of Europe, for example, Romania in 1996 (Tsai et al. 1998), Russia in 1999 (Platonov et al. 2001), France in 2000 (Murgue et al. 2001), Italy in 2002, 2008 and 2009 (Autorino et al. 2002; Macini et al. 2008; Barzon et al. 2009; Rizzo et al. 2009; Angelini et al. 2010) and Greece in 2010 and 2011 (Danis et al. 2011b, c).

WNV is widely distributed in Africa and Eurasia where a lot of mosquito species have been found infected (Hubálek and Halouzka 1999). Principal vectors in Europe are *Cx. pipiens*, *Cx. modestus* and *Cq. richardii* (Reiter 2010), with *Cx. p. pipiens* biotype *molestus* being the most important bridge vector transmitting the virus from birds to humans (Kilpatrick et al. 2005; Hamer et al. 2008, 2009). Transovarial transmission of WNV has been observed, for instance, in *Ae. albopictus* and *Cx. pipiens* (Baqar et al. 1993; Goddard et al. 2003), and overwintering was documented to occur in *Cx. pipiens* (Nasci et al. 2001).

Vertebrate reservoir hosts of WNV are mainly wild birds, with most of the species displaying no clinical signs after infection (Malkinson and Banet 2002). Migratory birds seem to play a particular role in virus transportation over long distances, for example, from Africa to the European continent (Rappole et al. 2000). WNF outbreaks are often connected to wetlands where huge numbers of migratory and non-migratory birds aggregate and come into contact with myriads of mosquitoes. In addition to this rural transmission cycle where amplification of the virus occurs and ornithophilic mosquitoes are involved, an urban cycle with less host-specific mosquito vector species appears to exist (Hubálek and Halouzka 1999; Koopmans et al. 2007).

While only exceptionally WNF has been diagnosed and virus could be isolated north of about latitude 49° in Europe (Hubálek 2008), there is evidence of more widespread virus circulation, for example, in Great Britain and Germany (Buckley et al. 2006; Linke et al. 2007), from serological studies on resident birds.

1.3.4 Dengue Virus

Contrasting the common assumption that dengue virus (DENV; Flaviviridae, *Flavivirus*, dengue antigenic complex) is associated with tropical regions, it used to be endemic in southern Europe as long as *Ae. aegypti* was present. In 1926–1927, DENV caused a major outbreak in Greece with more than 1,000 deaths due to dengue haemorrhagic fever among an estimated total of close to one million disease cases (Cardamatis 1929; Rosen 1986). With a low mortality rate, dengue fever probably also occurred in Spain at the beginning of the eighteenth century (Eritja et al. 2005). It disappeared from Europe together with its vector, and both here and elsewhere the absence of dengue epidemics in former endemic regions has been ascribed to successful mosquito/*Ae. aegypti* control (Gubler 1989).

On a global scale, dengue has been spreading and numbers of cases have been rising continuingly for decades, not the least because of the global spread of *Ae. albopictus* and ongoing urbanization (Guzman and Istúriz 2010). For this reason and for the re-emergence of *Ae. aegypti* in former endemic areas such as Madeira (Almeida et al. 2007) and the Georgian Black Sea coast (Iunicheva et al. 2008), concern about dengue resurgence is growing in Europe.

However, while the spread of *Ae. aegypti* is limited by its cold intolerance, this is obviously not the case with *Ae. albopictus* (Grist and Burgess 1994). Due to the continuing spread of *Ae. albopictus* in Europe, dengue is presently also regarded a risk to temperate climatic zones. Although located in warm regions of Europe, first autochthonous cases of DENV transmission, most likely caused by *Ae. albopictus*, after decades of absence of the disease have been reported from Croatia (Schmidt-Chanasit et al. 2010; Gjenero-Margan et al. 2011) and southern France (Gould et al. 2010; La Ruche et al. 2010).

Aedes aegypti is highly anthropophilic and must be accounted for the vast majority of cases of DENV transmission to humans. It is a particularly synanthropic mosquito species and prefers breeding in urban surroundings in small artificial water containers (Christophers 1960). Independent of this human urban dengue cycle driven by *Ae. aegypti*, a sylvatic cycle among wild primates exists in Southeast Asia and West Africa with other zoophilic *Aedes* mosquito species as vectors (Vasilakis et al. 2011). In rural areas, both cycles are connected by peridomestic *Aedes* species such as *Ae. albopictus* whose developmental stages are also frequently found in small man-made containers (Hawley 1988).

There is no DENV vertebrate reservoir other than humans and non-human primates.

Dengue can present with a broad spectrum of clinical illness, ranging from mild non-specific symptoms to severe and fatal disease (Halstead 2002). According to WHO case definitions, three clinical developments are possible: classical dengue, dengue haemorrhagic fever and dengue shock syndrome. Classical dengue occurs in most cases of infection when a host undergoes a primary infection with DENV. It is asymptomatic or causes a self-limited febrile syndrome with undifferentiated symptoms such as the sudden onset of fever, nausea, vomiting, headache, retro-orbital pain and rash, accompanied by severe joint and muscle pain ("breakbone fever"). In some cases, usually after a second DENV infection with another serotype (Halstead 1980), classical dengue proceeds as dengue haemorrhagic fever. This often happens around the time of defervescence and presents with haemorrhagic manifestations (petechiae, mucosal bleedings) and cavitary effusions. Up to about 10% of cases that develop shock (dengue shock syndrome) are fatal (WHO 2005).

1.3.5 Yellow Fever Virus

Yellow fever indeed is a disease of the tropics and subtropics, resulting from the thermophilic mosquito species *Ae. aegypti* being the principal vector of the causative agent, yellow fever virus (YFV; Flaviviridae, *Flavivirus*), in the human-to-human urban cycle. Similar to DENV, there is a jungle cycle in endemic regions in Africa and South America, respectively, where tree-hole-breeding mosquito species, such as *Ae. africanus* (Africa) and *Haemagogus* species (South America), transmit the virus between monkeys. In the moist savanna regions, the zones of emergence, a rural cycle takes place where the virus is transferred by various *Aedes* species, including *Ae. africanus* and *Ae. simpsoni*, from the jungle cycle monkey reservoirs to humans (Vainio and Cutts 1998). The virus is maintained in the mosquito population by vertical transmission via the eggs (Beaty et al. 1980; Fontenille et al. 1997).

Yellow fever may clinically present with a broad spectrum of symptoms ranging from non-specific flu-like illness to fatal haemorrhagic fever. Typical signs of a febrile illness, such as headache, myalgia, malaise, nausea, etc., appear during a first phase, the period of infection, which lasts several days. A short period of remission may follow in which the disease symptoms disappear. About a fourth of the people affected enter a more severe third phase, the period of intoxication, with fever, vomiting, jaundice, renal failure, haemorrhages, hypotension and coma. Up to 50% of the patients with hepatorenal dysfunction die within 1–2 weeks (Monath 2001).

In Europe, yellow fever outbreaks were recorded in Portugal, Spain, Italy and France in the eighteenth and nineteenth centuries (Strode 1951; Eritja et al. 2005). Although potential mosquito vectors spread and numbers of travellers returning from yellow fever endemic regions rise, the disease is not regarded a high priority risk in Europe since safe and efficacious vaccines are available (Roukens and Visser 2008).

1.3.6 Usutu Virus

Usutu virus (USUV; Flaviviridae, *Flavivirus*, Japanese encephalitis group) has its origin in sub-Saharan Africa where it occurs within a natural cycle between birds and mosquitoes, mainly species of the genus *Culex* (Nikolay et al. 2011). After an episode of increased mortality among blackbirds, it was detected for the first time outside of Africa in 2001 in Austria (Weissenböck et al. 2002). From then on, USUV infections were also demonstrated, either directly by isolation/immunohistochemistry/RNA amplification or indirectly by antibody detection, in birds in Great Britain (Buckley et al. 2003, 2006), Hungary (Bakonyi et al. 2007), Italy (Lelli et al. 2008), Czech Republic (Hubálek et al. 2008a), Poland (Hubálek et al. 2008b) and Switzerland (Steinmetz et al. 2011). The virus was found in *Cx. pipiens* mosquitoes from Spain in 2006 (Busquets et al. 2008), in *Cx. pipiens* and *Ae. albopictus* from Italy in 2009 (Busani et al. 2010; Tamba et al. 2010) and in *Cx. pipiens* from Germany in 2010 (Jöst et al. 2011a). Interestingly, 1 year after the demonstration of the virus in the German culicids, it could be isolated from blackbirds found dead in the same region (ProMED-Mail 2011). For Austria, it could be shown that USUV was able to overwinter and to establish a local transmission cycle (Weissenböck et al. 2003; Meister et al. 2008).

Being highly pathogenic for blackbirds, USUV is of unclear pathogenicity for humans (Vazquez et al. 2011). In addition to two benign infections in central Africa (Nikolay et al. 2011), two cases with involvement of the central nervous system in immunocompromised patients were reported from Italy (Cavrini et al. 2009; Pecorari et al. 2009).

1.3.7 Sindbis Virus

Sindbis virus (SINV; Togaviridae, *Alphavirus*, western equine encephalomyelitis virus complex) appears to have its main distribution in Fennoscandia and north-western Russia where outbreaks of disease with hundreds of clinical cases have been occurring since the early 1980s (Lundström 1999; Hubálek 2008). Interestingly, a 7-year periodicity could be observed in Finland (Brummer-Korvenkontio et al. 2002). Outbreaks have, however, also been reported from South Africa (McIntosh et al. 1976), and the virus has been demonstrated to also occur in other parts of Africa, the Middle East and even Australia (Niklasson 1989). In central and western Europe, specific antibodies were found in sentinel chickens in Great Britain (Buckley et al. 2006) and, most recently, the virus could be isolated from *Cx. torrentium*, *Cx. pipiens* and *An. maculipennis* s.l. mosquitoes collected in southern Germany (Jöst et al. 2010). While *Cx. torrentium*, *Cx. pipiens* and other ornithophilic species such as *Culiseta morsitans* must be considered to keep the transmission cycle among the passerine bird reservoir of the virus going, non-discriminative species such as *An. maculipennis* s.l. and *Ae. cinereus* are probably responsible for transmission to humans in Europe (Lundström 1999). In laboratory experiments, *Ae. albopictus* has also been proven to be susceptible for SINV (Dohm et al. 1995).

The disease associated with a SINV infection is called "Ockelbo fever" in Sweden (Skog and Espmark 1982), "Pogosta fever" in Finland (Brummer-Korvenkontio and Kuusisto 1981) and "Karelian fever" in north-western Russia (Lvov et al. 1982). It usually presents with headache, myalgia, arthralgia, malaise, conjunctivitis, pharyngitis and rash (Espmark and Niklasson 1984; Kurkela et al. 2005).

1.3.8 Chikungunya Virus

Chikungunya virus (CHIKV; Togaviridae, *Alphavirus*) is one of the most expansive viruses of modern times. Minor outbreaks have sporadically been reported since the 1960s in central and southern Africa and Southeast Asia. Starting in Kenya in 2004, however, major outbreaks almost continuously occurred until 2007 with hundreds of thousands of human cases and a geographical spread to hitherto unaffected islands of the Indian Ocean, India and parts of Southeast Asia (Powers and Logue 2007; Staples et al. 2009).

CHIKV is transmitted primarily by *Aedes* mosquitoes. Urban transmission with human involvement is usually mediated by the synanthropic species *Ae. aegypti* and *Ae. albopictus*, while other forest-dwelling *Aedes* species are responsible for a sylvatic transmission cycle among non-human primates in central and East Africa (Powers and Logue 2007).

Chikungunya has not remained restricted to tropical regions. Quite unexpectedly, an epidemic including more than 200 human clinical cases struck northern Italy in 2007 (Angelini et al. 2008). The virus had most likely been introduced by a viremic person from India who infected the established *Ae. albopictus* population, thus starting a local mosquito–human–mosquito transmission cycle (Rezza et al. 2007). More recently, two autochthonous cases of chikungunya were reported from southern France (Gould et al. 2010; Grandadam et al. 2011).

While up to ca. 25% of CHIKV infections remain asymptomatic (Sissoko et al. 2008), a symptomatic infection typically starts abruptly with fever (sometimes accompanied by rash), soon followed by severe polyarthralgias which may last for several months (Deller and Russell 1968; Queyriaux et al. 2008). Although increased mortality rates have been linked to chikungunya (Mavalankar et al. 2008), death caused directly by the disease seems to be rare.

1.3.9 Ťahyňa Virus

Ťahyňa virus (TAHV; Bunyaviridae, *Orthobunyavirus*, California serogroup) has been reported from most parts of Europe as well as from many Asian and African countries (Lundström 1999; Hubálek 2008). Numerous mosquito species of the genera *Aedes*, *Anopheles*, *Culiseta*, *Culex* and *Ochlerotatus* have been found virus positive, most frequently the anthropophilic species *Ae. vexans* (Gratz 2004; Hubálek 2008). This species and *Cs. annulata* are able to transmit TAHV transovarially to the next generation (Danielová and Ryba 1979; Bárdoš et al. 1975), while the virus may hibernate in *Cx. modestus* and *Cs. annulata* (Danielová and Minář 1969; Chippaux et al. 1970). Of major importance as vertebrate reservoir hosts are lagomorphs followed by pigs, rodents and insectivores (Aspöck 1996; Hubálek 2008). In humans, TAHV infection is generally either asymptomatic or connected with a mild febrile influenza-like illness, typically in summer, called "Valtice fever" (Sluka 1969a). Sometimes severe symptoms may occur, such as meningitis and atypical pneumonia (Bárdoš 1976; Bárdoš et al. 1980).

1.3.10 Batai/Čalovo Virus

Evidence for Batai virus (BATV; Bunyaviridae, *Orthobunyavirus*, Bunyamwera group) circulation has been found in various parts of Europe, Asia and central Africa. As to Europe, the virus was demonstrated in Scandinavia and central, eastern and southeastern Europe (Hubálek 2008). Here, it was repeatedly isolated from mosquitoes of the Maculipennis species complex (Bárdoš and Čupkova 1962; Smetana et al. 1967; Brudnjak et al. 1970; Jöst et al. 2011b) but also from *An. claviger* (Bárdoš and Čupkova 1962; Traavik et al. 1985), *Oc. communis* (Francy et al. 1989) and *Cq. richardii* (Aspöck and Kunz 1968; Aspöck et al. 1970). Experimentally infected *An. maculipennis* s.l. have been shown to enable BATV hibernation (Aspöck and Kunz 1970; Beletskaya and Alekseev 1988). Serological surveys carried out in several European countries contributed further data on the wide circulation of BATV in Europe (Lundström 1999). Primary vertebrate hosts of BATV appear to be ruminants and other domesticated livestock, resulting in an agroecosystem transmission cycle (Hubálek 2008). BATV infections usually take a mild course, often with clinical signs of an influenza-like illness (Sluka 1969b).

1.3.11 Inkoo Virus

Inkoo virus (INKV; Bunyaviridae, *Orthobunyavirus*, California serogroup) appears to be restricted in its distribution to northern Europe (Scandinavia, northern Russia) (Hubálek 2008). Infections have been demonstrated in *Oc. communis*, *Oc. punctor* and *Oc. hexodontus* mosquitoes (Brummer-Korvenkontio et al. 1973; Traavik et al. 1978, 1985; Francy et al. 1989; Mitchell et al. 1993) and in humans (Brummer-Korvenkontio et al. 1973; Lvov et al. 1989; Butenko et al. 1991; Lundström 1999; Vanlandingham et al. 2002). Although INKV is generally considered apathogenic or to cause mild influenza-like symptoms only (Karabatsos 1985; Kolobukhina et al. 1990), Demikhov (1995), Demikhov and Chaitsev (1995) and Lvov et al. (1996) reported severe illness including chronic neurological disease associated with INKV infection in Russia. Major vertebrate hosts of INKV seem to be cow, reindeer, moose, fox and snow hare (Brummer-Korvenkontio et al. 1973).

1.3.12 Lednice Virus

In Europe, Lednice virus (LEDV; Bunyaviridae, *Orthobunyavirus*, Turlock serogroup) has been shown to circulate in Austria, Romania and former Czechoslovakia (Malková et al. 1986). However, antibodies have primarily been detected in water birds including migratory birds (Wojta and Aspöck 1982), suggesting that the distribution area may extend to more southern regions. Antibodies have rarely been found in wild mammals and never in humans (Aspöck 1996). With regard to culicids, LEDV has so far been isolated from *Cx. modestus* only which probably does not only serve as a vector but also as a viral reservoir due to vertical transmission (Malková et al. 1974).

1.3.13 Rift Valley Fever Virus

Rift Valley fever virus (RVFV; Bunyaviridae, *Phlebovirus*) is an example of a mosquito-borne virus that has never before appeared in Europe, but raises concern of an introduction due to its geographic expansion in Africa and beyond (Martin et al. 2008; Chevalier et al. 2010). The virus is pathogenic to both humans and animals (ruminants). Mortality rates may reach 100% among neonatal sheep and cattle and 10–20% among adult ruminant livestock. Abortion in pregnant infected animals is rather the rule than the exception (Swanepoel 1994). Infections in humans are typically associated with self-limiting febrile illnesses. However, in 1–2% of affected individuals, RVF infections can progress to more severe disease including macular retinitis, encephalitis, liver necrosis and haemorrhage (Abdel-Wahad et al. 1978). Among severely affected persons who are hospitalized, the case fatality rate is approximately 10–20% (Madani et al. 2003).

RVFV was first discovered in the Rift Valley in Kenya in 1931 (Daubney et al. 1931), but since then, outbreaks have been reported from many sub-Saharan African countries, in particular on the east and south coasts (Meegan and Bailey 1988). Detection of specific antibodies indicates that the virus is much more widespread than clinical data suggest (Gerdes 2004). Major epidemics occurred in South Africa in 1950 with an estimated 100,000 deaths and 500,000 abortions in sheep (Gerdes 2004), in Kenya and Somalia in 1997–1998 with livestock losses of up to 70% in sheep and goats and 20–30% in cattle and camels (CDC 1998) and in Sudan in 2007–2008 with 230 deadly infections among the human population (Hassan et al. 2011). In 1977, RVF occurred for the first time as far north as in Egypt where, in addition to heavy losses among various ruminant species, more than 200,000 people fell ill and some 600 died (Meegan et al. 1979). At the Mauretania–Senegal border, a large RVF outbreak occurred among the human population between 1987 and 1988 with an exceptionally high frequency of neurological signs. Many thousands of people became sick and about one fifth of those died (Jouan et al. 1988). In 1993, RVF resurged in Egypt, this time with relatively little health consequences to animals and humans (Arthur et al. 1993), but in the year 2000, it was reported for the first time outside the African continent, in Saudi Arabia and Yemen, producing numerous human fatalities and heavy losses in the livestock population (Ahmad 2000; Balkhy and Memish 2003). Further future spreading tendencies are anticipated.

Humans are exposed to RVFV infection mainly by direct or indirect contact with the blood and organs of infected animals, for example, during slaughtering and butchering or veterinarian activities, while transmission through mosquito bites is considerably rare (WHO 2010). Infection occurs through inoculation, for example, via a wound from an infected knife or through contact with broken skin, or through inhalation of aerosols. In contrast, bites by infected mosquitoes are the main transmission route in ruminants. Among numerous mosquito species found infected with RVFV, floodwater *Aedes* species and some *Culex* species breeding in permanent freshwater bodies seem to be the most important vectors (Chevalier et al. 2010).

RVFV could be isolated from various unengorged *Aedes* species, such as *Ae. vexans*, which is also widely distributed in Europe, during interepizootic periods in Africa (Linthicum et al. 1985; Fontenille et al. 1995). This suggests that certain *Aedes* mosquito species, which are associated with temporary freshwater bodies, are not only vectors of the virus but reservoirs at the same time. Transovarially transmitted virus can be maintained and protected in the eggs of these mosquito species through adverse, i.e. dry, periods. The eggs hatch as a consequence of rising water levels and develop to adult females already infected with the virus before their first blood meal. Thus, a vertebrate reservoir host is not necessary for the maintenance of the natural cycle in regions where these mosquito species occur, and the virus can remain endemic without appearing in vertebrates (Linthicum et al. 1985). Not surprisingly from these considerations, RVF epidemics are usually linked to heavy rainfall, floods, the construction of dams and newly implemented irrigation schemes (Davies et al. 1985). In contrast to the "reservoir/maintenance" *Aedes* vectors, *Culex* species are the amplifying virus vectors during RVF epidemics (Pepin et al. 2010).

While of the mosquito species occurring in Europe, *Ae. vexans* as well as *Cx. theileri* and *Cx. tritaeniorhynchus* (the latter being recorded with single specimens from Albania and Greece; Samanidou and Harbach 2003) have been proven to be vector competent for RVFV (McIntosh et al. 1980; Jupp et al. 2002; Turell et al. 2008), *Ae. albopictus*, *Ae. caspius*, *Ae. detritus* and *Cx. pipiens* have been shown to be susceptible to virus infection and to disseminate the virus (Moutailler et al. 2008).

1.4 Conclusions

Of the mosquito species occurring in Europe, numerous have been found infected with disease agents of humans and animals. Whether they are actually able to transmit the pathogens and under which conditions is only clear in a few species that have been reared and used for transmission studies in the laboratory. The mosquito-borne disease agents presently circulating in Europe have led to severe outbreaks only exceptionally, such as in WNV, and most of them appear to be relatively harmless. Most critical are rather newly introduced mosquitoes and imported pathogens since these may lead to unprecedented combinations of imported and indigenous vectors and pathogens. In many European countries, it is recognized only now, i.e., after some recent transmission incidents, that old data on the geographic and seasonal distribution and on the biology of the indigenous mosquito species have to be updated and complemented for present-day risk assessments. The same is true for data on the incidence, dispersal and natural transmission cycles of pathogens vectored by mosquitoes. Classical mosquito research, both in the field and in the laboratory, has to be re-intensified to produce data on potential vectors to be able to adequately manage future outbreaks of mosquito-borne diseases which will definitely become more frequent as globalization and climate change continue.

References

Abdel-Wahad KSE, El Baz LM, El Tayeb EM, Omar H, Ossman MA, Yasin W (1978) Rift Valley fever virus infections in Egypt: pathological and virological findings in man. Trans R Soc Trop Med Hyg 72:392–396

Ahmad K (2000) More deaths from Rift Valley fever in Saudi Arabia and Yemen. Lancet 356:1422

Almeida AP, Gonçalves YM, Novo MT, Sousa CA, Melim M, Grácio AJ (2007) Vector monitoring of *Aedes aegypti* in the autonomous region of Madeira, Portugal. Euro Surveill 12, pi=3311

Alten B, Kampen H, Fontenille D (2007) Malaria in southern Europe: resurgence from the past? In: Takken W, Knols BGJ (eds) Emerging pests and vector-borne diseases in Europe, vol 1, Ecology and control of vector-borne diseases. Wageningen Academic, Wageningen, pp 35–57

Andreadis TG, Anderson JF, Vossbrinck CR, Main AJ (2004) Epidemiology of West Nile virus in Connecticut: a five-year analysis of mosquito data 1999–2003. Vector Borne Zoonotic Dis 4:360–378

Angelini P, Macini P, Finarelli AC, Pol C, Venturelli C, Bellini R, Dottori M (2008) Chikungunya epidemic outbreak in Emilia-Romagna (Italy) during summer 2007. Parassitologia 50:97–98

Angelini P, Tamba M, Finarelli AC, Bellini R, Albieri A, Bonilauri P, Cavrini F, Dottori M, Gaibani P, Martini E, Mattivi A, Pierro AM, Rugna G, Sambri V, Squintani G, Macini P (2010) West Nile virus circulation in Emilia-Romagna, Italy: the integrated surveillance system 2009. Euro Surveill 15, pii=19547

Arthur RR, el-Sharkawy MS, Cope SE, Botros BA, Oun S, Morrill JC, Shope RE, Hibbs RG, Darwish MA, Imam IZ (1993) Recurrence of Rift Valley fever in Egypt. Lancet 342: 1149–1150

Aspöck H (1996) Stechmücken als Virusüberträger in Mitteleuropa. Nova Acta Leopold 71:37–55

Aspöck H, Kunz C (1968) Isolierung des Čalovo- (=Batai- = Chitoor-)Virus aus Stechmücken in Österreich. Wien Med Wochenschr 118:497–498

Aspöck H, Kunz C (1970) Überwinterung des Čalovo-Virus in experimentell infizierten Weibchen von *Anopheles maculipennis messeae* Fall. Zentralbl Bakteriol 213:429–433

Aspöck H, Kunz C, Pretzmann G (1970) Phänologie und Abundanz der Stechmücken des östlichen Neusiedlersee-Gebietes (Ost-Österreich) in ihrer Beziehung zum Auftreten der durch Stechmücken übertragenen Arboviren. Zentralbl Bakteriol 214:160–173

Autorino GL, Battisti A, Deubel V, Ferrari G, Forletta R, Giovannini A, Lelli R, Murri S, Scicluna MT (2002) West Nile virus epidemic in horses, Tuscany region, Italy. Emerg Infect Dis 8:1372–1378

Bahnck CM, Fonseca DM (2006) Rapid assay to identify two genetic forms of *Culex* (*Culex*) *pipiens* L. (Diptera: Culicidae) and hybrid population. Am J Trop Med Hyg 75:251–255

Bailey CL, Eldridge BF, Hayes DE, Watts DM, Tammariello RF, Dalrymple JM (1978) Isolation of St. Louis encephalitis virus from overwintering *Culex pipiens* mosquitoes. Science 199:1346–1349

Bakonyi T, Erdélyi K, Ursu K, Ferenczi E, Csörgö T, Lussy H, Chvala S, Bukovsky C, Meister T, Weissenböck H, Nowotny N (2007) Emergence of Usutu virus in Hungary. J Clin Microbiol 45:3870–3874

Baldari M, Tamburro A, Sabatinelli G, Romi R, Severini C, Cuccagna G, Fiorilli G, Allegri MP, Buriani C, Toti M (1998) Malaria in Maremma, Italy. Lancet 351:1246–1247

Balenghien T, Vazeille M, Grandadam M, Schaffner F, Zeller H, Reiter P, Sabatier P, Fouque F, Bicout DJ (2008) Vector competence of some French *Culex* and *Aedes* mosquitoes for West Nile virus. Vector Borne Zoonotic Dis 8:589–895

Balkhy HH, Memish ZA (2003) Rift Valley Fever: an uninvited zoonosis in the Arabian peninsula. Int J Antimicrob Agents 21:153–157

Baqar S, Hayes CG, Murphy JR, Watts DM (1993) Vertical transmission of West Nile virus by *Culex* and *Aedes* species mosquitoes. Am J Trop Med Hyg 48:757–762

Bárdoš V (1976) The ecology and medical importance of Ťahyňa virus. Munch Med Wochenschr 118:1617–1620 [in German]

Bárdoš V, Čupkova E (1962) The Čalovo virus – the second virus isolated from mosquitoes in Czechoslovakia. J Hyg Epidemiol Microbiol Immunol 6:186–192

Bárdoš V, Ryba J, Hubálek Z (1975) Isolation of Ťahyňa virus from field-collected *Culiseta annulata* (Schrk.) larvae. Acta Virol 19:446

Bárdoš V, Medek M, Kania V, Hubálek Z, Juricová Z (1980) Das klinische Bild der Ťahyňa-Virus (California-Gruppe)-Infektionen bei Kindern. Pädiatrie 19:11–23

Barzon L, Squarzon L, Cattai M, Franchin E, Pagni S, Cusinato R, Palu G (2009) West Nile virus infection in Veneto region, Italy, 2008–2009. Euro Surveill 14, pii=19289

Beaty BJ, Tesh RB, Aitken THG (1980) Transovarial transmission of YF virus in *Stegomyia* mosquitoes. Am J Trop Med Hyg 29:125–132

Becker N (2008) Influence of climate change on mosquito development and mosquito-borne diseases in Europe. Parasitol Res 103(Suppl 1):S19–S28

Becker N, Petrić D, Zgomba M, Boase C, Madon M, Dahl C, Kaiser A (2010) Mosquitoes and their control, 2nd edn. Springer, Heidelberg, 577 pp

Beletskaya GV, Alekseev AA (1988) Experimental demonstration of the Batai virus maintenance in hibernating female *Anopheles maculipennis* mosquitoes. Med Parazitol 66:23–27 [in Russian]

Blacklock B, Carter HF (1920) Observations on *Anopheles (Coelodiazesis) plumbeus* Stephens, with special reference to its breeding places, occurrence in the Liverpool district and possible connection with the spread of malaria. Ann Trop Med Parasitol 13:421–444

Bruce-Chwatt LJ, de Zulueta J (1980) The rise and fall of malaria in Europe. Oxford University Press, Oxford, 239 pp

Brudnjak Z, Danielová V, Ryba J, Vesenjak-Hirjan J (1970) Isolation of Čalovo virus from *Anopheles maculipennis* s.l. mosquitoes in Yugoslavia. Folia Parasitol 17:323–324

Brummer-Korvenkontio M, Kuusisto P (1981) Has western Finland been spared the Pogosta? Suom Lääkäril 32:2606–2607 [in Finnish]

Brummer-Korvenkontio M, Saikku P, Korhonen P, Ulmanen I, Reunala T, Karvonen J (1973) Arboviruses in Finland. IV. Isolation and characterization of Inkoo virus, a Finnish representative of the California group. Am J Trop Med Hyg 22:404–413

Brummer-Korvenkontio M, Vapalahti O, Kuusisto P, Saikku P, Manni T, Koskela P, Nygren T, Brummer-Korvenkontio H, Vaheri A (2002) Epidemiology of Sindbis virus infections in Finland 1981–96: possible factors explaining a peculiar disease pattern. Epidemiol Infect 129:335–345

Bucklar H, Scheu U, Mossi R, Deplazes P (1998) Is dirofilariasis in dogs spreading to south Switzerland? Schweiz Arch Tierheilk 140:255–260 [in German]

Buckley A, Dawson A, Moss SR, Hinsley SA, Bellamy PE, Gould EA (2003) Serological evidence of West Nile virus, Usutu virus and Sindbis virus infection of birds in the UK. J Gen Virol 84:2807–2817

Buckley A, Dawson A, Gould EA (2006) Detection of seroconversion to West Nile virus, Usutu virus and Sindbis virus in UK sentinel chickens. Virol J 3:71

Busani L, Capelli G, Cecchinato M, Lorenzetto M, Savini G, Terregino C, Vio P, Bonfanti L, Pozza MD, Marangon S (2010) West Nile virus circulation in Veneto region in 2008–2009. Epidemiol Infect 139:818–825

Busquets N, Alba A, Allepuz A, Aranda C, Ignacio Nuñez J (2008) Usutu virus sequences in *Culex pipiens* (Diptera: Culicidae), Spain. Emerg Infect Dis 14:861–863

Butenko AM, Vladimirtseva EA, Lvov SD, Calisher CH, Karabatsos M (1991) California serogroup viruses from mosquitoes collected in the USSR. Am J Trop Med Hyg 45:366–370

Byrne K, Nichols RA (1999) *Culex pipiens* in London Underground tunnels: differentiation between surface and subterranean populations. Heredity 82:7–15

Cancrini G, Frangipane di Regalbono A, Ricci I, Tessarin C, Gabrielli S, Pietrobelli M (2003) *Aedes albopictus* is a natural vector of *Dirofilaria immitis* in Italy. Vet Parasitol 118:195–202

Cancrini G, Magi M, Gabrielli S, Arispici M, Tolari F, Dell'Omodarme M, Prati MC (2006) Natural vectors of dirofilariasis in rural and urban areas of the Tuscan region, central Italy. J Med Entomol 43:574–579

Cardamatis J (1929) La dengue en Grèce. Bull Soc Pathol Exot 22:272–292

Cavrini F, Gaibani P, Longo G, Pierro AM, Rossini G, Bonilauri P, Gerundi GE, Di Benedetto F, Pasetto A, Girardis M, Dottori M, Landini MP, Sambri V (2009) Usutu virus infection in a patient who underwent orthotropic liver transplantation, Italy, August-September 2009. Euro Surveill 14, pii=19448

CDC (Centers for Disease Control) (1998) Rift Valley Fever – East Africa, 1997–1998. Morb Mortal Wkly Rep 47:261–264

Chevalier V, Pépin M, Plée L, Lancelot R (2010) Rift Valley fever – a threat for Europe? Euro Surveill 15, pii=19506

Chippaux A, Rageau J, Mouchet J (1970) Hibernation de l'arbovirus Ťahyňa chez *Culex modestus* Fic. en France. C R Acad Sci D 270:1648–1650

Christophers R (1960) *Aedes aegypti* (L.), the yellow fever mosquito. Cambridge University Press, Cambridge, 739 pp

Cuadros J, Calvente MJ, Benito A, Arévalo J, Calero MA, Segura J, Rubio JM (2002) *Plasmodium ovale* malaria acquired in central Spain. Emerg Infect Dis 8:1506–1508

Danielová V (1972) The vector efficiency of *Culiseta annulata* mosquito in relation to Ťahyňa virus. Folia Parasitol 19:259–262

Danielová V, Holubová J (1977) Two more mosquito species proved as vectors of Ťahyňa virus in Czechoslovakia. Folia Parasitol 24:187–189

Danielová V, Minář J (1969) Experimental overwintering of the virus Ťahyňa in mosquitoes *Culiseta annulata* (Schrk.) (Diptera, Culicidae). Folia Parasitol 16:285–287

Danielová V, Ryba J (1979) Laboratory demonstration of transovarial transmission of Ťahyňa virus in *Aedes vexans* and the role of this mechanism in overwintering of this arbovirus. Folia Parasitol 26:361–368

Danielová V, Málková D, Minár J, Ryba J (1976) Dynamics of the natural focus of Ťahyňa virus in southern Moravia and species succession of its vectors, the mosquitoes of the genus *Aedes*. Folia Parasitol 23:243–249

Danis K, Baka A, Lenglet A, van Bortel W, Terzaki I, Tseroni M, Detsis M, Papanikolaou E, Balaska A, gewehr S, Dougas G, Sideroglou T, Economopoulou A, Vakalis N, Tsiodras S, Bonovas S, Kremastinou J (2011a) Autochthonous *Plasmodium vivax* malaria in Greece, 2011. Euro Surveill 16, pii=19993

Danis K, Papa A, Theocharopoulos G, Dougas G, Athanasiou M, Detsis M, Baka A, Lytras T, Mellou K, Bonovas S, Panagiotopoulos T (2011b) Outbreak of West Nile virus infection in Greece, 2010. Emerg Infect Dis 17:1868–1872

Danis K, Papa A, Papanikolaou E, Dougas G, Terzaki I, Baka A, Vrioni G, Kapsimali V, Tsakris A, Kansouzidou A, Tsiodras S, Vakalis N, Bonovas S, Kremastinou J (2011c) Ongoing outbreak of West Nile virus infection in humans, Greece, July to August, 2011. Euro Surveill 16, pii=19951

Daškova NG, Rasnicyn SP (1982) Review of data on susceptibility of mosquitoes in the USSR to imported strains of malaria parasites. Bull World Health Organ 60:893–897

Daubney RJ, Hudson JR, Garnham EC (1931) Enzootic hepatitis of Rift Valley fever: an undescribed virus disease of sheep, cattle and man from East Africa. J Pathol Bacteriol 34: 543–579

Davies FG, Linthicum KJ, James AD (1985) Rainfall and epizootic Rift Valley fever. Bull World Health Organ 63:941–943

Deller JJ, Russell PK Jr (1968) Chikungunya disease. Am J Trop Med Hyg 17:107–111

Demikhov VG (1995) Outcomes and prognosis of diseases caused by Inkoo and Ťahyňa viruses. Vopr Virusol 40:72–74 [in Russian]

Demikhov VG, Chaitsev VG (1995) Neurological characteristics of diseases caused by Inkoo and Ťahyňa viruses. Vopr Virusol 40:21–25 [in Russian]

Dohm DJ, Logan TM, Barth JF, Turell MJ (1995) Laboratory transmission of Sindbis virus by *Aedes albopictus*, *Aedes aegypti*, and *Culex pipiens* (Diptera: Culicidae). J Med Entomol 32:818–821

Eling W, van Gemert G-J, Akinpelu O, Curtis J, Curtis CF (2003) Production of *Plasmodium falciparum* sporozoites by *Anopheles plumbeus*. Eur Mosq Bull 15:12–13

Eritja R, Escosa R, Kucientes J, Marquès E, Molina R, Roiz D, Ruiz S (2005) Worldwide invasion of vector mosquitoes: present European distribution and challenges for Spain. Biol Invasions 7:87–97

Espmark A, Niklasson B (1984) Ockelbo disease in Sweden: epidemiological, clinical, and virological data from the 1982 outbreak. Am J Trop Med Hyg 33:1203–1211

Farajollahi A, Fonseca DM, Kramer LD, Marm Kilpatrick A (2011) "Bird biting" mosquitoes and human disease: a review of the role of *Culex pipiens* complex mosquitoes in epidemiology. Infect Genet Evol 11:1577–1585

Filipe AR (1972) Isolation in Portugal of West Nile virus from *Anopheles maculipennis* mosquitoes. Acta Virol 16:361

Fonseca DM, Keyghobadi N, Malcolm CA, Mehmet C, Schaffner F, Mogi M, Fleischer RC, Wilkerson RC (2004) Emerging vectors in the *Culex pipiens* complex. Science 303:1535–1538

Fontenille D, Traore-Lamizama M, Zeller H, Mondo M, Diallo M, Digoutte J-P (1995) Rift Valley fever in western Africa: isolations from *Aedes* mosquitoes during an interepizootic period. Am J Trop Med Hyg 52:403–404

Fontenille D, Diallo M, Mondo M, Ndiaye M, Thonnon J (1997) First evidence of natural vertical transmission of yellow fever virus in *Aedes aegypti*, its epidemic vector. Trans R Soc Trop Med Hyg 91:533–535

Francy DB, Jaenson TGT, Lundström JO, Schildt EB, Espmark Ö, Henriksson B, Niklasson B (1989) Ecologic studies of mosquitoes and birds as hosts of Ockelbo virus in Sweden and isolation of Inkoo and Batai virus from mosquitoes. Am J Trop Med Hyg 41:355–363

Genchi C, Rinaldi L, Cascone C, Mortarino M, Cringoli G (2005) Is heartworm disease really spreading in Europe? Vet Parasitol 133:137–148

Genchi C, Rinaldi L, Mortarino M, Genchi M, Cringoli G (2009) Climate and *Dirofilaria* infection in Europe. Vet Parasitol 163:286–292

Gerdes GH (2004) Rift Valley fever. Rev Sci Tech 23:613–623

Gjenero-Margan I, Aleraj B, Krajcar D, Lesnikar V, Klobučar A, Pem-Novosel I, Kurečić-Filipović S, Komparak S, Martić R, Duričić S, Betica-Radić L, Okmadžić J, Vilibić-Čavlek T, Babić-Erceg A, Turković B, Avsić-Županc T, Radić I, Ljubić M, Sarac K, Benić N, Mlinarić-Galinović G (2011) Autochthonous dengue fever in Croatia, August-September 2010. Euro Surveill 16, pii=19805

Goddard LB, Roth AE, Reisen WK, Scott TW (2003) Vertical transmission of West Nile Virus by three California *Culex* (Diptera: Culicidae) species. J Med Entomol 40:743–746

Gomes B, Sousa CA, Novo MT, Freitas FB, Alves R, Corte-Real AR, Salgueiro P, Donnelly MJ, Almeida AP, Pinto J (2009) Asymmetric introgression between sympatric *molestus* and *pipiens* forms of *Culex pipiens* (Diptera: Culicidae) in the Comporta region, Portugal. BMC Evol Biol 9:262–277

Gould EA, Gallian P, De Lamballerie X, Charrel RN (2010) First cases of autochthonous dengue fever and chikungunya fever in France: from bad dream to reality! Clin Microbiol Infect 16:1702–1704

Grandadam M, Caro V, Plumet S, Thiberge JM, Souarès Y, Failloux AB, Tolou HJ, Budelot M, Cosserat D, Leparc-Goffart I, Desprès P (2011) Chikungunya virus, southeastern France. Emerg Infect Dis 17:910–913

Gratz N (2004) The vector-borne human infections of Europe. Their distribution and burden on public health. WHO Regional Office for Europe, Copenhagen, 144 pp

Grist NR, Burgess NRH (1994) *Aedes* and dengue. Lancet 343:277

Gubler DJ (1989) *Aedes aegypti* and *Aedes aegypti*-borne disease control in the 1990s: top down or bottom up? Am J Trop Med Hyg 40:571–578

Gubler DJ (2007) The continuing spread of West Nile virus in the western hemisphere. Clin Infect Dis 45:1039–1046

Guzman A, Istúriz RE (2010) Update on the global spread of dengue. Int J Antimicrob Agents 36S: S40–S42

Halstead SB (1980) Immunological parameters of togavirus disease syndromes. In: Schlesinger RW (ed) The Togaviruses. Biology, structure, replication. Academic, New York, pp 107–173

Halstead SB (2002) Dengue. Curr Opin Infect Dis 15:471–476

Hamer GL, Kitron UD, Brawn JD, Loss SR, Ruiz MO, Goldberg TL, Walker ED (2008) *Culex pipiens* (Diptera: Culicidae): a bridge vector of West Nile virus to humans. J Med Entomol 45:125–128

Hamer GL, Kitron UD, Goldberg TL, Brawn JD, Loss SR, Ruiz MO, Hayes DB, Walker ED (2009) Host selection by *Culex pipiens* mosquitoes and West Nile virus amplification. Am J Trop Med Hyg 80:268–278

Hanke CA, Otten J-E, Berner R, Serr A, Splettstoesser W, von Schnakenburg C (2009) Ulceroglandular tularaemia in a toddler in Germany after a mosquito bite. Eur J Pediatr 168:937–940

Hannoun C, Panthier R, Mouchet J, Eouzan JP (1964) Isolement en France du virus West Nile à partir de malades et du vecteur *Culex modestus* Ficalbi. C R Acad Sci D 259:4170–4172

Hannoun C, Panthier R, Corniou B (1966) Isolation of Ťahyňa virus in the south of France. Acta Virol 10:362–364

Harbach RE, Dahl C, White GB (1985) *Culex* (*Culex*) *pipiens* Linnaeus (Diptera: Culicidae): concepts, type designations and description. Proc Entomol Soc Wash 87:1–24

Hassan OA, Ahlm C, Sang R, Evander M (2011) The 2007 Rift Valley fever outbreak in Sudan. PLoS Negl Trop Dis 5:e1229

Hawley WA (1988) Biology of *Aedes albopictus*. J Am Mosq Control Assoc 4:1–39

Hermosilla C, Pantchev N, Dyachenko V, Gutmann M, Bauer C (2006) First autochthonous case of canine ocular *Dirofilaria repens* infection in Germany. Vet Rec 158:134–135

Hesson JC, Lundström JO, Halvarsson P, Erixon P, Collado A (2010) A sensitive and reliable restriction enzyme assay to distinguish between the mosquitoes *Culex torrentium* and *Culex pipiens*. Med Vet Entomol 24:142–149

Hubálek Z (2008) Mosquito-borne viruses in Europe. Parasitol Res 103(Suppl 1):S29–S43

Hubálek Z, Halouzka J (1999) West Nile fever – e remerging mosquito-borne viral disease in Europe. Emerg Infect Dis 5:643–650

Hubálek Z, Halouzka J, Juricová Z, Sikutová S, Rudolf I, Honza M, Janková J, Chytil J, Marec F, Sitko J (2008a) Serologic survey of birds for West Nile flavivirus in southern Moravia (Czech Republic). Vector Borne Zoon Dis 8:659–666

Hubálek Z, Wegner E, Halouzka J, Tryjanowski P, Jerzak L, Šikutová S, Rudolf I, Kruszewicz AG, Jaworski Z, Włodarczyk R (2008b) Serologic survey of potential vertebrate hosts for West Nile virus in Poland. Viral Immunol 21:247–253

Huldén L, Huldén L, Heliövaara K (2005) Endemic malaria: an 'indoor' disease in northern Europe. Historical data analysed. Malar J 4:19

Iunicheva IV, Riabova TE, Markovich NI, Bezzhonova OV, Ganushkina LA, Semenov VB, Tarkhov GA, Vasilenko LE, Guzeeva TM, Shevereva TV, Sergiev VP (2008) First evidence for breeding *Aedes aegypti* L in the area of Greater Sochi and in some towns of Abkhasia. Med Parazitol 3:40–43 [in Russian]

Jöst H, Bialonski A, Storch V, Günther S, Becker N, Schmidt-Chanasit J (2010) Isolation and phylogenetic analysis of Sindbis viruses from mosquitoes in Germany. J Clin Microbiol 48:1900–1903

Jöst H, Bialonski A, Maus D, Sambri V, Eiden M, Groschup MH, Günther S, Becker N, Schmidt-Chanasit J (2011a) Isolation of Usutu virus in Germany. Am J Trop Med Hyg 85:551–553

Jöst H, Bialonski A, Schmetz C, Günther S, Becker N, Schmidt-Chanasit J (2011b) Isolation and phylogenetic analysis of Batai virus, Germany. Am J Trop Med Hyg 84:241–243

Jouan A, Leguenno B, Digoutte JP, Phillipe B, Riou O, Adam F (1988) A Rift Valley fever epidemic in southern Mauritania. Ann Virol 139:307–308

Jupp PG, Kemp A, Grobbelaar A, Leman P, Burt FJ, Alahmed AM, Al Mujalli D, Al Khamees M, Swanepoel R (2002) The 2000 epidemic of Rift Valley fever in Saudi Arabia. Mosquito vector studies. Med Vet Entomol 16:245–252

Kampen H (2004) Die Differenzierung von *Anopheles*-Zwillingsarten. In: Aspöck U (ed) Amöben, Bandwürmer, Zecken... Parasiten und parasitäre Erkrankungen des Menschen in Mitteleuropa, vol 13, Denisia. Biologiezentrum des Oberösterreichischen Landesmuseums, Linz, pp 497–513

Kampen H (2005) Integration of *Anopheles beklemishevi* (Diptera: Culicidae) in a PCR assay diagnostic for palaearctic *Anopheles maculipennis* sibling species. Parasitol Res 97:113–117

Kampen H (2009) What makes an insect a vector? Berl Münch Tierärztl Wochenschr 122:451–457

Kampen H, Werner D (2010) Three years of bluetongue disease in central Europe with special reference to Germany: what lessons can be learned? Wien Klin Wochenschr 122(Suppl 3): 31–39

Kampen H, Maltezos E, Pagonaki M, Hunfeld KP, Maier WA, Seitz HM (2002) Individual cases of autochthonous malaria in Evros Province, northern Greece: serological aspects. Parasitol Res 88:261–266

Kampen H, Proft J, Etti S, Maltezos E, Pagonaki M, Maier WA, Seitz HM (2003a) Individual cases of autochthonous malaria in Evros Province, northern Greece: entomological aspects. Parasitol Res 89:252–258

Kampen H, Sternberg A, Proft J, Bastian S, Schaffner F, Maier WA, Seitz HM (2003b) Polymerase chain reaction-based differentiation of the mosquito sibling species *Anopheles claviger* s.s. and *Anopheles petragnani* (Diptera: Culicidae). Am J Trop Med Hyg 69:195–199

Karabatsos N (ed) (1985) International catalogue of arboviruses, including certain other viruses of vertebrates, 3rd edn. American Society of Tropical Medicine and Hygiene, San Antonio, TX, 1041 pp

Kilpatrick AM, Kramer LD, Campbell SR, Alleyne LO, Dobson AP, Daszak P (2005) West Nile virus risk assessment and the bridge vector paradigm. Emerg Infect Dis 11:425–429

Kilpatrick AM, Kramer LD, Jones MJ, Marra PP, Daszak P, Fonseca DM (2007) Genetic influences on mosquito feeding behavior and the emergence of zoonotic pathogens. Am J Trop Med Hyg 77:667–671

Klee AL, Maidin B, Edwin B, Poshni I, Mostashari F, Fine A, Layton M, Nash D (2004) Long-term prognosis for clinical West Nile virus infection. Emerg Infect Dis 10:1405–1411

Kolobukhina LV, Lvov DK, Butenko AM, Nedyalkova MS, Kuznetsov AA, Galkina IV (1990) Signs and symptoms of infections caused by California serogroup viruses in humans in the USSR. Arch Virol Suppl 1:243–247

Koopmans M, Martina B, Reusken C, van Maanen K (2007) West Nile virus in Europe: waiting for the start of the epidemic? In: Takken W, Knols BGJ (eds) Emerging pests and vector-borne diseases in Europe, vol 1, Ecology and control of vector-borne diseases. Wageningen Academic, Wageningen, pp 123–151

Kramer LD, Li J, Shi PY (2003) West Nile virus. Lancet Neurol 6:171–178

Krüger A, Rech A, Su X-Z, Tannich E (2001) Two cases of autochthonous *Plasmodium falciparum* malaria in Germany with evidence for local transmission by indigenous *Anopheles plumbeus*. Trop Med Int Health 6:983–985

Kurkela S, Manni T, Myllynen J, Vaheri A, Vapalahti O (2005) Clinical and laboratory manifestations of Sindbis virus infection: prospective study, Finland, 2002–2003. J Infect Dis 191:1820–1829

La Ruche G, Souarès Y, Armengaud A, Peloux-Petiot F, Delaunay P, Desprès P, Lenglet A, Jourdain F, Leparc-Goffard L, Charlet F, Ollier L, Mantey K, Mollet T, Fournier JP, Torrents R, Leitmeyer K, Hilairet P, Zeller H, van Bortel W, Dejour-Salamanca D, Grandadam M, Gastellu-Etchegorry M (2010) First two autochthonous dengue virus infections in metropolitan France, September 2010. Euro Surveill 15, pii=19676

Labuda M, Kožuch O, Grešiková M (1974) Isolation of West Nile virus from *Aedes cantans* mosquitoes in west Slovakia. Acta Virol 18:429–433

Lelli R, Savini G, Teodori L, Filipponi G, Di Gennaro A, Leone A, Di Gialleonardo L, Venturi L, Caporale V (2008) Serological evidence of Usutu virus occurrence in north-eastern Italy. Zoon Publ. Health 55:361–367

Linke S, Niedrig M, Kaiser A, Ellerbrok H, Müller K, Müller T, Conraths FJ, Mühle R-U, Schmidt D, Köppen U, Bairlein F, Berthold P, Pauli G (2007) Serologic evidence of West Nile virus infections in wild birds captured in Germany. Am J Trop Med Hyg 77:358–364

Linthicum KJ, Davies FG, Kairo A, Bailey CL (1985) Rift Valley fever virus (family Bunyaviridae, genus Phlebovirus). Isolations from Diptera collected during an inter-epizootic period in Kenya. J Hyg (Lond) 95:197–209

Linton Y-M, Lee AS, Curtis C (2005) Discovery of a third member of the Maculipennis group in SW England. Eur Mosq Bull 19:5–9

Lundström JO (1999) Mosquito-borne viruses in Western Europe: a review. J Vector Ecol 24:1–39

Lundström JO, Andersson A-C, Bäckman S, Schäfer ML, Forsman M, Thelaus J (2011) Transstadial transmission of *Francisella tularensis holarctica* in mosquitoes, Sweden. Emerg Infect Dis 17:794–799

Lvov DK, Gromashevsky VL, Skvortsova TM, Berezin LK, Zakarian VA (1979) Kyzylagach virus (Togaviridae family, the genus Alphavirus), a new arbovirus isolated from *Culex modestus* mosquitoes collected in Azerbaijan SSR. Vopr Virusol 5:519–523 [in Russian]

Lvov DK, Skvortsova TM, Kondrashina NG, Vershinsky BV, Lesnikov AL, Derevyansky VS, Berezina LK, Gromashevsky VL, Adrianova DP, Yakovlev VI (1982) Etiology of Karelian fever, a new arboviral infection. Vopr Virusol 6:690–692 [in Russian]

Lvov DK, Skvortsova TM, Berezina LK, Gromashevsky VL, Yakovlev BI, Gushchin BV, Aristova VA, Sidorova GA, Gushchina EA, Klimenko SM, Lvov SD, Khutoretskaya NV, Myasnikova IA, Khiznyakova TM (1984) Isolation of Karelian fever agent from *Aedes communis* mosquitoes. Lancet 2:399–400

Lvov DK, Gromashevsky VL, Skvortsova TM, Sokolova NG, Andronova ON, Semenov VB, Makhlin PI, Kondrashina NG, Bystrova EA, Gushchina EA, Morozova TN, Kuznetsov AA, Galkina IV, Yamnikova SS, Aristova VA, Avershin AD, Berezin MV (1989) Circulation of California serogroup viruses in northwestern Russian plateau. Med Parazitol 6:74–77 [in Russian]

Lvov DK, Kolobukhina LV, Gromashevsky VL, Skvortsova TM, Morozova TN, Galkina IV, Nedyalkova MS (1996) Isolation of California antigenic group viruses from the patients with acute neuroinfection syndrome. Arbovirus Inf Exch June:16–18

Macini P, Squintani G, Finarelli AC, Angelini P, Martini E, Tamba M, Dottori M, Bellini R, Santi A, Loli Piccolomini L, Po C (2008) Detection of West Nile virus infection in horses, Italy, September 2008. Euro Surveill 13, pii=18990

Madani TA, Al-Mazrou YY, Al-Jeffri MH, Mishkhas AA, Al-Rabeah AM, Turkistani AM, Al-Sayed MO, Abodahish AA, Khan AS, Ksiazek TG, Shobokshi O (2003) Rift Valley fever epidemic in Saudi Arabia: epidemiological, clinical, and laboratory characteristics. Clin Infect Dis 37:1084–1092

Malkinson M, Banet C (2002) The role of birds in the ecology of West Nile virus in Europe and Africa. Curr Top Microbiol Immunol 267:309–322

Malková D, Danielová V, Minář J, Ryba J (1974) Virological investigation of mosquitoes in some biotopes of southern Moravia in summer season 1972. Folia Parasit 21:363–372

Malková D, Danielová V, Holubová J, Marhoul Z (1986) Less known arboviruses of central Europe: a new arbovirus Lednice. Academia Publishing House of the Czechoslovak Academy of Science, Prague, 75 pp

Marchant P, Eling W, van Gemert G-J, Leake CJ, Curtis CF (1998) Could British mosquitoes transmit *falciparum* malaria? Parasitol Today 14:344–345

Martin V, Chevalier V, Ceccato P, Anyamba A, De Simone L, Lubroth J, de La Rocque S, Domenech J (2008) The impact of climate change on the epidemiology and control of Rift Valley fever. Rev Sci Tech Off Int Epiz 27:413–426

Mavalankar D, Shastri P, Bandyopadhyay T, Parmar J, Ramani KV (2008) Increased mortality rate associated with chikungunya epidemic, Ahmedabad, India. Emerg Infect Dis 14:412–415

McCall JW, Genchi C, Kramer LH, Guerrero J, Venco L (2008) Heartworm disease in animals and humans. Adv Parasitol 66:193–285

McIntosh BM, Jupp PG, Dos Santos I, Meenehan GM (1976) Epidemics of West Nile and Sindbis viruses in South Africa with Culex (Culex) univittatus Theobald as vector. S Afr J Sci 72:295–300

McIntosh BM, Jupp PG, Dos Santos I, Barnard BJH (1980) Vector studies on Rift Valley fever virus in South Africa. S Afr Med J 58:127–132

Meegan JM, Bailey CL (1988) Rift Valley Fever. In: Monath TP (ed) The arboviruses: epidemiology and ecology, vol IV. CRC, Boca Raton, FL, pp 51–76

Meegan JM, Hoogstraal H, Moussa MI (1979) An epizootic of Rift Valley fever in Egypt in 1977. Vet Rec 105:124–125

Meister T, Lussy H, Bakonyi T, Sikutova S (2008) Serological evidence of continuing high Usutu virus (Flaviviridae) activity and establishment of herd immunity in wild birds in Austria. Vet Microbiol 127:237–248

Mitchell CJ, Lvov SD, Savage HM, Calisher CH, Smith GC, Lvov DK, Gubler DJ (1993) Vector and host relationships of California serogroup viruses in Western Siberia. Am J Trop Med Hyg 49:53–62

Monath TP (2001) Yellow fever: an update. Lancet Infect Dis 1:11–20

Moutailler S, Krida G, Schaffner F, Vazeille M, Failloux AB (2008) Potential vectors of Rift Valley fever virus in the Mediterranean region. Vector Borne Zoon Dis 8:749–753

Murgue B, Murri S, Zientara S, Durand B, Durand JP, Zeller H (2001) West Nile outbreak in southern France, 2000: the return after 35 years. Emerg Infect Dis 7:692–696

Muro A, Genchi C, Cordero M, Simón F (1999) Human dirofilariasis in the European Union. Parasitol Today 15:386–389

Nasci RS, Savage HM, White DJ, Miller JR, Cropp BC, Godsey MS, Kerst AJ, Bennett P, Gottfried K, Lanciotti RS (2001) West Nile virus in overwintering *Culex* mosquitoes, New York City, 2000. Emerg Infect Dis 7:742–744

Nicolescu G, Linton Y-M, Vladimirescu A, Howard TM, Harbach RE (2004) Mosquitoes of the *Anopheles maculipennis* group (Diptera; Culicidae) in Romania, with the discovery and formal recognition of a new species based on molecular and morphological evidence. Bull Entomol Res 94:525–535

Niklasson B (1989) Sindbis and Sindbis-like viruses. In: Monath TP (ed) The arboviruses: epidemiology and ecology. CRC, Boca Raton, FL, pp 167–176

Nikolaeva N (1996) Resurgence of malaria in the former Soviet Union (FSU). SOVE Newslett 27:10–11

Nikolay B, Diallo M, Bouh Boye CS, Sall AA (2011) Usutu virus in Africa. Vector Borne Zoonotic Dis 11:1417–1423

Odolini S, Parola P, Gkrania-Klotsas E, Caumes E, Schlagenhauf P, López-Vélez R, Burchard G-D, Santos-O'Connor F, Weld L, von Sonnenburg F, Field V, de Vries P, Jensenius M, Loutan L, Castelli F (2011) Travel-related imported infections in Europe, EuroTravNet 2009. Clin Microbiol Infect. doi:10.1111/j.1469-0691.2011.03596.x

Olin G (1942) Occurrence and mode of transmission of tularaemia in Sweden. Acta Pathol Microbiol Scand 19:220–247

Onyeka JOA (1982) The taxonomic value of pre-alar scales in the identification of *Culex* (*Culex*) *pipiens* L. and *Culex* (*Cx.*) *torrentium* Martini (Diptera: Culicidae). Mosq Syst 14:41–52

Overgaauw P, Van Dijk E (2009) Autochthonal case of *Dirofilaria repens* in a dog in The Netherlands. Vet Rec 164:158

Pampiglione S, Rivasi F (2000) Human dirofilariasis due to *Dirofilaria* (*Nochtiella*) repens: an update of world literature from 1995 to 2000. Parassitologia 42:231–254

Pampiglione S, Canestri Trotti G, Rivasi F (1995) Human dirofilariasis due to *Dirofilaria* (*Nochtiella*) repens: a review of world literature. Parassitologia 37:149–193

Pchelkina AA, Seledtsov II (1978) Experimental studies of the relationship between Ťahyňa virus and mosquitoes. Med Parazitol 47:59–63 [in Russian]

Pecorari M, Longo G, Gennari W, Grottola A, Sabbatini A, Tagliazucchi S, Savini G, Monaco F, Simone M, Lelli R, Rumpianesi F (2009) First human case of Usutu virus neuroinvasive infection, Italy, August-September 2009. Euro Surveill 17, pii=19446

Pepin M, Bouloy M, Bird BH, Kemp A, Paweska J (2010) Rift Valley fever virus (Bunyaviridae: Phlebovirus): an update on pathogenesis, molecular epidemiology, vectors, diagnostics and prevention. Vet Res 41:61

Pilaski J (1987) Contributions to the ecology of Ťahyňa virus in central Europe. Bull Soc Vector Ecol 12:544–553

Platonov AE, Shipulin GA, Shipulina OY, Tyutyunnik EN, Frolochkina TI, Lanciotti RS, Yazyshina S, Platonova OV, Obukhov IL, Zhukov AN, Vengerov YY, Pokrovskii VI (2001) Outbreak of West Nile virus infection, Volgograd Region, Russia, 1999. Emerg Infect Dis 7:128–132

Powers AM, Logue CH (2007) Changing patterns of chikungunya virus: re-emergence of a zoonotic arbovirus. J Gen Virol 88:2363–2377

Proft J, Maier WA, Kampen H (1999) Identification of six sibling species of the *Anopheles maculipennis* complex (Diptera: Culicidae) by a polymerase chain reaction assay. Parasitol Res 85:837–843

ProMED-Mail (2011) Usutu virus – Germany (02): birds, confirmed. ProMED-mail 2011; 16 Sept: 20110916.2827. http://www.promedmail.org. Accessed 28 Sept 2011

Queyriaux B, Simon F, Grandadam M, Michel R, Toulou H, Boutin JP (2008) Clinical burden of chikungunya virus infection. Lancet Infect Dis 8:2–3

Raccurt CP (1999) Dirofilariasis, an emerging and underestimated zoonoses in France. Med Trop 59:389–400 [in French]

Ramsdale C, Coluzzi M (1975) Studies on the infectivity of tropical strains of *Plasmodium falciparum* to some southern European vectors of malaria. Parassitologia 17:39–48

Ramsdale C, Snow K (2000) Distribution of the genus *Anopheles* in Europe. Eur Mosq Bull 7:1–26

Rappole JH, Derrickson SR, Hubálek Z (2000) Migratory birds and spread of West Nile virus in the western hemisphere. Emerg Infect Dis 6:319–328

Reiter P (2000) From Shakespeare to Defoe: malaria in England in the Little Ice Age. Emerg Infect Dis 6:1–11

Reiter P (2010) West Nile virus in Europe: understanding the present to gauge the future. Euro Surveill 15, pii=19508

Reusken C, De Vries A, Buijs J, Braks M, Den Hartog W, Scholte EJ (2010) First evidence for presence of *Culex pipiens* biotype *molestus* in the Netherlands, and of hybrid biotype *pipiens* and *molestus* in northern Europe. J Vector Ecol 35:210–212

Rezza G, Nicoletti L, Angelini R, Romi R, Finarelli AC, Panning M, Cordioli P, Fortuna C, Boros S, Magurano F, Silvi G, Angelini P, Dottori M, Ciufolini MG, Majori GC, Cassone A (2007) Infection with Chikungunya virus in Italy: an outbreak in a temperate region. Lancet 370:1840–1847

Rizzo C, Vescio F, Declich S, Finarelli AC, Macini P, Mattivi A, Rossini G, Piovesan C, Barzon L, Palù G, Gobbi F, Macchi L, Pavan A, Magurano F, Ciufolini MG, Nicoletti L, Salmaso S, Rezza G (2009) West Nile virus transmission with human cases in Italy, August-September 2009. Euro Surveill 14, pii=19353

Rosen L (1986) Dengue in Greece in 1927 and 1928 and the pathogenesis of dengue hemorrhagic fever: new data and a different conclusion. Am J Trop Med Hyg 35:642–653

Roukens AH, Visser LG (2008) Yellow fever vaccine: past, present and future. Expert Opin Biol Ther 8:1787–1795

Samanidou A, Harbach RE (2003) *Culex* (*Culex*) *tritaeniorhynchus* Giles, a newly discovered potential vector of arboviruses in Greece. Eur Mosq Bull 16:15–17

Santa-Ollala Peralta P, Vazquez-Torres MC, Latorre-Fandós E, Mairal-Claver P, Cortina-Solano P, Puy-Azón A, Adiego Sancho B, Leitmeyer K, Lucientes-Curdi J, Sierra-Moros MJ (2010) First autochthonous malaria case due to *Plasmodium vivax* since eradication, Spain, October 2010. Euro Surveill 15, pii=19648

Sartori M, La Terra G, Pozzoli G, Franzini C (1989) A case of autochthonous malaria in Italy. Scand J Infect Dis 21:357–358
Savage HM, Ceianu C, Nicolescu G, Karabatsos N, Lanciotti R, Vladimirescu A, Laiv L, Ungureanu A, Romanca C, Tsai TF (1999) Entomologic and avian investigations of an epidemic of West Nile fever in Romania in 1996, with serologic and molecular characterization of a virus isolate from mosquitoes. Am J Trop Med Hyg 61:600–611
Schmidt-Chanasit J, Haditsch M, Schöneberg I, Günther S, Stark K, Frank C (2010) Dengue virus infection in a traveller returning from Croatia to Germany. Euro Surveill 15, pii=19677
Shute PG (1940) Failure to infect English specimens of *Anopheles maculipennis* var. *atroparvus* with certain strains of *Plasmodium falciparum* of tropical origin. J Trop Med Hyg 43:175–178
Simková A, Danielová V, Bárdoš V (1960) Experimental transmission of the Ťahyňa virus by *Aedes vexans*. Acta Virol 4:341–347
Simón F, López-Belmonte J, Marcos-Atxutegi C, Morchón R, Martín-Pacho JR (2005) What is happening outside North America regarding human dirofilariasis? Vet Parasitol 133:181–189
Simón F, Morchón R, González-Miguel J, Marcos-Atxutegi C, Siles-Lucas M (2009) What is new about animal and human dirofilariosis? Trends Parasitol 25:404–409
Sissoko D, Moendandze A, Malvy D, Giry C, Ezzedine K, Solet JL, Pierre V (2008) Seroprevalence and risk factors of chikungunya virus infection in Mayotte, Indian Ocean, 2005–2006: a population-based survey. PLoS One 3:e3066
Skog M, Espmark A (1982) Ockelbo disease: epidemic arthritis-exanthema syndrome in Sweden caused by Sindbis-virus like agent. Lancet 1:795–796
Sluka F (1969a) Recognition of clinical forms of Valtice fever, a new arbovirus infection. Wien Med Wochenschr 119:765–769 [in German]
Sluka F (1969b) The clinical picture of Čalovo virus infection. In: Bárdoš V (ed) Arboviruses of the California complex and the Bunyamwera group. SAS, Bratislava, pp 337–339
Smetana A, Danielová V, Kolman JM, Málková D, Minář J (1967) The isolation of Čalovo virus from the group *Anopheles maculipennis* in southern Moravia. J Hyg Epidemiol Microbiol Immunol 11:55–59
Smith JL, Fonseca DM (2004) Rapid assay for identification of members of the *Culex* (*Culex*) *pipiens* complex, their hybrids, and other sibling species (Diptera: Culicidae). Am J Trop Med Hyg 70:339–345
Snow K, Ramsdale C (1999) Distribution chart for European mosquitoes. Eur Mosq Bull 3:14–31
Staples JE, Breiman RF, Powers AM (2009) Chikungunya fever: an epidemiological review of a re-emerging disease. Clin Infect Dis 49:942–948
Steinmetz HW, Bakonyi T, Weissenbock H, Hatt JM, Eulenberger U, Robert N, Hoop R, Nowotny N (2011) Emergence and establishment of Usutu virus infection in wild and captive avian species in and around Zurich, Switzerland – genomic and pathologic comparison to other central European outbreaks. Vet Microbiol 148:207–212
Strode GK (1951) Yellow fever. McGraw-Hill, New York, 725 pp
Swanepoel RCJ (1994) Rift Valley fever. In: Coetzer JAW, Thompson GR, Tustin RD (eds) Infectious diseases of livestock with special reference to Southern Africa. Oxford University Press, Cape Town, pp 688–717
Tamba M, Bonilauri P, Bellini R, Calzolari M, Albieri A, Sambri V, Dottori M, Angelini P (2010) Detection of Usutu virus within a West Nile virus surveillance program in Northern Italy. Vector Borne Zoonotic Dis 11:551–557
Tatem AJ, Hay SI, Rogers DJ (2006a) Global traffic and disease vector dispersal. Proc Natl Acad Sci USA 103:6242–6247
Tatem AJ, Rogers DJ, Hay SI (2006b) Estimating the malaria risk of African mosquito movement by air travel. Malar J 5:57
Tolle MA (2009) Mosquito-borne diseases. Curr Probl Pediatr Adolesc Health Care 39:97–140
Traavik T, Mehl R, Wiger R (1978) California encephalitis viruses isolated from mosquitoes collected in southern and arctic Norway. Acta Path Microbiol Scand B 86:335–341

Traavik T, Mehl R, Wiger R (1985) Mosquito-borne arboviruses in Norway: further isolations and detection of antibodies to California encephalitis viruses in human, sheep and wildlife sera. J Hyg 94:111–122

Triebenbach AN, Vogl S, Lotspeich-Cole L, Sikes D, Happ G, Hueffer K (2010) Detection of *Francisella tularensis* in Alaskan mosquitoes (Diptera: Culicidae) and assessment of a laboratory model for transmission. J Med Entomol 47:639–648

Tsai TF, Popovici F, Cernescu C, Campbell GL, Nedelcu NI (1998) West Nile encephalitis epidemic in southeastern Romania. Lancet 352:767–771

Turell MJ, Lundström JO, Niklasson B (1990) Transmission of Ockelbo virus by *Aedes cinereus*, *Ae. communis*, and *Ae. excrucians* (Diptera: Culicidae) collected in an enzootic area in central Sweden. J Med Entomol 27:266–268

Turell MJ, Linthicum KJ, Patrican LA, Davies FG, Kairo A, Bailey CL (2008) Vector competence of selected African mosquito (Diptera: Culicidae) species for Rift Valley fever virus. J Med Entomol 45:102–108

Vainio J, Cutts F (1998) Yellow fever. WHO document WHO/EPI/GEN/98.11. WHO, Geneva, 87 pp

Vanlandingham DL, Davis BS, Lvov DK, Samokhvalov EI, Lvov SD, Black WC, Higgs S, Beaty BJ (2002) Molecular characterization of California serogroup viruses isolated in Russia. Am J Trop Med Hyg 67:306–309

Vasilakis N, Cardosa J, Hanley KA, Holmes EC, Weaver SC (2011) Fever from the forest: prospects for the continued emergence of sylvatic dengue virus and its impact on public health. Nat Rev Microbiol 9:532–541

Vazquez A, Jimenez-Clavero M, Franco L, Donoso-Mantke O, Sambri V, Niedrig M, Zeller H, Tenorio A (2011) Usutu virus: potential risk of human disease in Europe. Euro Surveill 16, pii=19935

Weissenböck H, Kolodziejek J, Url A, Lussy H, Rebel-Bauder B, Nowotny N (2002) Emergence of Usutu virus, an African mosquito-borne flavivirus of the Japanese encephalitis virus group, central Europe. Emerg Infect Dis 8:652–656

Weissenböck H, Kolodziejek J, Fragner K, Kuhn R, Pfeffer M, Nowotny N (2003) Usutu virus activity in Austria, 2001–2002. Microbes Infect 5:1132–1136

Weissenböck H, Chvala-Mannsberger S, Bakonyi T, Nowotny N (2007) Emergence of Usutu virus in central Europe: diagnosis, surveillance and epizootiology. In: Takken W, Knols BGJ (eds) Emerging pests and vector-borne diseases in Europe, vol 1, Ecology and control of vector-borne diseases. Wageningen Academic, Wageningen, pp 153–168

WHO (World Health Organization) (2005) Dengue, dengue haemorrhagic fever and dengue shock syndrome in the context of the integrated management of childhood illness. WHO, Geneva, 40 pp

WHO (World Health Organization) (2010) Rift Valley fever, Fact sheet no. 207. http://www.who.int/mediacentre/factsheets/fs207/en/. Accessed 11 Nov 2011

Wojta J, Aspöck H (1982) Untersuchungen über die Möglichkeit der Einschleppung durch Stechmücken übertragener Arboviren durch Vögel nach Mitteleuropa. Mitt Österr Ges Tropenmed Parasitol 4:85–98

Zoller T, Naucke TJ, May J, Hoffmeister B, Flick H, Williams CJ, Frank C, Bergmann F, Suttorp N, Mockenhaupt FP (2009) Malaria transmission in non-endemic areas: case report, review of the literature and implications for public health management. Malar J 8:71

Chapter 2
Exotic Mosquitoes Conquer the World

Norbert Becker, Björn Pluskota, Achim Kaiser, and Francis Schaffner

Abstract Mosquitoes have inhabited the globe for more than 100 million years, long before *Homo sapiens* occurred on Earth. In the course of evolution, they were able to adjust their biology to a great variety of ecological conditions and reproduce in almost all aquatic habitats. Without the support of *Homo sapiens*, mosquitoes disperse passively by wind drift (up to ~25 km) or by active flight usually limited to <50 km per migration process. However, present-day human activities enable mosquitoes to be transported from one continent to another within a matter of hours to a few days. Increased transcontinental mobility of humans as well as the international trade, facilitate the dispersal and in some cases, the establishment of exotic mosquito species in other countries with favorable climatic conditions. The most remarkable ability of these species is that the eggs can survive desiccation and dryness for months or sometimes even years and can thus survive long periods with unfavorable living conditions. This ensures that e.g. eggs can survive in used tires or other small containers when these are shipped and consequently increase the probability of successful transport. In general, these species possess a high ecological potency and can rapidly adapt to new habitats due to their genetic plasticity. Within the about 30 species known to have established in new areas throughout the

N. Becker (✉)
German Mosquito Control Association (KABS), Ludwigstr. 99, 67165 Waldsee, Germany

Faculty of BioSciences, University of Heidelberg, Im Neuenheimer Feld 230, 69120 Heidelberg, Germany
e-mail: norbertfbecker@web.de

B. Pluskota • A. Kaiser
German Mosquito Control Association (KABS), Ludwigstr. 99, 67165 Waldsee, Germany

F. Schaffner
Vector Entomology Unit, Institute of Parasitology, University of Zürich, Winterthurerstr. 266a, 8057 Zürich, Switzerland

Avia-GIS Agriculture and Veterinary Information and Analysis, Risschottlei 33, 2980 Zoersel, Belgium

H. Mehlhorn (ed.), *Arthropods as Vectors of Emerging Diseases*,
Parasitology Research Monographs 3, DOI 10.1007/978-3-642-28842-5_2,

world, 3 species merit special recognition for their dispersal potential and also for their significance as vectors of human diseases: *Aedes (Stegomyia) aegypti* (Linnaeus 1762), *Ae. (Stegomyia) albopictus* (Skuse 1895) and *Ochlerotatus (Finlaya) japonicus* (Theobald 1901). In this chapter the taxonomy, biology, distribution and medical importance of the most successful invasive species are discussed, namely *Aedes aegypti*, *Aedes. albopictus*, *Ochlerotatus japonicus*, *Ochlerotatus koreicus*, *Ochlerotatus atropalpus* and *Ochlerotatus triseriatus*.

Keywords *Aedes aegypti* • *Aedes. albopictus* • Biology • Distribution • Exotic mosquitoes • Medical importance • *Ochlerotatus japonicus* • *Ochlerotatus koreicus* • *Ochlerotatus atropalpus* • *Ochlerotatus triseriatus* • Taxonomy

2.1 Introduction

Mosquitoes have inhabited the globe for more than 100 million years, long before *Homo sapiens* occurred on Earth. In the course of evolution, they were able to adjust their biology to a great variety of ecological conditions and reproduce in almost all aquatic habitats (Becker et al. 2010, 2011). Without the support of *Homo sapiens*, mosquitoes disperse passively by wind drift (up to ~25 km) or by active flight usually limited to <50 km per migration process (Bidlingmayer and Evans 1987; Becker et al. 2010). However, present-day human activities enable mosquitoes to be transported from one continent to another within a matter of hours to a few days. Increased transcontinental mobility of humans, as well as the international trade, facilitates the dispersal and, in some cases, the establishment of exotic mosquito species in other countries with favorable climatic conditions. A species is considered established after having produced at least three generations in a new territory (Nehring and Leuchs 1999). The term "invasive species" is used for introduced species that proliferate in great numbers and thus might have a negative impact on native species and the ecosystem or on human health and activities such as tourism or agriculture (Manguin and Christophe 2011). Established species which do not proliferate in great numbers and have no significant impact on the ecosystem are considered as "exotic species." Invasive species may have competitive advantages against native species or even against already established other invasive species, thus causing a reduction of the locally established species. They may also expand into new areas by filling an empty niche or exploiting unused resources. Nonnative species can benefit from reduced predation and parasitism in the new habitat as compared to the native range which also enhances their spread and proliferation (Juliano and Lounibos 2005).

Out of the more than 3,500 mosquito species worldwide, only a few species have begun to spread far beyond their original geographical borders. Particular characteristics of these species have favored their spread. Initially, these mosquitoes colonize small water bodies, such as water-filled tree holes, rock pools, coconut shells, dentrotelms, bromeliads, or other phytotelms. Usually, these small accumulations of water show an extremely temporary character, with large variations in temperature and other abiotic conditions. The survival in these temporary habitats requires special adaptations: the larval development has to be fast because of the risk of drying out, natural catastrophes which destroy all developing stages in a certain time period should not lead to an extinction, and the developmental stages have to tolerate and survive inhospitable living conditions.

The most remarkable adjustments to these specific habitats are the following: the embryos in the eggshells can survive desiccation and dryness for months or sometimes even years and can thus survive long periods with unfavorable living conditions. This ensures that, for example, eggs can survive in used tires or other small containers when these are shipped and consequently increase the probability of successful transport. Another phenomenon is "hatching in installments." Even within a batch of eggs laid by a single female and subjected to the same microclimate conditions, the larvae do not hatch simultaneously. If, for example, all of the larvae hatched at the same time under ideal conditions and if, as a consequence of a sudden dry spell, all of the breeding sites dry out before the brood could complete its developmental cycle, one single natural event could virtually wipe out the entire mosquito population. By "hatching in installments," the mosquito population can survive such potentially catastrophic events. After one hatching process, a large contingent of unhatched larvae still remains in the breeding site. At the next flooding, these larvae have the opportunity to successfully produce a new population without the necessity of new eggs being laid (Becker 1989; Becker et al. 2010).

Also the behavior of gravid females during oviposition is an adaptation to the living conditions in temporary breeding waters. A female of *Aedes/Ochlerotatus* lays about 100 eggs per oviposition. However, especially females of species breeding in small containers usually do not place all eggs in only one single breeding site, but rather distribute them in several breeding sites, ensuring that some of the developing stages survive even if one breeding site dries out or is destroyed before complete development of the brood.

There are further important behavioral factors that favor the spread of some container-breeding mosquitoes, i.e., they often have no strict host preference for blood meals, they are usually adapted to a wide climate range which allows the development in new areas, and they can develop in a short time in large numbers and quickly set up a stable population. In general, these species possess a high ecological potency and can rapidly adapt to new habitats due to their genetic plasticity. *Aedes albopictus* is an excellent example. This species has spread from tropical areas to areas with temperate climates which do not allow a constant follow-up of generations, for example, during winter periods. As a consequence, the species goes through a winter diapause during which the embryos are not able to hatch and remain in the eggshell until the living conditions allow a further development. Climate frequently limits the range of most invasive species; however, when the founder generation of a species in the new area originates from an area with similar climate conditions, the likelihood to establish in the new area is high. Overall, exotic species have a tremendous ability to adapt to new living conditions based on a very broad genetic plasticity, and they are easily disseminated by human activities.

Within the about 30 species known to have established in new areas throughout the world, three species merit special recognition for their dispersal potential and also for their significance as vectors of human diseases: *Aedes (Stegomyia) aegypti* (Linnaeus 1762), *Ae. (Stegomyia) albopictus* (Skuse 1895), and *Ochlerotatus (Finlaya) japonicus* (Theobald 1901). Due to the features discussed above, these species have been able to rapidly and successfully build up and establish stable populations in new geographic regions (Hawley 1988; Moore and Mitchell 1997; Pluskota et al. 2008; Schaffner et al. 2009; Medlock et al. in press).

All three species are characterized by their high vector competence for arboviruses. *Ae. aegypti* and *Ae. albopictus* are the primary and secondary vectors of dengue virus, causing fever and dengue hemorrhagic fever (DHF), with more than 40% of the human population worldwide at risk, especially in megacities of the tropics (Halstead 1980, 1982, 1992; Becker et al. 1991; Gratz 1999). *Aedes albopictus* has become the most important vector for the chikungunya virus and was recently involved in the transmission of this virus to humans in Italy in 2007 (Reiter et al. 2006; Beltrame et al. 2007; Angelini et al. 2007; Rezza et al. 2007). *Ochlerotatus j. japonicus* is a competent laboratory vector of several arboviruses, such as West Nile (WN) virus, and is considered a significant public health risk (Sardelis and Turell 2001).

2.2 Mode of Spreading

For most of the human history, continental populations have been relatively isolated from each other. Only relatively recently, there has been extensive contact between humans, flora, and fauna from both the Old and New World (Carey 2002). Several centuries ago, diseases like yellow fever and its mosquito vector *Ae. aegypti* have been transported on slave ships from West Africa to the New World. Nowadays, modern transport networks are the basis for the global growth of economy. Globalization of the world economy has resulted in a steady and rapid increase of airfreight and shipping traffic (Upham et al. 2003; Zachcial and Heideloff 2003).

Rapid transportation systems connect the world's biota more than any time in Earth's history. Within a couple of hours or days, organisms are transported from one continent to another. Besides economic activities, human migration and tourism is increasing the risk for spreading both disease vectors and diseases. Especially the international trade, mainly of used tires and occasionally of lucky bamboo (*Dracaena* spp.) cuttings, is the vehicle for the spread of most of the invasive mosquitoes.

Containers, for example, full of used tires, can harbor thousands of *Aedes/Ochlerotatus* eggs (Reiter and Sprenger 1987; Cornel and Hunt 1991; Gratz 2004). When the goods reach their region of destination, they are often stored outdoors, and an accumulation of rainwater allows the larvae to hatch and to develop to a stable population if the climate conditions are favorable. A good example is the introduction of *Ae. albopictus* in Italy by used tires (Sabatini et al.

1990). Within a few years, the species established from the north to Sicily in the south, supported by the wide-scale distribution of tires and dispersal by vehicles. In 2007, *Ae. albopictus* was the first time involved in a small outbreak of chikungunya fever in the Emilia-Romana region caused by the tropical virus which was introduced by an infected tourist from India (Beltrame et al. 2007; Pfeffer and Loescher 2006).

Once established in Italy, the species mainly spread by vehicles to neighboring countries such as France (French Riviera), Spain, Switzerland, and across the Alps into Germany by vehicles originating from Italy (Pluskota et al. 2008). Mosquitoes can also spread by trains and even by boats. It is assumed that the establishment of *Ae. albopictus* along the Adriatic coast in Croatia was supported by tourist boats from Italy that were infested with the species.

The international trade of plants shipped in containers from Asia (China) is another source for the introduction of exotic mosquitoes, particularly if the plants are transported in boxes holding water. A good example is the trade of lucky bamboo (*Dracaena* spp.), which leads to the introduction of *Aedes albopictus* to California (Madon et al. 2002, 2004) and the Netherlands (Scholte et al. 2007, 2010).

In general, the simultaneous distribution of large numbers of developing stages of an exotic species by trade may lead more rapidly and easily to a stable established population, as compared to the introduction of single individuals, even if gravid females are considered which may lay dozens of eggs. The faster a founder population reproduces in masses, the greater is the likelihood to become established.

However, not only the process of globalization favors the spread of exotic mosquitoes but also the phenomenon of global climate change if associated with higher temperatures and precipitation. Models considering IPCC of global climate scenarios applied to *Ae. albopictus* reveal suitable areas according to scenario and timescale (ECDC 2009; Becker 2009).

2.3 Important Exotic Mosquitoes

In the following, the most successful invasive species which are distributed by merchandise will be discussed, namely, *Aedes aegypti*, *Ae. albopictus*, *Ochlerotatus japonicus*, *Oc. koreicus*, *Oc. atropalpus*, and *Oc. triseriatus*.

2.3.1 Aedes (Stegomyia) aegypti *(Linnaeus) 1762*

[= *Stegomyia aegypti* sensu Reinert et al. 2004]

Female: A medium-sized dark species with contrasting silvery white ornamentation on the head, scutum, legs, and abdomen (Becker et al. 2010). *Aedes aegypti* is

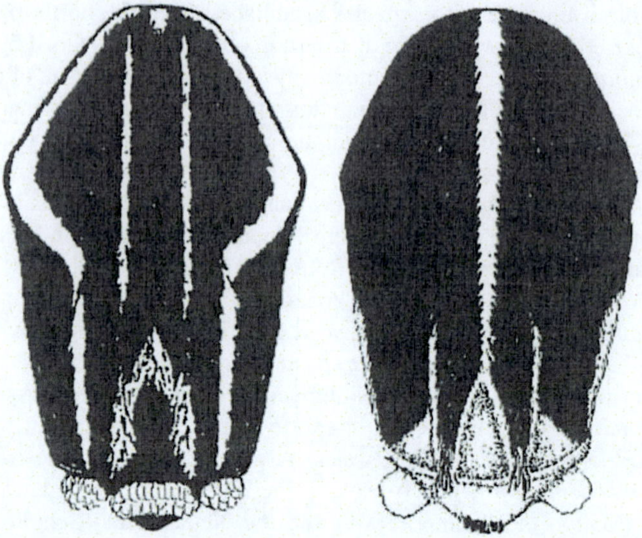

Fig. 2.1 Scutal patterns of female of *Aedes aegypti* and *Ae. albopictus*

easily recognized and distinguished from the other members of the subgenus *Stegomyia* by the white scutal markings which form a typical "lyre-shaped" pattern (Fig. 2.1). The hind tibia is entirely dark, the hind tarsus has broad basal white bands on tarsomeres I–IV, and tarsomere V is all white.

Larva: The antennal seta (1-A) is small and single, inserted slightly beyond the middle of the antennal shaft. The frontal setae (5-C to 7-C) are long and single, and the outer frontal seta (7-C) very rarely has two branches. The comb consists of a single irregular row of 6–12 scales, each scale with a long median spine and strong subapical spines forming a "trifid" appearance. The pecten has 8–22 teeth (usually 10–16), evenly spaced, or sometimes the distal most tooth is detached apically. The siphonal tuft (1-S) has three or four branches, usually inserted close to the distal pecten tooth and just beyond the middle of the siphon.

Biology: The "African tiger mosquito" ("yellow fever mosquito"), *Ae. aegypti*, is supposed to have its origin in Africa. An ancestor form (*Ae. aegypti formosus*) is still present on the African continent, whereas a domestic form (*Ae. aegypti aegypti*) has spread across almost all tropical and subtropical countries, during the past five centuries.

Its populations increase especially in areas where household water storage in containers is common and where solid waste disposal services are inadequate. In subtropical climates, the species is found almost always in the close vicinity of human settlements. The larvae occur in a wide variety of small artificial containers and water recipients of all kinds, both inside and outside of human habitations in gardens and within a circle of 500 m around dwellings, for example, in earthenware pots and water tanks for storing water, uncovered cisterns, rain-filled empty cans or flower pots, broken bottles, or discarded tires. The larvae may also be found in

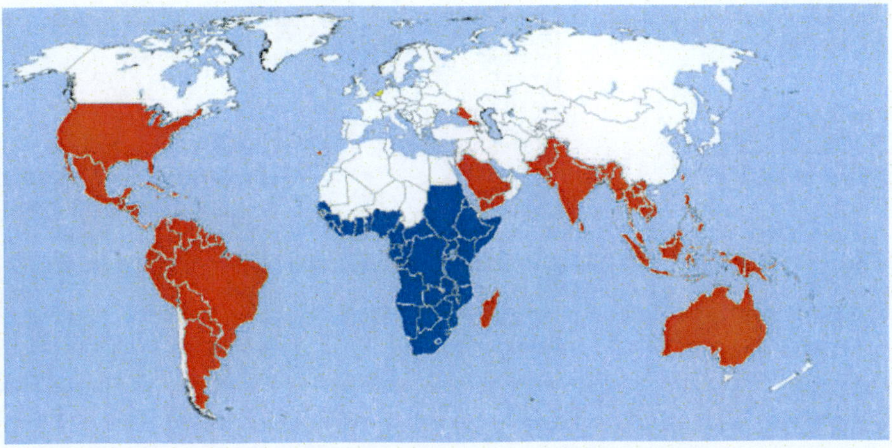

Map 2.1 Worldwide distribution of *Ae. aegypti* (blue = countries with native populations; red = countries with nonnative (established) populations; yellow = countries with sporadic occurrence – not established populations)

tree holes, leaf axils, bamboo stumps, or coconut shells after heavy rainfall. The breeding water is mostly clean or has a moderate content of organic matter. The larvae spend a long time under water feeding on the bottom of their breeding sites. The eggs are resistant to desiccation and are deposited close to the waterline in the mentioned recipients. At a temperature of 27–30°C, the larvae will hatch 2 days after egg deposition if flooded, pupation occurs after 8 days, and the adults emerge from the pupae 9–10 days after the eggs have been flooded. The females feed predominantly during the day in shaded places and only occasionally during the night in rooms. Human blood seems to be preferred to that of domestic animals (Carpenter and La-Casse 1955). The adults are frequently found resting indoors; they do not migrate over long distances and rarely fly more than several 100 m from their breeding sites. The available literature about *Ae. aegypti*, its biology, and medical significance is numerous, and a well-recognized monograph was published by Christophers in 1960.

Distribution: This cosmotropical species is distributed in the tropical, subtropical, and warm temperate regions of both hemispheres (Map 2.1). Its range is mainly limited by the 10°C cold-month isotherms allowing continuous breeding all year round (Christophers 1960). Certain populations may extend their summer range considerably north of this line, for example, in North America, specimens may be found up to 40° northern latitude in southern Illinois and Indiana, but they are not able to survive during cold winter months; this prevents the establishment of permanent populations. In Europe, prior to 1945, all Mediterranean countries and most major port cities had reported at least occasional introductions of *Ae. aegypti* (Mitchell 1995). It could be found in Portugal, Spain, France, the UK, Italy, former Yugoslavia, Greece, Turkey, Ukraine, and Russia, but has been eradicated from these countries. Recently, *Ae. aegypti* reestablished in Madeira and around the

Fig. 2.2 Female of *Aedes albopictus*

Black Sea in southern Russia, Abkhazia, and Georgia (Riabova et al. 2005; Almeida et al. 2007). Very recently, *Ae. aegypti* was even found in the Netherlands in used tires imported from the USA, associated with *Ae. atropalpus* and *Ae. albopictus* (Scholte et al. 2010). This is the first record of *Ae. aegypti* in northern Europe after decades (Hansford et al. 2010).

Medical importance: As the major vector of yellow fever virus, *Ae. aegypti* has long been notorious as the "yellow fever mosquito," but it is also an important vector of dengue, chikungunya, and several other viral infections.

2.3.2 Aedes (Stegomyia) albopictus *(Skuse) 1894*

[= *Stegomyia albopicta* sensu Reinert et al. 2004]

Female: The scutum is mainly covered with narrow dark scales, with a prominent longitudinal stripe of narrow white scales in the middle of the scutum (Figs. 2.1 and 2.2). The hind tibiae are entirely dark, hind tarsus has broad basal white bands on tarsomeres I–IV, and tarsomere V is all white.

Larva: The antennal seta (1-A) is simple and small and situated close to the middle of the antennal shaft. The median frontal seta (6-C) has 1–2 branches, and the inner frontal seta (5-C) is longer and simple. The comb consists of 6–13 (usually 8–10) long slender scales in a single row. The siphon is short and tapers distinctly from the middle; the siphonal index is 1.7–2.5. The number of pecten teeth is between 8 and 14; they are evenly spaced. The siphonal tuft (1-S) has 2–4 branches and is inserted beyond the distal most pecten tooth, slightly beyond the middle of the siphon.

Biology: The immature stages occur in a variety of small natural and artificial containers, for example, in tree holes, bamboo stumps, coconut shells, rock holes, plant axils, or palm fronds and in flower pots, tin cans, water jars, metal and wooden buckets or drums, broken glass bottles, or discarded tires (Huang 1972; Becker et al. 2010). The eggs are resistant to desiccation which facilitates their transport in used tire casings, even over long distances. Continuous breeding throughout the year takes place in tropical and subtropical areas, but in more temperate climatic zones, such as Europe, populations of *Ae. albopictus* are found which show embryonic diapause and overwinter in the egg stage. Several generations per year may occur. Adult females predominantly feed on humans but may also bite other mammals including rabbits, dogs, cows, and squirrels or occasionally avian hosts. This feeding behavior indicates that *Ae. albopictus* is well suited for transmitting of a variety of arboviruses that use mammals and birds as their main hosts (Mitchell 1995). To feed on humans, the females bite during daytime outside houses, mainly in shaded situations, and rarely enter dwellings during dusk and night (Delatte et al. 2010). *Aedes albopictus* is a mass species in East Asia causing great nuisance wherever it occurs and, although not present before 1990, has become a major pest species in some areas of the USA and Italy.

Distribution: The Asian tiger mosquito *Ae. albopictus*, originating from South-east Asia, has undergone a noteworthy expansion of its range within few decades (Hawley 1988). With the increase in international trade of used tires, *Ae. albopictus* has spread across very large distances and between continents (Reiter and Sprenger 1987; Reiter 1998; Scholte and Schaffner 2007).

The mosquito spread from its native range initially to the Pacific Islands (Gratz 2004) in the 1930s and then, by the end of the century, to other countries in both the Old and New World (Moore and Mitchell 1997; Gratz 2004). In the Palaearctic, it occurred in Japan and China. In 1985, it was discovered for the first time in the New World (Houston, Texas), and this was the beginning of a rapid spread of *Ae. albopictus* on the American continent as well as new introductions in many parts of the world (Mitchell 1995). It is now present in over 25 states of the USA and in several countries of South America and Africa (Map 2.2). Specimens have been found in Australia and New Zealand, but breeding populations have so far not become established there.

In Europe, *Ae. albopictus* has probably been present in Albania since at least 1979 (Adhami and Murati 1987; Adhami and Reiter 1998). In the early 1990s, it was passively introduced into Italy due to the international trade of used tires. The species was first detected in Genoa in September 1990 (Dalla Pozza and Majori 1992) followed by a rapid spread into other areas of northern and central Italy (Romi 1994). Since 1999, *Ae. albopictus* has been found in various southern and central European countries, including France (Schaffner et al. 2001), Bosnia and Herzegovina (Petrić et al. 2006), Montenegro (Petrić et al. 2001), Belgium (Schaffner et al. 2004), Switzerland (Flacio et al. 2004), Greece (Samanidou-Voyadjoglou et al. 2005), Malta (Gatt et al. 2009), Monaco (ECDC 2009), Croatia (Klobucar et al. 2006), San Marino (ECDC 2009), Slovenia (Petrić et al. 2006), Spain (Aranda et al. 2006), the Netherlands (Scholte et al. 2007), Vatican City (ECDC 2009), and Germany (Pluskota et al. 2008; Werner et al. 2012).

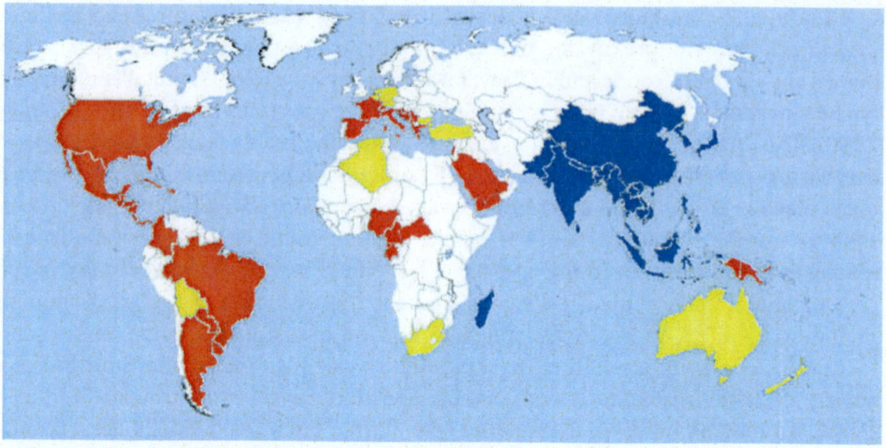

Map 2.2 Worldwide distribution of *Aedes albopictus* (blue = countries with native populations; red = countries with nonnative (established) populations; yellow = countries with sporadic occurrence – not established populations)

Another way of spreading the eggs and larvae of *Ae. albopictus* is the trade with the ornamental plant *Dracaena* sp. ("lucky bamboo"). These plants are packaged in standing water during shipment and permit an "ideal insectary in transit" which led to the introduction of *Ae. albopictus* from Asia to California and the Netherlands (Madon et al. 2004; Scholte et al. 2007). The species is not yet established in Belgium and Germany neither in the Netherlands where it is repeatedly introduced and controlled.

Medical importance: Aedes albopictus is a vector of at least 22 arboviruses, including chikungunya, West Nile, dengue, and yellow fever viruses as well as *Dirofilaria immitis* (dog heartworm) (Gubler and Rosen 1976; Gratz 2004; Becker et al. 2010). Although *Ae. aegypti* is the principal vector of dengue fever, recent outbreaks of dengue disease in the absence of *Ae. aegypti* have implicated *Ae. albopictus* as a competent vector (Gratz 2004; Effer et al. 2005). It is assumed that the Asian tiger mosquito is also involved in the autochthonous transmission of Dengue viruses in Europe, namely, in southern France and Croatia as well as in the outbreak of chikungunya fever in Italy 2007 (Bonilauri et al. 2008; La Ruche et al. 2010; Schmidt-Chanasit et al. 2010; Gjenero-Margan et al. 2011).

2.3.3 Ochlerotatus (Finlaya) japonicus japonicus *(Theobald 1901)*

[sensu Reinert et al. 2004 = *Aedes japonicus* sensu auctorum = *Hulecoeteomyia japonica* sensu Reinert et al. 2006]

There exist at least four morphologically similar subspecies (Tanaka et al. 1979) which occur in different parts of East Asia: *Oc. japonicus japonicus* (Palaearctic Japan and Korea), *Oc. japonicus shintiensis* (Taiwan), *Oc. japonicus yaeyamensis*, and

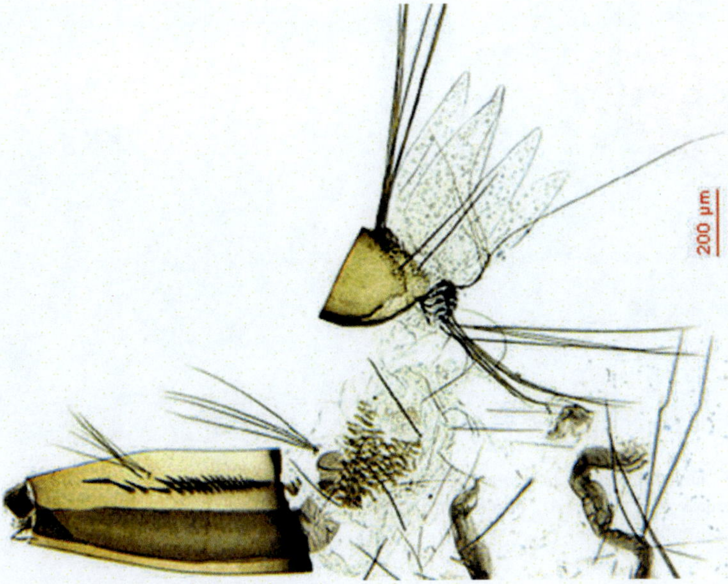

Fig. 2.3 Larva of *Oc. j. japonicus*

Oc. japonicus amamiensis (both on the Ryukyu Archipelago). The subspecies differ by morphology mainly in the ornamentation of the hind femora.

Female: The adult female of *Oc. japonicus* is a medium/large-sized mosquito of dark to blackish-brown appearance, with white scales on the body and legs. It is similar to the closely related *Oc. koreicus*. The pattern on the scutum is distinctive with five golden stripes of scales; the lateral scales are pale and white in a lyre shape. There are also broad pale basal bands on the hind tarsomeres I–III, and tarsomeres IV and V are entirely black. Subspiracular area bears only a few scales (contrary to *Oc. koreicus*). The proboscis is entirely dark. The head is dark and the abdomen is creamy colored (Miyagi 1971). Terga II–VII dark scaled with lateral patches of white scales at the base of each segment.

Larvae: Typical are the pecten with detached very strong teeth and the anal saddle with its highly spiculated distal margin; the inner and median frontal setae are multiple (tufts) and arranged in a straight line close to the apical margin of the head. The siphonal tuft has multiple branches (4–6), is very short, and inserted at the middle of the siphon and within the pecten. Comb scales are displayed in a patch, and the siphon index is 2.5 (Fig. 2.3).

Biology: Adult of *Oc. j. japonicus* prefer forested areas and bite usually during the daytime (Tanaka et al. 1979). They feed on a great variety of mammals and

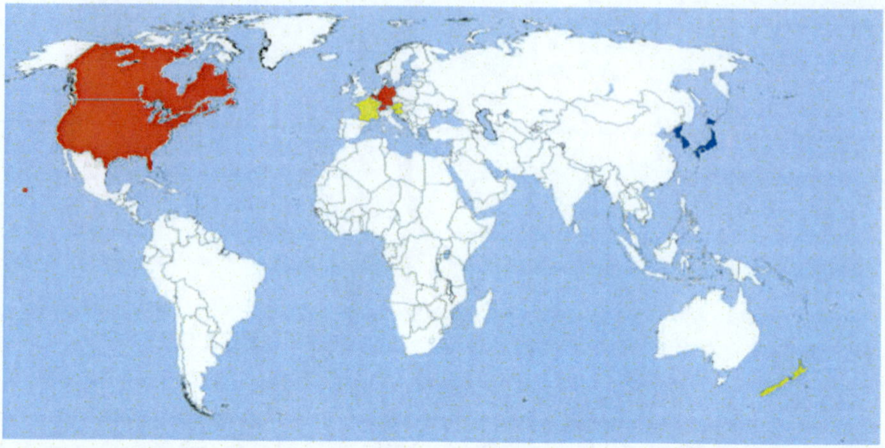

Map 2.3 Distribution of *Oc. japonicus* (blue = countries with native populations; red = countries with nonnative (established) populations; yellow = countries with sporadic occurrence – not established populations)

birds such as dogs, pigs, deers, rodents, or chicken (Scott 2003). They are more reluctant to bite humans and not as aggressive biters as floodwater mosquitoes like *Ae. vexans* or *Oc. sticticus*.

This multivoltine species has adapted to colder climates and can survive in their desiccation-resistant eggs even during cold winter times (Tanaka et al. 1979; Andreadis et al. 2001). Under moderate climate conditions, they can also be found throughout the winter as larvae. The larvae of *Oc. japonicus* are typically found in a great variety of usually small-volume natural water collections such as tree holes, bamboo stems, or rock pools or in artificial breeding sites including vases, tins, tires, drums, buckets, roof gutters, cement catch basins, or bird baths. They prefer breeding sites which are rich in organic matter, but not heavily polluted.

Distribution: *Ochlerotatus japonicus* is an Asian species originally found in Japan, Korea, South China, Taiwan, and the eastern part of the Russian Federation (Tanaka et al. 1979). It was intercepted several times in New Zealand in the 1990s (Laird et al. 1994; Fonseca et al. 2001). It has established for the first time outside its native range in the USA in 1998; it is now recorded in at least 29 other states of the country including Hawaii as well as in Canada (Map 2.3) (Andreadis et al. 2001; Saenz et al. 2006; Williges et al. 2008; Fonseca et al. 2010). The species was introduced via international transport mostly of used tires (Peyton et al. 1999; Thielman and Hunter 2006).

In 2000, *Oc. j. japonicus* was first recorded in Europe when larvae were found on a storage yard of imported used tires in France (Schaffner et al. 2003), from where it was successfully eradicated, thanks to immediate control measures. Since 2002, *Oc. j. japonicus* has repeatedly been observed on a secondhand tire company in southern Belgium (Versteirt et al. 2009a, b). Finally, in 2008, the species was detected in northern Switzerland spreading into bordering Germany (Schaffner et al. 2009). Since 2009, an intensive monitoring program of the distribution of *Oc. j. japonicus* has been conducted in Southern Germany (state of Baden-Württemberg). More than 6,500 water-filled containers in 291 municipalities distributed over the whole area of Baden-Württemberg have been investigated for *Oc. j. japonicus* developing stages. Out of 291 municipalities, 54 (18, 2%) were positive for *Oc. j. japonicus* covering an area of about 2,000 km^2 (Becker et al. 2011; Huber et al. 2012). Recently, a second focus localized 80 km apart from the known colonized area was reported (Schneider 2011). Finally, the species was reported from a large cross-bordering area of Austria and Slovenia (B. Seidel, unpublished).

Medical importance: *Ochlerotatus j. japonicus* is a competent laboratory vector of several arboviruses, such as West Nile (WN) virus and Japanese encephalitis (JE) virus, but it can also transmit St. Louis encephalitis, Eastern equine encephalitis, La Crosse, Dengue, and Chikunguny viruses (Schaffner et al. 2011) and is considered a significant public health risk (Sucharit et al. 1989; Sardelis and Turell 2001; Sardelis et al. 2002a, b, 2003).

2.3.4 Ochlerotatus (Finlaya) koreicus *(Edwards 1917)*

[sensu Reinert et al. 2004 = *Aedes koreicus* sensu auctorum = *Hulecoeteomyia koreica* sensu Reinert et al. 2006]

Females: The midscutal line is broad, and the submedian line is broken at the level of the scutal angle, the anterior end of the posterior part extending outward along the scutal angle to the scutal margin. The hind tarsi I–IV are basally banded, and the fifth tarsus bears some pale scales or even a basal band (*Oc. japonicus*: only hind tarsi I–III basally banded). The subspiracular area bears a distinct patch of broad white scales (Tanaka et al. 1979).

Larvae: The larva reminds those of *Oc. japonicus* but has no enlarged simple pecten tooth nor a spiculated distal margin of the anal saddle. The apical most teeth may be a little bit larger than the others, but never detached, as in *Oc. japonicus* (Miyagi 1971).

Biology: Adults bite during day and night and suck blood on humans and mammals. Immature stages can be found in natural breeding sites such as rock pools and tree holes but also in a great variety of artificial containers such as used tires or vases. The species overwinters in the egg stage, and the larvae are hatching after the snowmelt.

Distribution: *Ochlerotatus koreicus* is an Asian species native to Japan, China, Korea, and eastern Russia (Map 2.4). It was first found outside of its native range in Belgium in 2008 (Versteirt et al. 2009a, b). Very recently, in May 2011, developmental stages of *Oc. koreicus* were collected in the Veneto region from a manhole and other artificial containers (Capelli et al. 2011). The species is tolerant to cold

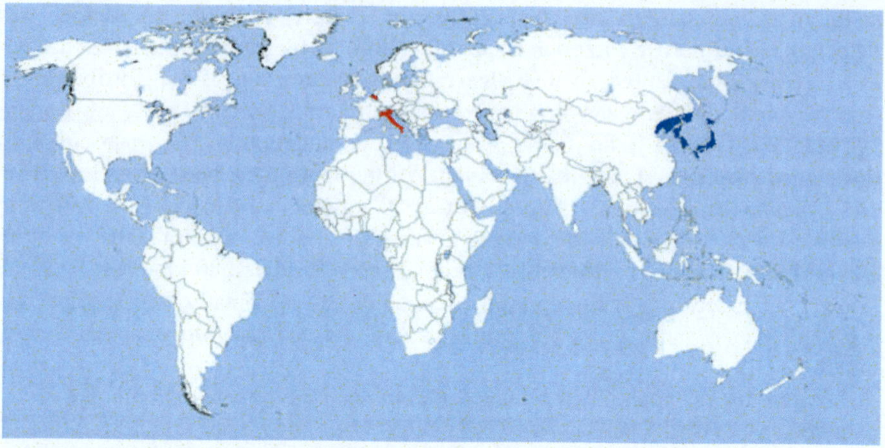

Map 2.4 Distribution of *Oc. koreicus* (blue = countries with native populations; red = countries with nonnative (established) populations; yellow = countries with sporadic occurrence – not established populations)

climates, making it capable of surviving and becoming established in Belgium and the hilly and pre-alpine areas of Italy. It seems that after *Oc. japonicus,* this species is starting to spread across the world (Cameron et al. 2010). No introduction route could be evidenced so far, neither direct from Asia nor from Belgium to Italy.

Medical importance: The species is considered a potential vector of arboviruses. It has been shown experimentally to be able to transmit Japanese encephalitis virus and the dog heartworm *Dirofilaria immitis*. However, vector competence studies are strongly required to better define the role of this mosquito in the transmission of arboviruses, such as West Nile and USUTU viruses (Capelli et al. 2011).

2.3.5 Ochlerotatus (Ochlerotatus) atropalpus *(Coquillett 1902)*

[sensu Reinert et al. 2004 = *Aedes atropalpus* sensu auctorum = *Georgecraigius atropalpus* sensu Reinert et al. 2006]

Females: Scutal scales are pale yellow, dark scales are mixed with light scales, and sometimes a darker medial stripe is present. Terga II–VII have narrow white basal bands of scales and dark brown apically. The hind tibia has pale scales at basis and apex. The hind tarsus has large basal and smaller apical ring of whitish scales at each tarsal joint; hind tarsomere V is entirely white (Wood et al. 1979).

Larvae: Ochlerotatus atropalpus resembles *Oc. japonicus* based on their detached pecten teeth and the siphonal tuft inserted within the pecten row; however, it can be separated by their inner and median frontal setae which are single and the difference in spiculation on the anal saddle. The pecten extends almost to the tip of the siphon, and the siphonal tuft inserts at the middle of the siphon within the row of

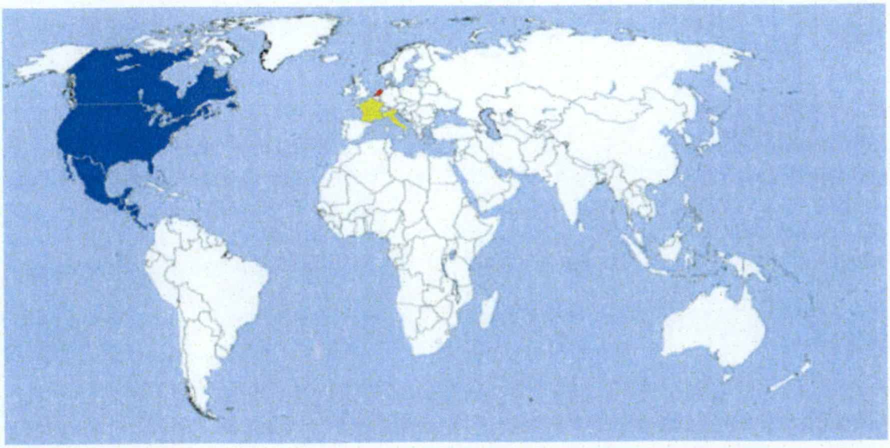

Map 2.5 Worldwide distribution of *Oc. atropalpus* (blue = countries with native populations; red = countries with nonnative (established) populations; yellow = countries with sporadic occurrence – not established populations)

pecten teeth. The comb scales are arranged in patch-like, irregular rows, contrary to *Ae. aegypti*, *Ae. albopictus*, and *Oc. triseriatus* with comb scales in a row (Wood et al. 1979).

Biology: Ochlerotatus atropalpus is a multivoltine species with a life cycle that resembles *Ae. triseriatus* in terms of larval habitats. It is most often associated with fresh water rock pool habitats along mountain streams in Northern America's inland areas (common name: American rock pool mosquito). However, it is known to breed in a variety of artificial containers especially in discarded tires and other man-made water collections such as in boats (King et al. 1960; Shaw and Maisey 1961; Berry and Craig 1984). The species bites human and other mammals outdoor during daytime.

Distribution: The species is native from North and Central America. It is found from Labrador, Canada to Panama, and from the Atlantic seaboard to Baja California (O'Meara and Craig 1970).

In Europe, it has been recorded from Italy (1996), France (2004), and the Netherlands (2009) (Map 2.5). In all cases, it was introduced via trade of used tires (Romi et al. 1997; Scholte and Schaffner 2007; Scholte et al. 2009).

Medical importance: In experiments, the mosquito was proven to be a vector of West Nile, Japanese encephalitis, La Crosse encephalitis, and some other arboviruses (Turell et al. 2001, 2005).

2.3.6 Ochlerotatus (Protomacleaya) triseriatus *(Say 1923)*

[sensu Reinert et al. 2004 = *Aedes triseriatus* sensu auctorum]

Female: Proboscis and palps dark scaled. Scutum with a middorsal longitudinal broad stripe of brown scales. Two stripes of silvery scales on the dorsal side of the

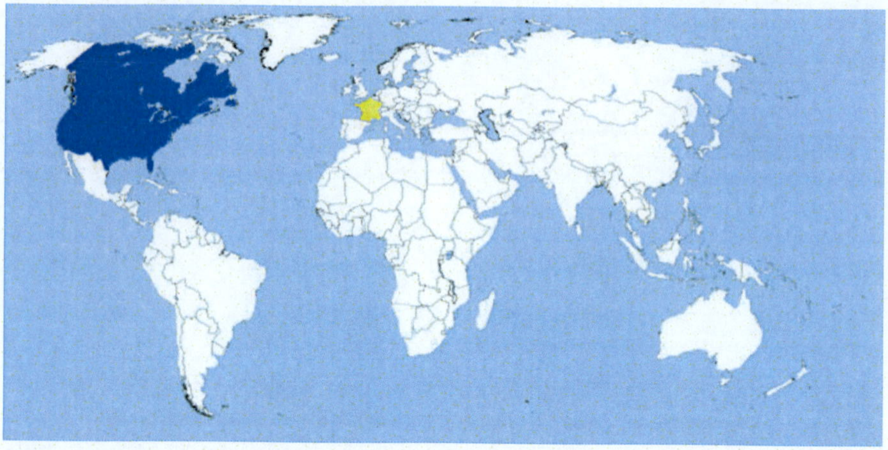

Map 2.6 Worldwide distribution of *Oc. triseriatus* (blue = countries with native populations; red = countries with nonnative (established) populations; yellow = countries with sporadic occurrence – not established populations)

thorax are the most diagnostic characteristic (Wood et al. 1979). The abdominal segments are dark and unbanded. Most of the tibiae and all tarsi are dark scaled (unbanded).

Larvae: Larvae have a characteristic gray coloration and swim with a characteristic serpentine motion. Frontal setae 5-C are single, and 6-C are 2–3 branched. The pecten teeth are evenly spaced, and the siphonal tuft inserts beyond the uppermost tooth (Wood et al. 1979). The comb has a single row of scales that are arranged in an extremely irregular fashion (comb scales of *Ae. aegypti* and *Ae. albopictus* are set in a concise single row). The ventral pair of anal gills is shorter than the dorsal pair (*Ae. albopictus* equal in size).

Biology: *Oc. triseriatus* is a multivoltine species which overwinters in the egg stage. The typical breeding sites are tree holes (common name: American eastern tree hole mosquito), but developmental stages can be found also in artificial breeding sites such as discarded tires, buckets, barrels, and cans. The eggs can withstand long periods (a year or more) without water. The females bite during day, especially in shaded areas. They take blood meals from humans, other mammals, and even birds. In tire yards, it can become a major pest species. However, the flight range is only a few 100 m from the breeding area (Wood et al. 1979).

Distribution: *Ochlerotatus triseriatus* is widely distributed in Canada and the USA (east of the Rocky Mountains) and inhabits all of the Southeastern United States, from Minnesota and Texas to Maine and Florida (Map 2.6). The species was found at one occasion in Greenland, maybe after accidental introduction (Messersmith 1971); it is not known if the species has persisted there. More recently, some females were intercepted at a used tire yard in France in 2004, freshly imported from the USA; immediate control measures prevent the species to establish (F. Schaffner in Medlock et al. 2012).

Medical importance: It can transfer a variety of viruses such as eastern equine encephalitis and is linked to the spread of La Crosse encephalitis virus (Watts et al. 1973).

2.4 Surveillance of Exotic Mosquitoes

The primary dispersal mode of exotic mosquitoes by human activity is through transport of desiccation-resistant eggs in cargo that previously contained stagnant water. The most important type of cargo is used tires that have been stored outdoors (Knudsen 1995). Businesses processing or trading used tires should be given high priority for monitoring populations. Another source of introduction is by ornamental plants, for example, "lucky bamboo" (*Dracaena* spp.) from Southeast Asia. For instance, "lucky bamboo," which is transported in containers with standing water, permits an "ideal insectary in transit." Thus, multiple introductions of the "Asian tiger mosquito" to the Netherlands in greenhouses of horticulture companies could be traced back to intensive trade of this plant (Scholte et al. 2007, 2010).

Due to high humidity and cool air temperature, refrigerated transoceanic containers offer ideal conditions suitable for the transport of living insects (Reiter and Darsie 1984). Therefore, container terminals at inland ports as well as terminals receiving trains with containers from infested countries should be routinely monitored, but surveillance may focus first and foremost at place where containers are open (e.g., used tire importation platforms).

Rest areas and parking lots along highways leading to areas infested with exotic species can also serve as sites of introduction (Flacio et al. 2004, 2006; Pluskota et al. 2008).

Cemeteries harbor usually many artificial containers such as flower vases and are frequently infested by exotic mosquitoes such as *Oc. japonicus* (Schaffner et al. 2009; Huber et al. 2012).

Appropriate tools for monitoring of presence and densities of vector populations to implement early warning systems are important for risk assessments. The data should be transformed into risk maps and finally entered into national and international information systems such as those developed and maintained by ECDC (http://www.vbornet.eu).

Numerous tools for monitoring the presence and/or to follow population densities of mosquitoes are available. The traps should be placed at hot spots or along most likely introduction pathways.

Suitable tools for surveillance or monitoring of exotic mosquitoes are ovitraps, gravid traps, carbon dioxide-baited CDC traps, BG-Sentinel traps, sticky traps, Mosquito Magnet traps, human bait catching, and sampling of larvae and pupae.

2.4.1 Ovitraps

Ovitraps are used to monitor the presence of a given mosquito species or to estimate the population density based on the number of deposited eggs (Bellini et al. 1996). The design of standard traps is simple. They consist of a dark plastic container with a total volume of 0.3–2.0 l. They are filled with water up to 3/4, and an oviposition support (i.e., wooden board or piece of polystyrene) is added in. The eggs will be deposited on the wooden board or polystyrene above the water surface. In order to prevent the potential development of larvae, it is advisable to add some larvicides to the water e.g. products based on *Bacillus thuringiensis israelensis*. The wooden boards should be retrieved and replaced at regular intervals, for example, every 2 weeks. The traps should be rinsed thoroughly with water, before adding a fresh wooden board. The wooden boards are taken to the laboratory and checked under the binocular for the presence of eggs.

2.4.2 Gravid Traps

Gravid traps are designed to attract gravid females. They are attracted by baits that release a mixture of odors. As soon as the females are near the trap, they are sucked into the trap by an air stream. Plastic buckets, wash tubes, or other receptacles containing several liters of infusion of hay, brewer's yeast, dog biscuit, alfalfa, or sewage water can be used as oviposition traps (Yasuno et al. 1973; Leiser and Beier 1982; Reiter 1983, 1986).

2.4.3 EVS Traps

The EVS (Encephalitis Vector Survey) trap is the most frequently used traps (Becker et al. 2010). The traps are used to monitor females searching for a blood meal. The attractant is CO_2 released from dry ice. Female mosquitoes, attracted mainly by the carbon dioxide, enter through an opening and are sucked downward by airflow created by a plastic fan attached to a small motor into a 30-cm-long nylon netting catch bag. In routine monitoring programs, the trap is baited with 0.8–1 kg dry ice, which is enough to catch mosquitoes during one night.

A carrying handle is provided, along with a metal chain to facilitate hanging from a tree branch or other object. The traps are set up 2 h before dusk and 2 h after down in the next morning. The catch, including the net, can be transferred into a container with dry ice to kill the mosquitoes, or kept alive in a cool box. In the laboratory, the species composition is determined. Catches at regular intervals (e.g., each fortnight) yield valuable information about the seasonality and population size of the adult mosquitoes.

2.4.4 BG-Sentinel Traps

The BG-Sentinel is a more recent monitoring trap. Without the need to add carbon dioxide, the BG-Sentinel is used to monitor *Aedes aegypti* and *Ae. albopictus*. The trap catches not only host-seeking female *Ae. albopictus* but also gravid females. The attractant contains artificial skin emanations which are found on human skin (ammonia, lactic acid, and caproic acid). With carbon dioxide (either from dry ice or from gas cylinders), the BG traps become a general mosquito trap equivalent to the CDC trap, with further improved catch rates for *Ae. albopictus* (Bhalala and Arias 2009; Becker et al. 2010).

2.4.5 Sticky Traps

These traps are, for example, like ovitraps but contain usually an adhesive film (sticky traps). After having been attracted, the mosquitoes stick to boards covered with glue (Facchinelli et al. 2007).

2.4.6 Mosquito Magnet

The commercially available "Mosquito Magnet" is also based on CO_2 produced by the burning of propane. Mosquito Magnets can attract mosquitoes within a radius of several meters. The attractiveness of Mosquito Magnets can be increased with short distance bite attractants based, for example, on octenol. Mosquito Magnets run without interruption for up to 1 month. They are very useful for long-term sampling (Becker et al. 2010).

2.4.7 Human Bait Catching

Determining the landing rates on legs or arms is quite an efficient method for monitoring anthropophilic mosquitoes. Location and time must carefully be chosen. Ethnic regulations must be taken into account (Becker et al. 2010).

2.4.8 Sampling of Larvae and Pupae

This is done with commercially available dippers of standard volumes. The white plastic dippers, which are fitted to a wooden or aluminum dowel, have a diameter of 11 cm and a volume of 350 ml (Dixon and Brust 1972). Dippers are used to check breeding sites for the presence of developing stages.

2.5 Control of Exotic Species

Control activities should focus first of all on the avoidance of the spread of exotic species, and if alien species are already introduced or established, on the control of immature and adult stages. It is advisable to control immature stages because they are usually defined and concentrated as populations in their breeding sites, whereas adults of container-breeding mosquitoes can spread several 100 m from their breeding grounds. However, when transmission of pathogens takes place, adults as a source for infections have to be controlled at least 100 m around the infection sites. This is the average migration rate of major invasive vector species. In general, the integrated vector management (IVM) should be the strategic approach for the control of mosquitoes (WHO 2009; Becker et al. 2010). It comprises usually the following elements and all appropriate tools should be implemented in an IVM program:

Physical control: Environmental management (e.g. source reduction) and environmental manipulation (e.g. improvement of the water supply); surface layers; reduction of human vector contact (e.g., use of bed nets based on community participation, repellents)

Chemical control: Larviciding and adulticide spraying

Biological control: Fish, invertebrates (e.g., copepods), microbial control agents (e.g., *Bacillus thuringiensis israelensis* and *Bacillus sphaericus, Saccharopolyspora spinosa*)

Genetical control: (e.g., sterile insect techniques (SIT) and population replacement, genetic engineering)

The implementation of each component of a control program should be based on a well-structured surveillance program which should be adapted to the situation in a particular area (e.g., surveillance for quick detection of an exotic species to avoid its establishment or to assess the efficacy of control measures and the risk for transmission of pathogens).

However, above all, the introduction of alien species should be avoided. Therefore, it is important to understand the pathways of the introduction. The goal should be first to avoid and second to stop further invasions of neobiota by governmental regulations in close cooperation with regulating agencies and trade companies. The spread of exotic mosquitoes is particularly associated with the trade of used tires (disposal sites of tires play a key role) and exotic plants. For instance, the state of Illinois (USA) has passed legislation to limit the storage of used tires and to require proper disposal and recycling (Novak 1995). The import of critical goods (container) should be regularly inspected when they arrive in the country of destination, and if exotic species are recorded, control measures have to be conducted immediately.

If the mosquitoes are already introduced or established, the following control measures could be conducted:

2.5.1 Environmental Management of Mosquitoes in Urban Areas

Urban areas provide a wide range of water bodies, ranging from flower vases at cemeteries, water barrels, buckets, cans, water catch basins, bird baths, and many more artificial and natural water bodies. Environmental management of mosquitoes may include a very wide range of interventions such as proper water storage containers, proper water management practices (including every possible effort to reduce stagnant water), the design of urban structures, to the provision of information to the public, for example, to cemetery visitors, to public information website, training of health department personnel and community activists, soliciting cooperation and support of local politicians, and the preparation and distribution of educational leaflets about mosquitoes to communities to increase public awareness. Such interventions often require a multidisciplinary approach, perhaps involving entomologists, civil engineers, agronomists, ecologists, behavioral scientists, economists, local politicians, and community representatives. Newton and Reiter (1992) used a deterministic model of dengue transmission to argue that techniques such as source reduction, which provided a long-term reduction of *Ae. aegypti* populations, were actually more effective and sustainable than ULV insecticide treatment in combating dengue outbreaks.

Some examples for environmental management (WHO 2009):

- Empty, clean, and refill drums or vases on a weekly basis.
- Use waterproof covers for drums, tanks, or cisterns.
- Storage under roofs (e.g., used tires) or discard containers.
- Modify design or repair (e.g., roof gutters, water catch basins, water storage tanks).
- Fill with sand (e.g., tree holes or tires).
- Replace natural flowers with plastic flowers and fill vases with sand.
- Collect, recycle, and dispose containers and waste.

Personal protection can be achieved by using repellents which contain for instance DEET or picaridin. Insecticide vaporizers or mosquito coils and insecticide-treated materials such as curtains, or long-lasting nets which cover water containers can reduce adult mosquito populations.

2.5.2 Community Participation

Environmental management measures are best implemented by the local community (Becker 1992; WHO 2009). Interventions intended to gain community support and cooperation may consist of any or all of the following: direct communication with residents, health center visitors, farmers, community leaders, etc., via public meetings, workshops, school visits, theater events, workplace meetings and practical demonstrations; distribution of brochures and literature to individual households; and mass media (posters, newspaper, radio, television, Internet). In Germany, each year millions of tablets based on *Bacillus thuringiensis israelensis* are distributed in city halls. The citizens get the tablets free of charge together with thorough information about the biology of mosquitoes and the use of the tablets to control their container-breeding mosquitoes by themselves. Bacterial control agents have a considerable safety advantage over synthetic insecticides because neither the operator nor the occupants of treated sites become exposed to potentially dangerous chemicals. For this reason, such preparations are particularly well suited for use by volunteers and in the fame of community participation.

Ovitraps which are commonly used for surveillance of *Ae. aegypti* or *Ae. albopictus* can be modified to lethal ovitraps (insecticide or glue is located on the oviposition site) or autocidal ovitraps which allow oviposition but contain an insecticide as *Bti* or IGRs to kill the offspring of the mosquitoes (Facchinelli et al. 2007). If high densities of traps are distributed, a significant reduction of the adult population can be achieved.

2.5.3 Biological Control

Biological control measures against container-breeding mosquitoes are mainly based on microbial control agents (e.g., formulations like tablets based on *Bacillus thuringiensis israelensis)*, invertebrates (e.g., copepods), or fish.

The application of *Bti* (VectoBac® DT/Culinex®) tablets can kill container-breeding mosquitoes over a period of several weeks (Becker et al. 1991). They are used in Italy against *Ae. albopictus* and in Germany against all container-breeding mosquitoes by millions. Also in tropical countries, they can contribute to the integrated control of dengue (Kroeger et al. 1995). The WHO/WHOPES recommends the use of *Bti* tablets in potable water containers because of their safety aspects (WHO 1999). The tablets are sterilized by gamma radiation before usage that only the protein crystals produced by the bacilli as active ingredient and not spores or bacilli are applied (Becker 2002).

The best-known aquatic predator of mosquitoes is the mosquito fish, *Gambusia affinis,* which is native to the Southeast United States, eastern Mexico, and the Caribbean, and the common guppy, *Poecilia reticulata,* which is native to tropical South America. Both fish are effective predators because their upward-facing mouth enables them to consume mosquito larvae living close to the water surface. In some Asian countries, such as in Malaysia, the fish are used in water containers for the control of *Ae. aegypti* (Seng et al. 2008).

Among the crustaceans, copepods are important predators of mosquito larvae (Miura and Takahashi 1985; Marten and Reid 2007). Hintz (1951) reported that *Cyclops* sp. caught about five first- and second-instar larvae of *Ae. aegypti* per day. The use of *Mesocyclops aspericornis,* which has been introduced into artificial containers, wells, and terrestrial crab burrows, has led to a reduction of >90% of *Ae. aegypti* and *Ae. polynesiensis* in Asia (Riviere et al. 1987; Kay et al. 1992). Similar results have been achieved by Vu et al. (1998) in northern Vietnam, where copepods of the genus *Mesocyclops* were used for the effective control of *Ae. aegypti* by inoculation into wells, large cement tanks, ceramic jars, and other domestic containers. The use of *Mesocyclops* sp. was complemented by community participation with respect to recycling to eliminate unused and discarded containers.

2.5.4 Chemical Control

Chemical control should be only complementary to environmental sanitation or biological control when disease transmission is threatening humans (WHO 2009). Besides biological products such as *Bti* formulations, chemical products such as methoprene, diflubenzuron, pyriproxifen (IGRs), or temephos (organophosphate) could be applied. Space spraying of adulticides should be only applied in case of emergencies (Reiter and Nathan 2001). The disease surveillance is important to detect cases in the early stage of the epidemic in order to conduct space spraying with malathion or pyrethroids (deltamethrin, permethrin, or cypermethrin). Formulations are mainly oil-based which could be a reason that residents reject spraying. Spraying should be done every 3 days in three rounds. Resistance to chemicals is always a severe problem in control programs.

2.5.5 Genetical Control

The potential success of the sterile insect technique hinges around the fact that female mosquitoes are monogamous, i.e., they mate once only, and use the sperm stored in their spermatheca to fertilize each successive batch of eggs. If that mating happens to be with a sterile male, then that female will lay only sterile or semisterile eggs and so will not contribute to the next generation of mosquitoes (Alphey et al. 2010). If sterilized male mosquitoes can be released in sufficient numbers and over a sufficiently extended period so that they outcompete the indigenous males in terms of mating with the females, then the population should decline theoretically to extinction. Genetical control is especially promising when mosquitoes do not migrate long distances to avoid invasion of nonsterile males or on define areas like islands.

Chemosterilization is typically carried out by immersion of the pupae for a fixed period of time in a standard solution of an alkylating aziridinyl compound. High levels of sterility may be achieved, with minimal loss of fitness (Seawright et al. 1977). However, use of large-scale chemosterilization has ceased now, partly due to concerns about safety to staff working with these potentially mutagenic chemicals.

Irradiation is now the most commonly used technique. It is typically the pupal stage that is exposed to the gamma rays. The gamma source used is typically cobalt-60, with a dose typically in the range 70–120 Gy. The exposure induces dominant lethal mutations in the sperm of the male mosquitoes. The lower doses cause partial sterilization, while higher doses induce a more complete level of sterilization, but are associated with damage to other cells within the male mosquito, causing a decrease in fitness and competiveness (Helinski and Knols 2009). It has been suggested that when using lower radiation doses, the benefits of higher fitness may outweigh the drawbacks of partial sterilization (Helinski et al. 2006).

SIT has been used in practice against mosquitoes on numbers of occasions (Benedict and Robinson 2003; Bellini 2005; Bellini et al. 2007), sometimes to explore and validate aspects of the technique and sometimes to attempt to control mosquito populations. Release of the sterilized males is ideally carried out at a time when the numbers of the indigenous insects are at a minimum. This may be achieved via prerelease use of conventional insect control techniques, or release may be timed to coincide with a natural seasonal depression in numbers.

Acknowledgments We acknowledge the valuable contributions of Prof. Alexander Mathis and Dr. Thin Thin Oo in the preparation of the manuscript.

References

Adhami J, Murati N (1987) Presence of the mosquito *Aedes albopictus* in Albania (in Albanian). Rev Mjekesore 1:13–16

Adhami J, Reiter P (1998) Introduction and establishment of *Aedes* (*Stegomyia*) *albopictus* Skuse (Diptera: Culicidae) in Albania. J Am Mosq Control Assoc 14:340–343

Almeida APG, Gonçalves YM, Novo MT, Sousa CA, Melim M, Grácio AJS (2007) Vector monitoring of *Aedes aegypti* in the autonomous region of Madeira, Portugal. Europ Surveill 12(46), http://www.Europsurveillance.org

Andreadis TG, Anderson JF, Munstermann LE, Wolfe RJ, Florin DA (2001) Discovery, distribution and abundance of the newly introduced mosquito *Ochlerotatus japonicus* (Diptera: Culicidae) in Connecticut, USA. J Med Ent 38:774–779

Angelini R, Finarelli AC, Angelini P, Po C et al (2007) Chikungunya in north-eastern Italy: a summing up of the outbreak. Euro Surveill 12(47), available online: http://www.eurosurveillance.org/ViewArticle.aspx?ArticleId=3313

Alphey L, Benedict M, Bellini R, Clark G, Dame D, Service M, Dobson S (2010) Sterile-Insect Methods for Control of Mosquito-Borne Diseases: An Analysis. Vector Borne Zoonotic Dis 10(3):295–311

Aranda C, Eritja R, Roiz D (2006) First record and establishment of *Aedes albopictus* in Spain. Med Vet Ent 20:150–152

Becker N (1992) Community participation in the operational use of microbial control agents in mosquito control programs. Bull Soc Vector Ecol 17(2):114–118

Becker N (1989) Life strategies of mosquitoes as an adaptation to their habitats. Bull Soc Vector Ecol 14(1):6–25

Becker N (2002) Sterilisation of *Bacillus thuringiensis israelensis* products by gamma radiation. J Am Mosq Control Assoc 18:57–62

Becker N (2009) Die Rolle der Globalisierung und Klimaveränderung auf die Entwicklung von Stechmücken und von ihnen übertragenen Krankheiten in Zentral-Europa. Umweltwiss Schadst Forsch 21:212–222

Becker N, Djakaria S, Kaiser A, Zulhasril O, Ludwig HW (1991) Efficacy of a new tablet formulation of an asporogenous strain of *Bacillus thuringiensis israelensis* against larvae of *Aedes aegypti*. Bull Soc Vector Ecol 16(1):176–182

Becker N, Petrić D, Zgomba M, Boase C, Madon M, Dahl C, Kaiser A (2010) Mosquitoes and their control. Springer, Heidelberg. Dordrecht, London, p 577

Becker N, Huber K, Pluskota B, Kaiser A (2011) *Ochlerotatus japonicus japonicus*—a newly established neozoan in Germany and a revised list of the German mosquito fauna. Europ Mosq Bull 29:88–102

Bellini R (2005) Applicazione della tecnica del maschio sterile nella lotta ad *Aedes albopictus* Tesi di Dottorato Entomologia Agraria, DiSTA-Università degli Studi di Bologna, p 82

Bellini R, Carrieri M, Burgio G, Bacchi M (1996) Efficacy of different ovitraps and binomial sampling in *Aedes albopictus* surveillance activity. J Am Mosq Control Assoc 12:632–636

Bellini R, Calvitti M, Medici A, Carrieri M, Celli G, Maini S (2007) Use of the sterile insect technique against *Aedes albopictus* in Italy:first results of a pilot trial. Area-wide control of insect pests: from research to field implementation, pp 505–515

Beltrame A, Angheben A, Bisoffi Z, Monteiro G, Marocco S, Calleri G (2007) Imported chikungunya infection, Italy. Emerg Infect Dis 13(8):1264–1266, http://www.cdc.gov/EID/content/13/8/1264.htm

Benedict MQ, Robinson AS (2003) The first releases of transgenic mosquitoes: an argument for the sterile insect technique. Trends Parasitol 19:349–355

Berry WJ, Craig GB Jr (1984) Bionomics of *Aedes atropalpus* breeding in scrap tires in northern Indiana. Mosq News 44:476–478

Bhalala H, Arias JR (2009) The Zumba™ mosquito trap and Bg-Sentinel™ trap: novel surveillance tools for host-seeking mosquitoes. J Am Mosq Control Assoc 25:134–139

Bidlingmayer WL, Evans DG (1987) The distribution of female mosquitoes about a flight barrier. J Am Mosq Control Assoc 3(3):369–377

Bonilauri P, Bellini R, Calzolari M, Angelini R, Venturi L, Fallacara F, Cordioli P, Angelini P, Venturelli C, Merialdi G, Dottori M (2008) Chikungunya virus in *Aedes albopictus*, Italy. Emerg Infect Dis 14(5):852–854

Cameron EC, Wilkerson RC, Mogi M, Miyagi I, Toma T, Kim HC, Fonseca DM (2010) Molecular phylogenetics of *Aedes japonicus*, a disease vector that recently invaded Western Europe, North America, and the Hawaiian islands. J Med Ent 47(4):527–535

Capelli G, Drago A, Martini S, Montarsi F, Soppelsa M, Delai N, Ravagnan S, Mazzon L, Schaffner F, Mathis A, Di Luca M, Romi R, Francesca Russo F (2011) First report in Italy of the exotic mosquito species *Aedes* (*Finlaya*) *koreicus*, a potential vector of arboviruses and filariae. Parasites Vectors 4:188

Carey A (2002) Globalization of Flora: inviting worldwide ecosystem disaster. J Renewable Resources 20(1):13–17

Carpenter SJ, La-Casse WJ (1955) Mosquitoes of North America (north of Mexico). University of Calif Press illus 127pls, p 360

Christophers SR (1960) *Aedes aegypti* (L), the yellow fever mosquito. Its life history, bionomics, and structure. Cambridge University Press, Cambridge, p 739

Cornel AJ, Hunt RH (1991) *Aedes albopictus* in Africa? First records of live specimens in imported tires in Cape Town. J Am Mosq Control Assoc 7:107–108

Dalla Pozza G, Majori G (1992) First record *of Aedes albopictus* establishment in Italy. J Am Mosq Control Assoc 8:318–320

Delatte H, Desvars A, Bouétard A, Bord S, Gimonneau G, Vourc'h G, Fontenille D (2010) Blood-feeding behavior of *Aedes albopictus*, a vector of chikungunya on La Réunion. Vector-Borne Zoon Dis 100:249–258

Dixon RO, Brust RA (1972) Mosquitoes of Manitoba III. Ecology of larvae in the Winnipeg area. Can Ent 104:961–968

ECDC (2009) Development of *Aedes albopictus* risk maps. http://www.ecdc.europa.eu/en/publications/Publications/0905 TER Development of *Aedes albopictus* Risk Maps.pdf

Effer PV, Pang L, Kitsutani P, Vorndam V, Nakata M, Ayers T, Elm J, Tom T, Reiter P, Rigau-Perez JG, Hayes JM, Mills K, Napier M, Clark GG, Gubler DJ (2005) Dengue fever, Hawaii, 2001–2002. Emerg Infect Dis 11:742–749

Facchinelli L, Valerio L, Pombi M, Reiter P, Costantini C, Della Torre A (2007) Development of a novel sticky trap for container-breeding mosquitoes and evaluation of its sampling properties to monitor urban populations of *Aedes albopictus*. Med Vet Entomol 21(2):183–195

Flacio E, Lüthy P, Patocchi N, Guidotti F, Tonolla M, Peduzzi R (2004) Primo ritrovamento di *Aedes albopictus* in Svizzera. Bollettino della Societa. Ticinese di Scienze Naturali 92: 141–142

Flacio E, Lüthy P, Patocchi N, Peduzzi R, Guidotti F, Radczuweit S (2006) Bericht 2006 zur Überwachung und Bekämpfung der asiatischen Tigermücke, *Aedes albopictus*, im Kanton Tessin. Jahresbericht 2006 Gruppo lavoro zanzare, Divisione della salute publica, 6501 Bellinzona, pp 1–10

Fonseca DM, Campbell S, Crans WJ, Mogi M, Miyagi I, Toma T, Bullians M, Andreadis TG, Berry RL, Pagac B, Sardelis MR, Wilkerson RC (2001) *Aedes (Finlaya) japonicus* (Diptera: Culicidae), a newly recognized mosquito in the United States: analyses of genetic variation in the United States and putative source populations. J Med Ent 38:135–146

Fonseca DM, Widdel AK, Hutchinson M, Spichiger SE, Kramer LD (2010) Fine-scale spatial and temporal population genetics of *Aedes japonicus*, a new US mosquito, reveal multiple introductions. Mol Ecol 19:1559–1572

Gatt P, Deeming JC, Schaffner F (2009) First records of *Aedes (Stegomyia) albopictus* in Malta. Euro Mos Bull 27:56–64

Gjenero-Margan I et al (2011) Autochthonous dengue fever in Croatia, August–September 2010. Europ Surveil 16(9):19805, http://www.Europsurveillance.org

Gratz NG (1999) Emerging and resurging vector-borne diseases. Annu Rev Ent 44:51–75

Gratz NG (2004) Critical review of the vector status of *Aedes albopictus*. Med Vet Ent 18:215–227

Gubler DJ, Rosen L (1976) Variation among geographic strains of *Aedes albopictus* in susceptibility to infection with dengue viruses. Am J Trop Med Hyg 25:318–325

Halstead SB (1980) Dengue haemorrhagic fever-a public health problem and a field for research. Bull WHO 58(1):1–21

Halstead SB (1982) WHO fights dengue haemorrhagic fever. WHO Chron 38(2):65–67

Halstead SB (1992) The XXth century dengue pandemic: need for surveillance and research. World Health Stat Q 45:292–298

Hansford K, Bennett E, Medlock JM (2010) Public health importance of the invasive mosquitoes of Europe. ECDC, VBORNET 2:10–13

Hawley WA (1988) The biology of *Aedes albopictus*. J Am Mosq Control Assoc Suppl 4:1–39

Helinski MEH, Knols BGJ (2009) Sperm quantity and size variation in un-irradiated and irradiated males of the malaria mosquito *Anopheles arabiensis* Patton. Acta Trop 109(1):64–69

Helinski MEH, Parker AG, Knols BGJ (2006) Radiation-induced sterility for pupal and adult stages of the malaria mosquito *Anopheles arabiensis*. Malariol J 5:41

Hintz HW (1951) The role of certain arthropods in reducing mosquito populations of permanent ponds in Ohio. Ohio J Sci 51(5):277–279

Huang HM (1972) The subgenus *Stegomyia* of *Aedes* in Southeast Asia. The Scutellaris group of species. Contrib Am Ent Inst 9:108

Huber K, Jöst A, Pluskota B, Hoffmann K, Becker N (2012) Status of the invasive species *Ochlerotatus japonicus japonicus (Diptera: Culicidae)* in South Germany. Europ Mosq Bull (in press)

Juliano SA, Lounibos Ph (2005) Ecology of invasive mosquitoes: effects on resident species and on human health. Ecol Lett 8(5):558–574

Kay BH, Cabral CP, Sleigh AC, Brown MD, Ribeiro ZM, Vasconcelos AW (1992) Laboratory evaluation of Brazilian Mesocyclops (Copepoda: Cyclopidae) for mosquito control. J Med Ent 29:599–602

King WV, Bradley GH, Smith CN, McDuffie WC (1960) A handbook of the mosquitoes of the southeastern United States. U.S. Department of Agriculture, Washington, DC Agri Handbk 173, p 188

Klobucar A, Merdic E, Benic N, Baklaic Z, Krcmar S (2006) First record of *Aedes albopictus* in Croatia. J Am Mosq Control Assoc 22(1):147–148

Knudsen AB (1995) Global distribution and continuing spread of *Aedes albopictus*. Parassitologia 37:91–97

Kroeger A, Dehlinger U, Burkhardt G, Anaya H, Becker N (1995) Community based dengue control in Columbia: people's knowledge and practice and the potential contribution of the biological larvicide *Bacillus thuringiensis israelensis*. Trop Med Parasitol 46:241–246

La Ruche G, Souarès Y, Armengaud A, Peloux-Petiot F, Delaunay P, Desprès P (2010) First two autochthonous dengue virus infections in metropolitan France, September 2010. Europ Surveill 15(39), http://www.Europsurveillance.org

Laird M, Calder L, Thornton RC, Syme R, Holder PW, Mogi M (1994) Japanese *Aedes albopictus* among four mosquito species reaching New Zealand in used tires. J Am Mosq Control Assoc 10:14–23

Leiser LB, Beier JC (1982) A comparison of oviposition traps and New Jersey light traps for *Culex* population surveillance. Mosq News 42:391–395

Madon MB, Mulla MS, Shaw MW, Hazelrigg JE (2002) Introduction and establishment of *Aedes albopictus* in Southern California. J Vector Ecol 27(1):149–154

Madon MB, Hazelrigg JE, Shaw MW, Kluh S, Mulla MS (2004) Has *Aedes albopictus* established in California? J Am Mosq Control Assoc 19:298

Manguin S, Christophe B (2011) Global Impact of Mosquito Biodiversity, Human Vector Borne Diseases and Environmental Change Tech, pp 27–50

Marten GG, Reid JW (2007) Cyclopoid copepods. J Am Mosq Control Assoc 23:65–92

Medlock J, Hansford KM, Schaffner F, Versteirt V, Hendrickx G, Zeller H, van Bortel W (2012) A review of the invasive mosquitoes in Europe: ecology, public health risks, and control options. Vector Borne Zoonotic Dis (in press)

Messersmith DH (1971) Extension of the distributional range of Aedes triseriatus (Say) to Greenland. Mosq Syst Newsletter 3(1):7–8

Mitchell CJ (1995) Geographic spread of *Aedes albopictus* and potential for involvement in arbovirus cycles in the Mediterranean basin. J Vect Ecol 20:44–58

Miura T, Takahashi RM (1985) A laboratory study of crustacean predation on mosquito larvae. Proc Calif Mosq Control Assoc 52:94–97

Miyagi I (1971) Notes on the *Aedes* (*Finlaya*) *chrysolineatus* Subgroup in Japan and Korea (Diptera: Culicidae). Trop Med 13(3):141–151

Moore CG, Mitchell CJ (1997) *Aedes albopictus* in the United States: ten year presence and public health implications. Emerg Infect Dis 3:329–334

Nehring S, Leuchs H (1999) Neozoa (Makrozoobenthos) an der deutschen Nordseeküste—eine Übersicht der Bundesanstalt für Gewässerkunde. Koblenz, BfG 1200:1–131

Newton EAC, Reiter P (1992) A model of the transmission of dengue fever with an evaluation of the impact of ultra-low volume (ULV) insecticide applications on dengue epidemics. Am J Trop Med Hyg 47(6):709–720

Novak R (1995) A North American model to contain the spread of *Aedes albopictus* through tire legislation. Parassitologia 37(2/3):129–139

O'Meara G, Craig GB Jr (1970) Geographical variation in *Aedes atropalpus* (Diptera: Culicidae). Ann Entomol Soc Am 63:1392–1400

Petrić D, Pajović I, Ignjatović-Ćupina A, Zgomba M (2001) *Aedes albopictus* (Skuse, 1894) new mosquito species (Diptera, Culicidae) in entomofauna of Yugoslavia. Plant Doctor, Novi Sad XXIX 6:457–458

Petrić D., Zgomba M., Ignjatović Ćupina A., Pajovic I., Merdić E., Boca I., and Landeka N. (2006). Invasion of the *Stegomyia albopicta* to a part of Europe. Presentation at the 15th European Society for Vector Ecology Meeting, 10–14 April 2006, Serres, Greece

Peyton EL, Campbell SR, Candeletti TM, Romanowski M, Crans WJ (1999) *Aedes (Finlaya) japonicus japonicus* (Theobald), a new introduction into the United States. J Am Mosq Control Assoc 15:238–241

Pfeffer M, Loescher T (2006) Cases of chikungunya imported into Europe. http://www.Europsurveillance.org

Pluskota B, Storch V, Braunbeck T, Beck M, Becker N (2008) First record of *Stegomyia albopicta* (Skuse) (Diptera: Culicidae) in Germany. Europ Mosq Bull 26:1–5

Reinert JF, Harbach RE, Kitching IJ (2004) Phylogeny and classification of Aedini (Diptera: *Culicidae*), based on morphological characters of all life stages. Zool J Linn Soc 142:289–368

Reinert JF, Harbach RE, Kitching IJ (2006) Phylogeny and classification of *Finlaya* and allied taxa (Diptera: Culicidae: *Aedini*) based on morphological data of all life stages. The linnean society of London. Zool J Linn Soc 148:1–101

Reiter P (1983) A portable, battery-powered trap for collecting gravid *Culex* mosquitoes. Mosq News 43:496–498

Reiter P (1986) A standardized procedure for the quantitative surveillance of certain *Culex* mosquitoes by egg raft collection. J Am Mosq Control Assoc 2:219–221

Reiter P (1998) *Aedes albopictus* and the world trade in used tires, 1988–1995: The shape of things to come. J Am Mosq Control Assoc 14:83–94

Reiter P, Darsie RF Jr (1984) *Aedes albopictus* in Memphis, Tennessee (USA): An achievement of modern transportation? J Am Mosq Control Assoc 44:396–399

Reiter P, Nathan MB (2001) Guidelines for assessing the efficacy of insecticidal space sprays for control of the dengue vector *Aedes aegypti*. WHO/CDS/PVC, pp 33

Reiter P, Sprenger D (1987) The used tire trade: a mechanism for the worldwide dispersal of container breeding mosquitoes. J Am Mosq Control Assoc 3:494–501

Reiter P, Fontenille D, Paupy C (2006) *Aedes albopictus* as an epidemic vector of chikungunya virus: another emerging problem? Lancet Infect Dis 6:463–464

Rezza G, Nicoletti L, Angelini R, Romi R et al (2007) Infection with CHIKV in Italy: an outbreak in a temperate region. Lancet 370(9602):1840–1846

Riabova TE, Iunicheva IY, Arkovich N, Ganusshkina LA, Orabei VG, Sergiev VP (2005) Detection of *Aedes* (*Stegomyia*) *aegypti* L. mosquitoes in Sochi city. Med Parasitol 3:3–5

Riviere F, Kay BH, Klein JM, Sechan Y (1987) *Mesocyclops aspericornis* (Copepoda) and *Bacillus thuringiensis* var. *israelensis* for the biological control of *Aedes* and *Culex* vectors (Diptera: Culicidae) breeding in crap holes, tree holes and artificial containers. J Med Ent 24:425–430

Romi R (1994) *Aedes albopictus* in Italia: problemi sanitari, strategie di controllo e aggiornamento della distribuzione al 30 settembre 1994. Notiziario ISS 7:7–11

Romi R, Sabatinelli G, Savelli LG, Raris M, Zago M, Malatesta R (1997) Identification of a North American mosquito species, *Aedes atropalpus* (Diptera: Culicidae) in Italy. J Am Mosq Control Assoc 13(3):245–246

Sabatini A, Raineri V, Trovato G, Coluzzi M (1990) *Aedes albopictus* in Italy and possible diffusion of the species into the Mediterranean area. Parassitologia 32:301–304

Saenz VL, Townsend LH, Vanderpool RM, Schardein MJ, Trout RT, Brown GC (2006) *Ochlerotatus japonicus japonicus* in the state of Kentucky. J Am Mosq Control Assoc 22:754–755

Samanidou-Voyadjoglou A, Patsoula E, Spanakos G, Vakalis NC (2005) Confirmation of *Aedes albopictus* (Skuse) (Diptera: Culicidae) in Greece. Europ Mosq Bull 19:10–12

Sardelis MR, Turell MJ (2001) *Ochlerotatus j. japonicus* in Frederick County, Maryland: discovery, distribution, and vector competence for West Nile virus. J Am Mosq Control Assoc 17:137–141

Sardelis MR, Dohm JD, Pagac B, Andre RG, Turell MJ (2002a) Experimental transmission of eastern equine encephalitis virus by *Ochlerotatus j. japonicus* (Diptera: Culicidae). J Med Ent 39:480–484

Sardelis MR, Turell MJ, Andre RG (2002b) Laboratory transmission of La Crosse virus by *Ochlerotatus j. japonicus* (Diptera: Culicidae). J Med Ent 39:635–639

Sardelis MR, Turell MJ, Andre RG (2003) Experimental transmission of St Louis encephalitis virus by *Ochlerotatus j. japonicus*. J Am Mosq Control Assoc 19:159–162

Schaffner F, Bouletreau B, Guillet B, Guilloteau J, Karch S (2001) *Aedes albopictus* established in metropolitan France. Europ Mosq Bull 9:1–3

Schaffner F, Chouin S, Guilloteau J (2003) First record of *Ochlerotatus (Finlaya) japonicus japonicas* (Theobald, 1901) in metropolitan France. J Am Mosq Control Assoc 19:1–5

Schaffner F, Van-Bortel W, Coosemans M (2004) First record of *Aedes* (*Stegomyia*) *albopictus* in Belgium. J Am Mosq Control Assoc 20:201–203

Schaffner F, Kaufmann C, Hegglin D, Mathis A (2009) The invasive mosquito *Aedes japonicus* in Central Europe. Med Vet Ent 23:448–451

Schaffner F, Vazeille M, Kaufmann C, Failloux A, Mathis A (2011) Vector competence of Aedes japonicus for chikungunya and dengue viruses. Europ Mosq Bull 29:141–142

Schmidt-Chanasit J, Haditsch M, Schöneberg I, Günther S, Stark K, Frank C (2010) Dengue virus infection in a traveller returning from Croatia to Germany. Europ Surveill 15(40)

Schneider K (2011) Breeding of *Ochlerotatus japonicus japonicus* 80 km north of its known range in southern Germany. Europ Mosq Bull 29:129–132

Scholte EJ and Schaffner F (2007) Waiting for the tiger: establishment and spread of the *Aedes albopictus* mosquito in Europe. In: Takken W, Knols BGJ (eds) Emerging pests and vector-borne diseases in Europe. Ecology and control of vector-borne diseases, vol 1, Wageningen Academic Publishers, Wageningen, The Netherlands, p 499 (241–260)

Scholte EJ, Jacobs F, Linton YM, Dijkstra E, Fransen J, Takken W (2007) First record of *Aedes* (*Stegomyia*) *albopictus* in the Netherlands. Europ Mosq Bull 22:5–9

Scholte EJ, Hartog WD, Braks M, Reusken C, Dik M, Hessels A (2009) First report of a North American invasive mosquito species *Ochlerotatus atropalpus* (Coquillett) in the Netherlands. Europ Surveill 14(45)

Scholte EJ, Hartog W, Dik M, Schoelitsz B, Brooks M (2010) Introduction and control of three invasive mosquito species in the Netherlands, July-October 2010. Europ Surveill 15:19710, http://www.Europsurveillance.org

Scott JJ (2003) The ecology of the exotic mosquito *Ochlerotatus* (Finlaya) *japonicus japonicus* (Theobald 1901) (Diptera: Culicidae) and an examination of its role in the West Nile virus cycle in New Jersey. Rutgers University, New Brunswick, NJ, p 180

Seawright JA, Kaiser PE, Dame DA (1977) Mating competitiveness of chemosterilised hybrid males of *Aedes aegypti* (L.) in field tests. Mosq News 37(4):615–619

Seng CM, Setha T, Nealon J, Socheat D, Chantha N, Nathan MB (2008) Community-based use of the larvivorous fish *Poecilia reticulata* to control the dengue vector *Aedes aegypti* in domestic water storage containers in rural Cambodia. J Vector Ecol 33(1):139–144

Shaw FR, Maisey SA (1961) The biology and distribution of the rockpool mosquito, *Aedes atropalpus* (Coq.). Mosq News 24:12–16

Sucharit S, Surathin K, Shrestha SR (1989) Vectors of Japanese encephalitis virus (JEV): species complexes of the vectors. Southeast Asian J Trop Med Publ Hlth 20:611–621

Tanaka K, Mizusawa K, Saugstad ES (1979) A revision of the adult and larval mosquitoes of Japan (including the Ryukyu Archipelago and the Ogasawara islands) and Korea (Diptera: Culicidae). Contr Am Ent Inst Ann Harbor 16:1–987

Thielman A, Hunter FF (2006) Establishment of *Ochlerotatus japonicus* (Diptera: Culicidae) in Ontario, Canada. J Med Ent 43:38–142

Turell MJ, O'Guinn ML, Dohm DJ, Jones JW (2001) Vector competence of North America mosquitoes (Diptera: Culicidae) for West Nile virus. J Med Ent 38:130–134

Turell MJ, Dohm DJ, Sardelis MR, O'Guinn ML, Andreadis TG, Blow JA (2005) An update on the potential of North American mosquitoes (Diptera: Culicidae) to transmit West Nile virus. J Med Ent 42:57–62

Upham P, Calum T, Gillingwater D, Raper D (2003) Environmental capacity and airport operations: Current issues and future prospects. J Air Transport Manag 9(3):145–152

Versteirt V, Schaffner F, Garros C, Dekoninck W, Coosemans M, Bortel WV (2009a) Introduction and establishment of the exotic mosquito species *Aedes japonicus japonicus* (Diptera: Culicidae) in Belgium. J Med Ent 46:1464–1467

Versteirt , et al: (2009) Arrival and acclimatisation of the exotic mosquito species *Aedes koreicus* in Belgium, European Mosquito vectors Disease: spatial biodiversity, drivers of change, and risk. Final Report. Brussels: Belgian Sci Policy, p 131

Vu SN, Nguyen TY, Kay BH, Marten GG, Reid JW (1998) Eradication of *Aedes aegypti* from a village in Vietnam, using copepods and community participation. Am J Trop Med Hyg 59 (4):657–660

Watts DM, Pantuwatana S, DeFoliart GR, Yuill TM, Thompson WH (1973) Transovarial transmission of La Crosse virus in the mosquito *Aedes triseriatus*. Science 182:1140–1141

Werner D, Kronefeld M, Schaffner F, Kampen H (2012) Two invasive mosquito species, *Aedes albopictus* and *Aedes japonicus japonicus*, trapped in south-west Germany, July to August 2011. Euro Surveill. 2012, 17(4) Available online: http://www.eurosurveillance.org/ViewArticle.aspx?ArticleId=20067

WHO (1999) *Bacillus thuringiensis*, Environ Health Criteria, p 217

WHO (2009) Dengue-Guidelines for Diagnosis, Treatment, Prevention and Control. WHO/TDR, p 146

Williges E, Farajollahi A, Scott JJ, McCuiston LJ, Crans WC, Gaugler R (2008) Laboratory colonization of *Aedes japonicus japonicus*. J Am Mosq Control Assoc 24:591–593

Wood DM, Dang PT, Ellis RA (1979) The mosquitoes of Canada (Diptera: Culicidae). Series: The insects and arachnidae of Canada. Biosystematics Res Inst Canada, Dept Agr Publ 1686, 6:390

Yasuno M, Kazmi SJ, LaBrecque GC, Rajagopalan PK (1973) Seasonal change in larval habitats and population density of *Culex fatigans* in Delhi Villages, WHO/VBC/73, 429:12

Zachcial M, Heideloff C (2003) ISL shipping statistics yearbook 2003. Institute of shipping economics and logistics, Bremen

Chapter 3
Future Strategies for European Pest Management

Reiner Pospischil

Abstract Global warming together with global trade and tourism gives foreign organisms many opportunities to enter Europe and to find suitable habitats to establish themselves. A number of these species already have the status of pests in their native countries and are now starting to be a major concern in Europe. However, particularly in industrial countries there is growing rejection by the public of the use of synthetic insecticides. The use of these insecticides is restricted, e.g., in food production, by new national and European directives, and pest management is limited to monitoring and the use of baits in the presence of foodstuffs. The consequences for scientific research and for the pest control industry are to invest in insecticides with a novel action with special focus on user-friendly and eco-friendly properties. Existing resistance of pests to frequently used insecticides should be overcome by the development of compounds which interfere with the metabolism and nervous system of arthropods in ways different from those of conventional insecticides. Additionally, new monitoring systems are required which will also detect alien species. Completely new control strategies without short-acting insecticides and baits which can be used without risk of contaminating the environment should be developed particularly for sensitive areas such as food production, pharmacies, and hospitals. Countries with a high diversity of vegetation (e.g., tropical rainforests) will be of major significance in the research and development of natural compounds for pest control in the future.

Keywords Alien species • Baits • Globalization • Natural compounds • Pheromones

R. Pospischil (✉)
Bayer Cropscience A6, Leverkusen, Germany (Retired), Im Tiergarten 9, 50129 Bergheim Erft, Germany
e-mail: reiner.pospischil@t-online.de

H. Mehlhorn (ed.), *Arthropods as Vectors of Emerging Diseases*,
Parasitology Research Monographs 3, DOI 10.1007/978-3-642-28842-5_3,

3.1 Introduction

Pests have lived with humans since humans started to settle more than 10,000 years ago. Some species even followed humans from East Africa on their way north beginning about 100,000 years ago, such as the larger house fly *Musca domestica*. Many strategies were developed to minimize the economic damage which was caused by these pests (Bodenheimer 1928; Levinson and Levinson 1990; Robinson 1996). During the nineteenth century and the first decades of the twentieth century, pest control relied mainly on inorganic compounds which were also highly toxic to mammals, persistent, and dangerous for the environment (Braness 2011).

With the development of the first synthetic organic pesticides, pest control suddenly started to be feasible without major efforts in the identification and biology of pests. Until 20 years ago, the control of cockroaches and other arthropods relied largely on synthetic insecticide sprays. With the development of baits which can be applied selectively where pests such as cockroaches, silverfish, ants, and flies occur, this situation has changed. With some knowledge of the identification and biology of pests, these baits can be applied in an eco-friendly way. Another user-friendly and environmentally friendly strategy is the use of larvicides (chitin synthesis inhibitors and juvenile-hormone analogue substances) which are applied as baits or sprays. A lot of research was done with natural compounds during the last two decades, and some have already reached the market.

Nevertheless, sprays retained their importance for pest management, and much effort was invested by the pest control industry to create more user-friendly sprays with fewer active ingredients but longer residual efficacy and that would pollute the environment less. With discussions on the use of synthetic insecticides on one side and the risk of vector-borne diseases and economic losses through the introduction of alien pest species on the other side, a controversial discussion on the future need for pest control products has started.

3.2 History

Transport of insects with goods on the commercial trade routes along the Mediterranean coast particularly by the Phoenicians has been known for more than 2,000 years (Bodenheimer 1928; Levinson and Levinson 1990). Exotic pests were transported hundreds of years ago by global trade. One example is the introduction of the American cockroach (*Periplaneta americana*) from tropical Africa to Central America via slave ships in the seventeenth century (Bell and Adiyodi 1982).

From the Middle Ages until the present, insects, mainly cockroaches and stored-product pests, have frequently arrived in Central European harbors on vessels from other continents. In the past, most goods were checked during unloading. Most infestations were recognized, and if at all, the pests could only survive the summer season in the harbor. Further distribution into the inner parts of Europe did not occur. This situation changed in the middle of the last century with strongly increasing global trade and tourism. The introduction of pest species to Europe now took place by aircraft, boats, and railways. Goods from other continents arrived directly at their destination inland, and the significance of harbors as filters against pests declined significantly (Pospischil 2001b).

Global trade and tourism now allow pests to travel undetected in aircraft between continents within a few days. Many tropical pest species withstand the cold temperature in the hold during the flight well. Nowadays, many goods such as fruits, vegetables, and other foodstuffs arrive by airfreight directly at their destination in the interior of the continent, or they arrive as container freight and are put on a truck directly without inspection after unloading in the harbor. Infestation of goods is often first recognized at the final destination, and the local pest management professional has to identify the pest and decide what measures to take. The pest control business has to consider these changing conditions and establish training schemes for pest management professionals to provide them with the right answers to the new situation in pest control (Pospischil 2001b).

3.2.1 *Urbanization*

Worldwide urbanization has provided many pest species with excellent conditions for development and propagation. At the beginning of the nineteenth century, only 2% of the world's population lived in urban areas. The vast majority of the population was distributed in rural areas. This situation changed after World War II. Twenty-eight percent of the world's population was now situated in large towns, and by 1985 this had increased to 42% (Robinson 1996; Watson 1993). Until the beginning of the twenty-first century, urbanization did grow rapidly, and in 2008 more than 50% of the world's population was estimated to live in large cities (United Nations Population Fund 2008). Large cities and industrial conurbations have a strong impact on climate conditions. The median temperature is up to 5°C higher in urban centers than in rural areas. Owing to the discharge of warm water, sewer systems may reach a temperature of more than 25°C, which provides convenient conditions for tropical species, e.g., *Periplaneta americana* (Stellmacher 1996). Growing prosperity and economic progress, modern buildings with central heating, and constantly warm climate conditions are achievements of the last 60 years. Extensive food processing and handling in industrial facilities gives tropical pests shelter over the whole year. The increasing number of buildings with tropical climate conditions in zoological and botanical gardens and tropical indoor swimming pools also offer excellent conditions for tropical invaders (Holway et al. 2002; Pospischil 2001b, 2004).

A strong increase of arthropod pests was observed in large Central European livestock facilities with a constantly warm climate. Examples are synanthropic flies which build up high population densities even in winter (e.g., *Musca domestica*, *Stomoxys calcitrans*, *Hydrotaea* spp., *Drosophila* spp.), and also litter beetles

(*Alphitobius diaperinus*) and red poultry mites (*Dermanyssus gallinae*). The strong population increase of *Alphitobius diaperinus*, a species of tropical origin, demonstrates this development well (Pospischil 2001b). The species frequently arrived with goods in European harbors but could only survive for some months during summer. With the establishment of large layer units and industrial pig farming with a constantly warm climate over the whole year, *Alphitobius diaperinus* established itself in the 1970s. Now high populations are found in livestock and cause enormous damage (Fiocre 1991). The control of this species is difficult because of the high insecticide resistance which is now present in most populations.

The significance of global warming for these pests to establish themselves outdoors in urban areas should not be underestimated. The Oriental cockroach *Blatta orientalis*, which was restricted in Central Europe to constantly warm indoor habitats until the end of the last century, is now found outdoors in urban areas during warm summer periods (Freise and Röttgers 2006).

3.2.2 Invasive Species

The number of alien species which arrive and establish themselves in Europe as well as on other continents is growing steadily, triggered by growing world trade and short transportation routes. Through the increasing trade with aircraft over long distances between continents, even sensitive species such as mosquitoes can arrive safely in Central Europe, e.g., *Aedes japonicus* (Switzerland) or *Anopheles* species, which cause the so-called airport malaria (Anderson et al. 2005). Global warming allows arthropods which have already reached southern Europe (e.g., *Aedes albopictus*) to enter Central Europe and to extend their habitat.

The introduction of exotic ants to Central Europe has played an increasing role for the last 20 years, caused by growing world trade and short transportation routes. Particularly, tropical greenhouses offer ideal conditions for ant species with high temperature and humidity preferences and with the ability to withstand cold and dry conditions during transportation (Pospischil 2011a).

As social insects, ants need to fulfill special requirements for successful propagation and adaptation after entering a new environment (Klotz et al. 2008; McGlynn 1999). These species live in close association with humans and are primarily dispersed mostly by human activity. They readily adapt to new nesting sites and to new nutrition sources. Both workers and queens have high foraging mobility. Their colonies are polygynous and polydomous. Nuptial flights occur only seldom, and reproduction is done by budding. Aggressiveness does not occur between members of different colonies. Some of these species build supercolonies. So-called tramp ants live close to people and do not disperse actively in their new environment. Invasive ant species spread actively after their introduction in urban and rural areas and displace native species.

During the last 20 years, a strong increase in the number of species has been observed in Central Europe; this increase has caused tremendous problems in urban areas and tropical greenhouses (Boer and Vierbergen 2008; Giraud et al. 2002; Heinze et al. 2006; Pospischil 2010, 2011a; Ugelvig et al. 2008). In 1999, 147 species of ants were recorded living in nonnative habitats worldwide (McGlynn 1999). Five ant species are now mentioned on the list of the 100 worldwide most important aliens (the red imported fire ant *Solenopsis invicta*, the little fire ant *Wasmannia auropunctata*, *Anoplolepis gracilipes*, the bigheaded ant *Pheidole megacephala*, and the Argentine ant *Linepithema humile*) (http://www.issg.org/database).

Global tourism and business trips bring pests to Europe which were thought to have been more or less eradicated 60 years ago. Bedbugs have lived together with humans for more than 10,000 years, but their numbers declined in many industrial countries to a level where they were unimportant in the middle of the twentieth century. The resurgence of bedbugs in the last 15 years has resulted in immense efforts in research and development of innovative detection and control methods (Boase 2007). This resurgence, which has been accompanied by the detection of highly insecticide resistant populations (Romero et al. 2007), shows the urgent need for new control strategies with good monitoring systems, kits for resistance detection, resistance-breaking products, and physical control strategies.

The introduction of pests is only one challenge for the pest control industry. The arrival of diseases from tropical areas with infested people or animals in Europe must not be underestimated. In the case of blue tongue virus, which was introduced to Central Europe in 2006 via infested animals, native biting midges of the genus *Culicoides* started to transmit the virus instead of the original vector *Culicoides imicola*, which does not occur in Central Europe. The outbreak of blue tongue disease in several Central Europe countries caused high economic loss in terms of livestock, but strategies to overcome this disease were not available and had to be developed first (Mehlhorn et al. 2009). Efforts have to be undertaken to overcome hazards caused by these invaders to people and animals (livestock and pets) and their adverse effects on nature (Holway et al. 2002).

3.2.3 History of Pest Control

Many strategies against pests were developed in the past from the time people started to settle and to store grain and other products to minimize the economic damage caused by these pests. Inorganic compounds were used prior to World War II and were then replaced by organic insecticides. Some frequently used inorganic products were boron, mercury, and arsenic compounds and sodium fluoride (Braness 2011). The first synthetic organic insecticides were synthesized at the beginning of the twentieth century, and the first pest control products with organic insecticides reached the market in about 1940. These first compounds belonged to the class of organochlorines. They were followed by carbamates and organophosphates, which are inhibitors of acetylcholinesterase, and later by pyrethroids.

Insect growth regulators were first synthesized in the middle of the 1970s and affect metabolic processes in insects, but have low toxicity for mammals. The compounds belong either to the inhibitors of chitin biosynthesis (benzoylphenylureas) or to the juvenile-hormone mimics. The benzoylphenylureas inhibit chitin synthetase and interfere in the molting process of the larvae. Juvenile-hormone mimics prevent the development of larvae or nymphs to adult insects (Braness 2011).

Microbial disrupters of insect midgut membranes (e.g., *Bacillus thuringiensis*) are particularly used against mosquito larvae (Culicidae). Avermectins were isolated in the late 1970s from the actinomycete fungus *Streptomyces avermitilis*. They consist of four separate mixtures. One of them, abamectin, has good efficacy against household pests at low rates but low mammalian toxicity (Braness 2011; Quarles 2006). Inert materials such as silica gel and diatomaceous earth have a physical effect on arthropods as desiccants. They are used alone as dust or together with organic insecticides (Quarles 2007). Pyrethrum and neem tree compounds with insecticidal action are natural compounds with a long history of use (Casida and Quistad 1995).

These examples demonstrate well that much effort has been made to find new active ingredients which are less toxic to mammals and to the environment. The outcome of this research is an excellent source for the development of novel insecticides.

3.2.4 Formulation Development, a Key for Successful Pest Control

Another important key for successful pest control is the development of formulations which enhance the efficacy of the pesticide, and improve its efficacy, dispersion in water, application, safety, and storage stability. These properties are achieved with special adjuvants, buffering agents, carriers, emulsifiers, wetting agents, etc. (Braness 2011). The aim of formulation development is to reduce the amount of the active ingredient, to make the product more user-friendly, and to reduce pollution of the environment. Many different formulation types have been developed to fulfill these requirements.

The first formulations were dustible powders, followed by wettable powders which were dispersed in water. Emulsifiable concentrates could be mixed with water or organic solvents. They were later replaced by highly sophisticated formulations which were evaluated for the special use of the product and according to the special physical properties of the insecticide. Emulsion concentrates, suspension concentrates, capsule suspensions, and water-dispersible granules are examples of concentrates which are now frequently used (Braness 2011).

3.2.5 *Baits*

The development of highly sophisticated baiting systems particularly for control of cockroaches, ants, and flies started at the beginning of the 1990s with the development of novel compounds with oral action which belong to the chloronicotinyls, aryl pyrazoles, and later spinosyns. Cockroach gel baits which can be applied selectively in small portions where the insects are living entered the market nearly 20 years ago and have gained steadily in importance compared with conventional sprays. They are particularly applied in premises which are used for the preparation of food, in hospitals, in kindergartens, and in other sensitive areas where sprays are not welcome or are banned (Benson and Zungoli 1997; Pospischil et al. 2000). This development was triggered by the synthesis of new compounds which exert their efficacy after oral ingestion (Krämer and Mencke 2001; Londershausen 1996).

More than 90% of these highly sophisticated baiting systems consist of nutrition mixtures which represent an optimal compromise between attractiveness for the pest species and the physicochemical properties of the baits for long-lasting control. The attractiveness of these baits is enhanced by pheromones, attractants, and special colors. The baits are applied with special application devices in small portions of 30–250 mg and well hidden behind equipment, near the hiding places of the pest species. Application is recommended in electronic equipment in food production facilities, restaurants, and even aircraft. Some baits have demonstrated long-lasting efficacy and palatability for more than 1 year under adverse conditions of high temperature and humidity (Pospischil 2001a; Pospischil et al. 2000).

3.3 Pest Control Companies: A Changing Business

In the past, a pest control operator was called to solve an already existing pest problem by treating the infested area. Nowadays, the use of pesticides is restricted in sensitive areas (e.g., in food production) or people do not accept insecticides in their environment. The pest control business is changing to pest management, and the original pest control operator now meets the criteria of a pest management professional, a business description which is already established in the USA. Pest monitoring and giving advice to clients are now important tasks in pest management, and their importance will increase in the future.

3.3.1 *Decreasing Efficacy of Commercial Products*

3.3.1.1 Resistance

Resistance of invertebrates started to be of major concern a few years after the first synthetic insecticides reached the market. Particularly pests with a high reproduction rate and short development cycle started to develop resistance only a few years after the first contact with a new compound (e.g., *Musca domestica*). Resistance management strategies have existed for several decades and are now implemented by the Insecticide Resistance Action Committee, which started in 1984 as an intercompany organization to coordinate strategies to prevent or delay the development of resistance in arthropods (McCaffery and Nauen 2006). There is now a great challenge to develop new compounds and synergists to overcome the existing resistances of pests to frequently used insecticides.

3.3.1.2 Bait Aversion

Bait aversion occurred several years after the introduction of cockroach gel baits in the USA. It is a genetically triggered aversion to D-glucose, a nutrition component in many cockroach gels. After replacement of D-glucose with fructose, the baits were again well accepted by the German cockroach *Blattella germanica* (Bao and Macom 2005; Silverman and Bieman 1993).

3.4 Pest Management in the Future

A world without pests will never be achieved, and pest control will be an even more important task in the future. The development of new, effective pest management strategies has become a global challenge, because global trade and tourism allow aliens from tropical areas to travel undetected between continents within a few days. Global warming is an important prerequisite for these species to establish themselves even in temperate zones of the northern or southern hemisphere, where the low temperatures did not allow these species to survive outdoors in the past. People's expectations of better housing, health care, and vector control and their tolerance of pest control measures have changed, particularly in urban areas. New pest management strategies are required which meet the requirements for new compounds with user-friendly and eco-friendly characteristics to be accepted by the public, and the pest control industry has the responsibility to provide pest management companies with these products. The pest control business has to consider these changing conditions and establish a training scheme for pest management professionals to provide them with the right answers to the new situation in pest control (Pospischil 2001b, 2011b).

3.4.1 Research and Development of New Compounds

New compound groups with novel modes of action are urgently needed to overcome resistance problems. Other important requirements of these new pesticides are user-friendly and eco-friendly characteristics. The challenges in the search for new compounds are the increasing time and costs for the development and the high toxicology hurdles. The cost of developing a new compound for the pest control market now exceeds $240 million (Braness 2011). Most of the 8–10 years' development time for a new compound or product is now spent on studying properties related to health, ecological effects, and environmental fate (Braness 2011). Owing to the long time from the discovery of a new compound class until delivery to the market, novel formulation techniques are needed to make compounds which are already on the market more user-friendly and eco-friendly, and more effective against pest species.

The public place great emphasis on natural or "green" products, and the pest control industry needs to think about new compounds which mainly derive from nature. A large number of extracts from plants have already been tested in scientific institutes, and the results have been presented in a strongly increasing number of publications. Plants have developed many strategies and toxic substances during their evolution to defend themselves against predators. Natural products will therefore be important components in pest management in the future. Countries with a high diversity of vegetation (e.g., tropical rainforests) are welcome resources of natural active ingredients. These natural pesticides or biopesticides are derived from animals, plants, bacteria, fungi, or certain minerals. Biochemical pesticides such as essential oils from highly aromatic plants and insect pheromones are subgroups of the natural compounds. Global cooperation in the research and development of these natural compounds and their use in pest control will be an important objective for the pest control industry in the future.

One well-known example is the neem tree (*Azadirachta indica*), which has more than 100 active ingredients (e.g., azadirachtin, salannin, meliantriol, and nimbidin). The insecticidal properties of the tree are well known by the people in its natural habitat, and research on different neem extracts started after the middle of the last century (Ahmed and Grainge 1986). Pyrethrum has been used for a long time as a natural insecticide. It belongs to the chemical class of pyrethrins. These compounds are extracted from *Tanacetum* spp. (Asteraceae) (Glynne-Jones 2001; Casida and Quistad 1995). The entomopathogenic fungi *Beauveria bassiana* (*Cordycipitaceae/Ascomycota*) and *Metarhizium anisopliae* (*Clavicipitaceae/Ascomycota*) grow naturally in soils throughout the world and act against a number of arthropods, such as root weevils (e.g., *Othiorhynchus sulcatus*), different termite species, flies, gnats, and whiteflies, depending on the strains of the fungi.

D-Limonene, an essential oil from the peel of citrus fruits, eucalyptus oil, linalool extracted from lavender, thyme oil, rosemary oil, and lemongrass oil are other natural products which are described in the literature (Hink and Fee 1986; Hollingsworth 2005; Isman 2006; Karr and Coats 1988; Sheppard 1984).

3.4.2 Screening

Screening for new compounds is an important tool to recognize special effects of new compounds on the metabolism or behavior of pests. The search for new components starts with biochemical in vitro screening to assess their action against important receptors of pest insects and the properties of these new substances as inhibitors of important metabolic enzymes such as the cytochrome P450 complex. Synergists are already important factors in insect control. Their significance will increase strongly, and the potency of new compounds as synergists should be identified in an early screening stage. Recently introduced pest species should be included in the following in vivo screening.

3.4.3 Formulation Techniques

In the past, the toxicological effects of active ingredients on humans and companion animals or ecosystems had a stronger impact on the decision for development of a new compound than their effects on arthropod pests. However, it is also a fact that an excellent formulation may overcome weak efficacy of the active ingredient itself. Highly sophisticated formulations will therefore be more important in the future for the development of pest control products than the active ingredient itself.

Innovative formulation and application techniques should start with already existing pesticides to make them more user-friendly and eco-friendly. These new techniques are required to apply the product directly on the surfaces to be treated without contamination of room air or the environment (e.g., against flies in livestock). Nanotechnology is implemented in many technical innovations, and there is need to discuss the suitability of this technology in pest management. One possible benefit of nanotechnology would be better spreading behavior of sprays on surfaces, with fewer active ingredient being needed. Insecticide-impregnated nets are already used commercially against malaria mosquitoes. The development of nets which are impregnated with natural products (e.g., repellents, or spores from entomopathogenic fungi) which act against other pest species is a task for the pest control industry.

3.4.4 New Bait Formulations

Bait techniques open up many possibilities for innovative and targeted pest control for the future, not only for common pests, but also for exotic invaders. Novel baiting techniques which have selective action against special pests but are not dangerous to beneficial species are urgently needed for pests in greenhouses and in sensitive areas such as the food processing industry. Exotic invaders should be included in this research. Urban farming is a strategy to produce food in the direct neighborhood of people in large towns and megacities and is now starting to become economically viable. However, this development also needs innovation in pest management strategies with selectively acting formulations which do not harm beneficial organisms.

Our environment is driven by ongoing changes. Pest organisms adapt fast to new conditions. The modern pest control industry needs to realize new situations which will be adopted by pests as a new habitat at an early stage to give advice to pest management professionals and to put suitable monitoring or control techniques at the technician's disposal.

3.4.5 Effective Monitoring Devices

In the past, monitoring and preventive measures in pest management were far behind the development of new pesticides in importance. In the future, more attention should be given to monitoring systems which are equipped with attractive lures, pheromones, etc., for annoying pests such as cockroaches, ectoparasites such as mosquitoes, bedbugs, and fleas, and stored-product pests. Monitoring devices with the ability to detect specific attractants and pheromones which are emitted by pests such as bedbugs, triatomine bugs, cockroaches, and other insects would be highly appreciated by pest management companies.

Detection kits for frequently used insecticides are required to recognize starting resistance at an early stage and avoid unsuccessful treatments, with special regard to stored-product protection, fly control, and cockroach control. Devices for the detection of allergens which derive from cockroaches, house-dust mites, molt, or other organisms indoors are urgently needed.

3.5 Strategies to Avoid Introduction of Arthropods into Buildings

Pests may enter buildings with goods and transportation equipment but also through ventilation joints, along water and electricity pipes, from the canalization or the garbage management near the foundations, etc. Inside the building they may find shelter in cracks and crevices, in hollow spaces, inside work tools and machines, and in other available hiding places. One task of modern pest management is the prevention of pests from entering buildings. Future pest management should therefore start with the construction of a new building, and a pest management professional should be included in the planning team to give advice on pest-proof construction.

The pest control industry should provide the construction industry with pest-proof products for the construction and equipment of new buildings which are "unfriendly" to pests.

Nanotechnology may help to create vertical surfaces which are smooth and prevent pests from climbing up them. Insect-proof air ventilation systems may be used to avoid penetration of arthropods into buildings to hibernate there. Sealed cable ducts and water pipes and sealing materials should be made insect-proof with repellents or long-lasting insecticides. Electric barriers can be installed around ventilation joints and other possible entrances to stop invaders, including rodents.

3.6 Pest Control Industry and Academia: An Important Symbiosis

Pest management in the twenty-first century needs to be environmentally friendly, considerate to natural resources, and sustainable. The development of innovative technologies and products for pest management is a global challenge and needs interdisciplinary research with international teams of specialists with a different emphasis. Global cooperation between universities, federal institutes, and the pest control industry has already existed for a long time, but has to be intensified.

The pest control industry has excellent research and screening facilities to assess the efficacy of new synthetic and natural compounds against different pest species. The pest control industry has extensive experience in the development of special highly sophisticate formulations to enhance the efficacy of new active ingredients, but decrease their environmental and toxicological effects. This is another important task for the pest control industry.

Scientific institutes provide the pest control business with important information on the identification, biology, and behavior of new pest organisms. Physiological research on pheromones, lures, synergists, biochemical control strategies, biological control of pests, etc., is a broad field with strong competence by universities. And together with pest control industry they play an important role in research on natural compounds. New results on metabolic or physiological mechanisms in arthropods can be used for further resistance management by the pest control industry and the pest control business. The scientific literature on these results is an important source of information for the pest control industry and the pest control business. Networks on the introduction and impact of aliens are an important source for the pest control industry to be prepared for new challenges (Hendrickx et al. 2011). Technical advice, resistance management, and new control strategies should be evaluated by scientific and federal institutes together with the pest control industry.

3.7 Conclusion

The pest control business is triggered by fast global trade, changing living conditions in urban areas, large modern livestock facilities, and environmental factors such as global warming. People's expectations of better housing, health care, and vector control and their tolerance of pest control measures have changed, particularly in urban areas. Short transportation periods in aircraft give sensitive species the opportunity to reach Central Europe with trade, and to establish themselves! A rapid increase in the number of introduced species has been observed during the last 20 years. Even tropical pest species now profit from urbanization and the changing environmental factors. The rapidly changing situation in pest management has to be considered by the pest control industry. Innovative pest management concepts have to be provided by the pest control industry.

Future research in the pest control industry has clear prerequisites. New compounds must not be hazardous to the environment, must not accumulate in the environment, and must be quickly (bio)degradable. They must be effective against pest species, but should be less dangerous to nontarget species. Weak points in the efficacy can be overcome by special formulations. However, new products should derive from renewable resources and should be sustainable.

History has shown that development of new compounds and strategies cannot be done in a few years, but needs up to 10 years and even more. This should be considered in the future. In the meantime, novel and intelligent formulations may provide already-used compounds with better efficacy in lower amounts and decrease environmental risks. Resistance of pests to frequently used products is now a great challenge to develop new technologies and synergists to overcome the existing resistance of pests against frequently used compounds until new compound classes are developed.

Natural products are important resources for effective but eco-friendly compounds for future pest management. Many innovative products and strategies have already been developed or are planned for use soon. They should have a nontoxic mode of action and they should occur in nature. Tropical rainforests, which have vegetation with the highest diversity worldwide, are the most important resource of natural compounds. The protection of these forests is an important task to preserve this valuable resource for the future. Because of these facts, continuous changes in pest management are inevitable, and the pest control industry, public health authorities, the food processing industry, pest management professionals, physicians, and veterinarians must adjust themselves to them (Pospischil 2001b).

Most alien arthropods have a way of life different from that of native species and require special control strategies. Detailed descriptions of economically important species are often found in their countries of origin or in the American and Australian literature. Modern pest management therefore requires global knowledge of pest control strategies and study of the global pest control literature. Some control advice may also be found on the Internet when the respective alien species has been identified.

The pest control industry should realize changes in pest control even earlier than pest control companies to develop innovative products and strategies before they are really needed and to provide pest management professionals with these tools at an early stage. Eco-friendly strategies such as natural compounds, impregnated tissues, innovative baits, physical control methods, and effective monitoring devices are required additionally to the already existing product types. This industry sector is therefore occupied no longer only with the development of new products (mostly pesticides) for the control of pests, but also or even more so with the development of monitoring systems, prevention of infestation, and development of materials which are "unfriendly" to pests for the construction of new buildings. The pest control industry and scientific institutes should collaborate as a global team to overcome the future challenges in pest management. The future task of the pest management business and the pest control industry is to learn from invasive pests, from the environment, and from nature.

References

Ahmed S, Grainge M (1986) Potential of the neem tree (*Azadirachta indica*) for pest control. Econ Bot 40:201–209

Anderson M, Ware R, Murphy RG (2005) Population survey of mosquito species in proximity to an international airport and other urban sites in the UK. In: Lee C-Y, Robinson WH (eds) Proceedings of the 5th international conference on urban pests, Perniagaan Ph'ng @ P&Y Design Network, Malaysia, pp 271–274

Bao N, Macom T (2005) Resurrection of bait aversion and management strategies for the German cockroach (Blattodea: Blattellidae). In: Lee C-Y, Robinson WH (eds) Proceedings of the 5th international conference on urban pests, Perniagaan Ph'ng @ P&Y Design Network, Malaysia, pp 73–79

Bell WJ, Adiyodi KG (1982) The American cockroach. Chapman and Hall, London

Benson EP, Zungoli PA (1997) Cockroaches. In: Mallis A (ed) Handbook of pest control, 8th edn. Mallis, Cleveland, pp 123–202

Boase CJ (2007) Bed bugs: research and resurgence. In: Takken W, Knols BGJ (eds) Ecology and control of vector-borne diseases, vol 1. Wageningen Academic, Wageningen, pp 261–280

Bodenheimer FS (1928) Materialien zur Geschichte der Entomologie, vol 1. Junk W, Berlin

Boer R, Vierbergen B (2008) Exotic ants in the Netherlands (Hymenoptera: Formicidae). Entomologische Berichten 68:121–129

Braness G (2011) Insecticide & pesticide safety. In: Moreland D (ed) Mallis handbook of pest control, 10th edn. G.I.E. Media, Cleveland, pp 1262–1339

Casida JE, Quistad GB (1995) Pyrethrum flowers: production, chemistry, toxicology, and uses. Oxford University Press, New York

Fiocre O (1991) Un Nuisible des Locaux d'Elevage hors Sol: Alphitobius diaperinus (Coleoptère, Ténébrionidae). These de l'Ecole Nationale Veterinaire de Nantes

Freise JF, Röttgers HR (2006) Großräumiger Schabenbefall in Damme. Der Praktische Schädlingsbekämpfer 58:18–19

Giraud T, Pedersen JS, Keller L (2002) Evolution of supercolonies: the Argentine ants of southern Europe. Proc Natl Acad Sci USA 99:6075–6079

Global Invasive Species Database. http://www.issg.org/database

Glynne-Jones A (2001) Pyrethrum. Pestic Outlook 10:195–198

Heinze J, Cremer S, Eckl N, Schrempf A (2006) Stealthy invaders: the biology of Cardiocondyla tramp ants. Insectes Soc 53:1–7

Hendrickx G, Wint W, Bastier S, Kruijff M, Schaffner F, Ducheyne E (2011) Vecmap, a one-stop-shop to vector mapping. In: Carvalho Campos AE, Robinson WH (eds) Proceedings of the 7th international conference on urban pests, Instituto Biologico, Sao Paulo, p 390

Hink WF, Fee BJ (1986) Toxicity of d-limonene, the major component of citrus peel oil, to all life stages of the cat flea, *Ctemocephalides felis* (Siphonaptera: Pulicidae). J Med Entomol 23:400–404

Hollingsworth RG (2005) Limonene, a citrus extract, for control of mealybugs and scale insects. J Econ Entomol 98:772–779

Holway DA, Lach L, Suarez AV, Tsutsui ND, Case TJ (2002) The causes and consequences of ant invasions. Annu Rev Entomol 33:181–233

Isman MB (2006) Botanical insecticides, deterrents and repellents in modern agriculture and an increasingly regulated world. Annu Rev Entomol 51:45–66

Karr LL, Coats JR (1988) Insecticidal properties of d-limonene. J Pestic Sci 13:287–290

Klotz JH, Hansen LD, Pospischil R, Rust M (2008) Ants and their management in urban environments. Cornell University Press, Ithaca

Krämer F, Mencke N (2001) Flea biology and control. Springer, Berlin

Levinson H, Levinson A (1990) Die Ungezieferplagen und Anfänge der Schädlingsbekämpfung im Alten Orient. Anz Schaedlingskde Pflanzenschutz Umweltschutz 63:81–96

Londershausen M (1996) Approaches to new parasiticides. Pestic Sci 48:269–292

McCaffery A, Nauen R (2006) The Insecticide Resistance Action Committee (IRAC): public responsibility and enlightened industrial self interest. Outlooks Pest Manag 2:11–14

McGlynn TP (1999) The worldwide transfer of ants: geographical distribution and ecological invasions. J Biogeogr 26:535–548

Mehlhorn H, Walldorf V, Klimpel S, Schaub G, Kiel E, Focke R, Liebisch G, Liebisch A, Werner D, Bauer C, Clausen H, Bauer B, Geier M, Hörbrand T, Bätza H-J, Conraths FJ, Hoffmann B, Beer M (2009) Bluetongue disease in Germany (2007–2008): monitoring and entomological aspects. Parasitol Res 105:313–319

Pospischil R (2001a) Neue Entwicklungen in der Schabenbekämpfung. Mitt Dtsch Ges Allg Angew Ent 13:557–560

Pospischil R (2001b) Veränderungen der Schädlingsfauna im menschlichen Umfeld seit 1950. Verh. Westd. Entom. Tag 2000. Löbbecke Museum Düsseldorf, Düsseldorf, pp 319–324

Pospischil R (2004) Schaben – Dictyoptera. Ein Beispiel für die Einschleppung und Einbürgerung von Insekten in Europa. Mitt Dtsch Ges Allg Angew Ent 14:93–100

Pospischil R (2010) Ameisen (Hymenoptera, Formicidae) – ihre Lebensweise und medizinische Bedeutung. In: Aspöck H (Ed) Krank durch Arthropoden, Denisia 30. Biologiezentrum/ Oberösterreichische Landesmuseen. Linz, Austria, pp 213–224

Pospischil R (2011a) The role of tropical greenhouses for introduction and establishment of foreign ant species (Hymenoptera: Formicidae) in Central Europe. In: Carvalho Campos AE, Robinson WH (eds) Proceedings of the 7th international conference on urban pests, Instituto Biologico, Sao Paulo, pp 59–66

Pospischil R (2011b) Development of future strategies for pest management in Europe. In: Carvalho Campos AE, Robinson WH (eds) Proceedings of the 7th international conference on urban pests, Instituto Biologico, Sao Paulo, pp 351–354

Pospischil R, Schneider U, Böcker T, Junkersdorf J, Nentwig G, Smith G, Sonneck R (2000) Imidacloprid – a new active ingredient for control of cockroaches in a gel formulation. Pflanzenschutz Nachrichten Bayer 52:376–390

Quarles W (2006) The avermectins: successful biopesticides. IPM Pract 13:6–15

Quarles W (2007) Diatomaceous earth the 20-million-year-old pesticide. Pest Control Technol 35 (64–66):68

Robinson W (1996) Urban entomology. Chapman and Hall, London

Romero A, Potter MF, Potter DA, Haynes KF (2007) Insecticide resistance in the bed bug: a factor in the pest's sudden resurgence? J Med Entomol 44:175–178

Sheppard DC (1984) Toxicity of citrus peel liquids to the house fly and red important fire ant. J Agric Entomol 1:95–100

Silverman J, Bieman DN (1993) Glucose aversion in the German cockroach, Blattella germanica. J Insect Physiol 39:925–933

Stellmacher A (1996) Amerikanische Schaben in Deutschland. Der Praktische Schädlingsbekämpfer 48:19–22

Ugelvig LV, Drijfhout FP, Kronauer DJC, Boomsma JJ, Pedersen JS, Cremer S (2008) The introduction history of invasive garden ants in Europe: integrating genetic, chemical and behavioural approaches. BMC Biol 6:11

United Nations Population Fund (2008) http://www.unfpa.org/swp/

Watson C (1993) Trends in World Urbanisation. In: Wilday KB, Robinson WH (eds) Proceedings of the 1st international conference on urban pests, BPCC Wheatons Ltd, Exeter, pp 1–7

Chapter 4
Assessing Diversity and Abundance of Vector Populations at a National Scale: Example of *Culicoides* Surveillance in France After Bluetongue Virus Emergence

R. Venail, T. Balenghien, H. Guis, A. Tran, M.-L. Setier-Rio, J.-C. Delécolle, B. Mathieu, C. Cêtre-Sossah, D. Martinez, J. Languille, T. Baldet, and C. Garros

Abstract Biting midges of the genus *Culicoides* (Diptera: Ceratopogonidae) are small hematophagous dipterans responsible for transmitting major viruses to livestock such as African horse sickness virus in equids, and epizootic haemorrhagic disease and bluetongue viruses (BTV) in ruminants. Between 1998 and 2005, BTV outbreaks occurred into many countries around the Mediterranean basin associated with the northward extension of *C. imicola*, which colonized new Mediterranean territories in the past decades due to the global increase of temperatures. In August 2006, BTV serotype 8 (BTV-8) was unexpectedly introduced in the Netherlands and was intensively transmitted by autochthonous Palaearctic *Culicoides* species, leading in 2007 and 2008 to a major sanitary crisis in whole continental Europe with huge economic losses due to animal movement restrictions. This chapter presents a synthesis of the data gathered through the different entomological surveillance networks implemented in France since 2000 when *C. imicola* was recorded for

R. Venail (✉) • M.-L. Setier-Rio
EID Méditerranée, 34184 Montpellier, France
e-mail: rvenail@eid-med.org

T. Balenghien • H. Guis • C. Cêtre-Sossah • D. Martinez • T. Baldet • C. Garros (✉)
Cirad, UMR Contrôle des Maladies, 34398 Montpellier, France
e-mail: claire.garros@cirad.fr

A. Tran
Cirad, UPR Animal et Gestion Intégrée des Risques, 34398 Montpellier, France

J.-C. Delécolle,
UdS, IPPTS, Faculté de Médecine, 67000 Strasbourg, France

B. Mathieu
EID Méditerranée, 34184 Montpellier, France
UdS, IPPTS, Faculté de Médecine, 67000 Strasbourg, France

J. Languille
Ministère de l'Agriculture, de l'Alimentation, de la Pêche, de la Ruralité et de l'Aménagement du territoire, Paris, France

H. Mehlhorn (ed.), *Arthropods as Vectors of Emerging Diseases*,
Parasitology Research Monographs 3, DOI 10.1007/978-3-642-28842-5_4,

the first time in France. These networks aimed at first to monitor the spread and establishment of the invasive *C. imicola* and then from 2006 onwards, to study the diversity and dynamics of autochthonous *Culicoides* species. The data collected enable to describe for the first time the spread of *C. imicola* into French mainland, propose an updated list of *Culicoides* species for France, describe the *Culicoides* species distribution and seasonal dynamics, and report assays to identify BTV in field-collected *Culicoides* during outbreaks.

Keywords Bluetongue • *Culicoides* • *Culicoides imicola* • Entomological surveillance • France

Assessing accurately diversity and abundance of organisms is one of the most difficult challenges in ecological studies, especially for taxonomic groups such as insects, which can be highly diverse and abundant (Southwood and Henderson 2000). Efficiency of this assessment has decisive sanitary interests when target groups are vectors of pathogens. Surveillance of insect vector populations may aim to (1) detect introduction of invasive species into a new territory, (2) detect the presence and circulation of a given pathogen in insect population, before appearance of vertebrate cases and (3) follow up characteristics of vector populations to help assess pathogen transmission risk. Sampling strategies depend on the objectives as illustrated by the changes in surveillance system of *Culicoides* populations in France from 2001 to 2009 due to changes in bluetongue virus (BTV) transmission.

Biting midges of the genus *Culicoides* (Diptera: Ceratopogonidae) are small hematophagous dipterans responsible for nuisance limiting touristic industry, causing acute allergenic dermatitis in horses and transmitting major viruses to livestock such as African horse sickness virus in equids, and epizootic haemorrhagic disease and bluetongue viruses in ruminants (Mellor et al. 2000, 2009). Bluetongue was reported for the first time in the Europe in 1924 in Cyprus, which was until 1998 the only European area where bluetongue was endemic (Gambles 1949). Elsewhere, BTV presence was limited to periodic incursions in the south of the Iberian Peninsula and in some Greek islands. These zones, where incursions occurred, coincided with the distribution area of *C. imicola*, which is considered the main BTV vector in Africa and in the Middle East (Du Toit 1944). Indeed, *C. imicola* was first reported in the European continent in southern Spain and in southern Portugal (Mellor et al. 1983, 1985), and this species was not recorded outside the South of Iberian Peninsula despite *Culicoides* investigation in the 1970s (Mellor et al. 2000). At that time, some Palaearctic species were considered able to act as secondary vector such as *C. obsoletus*, based on virus isolations (Jennings and Mellor 1988). Between 1998 and 2005, BTV outbreaks occurred into many countries around the Mediterranean basin (Mellor and Wittmann 2002). This transmission was associated with the northward extension of *C. imicola*, which colonized new Mediterranean territories in the past decades favoured by the global increase of temperatures (Guis et al. 2012; Purse et al. 2005). In August 2006, BTV serotype

8 (BTV-8) was unexpectedly introduced near Maastricht (the Netherlands) and was intensively transmitted by autochthonous Palaearctic *Culicoides* species (Enserink 2006; Meiswinkel et al. 2008; Mellor et al. 2009; Saegerman et al. 2008), leading in 2007 and 2008 to a major sanitary crisis in whole continental Europe with huge economic losses due to animal movement restrictions (Velthuis et al. 2010).

This chapter presents a synthesis of the data gathered through the different entomological surveillance networks implemented in France since 2000 when *C. imicola* was recorded for the first time in France. These networks aimed at first to monitor the spread and establishment of the invasive *C. imicola* and then from 2006 onwards, to study the diversity and dynamics of autochthonous *Culicoides* species. The data collected enable to describe for the first time the spread of *C. imicola* into French mainland, assess the spatio-temporal variations in species distribution and dynamics, propose an updated list of *Culicoides* species for France, and report assays to identify BTV in field-collected *Culicoides* during outbreaks.

4.1 Overview of *Culicoides* Population Surveillance in France

Culicoides imicola was recorded for the first time in France in September 2000 in Corsica (Delécolle and de La Rocque 2002), a Mediterranean island close to the Italian island of Sardinia, just before the emergence of BTV serotype 2 in October 2000. Since then, the French ministry in charge of agriculture mandated the Centre international de coopération en recherche agronomique pour le développement (Cirad), and its associated partners (Institut de Parasitologie et de Pathologie Tropicale de Strasbourg (IPPTS), and l'Entente Interdépartementale pour la Démoustication du littoral méditérranéen (EID-Med)), to implement *Culicoides* trappings in Corsica and in the Mediterranean littoral of French mainland aiming to (1) establish *C. imicola* distribution and dynamics in Corsica and to (2) detect an eventual introduction and establishment of *C. imicola* to French mainland. Then, from 2002 to 2008, surveillance of *C. imicola* populations was implemented in 12 sheep or cattle farms in Corsica with one night trapping every 3 weeks with outdoor traps (Figs. 4.1 and 4.2).

Moreover, from 2002, 19 sites along the Mediterranean coast of the French mainland (involving seven departments (administrative units): Pyrénées-Orientales, Aude, Hérault, Gard, Bouches-du-Rhône, Var and Alpes-Maritimes) (Fig. 4.1) were sampled, one night collection per site every month (Baldet et al. 2004). Additional traps were used in departments close to known established *C. imicola* populations in countries bordering France: in Pyrénées-Orientales and Alpes-Maritimes and in Pyrénées-Atlantiques in 2005 (Fig. 4.1). Indeed, in neighbouring countries, the closest populations of *C. imicola* are present in Spain, in Catalonia Peralda (Girona, Spain) (Sarto i Monteys et al. 2005) and in the Basque Country (Goldarazena et al. 2006), and in Italy (Torina et al. 2004). The Mediterranean littoral surveillance system is largely developed below in the section entitled "First Record of *Culicoides imicola* in Mainland France and Surveillance of Its Extension".

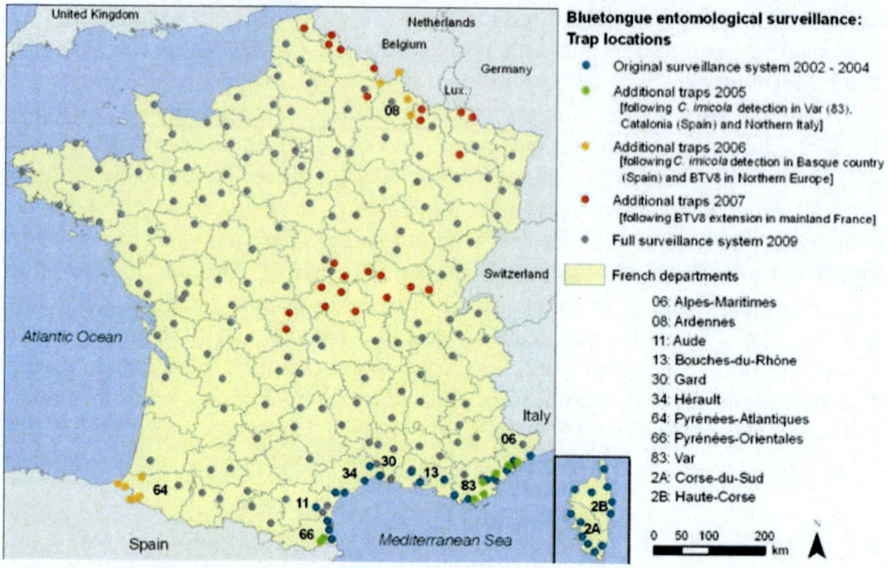

Fig. 4.1 Localization of *Culicoides* surveillance sites in France, 2002–2009

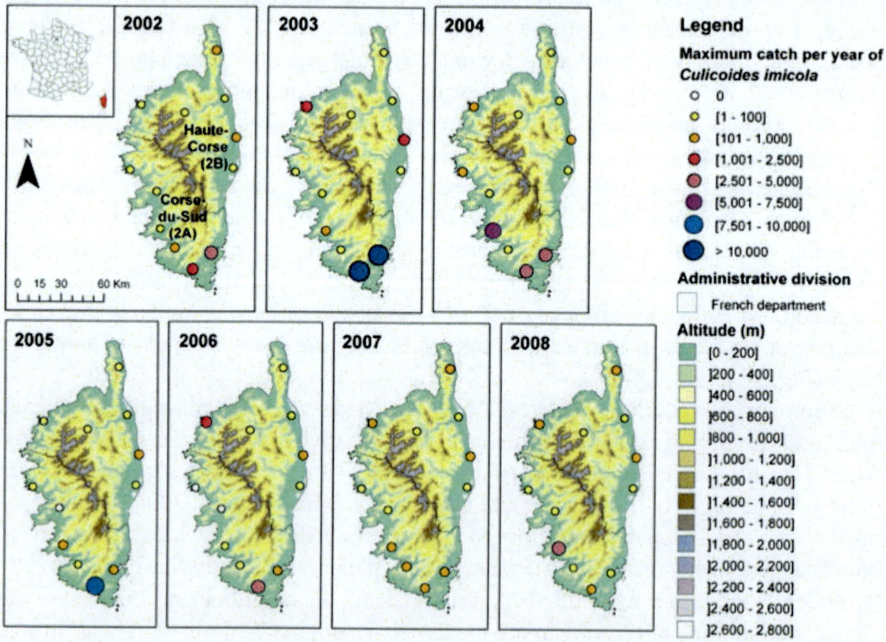

Fig. 4.2 Collection sites in Corsica with maximum catches of *Culicoides imicola* per year from 2002 to 2008

In 2006, when BTV-8 emerged in northern Europe, France was initially marginally affected with only seven outbreaks along the Belgian border. In late September 2006, four cattle farms in the Ardennes department were thus sampled one night weekly to confirm the absence of *C. imicola*, describe *Culicoides* diversity and the end of their activity period (Baldet et al. 2008). Due to the large BTV spread in Belgium, the Netherlands and Germany in late 2006, the surveillance network was extended early 2007 to 15 farms along the northern border, with outdoor and indoor traps (Fig. 4.1). In late 2007, following the disease front evolution, the surveillance included 13 farms in the centre of France (Fig. 4.1). The aim of this network was (1) to determine the start and the end of the active period of *Culicoides* populations by trapping one night per week from March to April and from October to December and (2) to follow up *Culicoides* dynamics via monthly collections from May to September. At that time, the transmission of BTV by autochthonous *Culicoides* species occurred massively throughout the whole French territory (15,253 infected farms with BTV-8 in 2007, 23,959 infected farms with BTV-8 in 2008 and 4,339 infected farms with BTV serotype 1 in 2008). Indeed, BTV-1 was introduced in the Basque country from southern Spain and spread mainly in south-western France. The trapping frequency was standardized in 2008 for all traps of French mainland, with weekly collections at the start or end of *Culicoides* activity (from February to April and from October to December) and monthly collections the rest of the year. Trap localizations remained unchanged except in the Mediterranean littoral where the number of traps was reduced to 13 traps. Finally, from 2009 to date, in order to obtain information on the whole territory, the number of traps was increased to 160 distributed in the mainland and Corsica with 1–2 traps per department monitored at the same frequency as described just above (Balenghien et al. 2010, 2011).

In all surveillance catches, all *Culicoides* specimens were sorted by species, sex and physiological stages for females (nulliparous, parous or engorged). Subsampling following a standardized procedure was applied when abundance was high.

4.2 First Record of *Culicoides imicola* in Mainland France and Surveillance of Its Extension

Following the first record of *C. imicola* in August 2000 in Sardinia, Italy (Goffredo et al. 2001), and given the short distance between Sardinia and Corsica, France (12 km), French authorities mandated an entomological surveillance of the island early in October in order to establish whether *Culicoides imicola* was present (and established) on the island and thereafter assess the risk of BTV spreading to French mainland (Fig. 4.1). Early October 2000, 2 weeks before the first BTV outbreaks in Corsica, an entomological survey was conducted in nine farms during a week, with one night trapping using UV-light suction traps (Rieb trap) (Fig. 4.3).

Fig. 4.3 The two models of UV-light trap used during *Culicoides* surveillance in France. Rieb traps (picture from IPPTS) were used from 2000 to 2006 in Corsica, from 2002 to 2005 on the Mediterranean littoral, and in 2006 in the southwest France (**a**). OVI traps (picture from Vet Services, department Indre) were used in 2006 onwards in all trapping sites except Corsica where they were implemented in 2007 (**b**)

This survey recorded for the first time *C. imicola* in Corsica (October 2000) (Delécolle and de La Rocque 2002). It also highlighted that the species could be very abundant (with a single-night catch of 12,000 individuals at Porto-Vecchio in the south-east) and distributed in almost all the studied sites throughout the island, notably on the littoral (Fig. 4.2). Given the proximity of Corsica to the French mainland (180 km), the authorities dread the establishment of these populations into the French mainland. Therefore, an entomological survey was carried out with one night trapping in eight locations for three departments (Hérault, Gard and Bouches-du-Rhône) along the Mediterranean littoral in May 2001 and in the Var department in October 2011 (Fig. 4.1). A total of 419 *Culicoides* were trapped, and no *C. imicola* individuals were captured. In Corsica, in June 2001, 12 farms were prospected during a week confirming the presence of *C. imicola* in high numbers. In 2002, a wider surveillance programme was implemented in Corsica (12 sites) and in the French mainland (19 sites) (Figs. 4.1 and 4.2). This network enabled to trap the first *C. imicola* individuals on the mainland in 2003: one male the 22nd of May, in the Var department, and one nulliparous female the 24th of September in the Alpes-Maritimes department (Fig. 4.4), confirming that individuals could spread passively from Corsica. Additional nights of trapping conducted in and around both positive sites failed to confirm the presence of established populations.

These catches coincided with a year of high abundance of *C. imicola* within Corsica and, particularly, in the north-east of the island in 2003 (Figs. 4.2 and 4.8). The next year, in June 2004, a gravid female was collected by the surveillance network in the valley of Argens (Var department), followed by one parous female late September, one nulliparous and two parous females early October. An expected critical zone of establishment was defined in the valley of Argens. Thus, additional trappings were carried out in October 2004 in this area (Table 4.1, Fig. 4.4).

This extensive trapping during 2 consecutive nights in 16 sites localized in the critical zone revealed that *C. imicola* was present in 10 sites, with a total of 240

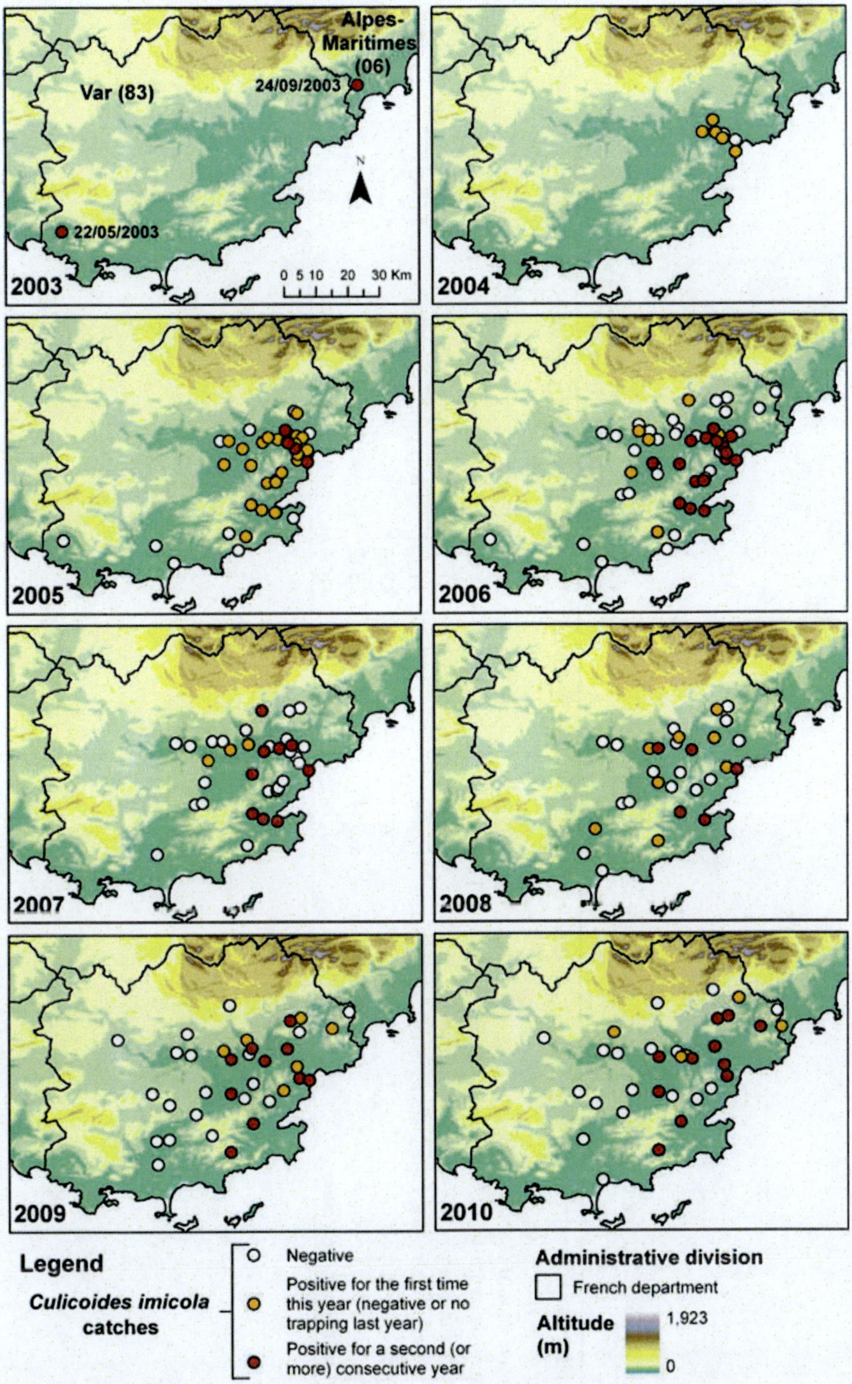

Fig. 4.4 Distribution area of *Culicoides imicola* in the Var department from 2003 to 2010

Table 4.1 Summary of *Culicoides* catches in the Var department, 2004–2010

Year	2004	2005	2006	2006	2007	2008	2009	2010
Sampling dates	19-20/10	26/09-10/10	19-23/09	26-29/09	01-04/10	29/09-04/10	28/09-02/10	06-10/09
Number of sites	16	42	51	45	34	29	35	34
Negative sites (0 *C. imicola*) [% total]	6 [37.5]	12 [28.6]	40 [78.4]	22 [48.9]	22 [64.7]	16 [55.2]	19 [54.3]	17 [50]
Positive sites with 1–10 *C. imicola* [% total]	7 [43.75]	21 [50]	8 [15.7]	13 [28.9]	6 [17.6]	7 [24.2]	11 [31.4]	11 [32.3]
Positive sites with 11–100 *C. imicola* [% total]	2 [12.5]	5 [11.9]	2 [3.9]	7 [15.5]	2 [5.9]	3 [10.3]	2 [5.7]	5 [14.7]
Positive sites with >100 *C. imicola* [% total]	1 [6.25]	4 [9.5]	1 [2]	3 [6.7]	4 [11.8]	3 [10.3]	3 [8.5]	1 [2.9]
Number of *C. imicola*	240	4,636	257	8,935	948	1,348	546	1,497
Number of other *Culicoides*	379	6,954	27,056	2,112	3,420	1,264	2,159	3,270
Total number of *Culicoides*	619	11,590	27,313	11,047	4,368	2,612	2,705	4,770

Table 4.2 Maximal spread of *Culicoides imicola* over the years in the Var department

Year	2005	2006	2007	2008	2009	2010	2004–2010
Number of traps positive for the first time	25	7	2	4	4	6	Total = 48
Maximal distance between a trap positive for the first time that year and its nearest positive trap of the preceding years (in km)	30.66	11.60	5.52	18.89	10.44	9.84	Max = 30.66 Mean = 14.49

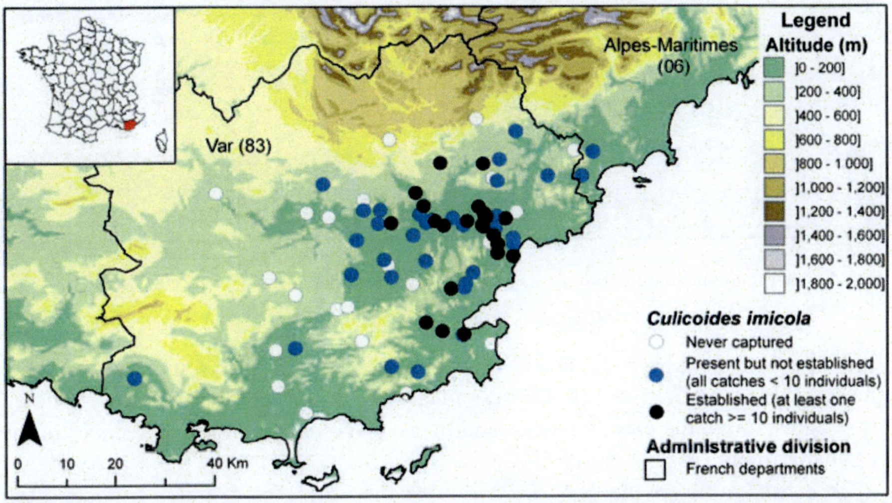

Fig. 4.5 Distribution area of *Culicoides imicola* in the Var department based on catches from 2003 to 2010

C. imicola trapped out of 619 *Culicoides*. It thus confirmed the presence and lasting establishment of *C. imicola* on the French mainland (Fig. 4.4). Since 2004, the different *C. imicola* populations established in the valley of Argens have been surveyed each year by extensive trapping conducted during one night in several sites at the end of the summer to follow their geographical extension (Table 4.1). From 2003 to 2010, a gradual extension of the distribution area of *C. imicola* in the Var department was observed (Table 4.1, Fig. 4.4).

In 2007, out of the 34 studied sites, 12 were positive for *C. imicola,* with four sites separated by a few kilometers collecting 90% of the total number of *C. imicola* individuals (Fig. 4.4). One isolated population was detected at Grimaud, south of the Argens valley. In 2008, *C. imicola* populations remained restricted to some areas: the valley of Argens, the coastal plains of Fréjus and St Tropez golf, and some sites such as Cannet-des-Maures, Grimaud, Bormes-les-Mimosas and Pierrefeu-du-Var. In 2009, despite important floods in June in the known distribution area of *C. imicola*, new positive sites were recorded eastward (Mandelieu-la Napoule, département Alpes-Maritimes, Callian, département Var), highlighting an eastward expansion. Up to now, the distribution area of *C. imicola* seemed to be

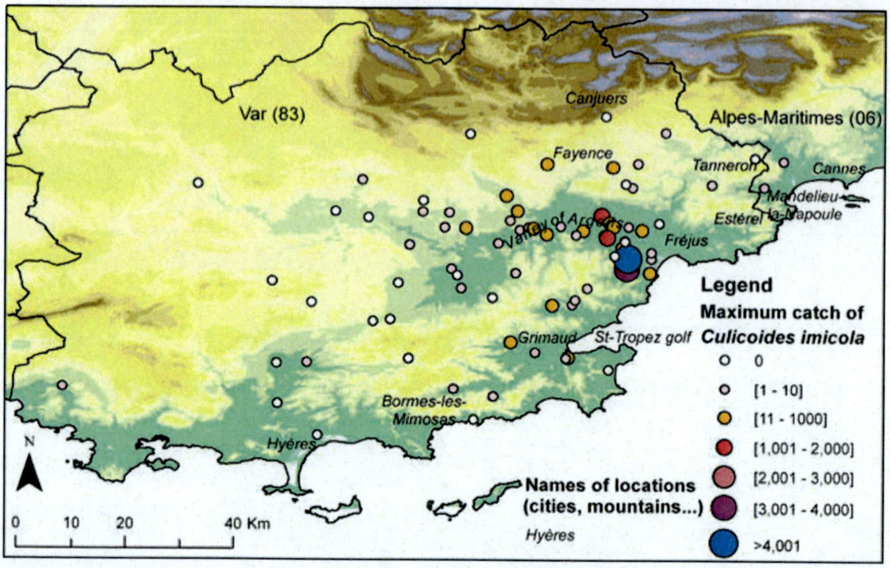

Fig. 4.6 Maximum catch of *Culicoides imicola* in the Var department from 2003 to 2010

limited by physical barriers such as the massifs of Estérel and Tanneron (600 m) to the east, planes of Canjuers (1,000 m) and la Colle du Rouet (500 m), slowing down its progression to the east (Cannes), south-east (Hyères) and north (Fayence) of the Var department (Figs. 4.5 and 4.6).

To estimate the spread of *C. imicola* over the years, the distance between each trap site which was positive in a given year and the closest nearest neighbouring site which was found positive at least once during the preceding years was computed. The maximal observed spread of *C. imicola* per year corresponds to the maximal distance between two nearest neighbours for that year. This relies on the hypothesis that *C. imicola* has spread between sites following the shortest possible path. The sites positive in 2003 were not included as they did not lead to the establishment of *C. imicola* populations the following year.

Over the 2004–2010 period, the average spread was 14.5 km/year. The maximal observed spread was 30.7 km/year and occurred in 2005. That year, the spread may have been overestimated as in 2004, the trapping campaign was not large enough to encompass a ring of negative trapping sites around positive sites, indicating that *C. imicola* may have spread further than what was observed in 2004. The speed of colonization of this species over land seems to be limited by topography and vegetation cover (Tran et al. 2007) (Figs. 4.5 and 4.6).

The potential extension of *C. imicola* from Spain to the Pyrénées-Orientales department was also surveyed starting 2009 since this species is permanently established in the Spanish bordering area (Catalonia). In October 2008, three erratic individuals were collected by the surveillance network at the lowest point along the

France-Spain border in the Pyrénées-Orientales department. To check whether *C. imicola* was absent from this region, additional trappings were carried out in October 2009 but did not collect any *Culicoides* because of heavy rains and wind. The following year, in October 2010, 11 sites were sampled for a week with one night trapping, and one site was positive with one parous female. The site was localized closed to the highway, suggesting that erratic individuals may passively be transported through the Pertus pass. Surveillance in this area is maintained, as this area is identified as risk of *C. imicola* establishment despite the efficient natural barrier of the Pyrenees Mountains and the North to South dominant winds.

4.3 Species Diversity

Through these surveys, the fauna list of *Culicoides* of France was updated. For most of the collections, traps were set up close to the hosts of interest (cattle and/or sheep); therefore, the checklist is probably not exhaustive and only related to the biting midge species related to domestic ruminants and therefore with a veterinary interest. Establishment of faunistic inventories is important to list the biodiversity, to analyse the possible spread of species into new territories and to analyse species communities in relation to the environment. Species diversity varies depending on the environment, reflecting differences in soil composition and meteorological conditions, and locally different breeding practices. When assessing the diversity of species to record rare or locally distributed species, it is important to keep in mind that collecting efforts often vary over time and that the type of trap used may bias the diversity of species sampled; for example, light traps are not appropriate to collect diurnal species.

Before bluetongue emergence in France, Delécolle (1985) recorded 65 species over the mainland territory with 45 recognized species in the Northeast region. A first checklist was published by Kremer et al. (1971) for Corsica recording 11 *Culicoides* species. However, the first bluetongue outbreaks in 2000, the first record of *C. imicola* on the island (Delécolle and de La Rocque 2002) and the resulting entomological surveillance increased the number of species records. Zientara et al. (2000) reported 21 species, and Delécolle and de La Rocque (2002) later recognized 37 species with 7 new species for Corsica. In 2005, Delécolle et al. (2005) updated the list of *Culicoides* in Corsica to 58 species with the first description of *C. riebi*. Today, the species list for French biting midges includes 83 species for mainland and 61 species for Corsica with recently three new records for mainland France: *C. manchuriensis*, *C. abchazicus* and *C. saevus* (Table 4.3). This list provides all the species collected through the surveillance networks plus some species recorded by experts but rarely collected (Delécolle and de La Rocque 2002; Delécolle et al. 2005). Whereas *C. imicola* can be dominant in Corsica, the common species collected in mainland France are the two sibling species of the Obsoletus Complex: *C. obsoletus* and *C. scoticus* usually referred as *C. obsoletus*/*C. scoticus* because females are morphologically indistinguishable, *C. chiopterus*,

Table 4.3 Updated list of *Culicoides* species recorded in France, and their distribution areas

Species name	Synonyms following the world biting midge catalogue	Distribution			
		Mainland France			
		Atlantic coast	centre and northeast	Mediteranean littoral	Corsica
C. abchazicus Dzhafaraov, 1964			x (SR)		
C. achrayi Kettle and Lawson, 1955		x	x	x	x
C. alazanicus Dzhafarov, 1961	*C. musilator* Kremer and Callot, 1961	x	x	x	x
C. albicans (Winnertz), 1852			x		
C. albihalteratus Goetghebuer, 1935[a]		x (SR)	x (SR)		
C. begueti Clastrier, 1957		x	x (SR)	x	x
C. brunnicans Edwards, 1939		x	x	x	x
C. cameroni Campbell and Pelham-Clinton, 1960				x	x (SR)
C. cataneii Clastrier, 1957		x	x	x	x
C. caucoliberensis Callot, Kremer, Rioux and Descous, 1967					X (SR)
C. chiopterus (Meigen), 1830	*C. amoenus* (Winnertz), 1852; *C. similis* (Goetghebuer), 1927; *C. dobyi* Callot and Kremer, 1969	x	x	x	x
C. circumscriptus Kieffer, 1918	*C. nadayanus* Kieffer, 1918; *C. edwardsi* Goetghebuer, 1921; *C. algarum* Kieffer, 1924; *C. salicola* Kieffer, 1924; *C. pictidorsum* Kieffer, 1924; *C. albonotatus* Vimmer, 1932; *C. albosignatus* Vimmer, 1932; *C. polymaculatus* Vimmer, 1932; *C. pulcher* Zilahi-Sebess, 1934; *C. kirovabadicus* Dzhafarov 1964; *C. matsuensis* Lien, Weng and Lin, 1996; *C. meridionalis* Xue, Liu and Yu, 2003	x	x	x	x
C. clastrieri Callot, Kremer and Déduit, 1962		x	x	x (SR)	x
C. clintoni Boorman, 1984			x		
C. comosioculatus Tokunaga, 1956	*C. chaetophthalmus* Amosova, 1957; *C. caucasicus* Sergejev, 1959; *C. setosus* Gutsevich, 1960		x		

C. corsicus Kremer, Leberre and Beaucournu-Saguez, 1971					x
C. dzhafarovi Remm, 1967	*C. dzhafarovi* Callot, Kremer, Molet and Bach, 1968		x		
C. deltus Edwards, 1939[a]	*C. lupicaris* Downes and Kettle, 1952	x	x	x	x
C. derisor Callot and Kremer, 1965		x			x (SR)
C. dewulfi Goetghebuer, 1936	*C. pseudochiopterus* Downes and Kettle, 1952	x	x	x	x
C. duddingstoni Kettle and Lawson, 1955		x	x	x	x
C. fagineus Edwards, 1939		x	x	x	x
C. fascipennis (Staeger), 1839	*C. distictus* Kieffer, 1916; *C. albonotatus* Kieffer, 1918; *C. turficola* Kieffer, 1925	x	x	x	x
C. festivipennis Kieffer, 1914	*C. odibilis* Austen, 1921; *C. winnertzi* Edwards, 1926	x	x	x	x
C. flavipulicaris Dzhafarov, 1964		x	x	x	x
C. furcillatus Callot, Kremer and Paradis, 1962		x	x	x	x
C. gejgelensis Dzhafarov, 1964	*C. bifasciatus* Tokunaga, 1951	x	x	x	x
C. griseidorsum Kieffer, 1918		x		x	x
C. grisescens Edwards, 1939	*C. remmi* Damian-Georgescu, 1972; *C. arschanicus* Mirzaeva, 1984	x	x		
C. haranti Rioux, Descous and Pech, 1959		x	x (SR)	x	x
C. heliophilus Edwards, 1921	*C. latifrontis* Shakirzjanova, 1962; *C. kobachidzei* Dzhafarov, 1964	x	x		
C. helveticus Callor, Kremer and Deduit, 1962			x		
C. heteroclitus Kremer and Callot, 1965		x (SR)		x	
C. ibericus Dzhafarov, 1963		x (SR)			
C. imicola Kieffer, 1913	*C. pallidipennis* Carter, Ingram and Macfie, 1920; *C. iraqensis* Khalaf, 1957; *C. minutus* Sen and Das Gupta, 1959; *C. pseudoturgidus* Das Gupta, 1962			x	x

(continued)

Table 4.3 (continued)

Species name	Synonyms following the world biting midge catalogue	Distribution			
		Mainland France			
		Atlantic coast	centre and northeast	Mediteranean littoral	Corsica
C. impunctatus Goetghebuer, 1920	*C. minor* Tokunaga, 1941	x	x	x	
C. indistinctus Khalaf, 1961[a]		x	x		x
C. jumineri Callot and Kremer, 1969		x		x	x
C. jurensis Callor, Kremer and Deduit, 1962			x		
C. kibunensis Tokunaga, 1937	*C. cubitalis* Edwards, 1939; *C. ponkikiri* Kono et Takahasi, 1940; *C. sitinohensis* Okada, 1941	x	x	x	x
C. kurensis Dzhafarov, 1960		x (SR)	x	x	x
C. longipennis Khalaf, 1957	*C. flavisimilis* Dzhafarov, 1964	x	x	x	x
C. lupicaris Downes et Kettle, 1952[a]		x	x	x	x
C. malevillei Kremer and Coluzzi, 1971				x (SR)	x
C. manchuriensis Tokunaga, 1941	*C. setiger* Goetghebuer, 1938; *C. goetghebueri* Arnaud, 1956; *C. machardyi* Campbell and Pelham-Clinton, 1960; *C. ochraceimaculatus* Shevchenko, 1970; *C. ochraceipennis* Shevchenko, 1970; *C. mesostigma* Remm, 1971; *C. vistulensis* Skierska, 1973	x	x (SR)		
C. maritimus Kieffer, 1924[a]	*C. submaritimus* Dzhafarov, 1962	x	x	x	x
C. maritimus paucisensillatus nv. sp.[a]			x (SR)	x	
C. minutissimus (Zetterstedt), 1855	*C. nanulus* Kieffer, 1919; *C. albihalter* Kieffer, 1919; *C. bychowskyi* Dzhafarov, 1964; *C. tugaicus* Dzhafarov, 1960	x	x		x
C. montanus Shakirzjanova, 1962					x (SR)
C. newsteadi Austen, 1921	*C. biclavatus* Kieffer, 1924; *C. halophilus* Kieffer, 1924; *C. edwardsi* Goetghebuer, 1933; *C. edwardsianus* Goetghebuer, 1933	x	x	x	x

C. nubeculosus (Meigen), 1830	*C. puncticollis* Goetghebuer, 1912; *C. punctaticollis* Goetghebuer, 1920	x	x	x	
C. obsoletus (Meigen), 1818	*C. varius* (Winnertz), 1852; *C. yezoensis* (Matsumura), 1911; *C. obscuripes* Santos Abreu, 1918; *C. lacteinervis* Kieffer, 1919; *C. rivicola* Kieffer, 1921; *C. clavatus* Kieffer, 1921; *C. heterocerus* Kieffer, 1921; *C. pegobius* Kieffer, 1922; *C. kabyliensis* Kieffer, 1922; *C. concitus* Kieffer, 1922; *C. intermedius* Okada, 1941; *C. sintrensis* Cambournac, 1956; *C. seimi* Shevchenko, 1967	x	x	x	x
C. odiatus Austen, 1921[a]	*C. niger* Dzhafarov, 1960; *C. indistinctus* Khalaf, 1961; *C. lailae* Khalaf, 1961; *C. kurekshaicus* Dzhafarov, 1962; *C. conicus* Remm, 1968	x	x	x	x
C. pallidicornis Kieffer, 1919	*C. susae* Kieffer, 1919; *C. dileucus* Kieffer, 1921; *C. brunneiscutellatus* Zilahi-Sebess, 1933; *C. bruneoscutellatus* Zilahi-Sebess, 1934;*C. niger* Root and Hoffman, 1937	x	x	x	x
C. paolae Boorman, 1996					x
C. paradisionensis Boorman, 1988					x
C. parroti Kieffer, 1922		x	x	x	x
C. pictipennis (Staeger), 1839	*C. arcuatus* (Winnertz), 1852; *C. guttularis* Kieffer, 1919; *C. maculatus* Zilahi-Sebess, 1936; *C. achkamalicus* Dzhafarov, 1964; *C. luganicus* Shevchenko, 1972	x	x	x	x
C. picturatus Kremer and Déduit, 1961		x	x	x	x
C. poperinghensis Goetghebuer, 1953		x	x	x	x
C. pseudopallidus Khalaf, 1961		x			
C. pulicaris (Linné), 1758	*C. setosinervis* Kieffer, 1913; *C. pullatus* Kieffer, 1915; *C. stephensi* Carter, 1916; *C. cinerellus* Kieffer, 1919; *C. quinquepunctatus* Goetghebuer, 1921;*C. flaviplumus* Kieffer, 1924; *C. sawamotoi* Kono and Takahasi, 1940	x	x	x	x
C. punctatus (Meigen), 1804	*C. punctatus* Latreille, 1809; *C. ocellaris* Kieffer, 1921; *C. kasachstanicus* Shakirzjanova, 1963	x	x	x	x

(continued)

Table 4.3 (continued)

Species name	Synonyms following the world biting midge catalogue	Distribution			
		Mainland France			
		Atlantic coast	centre and northeast	Mediteranean littoral	Corsica
C. puncticollis (Becker), 1903	*C. algecirensis* (Strobl), 1900; *C. impressus* Kieffer, 1918; *C. distigma* Kieffer, 1922; *C. donatieni* Kieffer, 1922; *C. sciniphes* Kieffer, 1925; *C. bipunctatus* Vimmer, 1932; *C. tripunctatus* Vimmer, 1932; *C. wenigi* Vimmer, 1932; *C. flavitarsis* Vimmer, 1932; *C. griseovittatus* Vimmer, 1932; *C. luteosignatus* Vimmer, 1932; *C. vavrai* Vimmer, 1932	x	x	x	x
C. reconditus Campbell and Pelham-Clinton, 1960			x	x	
C. riebi Delécolle, Mathieu and Baldet, 2005		x			x
C. riethi Kieffer, 1914	*C. cordatus* Kieffer, 1921; *C. crassiforceps* Kieffer, 1924	x	x		
C. riouxi Callot and Kremer, 1961		x	x	x	x (SR)
C. saevus Kieffer, 1922	*C. drenskii* Zilahi-Sebess, 1934; *C. puncticeps* Goetghebuer, 1934; *C. micromaculithorax* Khalaf, 1957		x	x	
C. sahariensis Kieffer, 1923	*C. baghdadensis* Khalaf, 1957; *C. coluzzii* Callot, Kremer and Bailly-Choumara, 1970		x	x	
C. salinarius Kieffer, 1914	*C. halobius* Kieffer, 1914; *C. meinerti* Kieffer, 1915; *C. punctatidorsum* Kieffer, 1924	x	x		x (SR)
C. santonicus Callot, Kremer, Rault and Bach, 1966		x	x	x	x
C. scoticus Downes and Kettle, 1952		x	x	x	x
C. segnis Campbell and Pelham-Clinton, 1960		x	x	x	
C. semimaculatus Clastrier, 1958	*C. karajevi* Dzhafarov, 1961	x		x	x
C. shaklawensis Khalaf, 1957	*C. caspius* Gutsevich, 1959	x	x (SR)	x (SR)	x

C. simulator Edwards, 1939		x	x	x	
C. sphagnumensis Williams, 1955	*C. lacicola* Arnaud, 1956; *C. carjalaensis* Glukhova, 1957		x		
C. stigma (Meigen), 1818	*C. kiefferi* Goetghebuer, 1910;*C. cordiformitarsis* Carter, 1916; *C. unimaculatus* Goetghebuer, 1920; *C. stigmoides* Callot, Kremer and Deduit, 1962	x	x		
C. subfagineus Delécolle and Ortega, 1998		x	x	x	x
C. subfasciipennis Kieffer, 1919	*C. analis* Kieffer, 1925	x	x	x	x
C. submaritimus Dzhafarov, 1962[a]			x (SR)	x	x
C. tauricus Gutsevich, 1959		x	x	x	
C. tbilisicus Dhafarov, 1964	*C. dendriticus* Boorman, 1976	x	x	x	x
C. truncorum Edwards, 1939	*C. sylvarum* Callot and Kremer, 1961	x (SR)			x
C. univittatus Vimmer, 1932	*C. agathensis* Callot, Kremer and Rioux, 1963	x	x (SR)	x	x
C. vexans (Staeger), 1839	*C. pungens* (Kieffer), 1901; *C. perpungens* Kieffer, 1919 ; *C. ajbassovi* Shakirzjanova, 1962	x	x	x	
C. vidourlensis Callot, Kremer, Molet and Bach, 1966				x	x
Total number of species		66	70	61	61

SR single record

[a]Means that there are some taxonomic discrepancies between the world biting midge catalogue (http://www.inhs.uiuc.edu/research/FLYTREE/CeratopogonidaeCatalog.pdf) and experts from the Palaearctic region about the species status

C. dewulfi, *C. pulicaris*, *C. newsteadi* and *C. punctatus* (Baldet et al. 2008; Balenghien et al. 2010, 2011) with some species locally being widely collected, as *C. brunnicans* in western France (Viennet et al. 2011). *Culicoides chiopterus* and *C. dewulfi* are rarely collected in Corsica.

Species identification based on morphological features is difficult for *Culicoides* midges. Most often slide mounting is required to validate identification by experts and to record new species. Moreover, cryptic diversity is suspected for important species groups (Pages et al. 2009), and numerous complexes are still in need of revision. Future works with combined molecular and morphological approaches may increase the fauna list for France.

4.4 Species Distribution and Seasonal Dynamics

The current species distribution results from the influence of intrinsic factors (spread ability, ecological amplitude and potential of evolution) and extrinsic factors (geographic, climatic, geologic and biotic). Moreover, seasonal dynamics of insect populations is driven mainly by climatic factors, i.e. rainfalls and temperature. France has a wide range of climate and ecogeographic zones which can explain the observed differences in species abundance and dynamics.

Corsica presents a Mediterranean climate characterized by dry and hot summers, and wet and mild winters. *Culicoides newsteadi*, *C. imicola*, *C. obsoletus*/*C. scoticus* and *C. pulicaris* were the dominant species. The three former could represent 80–90% of the total annual catch according to the sites (Fig. 4.7). Only six *C. dewulfi* and no *C. chiopterus* were collected by the surveillance network in Corsica from 2002 to 2008, both species remaining rare in Corsica.

Culicoides imicola and *C. obsoletus*/*C. scoticus* showed different seasonal dynamics in Corsica (Fig. 4.8). *Culicoides imicola* was present mainly from May to November, with highest abundance in August, September or October. *Culicoides imicola* populations progressively increased with monthly temperatures regardless of rainfall. Indeed, highest abundances were observed in 2003, considered as a dry and hot year. On the contrary, *C. obsoletus*/*C. scoticus* populations seemed to be limited by dryness. *Culicoides obsoletus*/*C. scoticus* populations were abundant around May and declined during summer dry months. This phenomenon was particularly marked in 2003. Nevertheless, populations could increase again in case of rainy summer as observed in 2002. In Corsica, *C. obsoletus*/*C. scoticus* populations became usually rare in January except in some year when populations were present all the year round as in winter 2002/2003.

South-western France presents an oceanic climate characterized by warm, but not hot summers and cool, but not cold winters, with a narrow annual temperature range. Rainfall is more abundant during winter. Due to low latitude and proximity to the ocean, this region presents particularly mild winter. The two species of the Obsoletus Complex were dominant, representing 30–95% of the total annual catches (Fig. 4.7). In sites where *C. obsoletus*/*C. scoticus* were not ultra-dominant, secondary species

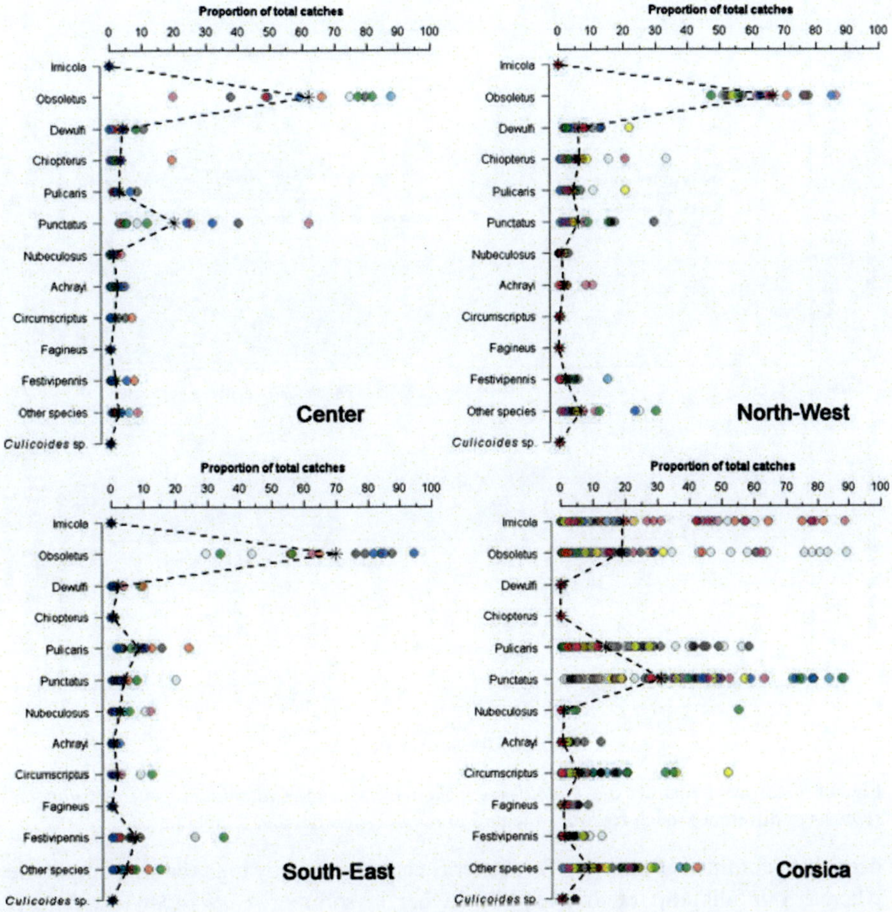

Fig. 4.7 Diversity of *Culicoides* species collected in different French regions, as the proportion of total annual catches. Each colour represents one site, which could be sampled different years. Stars represent the average proportion (whatever the site or the year). For convenience, species are grouped without any taxonomic validity. Imicola: *C. imicola*; Obsoletus: *C. obsoletus* and *C. scoticus*; Dewulfi: *C. dewulfi*; Chiopterus: *C. chiopterus*; Pulicaris: *C. flavipulicaris*, *C. lupicaris* and *C. pulicaris*; Punctatus: *C. punctatus* and *C. newsteadi*, Nubeculosus: *C. nubeculosus*, *C. puncticollis* and *C. riethi*; Achrayi: *C. achrayi*, *C. fascipennis*, *C. pallidicornis*, *C. picturatus* and *C. subfasciipennis*; Circumscriptus: *C. circumscriptus*, *C. salinarius* and *C. sphagnumensis*; Fagineus: *C. fagineus* and *C. subfagineus*; Festivipennis: *C. clastrieri*, *C. festivipennis*, *C. paloae* and *C. shaklawensis*; *Culicoides* sp.: unidentified specimens

were *C. festivipennis*, *C. pulicaris*, *C. lupicaris* and *C. nubeculosus*. *Culicoides dewulfi* represented up to 10% of the annual catch, whereas *C. chiopterus* was very rare. Centre of France presents a gradient from continental to mountainous climate characterized by cold winter and mild and wet summer. The dominant species were *C. obsoletus*/*C. scoticus* or *C. punctatus* depending on the sites. *Culicoides dewulfi*, and more rarely *C. chiopterus*, could be locally abundant representing up to 10% and 16%. The north-eastern France presents a west-eastern gradient of climate: from

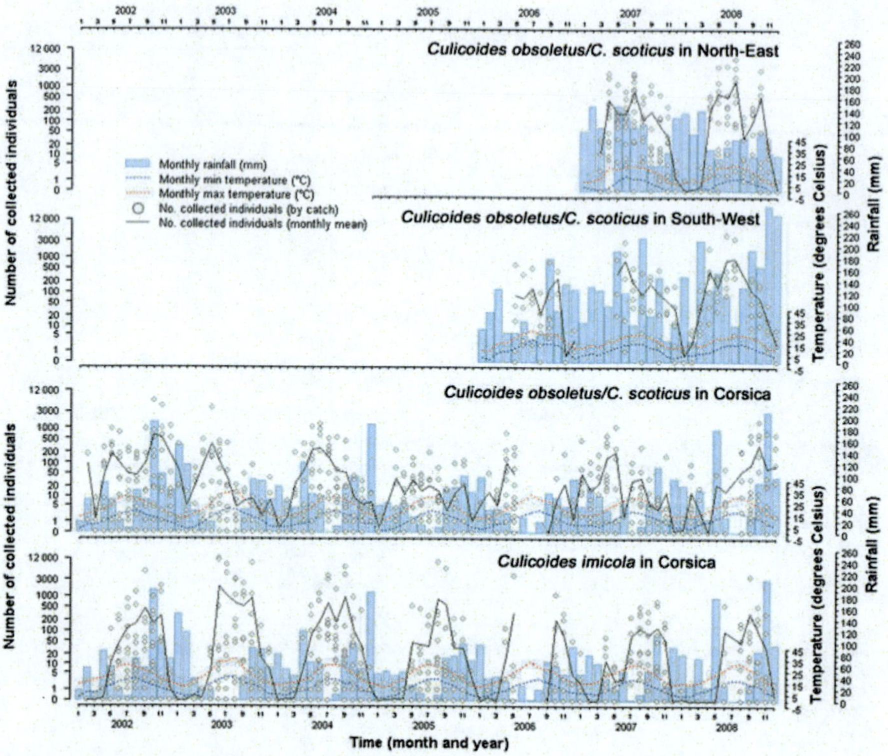

Fig. 4.8 Seasonal dynamics of *Culicoides imicola* in Corsica and *Culicoides obsoletus*/*Culicoides scoticus* in different French regions in regard to monthly rainfall and temperatures

oceanic to continental climate. This latter is characterized by high thermal amplitude, whereas rainfalls are regular, slightly higher in summer. *Culicoides obsoletus*/*C. scoticus* were clearly dominant, representing 45–90% of the total annual catches. *Culicoides dewulfi* and *C. chiopterus* may be abundant depending on the site: up to 20% and 35%. The other abundant species were *C. lupicaris*, *C. pulicaris*, *C. punctatus*, *C. achrayi*, *C. festivipennis*, *C. impunctatus*, *C. simulator* and *C. vexans*.

In non-Mediterranean mainland regions, *C. obsoletus*/*C. scoticus* dynamics were mainly unimodal, summer months being sufficiently wet. Nevertheless, populations may decrease during August in some years, as observed in 2006 in south-western France or in 2008 in north-eastern France, probably due to lack of rainfall in some sites.

4.5 Indoor and Outdoor Activity

Endophagous behaviour of Palaearctic species was reported by Overgaard Nielsen and Christensen (1975) in Denmark. In late 2006, comparing UV-light trap collections inside and outside sheds in the Netherlands, Meiswinkel et al. (2008)

Table 4.4 Mean number observed (max) and predicted *Culicoides* for the most abundant species from May to October 2007 and 2008 depending on the trap location and the animal presence (plus their interactions)

Effect	Value	*C. obsoletus*/*C. scoticus*		*C. chiopterus*		*C. dewulfi*	
		Observed	Predicted[b]	Observed	Predicted	Observed	Predicted
Trap location	Indoor	570 (3,452)	**125*****	37 (362)	7	37 (634)	**7*****
	Outdoor	156 (2,532)	**31**	11 (107)	4	23 (271)	**6**
Animal	Presence	377 (3,452)	73	28 (362)	**9*****	33 (634)	**6*****
	Absence	164 (2,532)	83	8 (67)	**2**	18 (108)	**7**
Year	2007	210 (2,155)	51	14 (204)	**2****	28 (271)	3
	2008	420 (3,452)	105	27 (362)	**9**	29 (634)	10
Loc × Pres[a]	–	$p < 0.001$		$p < 0.01$		$p < 0.001$	

[a]Interaction between trap location and animal presence
[b]Predicted values are in *bold* if $p < 0.05$
p values were given for effect modalities compared to a reference value, i.e. "indoor" for trap location or "absence" for animal presence ($^{***}p < 0.001$; $^{**}p < 0.01$ and $^{*}p < 0.05$)

collected threefold more *Culicoides* outside than inside, where cattle are kept for the night. Using the same methodology in France at the same period, Baldet et al. (2008) collected *Culicoides* in greater number inside than outside. Authors conceded that results are difficult to interpret because of the variation in building opening and in cattle abundance close to trap in the different collection sites (Baldet et al. 2008). With a standardized approach, Baylis et al. (2010) identified that the presence of animals and the opening of stable increased the indoor number of *Culicoides* collected by UV-light trap.

In north-eastern France, two OVI light-traps were placed in eight sites, one trap inside stables and the other outside in 2007 and 2008. Presence of animals indoor or outdoor was recorded. We used these surveillance data to test the influence of trap location (indoor vs. outdoor) and animal presence on *Culicoides* abundance during favourable months (from May to October). We modelled the *Culicoides* abundance using a generalized linear model with a Poisson distribution, with trap location, animal presence (plus their interaction) and year as fixed effects and site and date as random effects.

It seemed clear that it was not possible to interpret effect of trap location (indoor vs. outdoor) on *Culicoides* abundance, without taking into account animal presence. Indeed, interactions between trap location and animal presence had influence on *Culicoides* abundance ($p < 0.01$) for *C. obsoletus*/*C. scoticus*, *C. chiopterus* and *C. dewulfi* (Table 4.4). UV-light traps collected only few *Culicoides* indoor in the absence of animal (Fig. 4.9). On the contrary, indoor collections in the presence of animals were more abundant than outdoor (Fig. 4.9), maybe due to a higher efficiency of the trap indoor than outdoor or due to a higher density of animals around the trap indoor than outdoor (Garcia-Saenz et al. 2011). Outdoor collections may be abundant even if animals were not present (Fig. 4.9), but animal presence increased the number of collected *Culicoides*. These results confirmed the importance of presence of animals in increasing the indoor number of *Culicoides* collected by UV-light trap (Baylis et al. 2010). Future investigations should include the

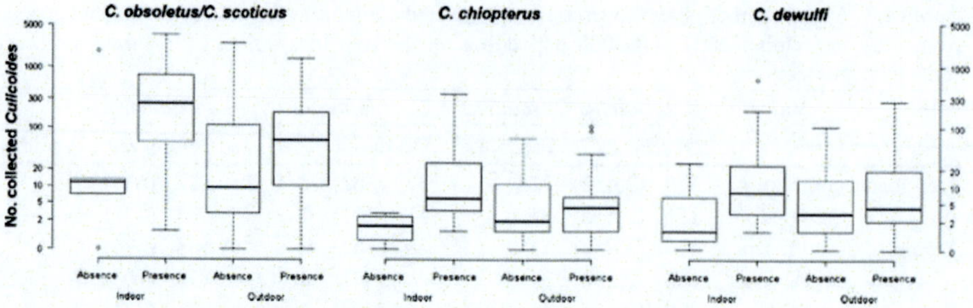

Fig. 4.9 Boxplots of main *Culicoides* species of veterinary interest abundance collected in eight sites in north-eastern France from 2007 to 2008 depending on trap location (*indoor and outdoor*) and on animal presence

opening of stable to untangle complex influences of animal presence and opening of stable on indoor abundance of *Culicoides*.

4.6 Field-Collected Infected Individuals

Reliable detection of bluetongue virus from field-collected specimens is difficult due to several constraints related to the genus *Culicoides* itself and to laboratory procedures. A species can be incriminated as a confirmed vector species when four criterions are met: biology compatible with vector/host contacts (abundance of species, contact with host of interest), ability of species to be infected after infectious blood meal, ability of the species to transmit the virus by bite, and detection or isolation of the virus within the field-collected species specimens (WHO 1961). This last criterion is challenging and requires insects to be collected at the moment of intense transmission to maximize the chances of collecting infected *Culicoides*. Virus detection is usually achieved with real-time reverse-transcription polymerase chain reaction (real-time RT-PCR) and virus isolation with cell cultures or embryonated chicken eggs. However, these results are meaningless in terms of vector implication if biting midge pools are not sorted out by species, sex and physiological status and if engorged females are not excluded from the pools tested.

Among Palaearctic species, BTV was isolated from field-collected pools of *C. obsoletus* (Mellor and Pitzolis 1979; Savini et al. 2004) and *C. pulicaris* (Caracappa et al. 2003; De Liberato et al. 2005). BTV-8 viral genome from *C. dewulfi* and *C. chiopterus* field individuals has been identified by real-time RT-PCR in the Netherlands (Dijkstra et al. 2008; Meiswinkel et al. 2007) and in France (see below). In Basque country, BTV-1 was detected by real-time RT-PCR from *C. obsoletus*/*C. scoticus*, *C. pulicaris* and *C. lupicaris* parous females (Romon et al. 2011). *Culicoides obsoletus* and *C. scoticus* from United Kingdom have been experimentally infected by serotypes 8 and 9 of bluetongue virus, *C. scoticus* having higher viral titers (Carpenter et al. 2008). In Sicily, pools of *C. pulicaris* were found infected with serotype 2 (Torina et al. 2004), but the taxonomic issues around this species and related species such as *C. lupicaris* make difficult the understanding of the involvement of this species in BTV transmission in northern Europe. This list of (potential) vector species is probably not exhaustive. It is therefore important to complete this list of potential vectors with information from other BTV-infected countries and with future vector competence experimentations with Palaearctic *Culicoides* species.

To identify which species of *Culicoides* species were involved in BTV-8 transmission in France, individuals collected in infected farms were tested for the presence of BTV in France at two occasions in 2007. In mid-September 2007, a single-night collection was carried out using two UV-light suction traps (OVI trap) in a cattle farm localized in north-eastern France (Ardennes department). This cattle farm was chosen because animals were developing clinical signs of bluetongue. After morphological identification, *Culicoides* were divided into pools from 1 to 57 individuals of the same species, sex and physiological stage. Among collected individuals, 973 unfed parous females were tested for the presence of BTV by real-time RT-PCR: 731 *C. chiopterus* (15 pools), 159 *C. obsoletus*/*scoticus* (3 pools), 72 *C. dewulfi* (2 pools), 3 *C. lupicaris* (1 pool), 2 *C. pulicaris* (1 pool), 2 *C. nubeculosus* (1 pool), 2 *C. pallidicornis* (1 pool), 1 *C. achrayi* (1 pool) and 1 *C. salinarius* (1 pool). Among the 15 pools of *C. chiopterus*, 2 were positive for BTV-8, leading to an estimate of the minimum infection rate (MIR) of 2.7% for this species. This result confirms the role as vector species of *C. chiopterus* (Dijkstra et al. 2008).

In October 2007, an extensive biting midge collection (three consecutive days) in a sheep farm reporting BTV-8 clinical cases was carried out. The farm was located in north-eastern France (Alsace department) and composed by a central building surrounded by fields hosting about 220 animals (2.3% with clinical signs and 6.3% [1.3–17.2] positive by ELISA). *Culicoides* were collected using (1) six UV-light traps dispersed in the farm closed to the animals, including four traps baited with CO_2 and operated continuously, (2) mouth aspirator on the abdomen of a sheep from 10 to 14 h and from 15 to 18 h (30 min sessions) and (3) a backpack aspirator during afternoons (30 min sessions) around haystacks covered by a black plastic tarpaulin. UV-light traps collected 383 *Culicoides*, with 71% of *C. obsoletus*/*scoticus* (21.3 *Culicoides*/trap/day). On the sheep, 33 unfed females (23 *C. obsoletus*/*scoticus*, 9 *C. dewulfi* and 1 *C. chiopterus*) mainly (32/33) from 10 to 11 h (17°C at 11 h) were collected in 1 day. With the backpack aspirator, 1,076 *Culicoides* were collected (135 *Culicoides*/session) with 99% of *C. obsoletus*/*scoticus* females, mainly parous (92%) and gravid (80%). We did not detect BTV-8 from the 1,377 parous females tested. This reflects the low sensitivity of the underlying method as very small numbers of vectors are infected, even during high transmission phases, and the necessity to collect *Culicoides* at the exact moment of the transmission.

4.7 Conclusions

The entomological surveillance implemented since 2000 in France because of the northward expansion of *C. imicola* together with the emergence of bluetongue in Western Europe helps to understand the *Culicoides* dynamics and diversity and species role in BTV transmission.

The principal findings of the *Culicoides* surveillance in France can be summarized as follows: 1. *C. imicola*, the main southern European/African BTV vector which was involved in BTV outbreaks in Corsica, has settled in mainland France but is restricted currently to the littoral of the Var department. 2. Several Palaearctic *Culicoides* species were considered potential BTV vectors in French mainland, including *C. obsoletus*/*C. scoticus*, *C. chiopterus* and *C. dewulfi* in the absence of *C. imicola*. 3. The spatio-temporal dynamics of *Culicoides* in farms depends on environment and climatic factors and at a local scale on husbandry practices, but generally *C. imicola* is abundant during summer in the littoral of Corsica and *C. obsoletus*/*C. scoticus* occurred widely and abundantly from spring to autumn on sheep and cattle holdings across the whole French mainland.

The favourable climatic conditions that prevailed in Europe at the time of the emergence and expansion of BTV-8 from 2006 to 2008 appear to have promoted the successful transmission by several Palaearctic *Culicoides* species (Guis et al. 2012). BTV vectors are known worldwide to transmit also other viral pathogens to livestock, such as African horse sickness, Akabane and epizootic haemorrhagic disease viruses. It is thus probable that these pathogens could be transmitted if introduced into Europe during climatically favourable periods. Future temperature increase may enhance the vector capacity of Palaearctic *Culicoides* species and could favour the settlement of new exotic species in French territory such as *C. imicola*. For these reasons, the French veterinary authorities are advised to maintain exhaustive *Culicoides* surveillance network to follow the expansion of *C. imicola*—and the settlement of new exotic species—and to determine precisely the abundance and seasonality of *Culicoides* species in livestock premises. Entomological surveillance of *Culicoides*, potential vectors of pathogens to livestock, should be harmonized at the European level leading to more integrated approaches and, hence, comparability between datasets. These surveys should improve the predictive models developed such as those based on the basic reproductive ratio (R_0) and would finally help to identify areas and periods at risk of transmission (Guis et al. 2012).

References

Baldet T, Delécolle J-C, Mathieu B, de La Rocque S, Roger F (2004) Entomological surveillance of bluetongue in France in 2002. Vet Ital 40:226–231

Baldet T, Delécolle JC, Cêtre-Sossah C, Mathieu B, Meiswinkel R, Gerbier G (2008) Indoor activity of *Culicoides* associated with livestock in the bluetongue virus (BTV) affected region of northern France during autumn 2006. Prevent Vet Med 87:84–97

Balenghien T, Garros C, Mathieu B, Setier-Rio ML, Allène X, Gardès L, Rakotoarivony I, Venail R, Akadar A, Drouet M, Delécolle J-C (2010) La surveillance des *Culicoides* en France. Bull Epidémiologique 35:8–9

Balenghien T, Delécolle JC, Setier-Rio ML, Rakotoarivony I, Allène X, Venail R, Delécolle D, Lhoir J, Gardès L, Chavernac D, Mathieu B, Languille J, Baldet T, Garros C (2011) Fièvre catarrhale ovine: bilan de la surveillance entomologique en 2010 en France. Bull Epidémiologique 46:26–31

Baylis M, Parkin H, Kreppel K, Carpenter S, Mellor PS, McIntyre KM (2010) Evaluation of housing as a means to protect cattle from *Culicoides* biting midges, the vectors of bluetongue virus. Med Vet Entomol 24:38–45

Caracappa S, Torina A, Guercio A, Vitale F, Calabro A, Purpari G, Ferrantelli V, Vitale M, Mellor PS (2003) Identification of a novel bluetongue virus vector species of *Culicoides* in Sicily. Vet Rec 153:71–74

Carpenter S, McArthur C, Selby R, Ward R, Nolan DV, Luntz AJ, Dallas JF, Tripet F, Mellor PS (2008) Experimental infection studies of UK *Culicoides* species midges with bluetongue virus serotypes 8 and 9. Vet Rec 163:589–592

De Liberato C, Scavia G, Lorenzetti R, Scaramozzino P, Amaddeo D, Cardeti G, Scicluna M, Ferrari G, Autorino GL (2005) Identification of *Culicoides obsoletus* (Diptera: Ceratopogonidae) as a vector of bluetongue virus in central Italy. Vet Rec 156:301–304

Delécolle J-C (1985) Nouvelle contribution à l'étude systématique et iconographique des espèces du genre *Culicoides* du Nord-Est de la France thése d'université, Université de Strasbourg, France

Delécolle J-C, de La Rocque S (2002) Contribution à l'étude des *Culicoides* de Corse. Liste des espèces recensées en 2000/2001 et redescription du principal vecteur de la fièvre cattarhale ovine: *C. imicola* Kieffer, 1913 (Diptera: Ceratopogonidae). Bull Soc Entomo France 107:371–379

Delécolle J-C, Mathieu B, Baldet T (2005) Nouvelle contribution à l'étude des *Culicoides* de Corse. II. Mise à jour de la liste des espèces du genre, description de *Culicoides riebi* n. sp. et redescription de *Culicoides paradisionensis* Boorman, 1988 (Diptera: Ceratopogonidae). Bull Soc Entomol France 110:69–76

Dijkstra E, van der Ven IJ, Hölzel DR, Van Rijn PA, Meiswinkel R (2008) *Culicoides chiopterus* as a potential vector of bluetongue virus in Europe. Vet Rec 162:422

Du Toit RM (1944) The transmission of blue-tongue and horse sickness by *Culicoides*. Onderstepoort J Vet Sci 19:7–16

Enserink M (2006) Emerging infectious diseases. During a hot summer, bluetongue virus invades northern Europe. Science 313:1218–1219

Gambles RM (1949) Bluetongue of sheep in Cyprus. J Comp Pathol Therapeut 59:176–190

Garcia-Saenz A, McCarter P, Baylis M (2011) The influence of host number on the attraction of biting midges, *Culicoides* spp., to light traps. Med Vet Entomol 25:113–115

Goffredo, M., Satta, G., Torina, A., Federico, G., Scaramozzino, P., Cafiero, M.A., Lelli, R., Meiswinkel, R., 2001. The 2000 bluetongue virus (BTV) outbreak in Italy : distribution and abundance of the principle vector *Culicoides imicola* Kieffer. In: Proc Tenth Int Symp Amer Assoc Vet Lab Diagnosticians, Salsomaggiore, Parma, pp 308–309

Goldarazena A, Romon P, Aduriz G, Balenghien T, Baldet T, Delécolle J-C (2006) First record of *Culicoides imicola*, the main vector of bluetongue virus in Europe, in the Basque Country (northern Spain). Vet Rec 162:820–821

Guis H, Caminade C, Calvete C, Morse AP, Tran A, Baylis M (2012) Modelling the effects of past and future climate on the risk of bluetongue emergence in Europe. J R Soc Interface R Soc 9:339–350

Jennings DM, Mellor PS (1988) The vector potential of British *Culicoides* species for bluetongue virus. Vet Microbiol 17:1–10

Kremer M, Leberre G, Beaucournu-Saguez F (1971) Note sur les *Culicoides* (Diptera: Ceratopogonidae) de Corse. Description de *C. corsicus* n. sp. Ann Parasitol 46:653–660

Meiswinkel R, van Rijn P, Leijs P, Goffredo M (2007) Potential new *Culicoides* vector of bluetongue virus in northern Europe. Vet Rec 161:564–565

Meiswinkel R, Baldet T, de Deken R, Takken W, Delécolle JC, Mellor PS (2008) The 2006 outbreak of bluetongue in northern Europe–the entomological perspective. Prevent Vet Med 87:55–63

Mellor P, Pitzolis G (1979) Observations on breeding sites and light trap of *Culicoides* during an outbreak of bluetongue in Cyprus. Bull Entomol Res 69:229–234

Mellor PS, Wittmann EJ (2002) Bluetongue virus in the Mediterranean basin 1998–2001. Vet J 164:20–37

Mellor PS, Boorman JPT, Wilkinson PJ, Martinez-Gomez F (1983) Potential vectors of bluetongue and African horse sickness viruses in Spain. Vet Rec 35:445–466

Mellor PS, Jennings DM, Wilkinson PJ, Boorman JPT (1985) *Culicoides imicola* a bluetongue virus vector in Spain and Portugal. Vet Rec 116:589–590

Mellor PS, Boorman J, Baylis M (2000) *Culicoides* biting midges: their role as arbovirus vectors. Annu Rev Entomol 45:307–340

Mellor PS, Baylis M, Mertens PP (2009) Bluetongue. Elsevier, London

Overgaard Nielsen B, Christensen O (1975) A mass attack by biting midge *Culicoides nubeculosus* (Mg.) (Diptera, Ceratopogonidae) on grazing cattle in Denmark a new aspect of sewage discharge. Nordisk Veterinaermedicin 27:365–372

Pagès N, Munoz-Munoz F, Talavera S, Sarto V, Lorca C, Nunez JI (2009) Identification of cryptic species of *Culicoides* (Diptera: Ceratopogonidae) in the subgenus *Culicoides* and development of species-specific PCR assays based on barcode regions. Vet Parasitol 165:298–310

Purse BV, Mellor PS, Rogers DJ, Samuel AR, Mertens PP, Baylis M (2005) Climate change and the recent emergence of bluetongue in Europe. Nat Rev Microbiol 3:171–181

Romon P, Higuera M, Delecolle JC, Baldet T, Aduriz G, Goldarazena A (2012) Phenology and attraction of potential *Culicoides* vectors of bluetongue virus in Basque Country (northern Spain). Vet Parasitol 186:415–424

Saegerman C, Mellor P, Uyttenhoef A, Hanon JB, Kirschvink N, Haubruge E, Delcroix P, Houtain JY, Pourquier P, Vandenbussche F, Verheyden B, De Clercq K, Czaplicki G (2008) The most likely time and place of introduction of BTV8 into Belgian ruminants. PLoS One 5:e9405

Sarto i Monteys V, Ventura D, Pagès N, Aranda C, Escosa R (2005) Expansion of *Culicoides imicola*, the main bluetongue virus vector in Europe, into Catalonia, Spain. Vet Rec 156:415–417

Savini G, Goffredo M, Monaco F, Di Gennaro A, de Santis P, Meiswinkel R, Caporale V (2004) The isolation of bluetongue virus from field populations of the Obsoletus Complex in central Italy. Vet Ital 40:286–291

Southwood TRE, Henderson PA (2000) Ecological methods, 3rd edn. Blackwell Publishing, Oxford, UK

Torina A, Caracappa S, Mellor PS, Baylis M, Purse BV (2004) Spatial distribution of bluetongue virus and its *Culicoides* vectors in Sicily. Med Vet Entomol 18:81–89

Tran A, Guis H, Barragué B, Mathieu B, Setier-Rio M-L, Gerbier G, Roger F, Baldet T (2007) Use of remote sensing to assess the risk of emergence for a vector-borne disease: Introduction and diffusion of *Culicoides imicola*, bluetongue vector in Southern France. Télédetection 7: 419–432

Velthuis AG, Saatkamp HW, Mourits MC, de Koeijer AA, Elbers AR (2010) Financial consequences of the Dutch bluetongue serotype 8 epidemics of 2006 and 2007. Prevent Vet Med 93:294–304

Viennet E, Garros C, Lancelot R, Allène X, Gardès L, Rakotoarivony I, Crochet D, Delécolle JC, Moulia C, Baldet T, Balenghien T (2011) Assessment of vector/host contact: comparison of animal-baited traps and UV-light/suction trap for collecting *Culicoides* biting midges (Diptera: Ceratopogonidae), vectors of Orbiviruses. Parasit Vectors 4:119

WHO (1961) Arthropod-borne viruses: report of a study group. In World Health Organization Technical Report Series (Geneva), p. 68.

Zientara S, De La Rocque S, Gourreau JM, Grégory M, Diallo A, Hendrickx P, Libeau G, Sailleau C, Delécolle J-C (2000) La fièvre catarrhale ovine en Corse en 2000. Epidémiol et santé anim 40:129–134

Chapter 5
Bluetongue Disease: An Analysis of the Epidemic in Germany 2006–2009

Franz J. Conraths, Michael Eschbaumer, Conrad Freuling, Jörn Gethmann, Bernd Hoffmann, Matthias Kramer, Carolina Probst, Christoph Staubach, and Martin Beer

Abstract In August 2006, bluetongue virus of serotype 8 (BTV-8), which had occurred before in the sub-Saharan region, Asia and South America, was introduced into Central Europe. The virus hit an area with a high population density of BTV-naive ruminants, suitable vectors (*Culicoides* spp.) and climatic conditions favourable for virogenesis and transmission. In 2006 and 2007, the disease spread over wide parts of western Germany and had a high economic impact on sheep and cattle farms. To reduce animal losses, mitigate the clinical symptoms and stop the further spread of the disease, Germany decided to implement a compulsory vaccination programme with a monovalent, inactivated vaccine against BTV-8 in May 2008 which has apparently led to the eradication of the disease. This chapter reviews the pathogenesis of bluetongue disease, the clinical signs, diagnosis, the course of the epidemic, control measures and the economic impact of the BTV-8 epidemic in Germany.

Keywords Bluetongue disease • Culicoides • Epidemiology • Germany • Vector

5.1 Introduction

Bluetongue disease (BT) is a non-contagious vector-borne disease that mainly affects ruminants but also camelids. It is caused by the BT virus (BTV), which belongs to the genus *Orbivirus* within the family *Reoviridae* and of which 24

F.J. Conraths (✉) • C. Freuling • J. Gethmann • M. Kramer • C. Probst • C. Staubach
Institute of Epidemiology, Friedrich-Loeffler-Institut, Seestraße 55, 16868 Wusterhausen, Germany
e-mail: franz.conraths@fli.bund.de

M. Eschbaumer • B. Hoffmann • M. Beer
Institute of Diagnostic Virology, Friedrich-Loeffler-Institut, Südufer 10, 17493 Greifswald-Insel Riems, Germany

H. Mehlhorn (ed.), *Arthropods as Vectors of Emerging Diseases*,
Parasitology Research Monographs 3, DOI 10.1007/978-3-642-28842-5_5,

serotypes are known; further serotypes awaiting confirmation were described in Switzerland (Toggenburg virus) (Hofmann et al. 2008) and Kuwait (Maan et al. 2011a, b). BTV is transmitted between hosts almost exclusively through the bites of female *Culicoides* biting midges. BT is a notifiable disease under the German Animal Disease Act, reportable in the European Union via the Animal Disease Notification System (ADNS) and at the global level notifiable to the World Organisation for Animal Health (OIE).

BT had never been reported in Germany before it occurred in the region of Aachen in North Rhine-Westphalia in August 2006, almost simultaneously with outbreaks in Belgium and the Netherlands (EU Rapid Press release Nr. IP/06/1112, ProMedMail, 20060828.2448, Mehlhorn et al. 2007; Toussaint et al. 2007).

5.2 Pathogenesis

BTV is spread by infected haematophagous insects, mainly biting midges (*Culicoides* spp.) that excrete the virus in their saliva. In addition to the transmission itself, it has been suggested that immunomodulatory proteins in the midge saliva aid in the initial infection of the host (Darpel et al. 2009). During a blood meal, BTV is inoculated into the skin, which may be both an important site for replication and a source of virus for blood-feeding vectors (Darpel et al. 2009). After inoculation, migrating dendritic cells transport the virus to the draining lymph node (Barratt-Boyes et al. 1995; Hemati et al. 2009), from where it disseminates to other lymphoid tissues (Pini 1976; Barratt-Boyes and Maclachlan 1994). Replication occurs principally in mononuclear phagocytic cells, proliferating lymphocytes and endothelial cells (Mahrt and Osburn 1986; Maclachlan et al. 1990; Barratt-Boyes et al. 1992; DeMaula et al. 2001; Drew et al. 2010b). Accordingly, bluetongue pathogenesis is characterized by virus-mediated immune suppression, endothelial injury and dysfunction (DeMaula et al. 2002a; Maclachlan 2004; Maclachlan et al. 2009; Umeshappa et al. 2010).

Virus replication in endothelial cells causes direct cell injury and necrosis. Vascular blockage leads to haemorrhage and tissue infarction (Mahrt and Osburn 1986) that can manifest as myonecrosis and mucosal ulceration (Drew et al. 2010b). The activation of pulmonary endothelial cells and macrophages (DeMaula et al. 2002a, b; Drew et al. 2010b), on the other hand, and the subsequent release of host-derived inflammatory and vasoactive mediators (Hemati et al. 2009) increase vascular permeability, potentially leading to the widespread oedema often seen in fatal BT, African horse sickness and other virus-induced haemorrhagic fevers (Maclachlan et al. 2009; Drew et al. 2010a; Maclachlan 2011). A confirmation of this hypothesis, however, will require a deeper understanding of the highly complex interplay of cytokines in infected ruminants beyond the isolated findings of groundbreaking in vitro studies.

5.3 Clinical Signs

Before the first outbreaks of BT in Central Europe in 2006, clinical signs of the disease were mainly described for sheep (Erasmus 1990). However, it is known since many years that BTV can infect several domestic and wild ruminant species (Tabachnick 1996; Darpel et al. 2007). The severity of the disease in sheep may depend to a marked extent on environmental conditions, most notably exposure to sunlight, a frequently ignored fact (Erasmus 1990). Other authors doubt whether there is an ill-defined interaction with the environment (Mellor and Wittmann 2002). The clinical picture also varies depending on the strain of the virus as well as the breed and age of the infected animals with older age groups being more susceptible (Tabachnick 1996; Mellor and Wittmann 2002). In general, the clinical picture of BT can be extremely variable (EFSA 2007).

5.3.1 Sheep

Irrespective of the region of the world from where BTV serotypes originate, clinical disease of sheep follows a similar pattern (Parsonson 1992). The severity of clinical signs depends on both the breed of sheep and the strain of virus (Darpel et al. 2007). All breeds of sheep are susceptible to BTV infection, although the clinical outcome may vary remarkably. Especially indigenous African breeds have been reported as resistant (Erasmus 1990). However, not only African but also other indigenous breeds seem to be less susceptible than introduced European breeds and Merino sheep (Parsonson 1992) or may only be affected subclinically (Darpel et al. 2007). Febrile reactions with fever exceeding 41°C are common (Erasmus 1990). This could also be reproduced with a BTV-8 strain in experimental infections (Darpel et al. 2007). Within 1–2 days after the onset of the disease, the skin of the muzzle and lips as well as the oral mucosa became hyperaemic and oedematous (Erasmus 1990). This was also frequently observed during the BTV-8 epidemic in Central Europe between 2006 and 2009. Figure 5.1 shows a ewe from a backyard farm in North Rhine-Westphalia in September 2006. During the BTV-8 outbreak in Germany 2006–2009, clinical disease was often very severe in sheep.

Nasal discharge, which became later sometimes mucopurulent, and resulting dyspnoea could often be observed.

Foot lesions developing with the subsidence of fever represented a frequently detected disease manifestation. The coronary band is hyperaemic often combined with petechial haemorrhages.

An investigation in sheep farms in 2007 revealed that stillborn lambs showed clinical signs such as crusts and lesions of the oral mucosa (Fig. 5.2a). One of the farmers reported that the newborn lambs were physically and motorically retarded for several weeks. Feet lesions were apparently so painful that some of the sheep were reluctant to walk for weeks even after the acute symptoms had healed. Instead, they crawled on their carpal joints (Fig. 5.2b). When the animals were supported or

Fig. 5.1 Ewe with hyperaemic muzzle and shallow erosions and crusts on the nostril

when they got shooed, they walked on their feet for a little while; however, they immediately fell back to their previous behaviour when they were left alone.

5.3.2 Goats

Reports on clinical signs with BTV-8 virus strains in goats are rare. During experimental studies with BTV-8 in Dutch dairy goats, fever, signs of general illness, apathy, dysphagia, diarrhoea and lameness were observed (Backx et al. 2007). The discrepancy between the observations in the field and in laboratory experiments might be explained by different routes of infection, e.g. intravenous injection vs. the natural route of infection via biting midges or by vector preferences or the types of husbandry systems (Backx et al. 2007). During the BTV-8 epidemic in Germany in 2006–2009, a total of 26,954 BTV-8-infected premises were recorded (TSN database; 15.11.2011, 1135 hours), among which there were 132 holdings where goats were reported as clinically affected (TSN database; 15.11.2011, 1135 hours). This shows that goats were affected by this epidemic in Germany, but only to a limited extent.

5.3.3 Cattle

It is thought that cattle have now largely replaced antelopes as a maintenance host of the virus in Africa (Gerdes 2004). However, before the BTV-8 epidemic in northern

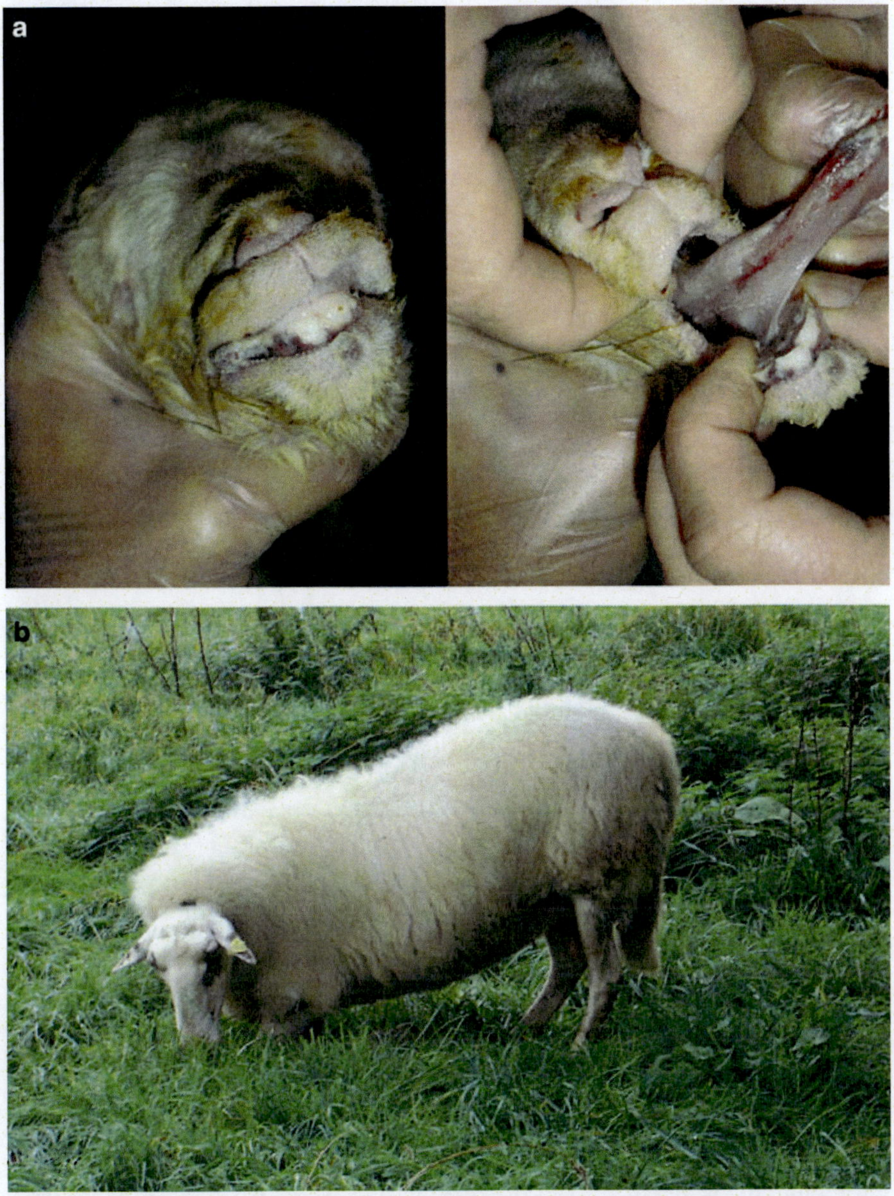

Fig. 5.2 Clinical signs of BT in sheep; (**a**) lips and tongue of a stillborn lamb with crusts and lesions in the oral mucosa; (**b**) sheep walking on carpal joints

Europe, natural and experimental BTV infection of cattle was considered asymptomatic in the vast majority of cases (Maclachlan et al. 1992; Maclachlan 2011). A transient febrile response, lacrimation and salivation were occasionally observed in infected animals (Erasmus 1990).

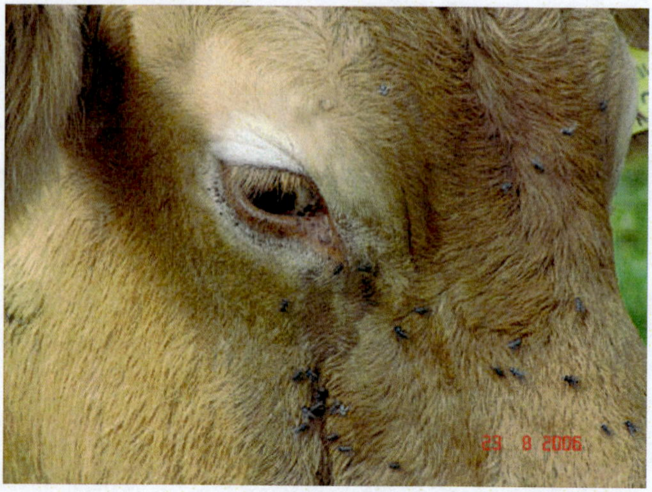

Fig. 5.3 Limousin cow with swollen eyelids, lacrimation and petechial haemorrhages

At the beginning of the BTV-8 epidemic in Central Europe, clinical symptoms such as fever (40–41°C) for 2–14 days, severe nasal discharge, lacrimation, facial oedema and nasal excoriations were described in adult cattle (Mehlhorn et al. 2007). However, using a field virus strain that originated from a BT outbreak in the Netherlands in an experimental infection, no pyrexia was recorded in any of the cattle (four 6-month-old male Holstein-Friesian calves) at any stage of the experiment (Darpel et al. 2007). One of the first German BTV-8 outbreaks was detected in a beef suckler herd (57 cattle, Limousin breed) in North Rhine-Westphalia. One cow showed a bilateral conjunctivitis, oedema of the eyelids combined with petechial haemorrhages on the swollen mucous membranes of the eyelids. Additionally, strong lacrimation was obvious as a predominant clinical sign (Fig. 5.3). The local tissue damage was complicated by bacterial infection.

Ulcers and erosions may occur in the oral mucosa (Fig. 5.4). The skin of the muzzle was inflamed at the beginning of the disease; later, cracks and peels could be observed (Fig. 5.5).

Large-scale erosions or haemorrhagic lesions in the skin of the teats were quite often reported in dairy cattle (Fig. 5.6). These lesions sometimes resulted in detachments of bigger parts of the skin of the affected teats. This caused pain during milking or nursing. Later on, these lesions were also affected by bacterial infection or characterized by the formation of crusts.

Lesions involving coronitis and inflammation of the whole foot region were also frequently reported. The coronary bands were hyperaemic, swollen and inflamed. Therefore, stiffness or lameness is common in BTV-infected cattle (Erasmus 1990).

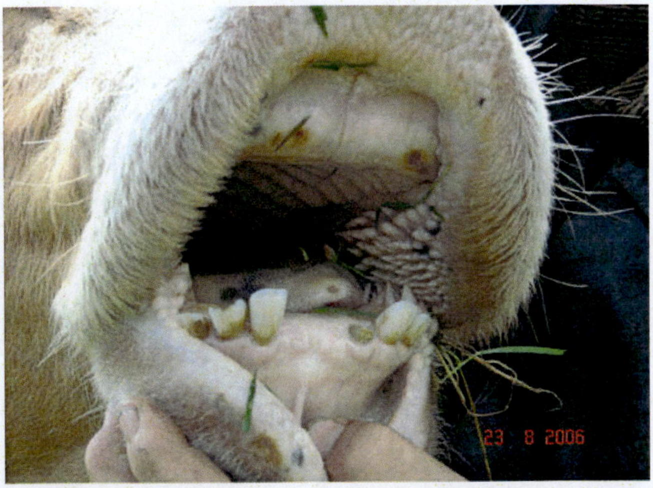

Fig. 5.4 Erosion on the oral mucosa in healing after BTV-8 infection

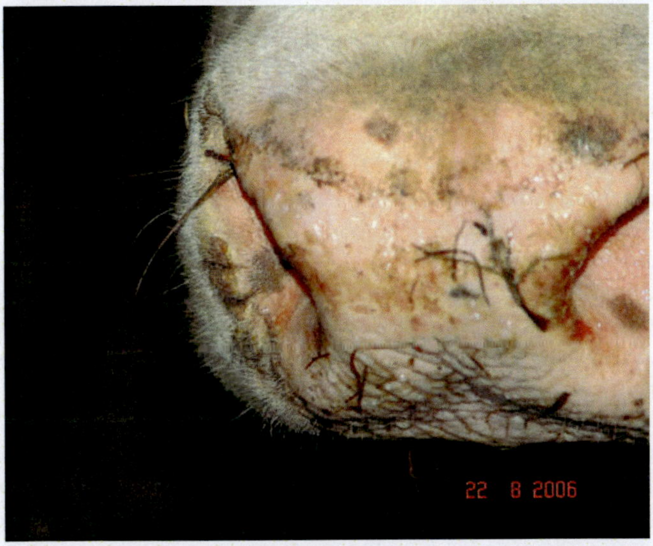

Fig. 5.5 Erosions, hemorrhagic lesions accompanied by bacterial superinfections and inflammatory foci of on the muzzle of a BTV-infected cattle

5.4 Diagnosis

5.4.1 BTV Isolation

For the direct detection of bluetongue virus, antigen or genome, whole blood is preferred over plasma because BTV is closely associated with red blood cells

Fig. 5.6 Large-scale hemorrhagic lesions on the teats of BTV-8 infected cattle

(Brewer and Maclachlan 1992; Nunamaker et al. 1992). Recommended specimens for the isolation of bluetongue virus are blood of live animals or spleen samples collected at necropsy. Heparinized or EDTA-treated blood that has been washed several times with phosphate-buffered saline to remove BTV antibodies is best suited for virus isolation. The release of virus particles from the erythrocyte membrane by sonication after washing can increase the success of the isolation. BTV-positive spleen samples are homogenized in cell culture medium and cleared by centrifugation. The purified supernatants can then be used for the inoculation of cultured cells or embryonated chicken eggs. For the latter, most laboratories use the intravenous inoculation method published by Goldsmit and Barzilai (1985). BTV-infected chicken embryos usually die within 2 and 7 days, and appear cherry red as a result of massive haemorrhages.

Besides mammalian cell lines (e.g. baby hamster kidney cells [BHK-21], African green monkey kidney cells [Vero] or bovine aorta endothelial cells), several insect-derived cell lines (e.g. *Aedes albopictus* clone C6/36 or *Culicoides variipennis* larval [KC] cells) can be used for isolation and propagation of BTV (Clavijo et al. 2000). The passage of virus in cell culture typically results in an adaptation of the virus to the in vitro conditions (Gould et al. 1989).

A highly sensitive and reliable method for the isolation of BTV is the inoculation of susceptible animals, mainly sheep or cattle. Although animal experiments are very expensive and ethical considerations have to be taken into account, an important advantage can justify this approach. Based on the large volume of inocula (up to 500 ml sample material are tolerated during intravenous injection), the sensitivity is very high. For samples with borderline infectivity (blood from animals in a late stage of infection or semen samples with a low viral load), the inoculation of susceptible animals can be the only way to propagate the virus (Hourrigan and Klingsporn 1975; Eschbaumer et al. 2010a, b).

Laboratory animals such as mice are not regularly used for BTV diagnosis and research. Nevertheless, the intracerebral inoculation of BTV isolates in 2–3 days old suckling mice causes clinical signs and death within 2–5 days after inoculation. Recently, a novel mouse model for BTV using type I interferon-receptor-deficient (IFNAR−/−) mice was developed (Calvo-Pinilla et al. 2009b; Eschbaumer et al. 2010b). Owing to the high susceptibility of these mice to fatal infection with BTV, they can also be used for virus isolation from samples harbouring a small viral load.

5.4.2 Molecular Diagnosis

Historically, confirmation and classification of BTV isolates have been based on immunological assays such as the indirect immune fluorescence test (Ruckerbauer et al. 1967; Jochim and Jones 1983), the complement fixation test (Shone et al. 1956), the haemagglutination assay (van der Walt 1980; Hübschle 1980), electron microscopy (Gould et al. 1989; Nunamaker et al. 1992), virus neutralization assays (Howell 1960) and competitive antigen-capture ELISAs (Mecham 1993; Mecham and Wilson 2004). With the development of nucleic acid detection methods in the late 1980s, cloned cDNA segments of several BTV serotypes were used to define the genetic relationships between and within the BTV serotypes (Unger et al. 1988; Ritter and Roy 1988). The introduction of the polymerase chain reaction (PCR) in the beginning of the 1990s revolutionized the molecular diagnosis of BTV (Wade-Evans et al. 1990; McColl and Gould 1991). In the following years, several improvements of BTV genome detection by PCR were published. Besides one-step RT-PCR assays, nested PCR systems and multiplex PCR systems for the identification of different BTV serotypes circulating in one region were developed (Wilson and Chase 1993; Katz et al. 1994; Aradaib et al. 1998; Zientara et al. 2004). Primers of different segments were used for the amplification and characterization of BTV strains (see Hoffmann et al. 2009a, for a review). For group-specific assays, conserved regions of the segments 5, 6, 7 and 10 were identified (Aradaib et al. 1998; Pierce et al. 1998; Bandyopadhyay et al. 1998; Anthony et al. 2007). Specific primers for the VP2 gene were developed and used in singleplex or multiplex assays to determine the BTV serotype (Wilson and Chase 1993; Eschbaumer et al. 2011b).

A new era for the molecular diagnosis of BTV began with the development of the real-time RT-PCR technology in the 1990s (Higuchi et al. 1993; Wittwer et al. 1997). Prior to the BTV-8 outbreak in Europe in the summer of 2006, only very few real-time quantitative RT-PCR (RT-qPCR) assays for the detection of BTV had been published. The first used primers were designed for the detection of the NS1 gene (Seg-5) (Wilson et al. 2004). However, this assay detected only 11 out of the 19 serotypes tested. The same year, another RT-qPCR was published using Förster resonance energy transfer (FRET) probe technology targeting genome segment

2 (VP2) (Orru et al. 2004). In 2006, an RT-qPCR assay was developed using a conserved region in RNA segment 5 of BTV-2 and BTV-4 (Jimenez-Clavero et al. 2006). This assay detected all of the recent Mediterranean isolates that were tested, BTV vaccine strains for serotypes 2 and 4 as well as 15 out of the 24 BTV reference strains. In the European outbreak of BTV-8, however, this assay showed a reduced sensitivity for the field strain of BTV-8 compared to other assays (Batten et al. 2008a). In the same year, an RT-qPCR was developed using a molecular beacon (MB) fluorescent probe designed within the NS3 conserved region of segment 10 (Orru et al. 2006).

Since the start of the northern European outbreak in August 2006, many RT-qPCR assays have been developed. Most of these were assays for the detection of all 24 serotypes identified at this time (Toussaint et al. 2007; Shaw et al. 2007), using conserved regions of the VP1 (seg-1) and NS1 (seg-5) genes. Nevertheless, the reduced sensitivity for the novel BTV serotypes 25 and 26 (Hofmann et al. 2008; Maan et al. 2011a, b) confirms the necessity for regular verification of the functionality of such pan-BTV assays. The parallel use of independent assays with equivalent diagnostic sensitivity and specificity can overcome the limitations of one assay. Another option is the application of pan-orbivirus assays, which use primer binding sites in the polymerase gene (seg-1) that are conserved among all currently known orbiviruses (Palacios et al. 2011). In addition to the group-specific (pan-BTV) real-time RT-PCR assays, serotype-specific BTV assays were developed and validated by diagnostic laboratories (Mertens et al. 2007; Hoffmann et al. 2009a, b). Furthermore, the advantages of the real-time PCR technology can be used for high-throughput analyses (Vandemeulebroucke et al. 2010). The use of robotics for automated extraction of BTV RNA combined with the co-amplification of the BTV target and an internal control RNA ensures a high diagnostic reliability during the investigation of large sample batches (Toussaint et al. 2007; Vandenbussche et al. 2010). These can be blood and tissue specimens in outbreak situations or insect samples from entomological monitoring programmes. High-throughput RNA extraction and real-time RT-PCR are particularly indispensable for the BTV analysis of extensive numbers of midges (Hoffmann et al. 2009c; Vanbinst et al. 2009).

5.4.3 Serology

Historically, BTV antibody detection in serum relied on complement fixation and agar gel immunodiffusion. Both, however, proved inferior to enzyme-linked immunosorbent assays (ELISAs) and were eventually replaced (Afshar 1994; Hamblin 2004). Highly sensitive ELISAs can pick up the humoral immune response to BTV as early as 1 week after infection (Batten et al. 2008a; Oura et al. 2009). Several systems are commercially available, mostly detecting antibodies against VP7, a BTV structural protein that is largely conserved across all serotypes. Three competitive ELISA kits (by ID VET, IDEXX and VMRD) are currently licensed for use in Germany. A latex agglutination test (Yang et al. 2010a) and immunochromatographic strips

(Yang et al. 2010b) have recently been proposed, promising faster sample turnover and the possibility of pen-side testing.

Apart from the competitive ELISAs, an indirect assay is available for individual and bulk milk samples (Kramps et al. 2008; Chaignat et al. 2010; Mars et al. 2010). Double-antigen sandwich ELISAs, which use peroxidase-labelled antigen to detect captured antibody (Laman et al. 1991), display superior sensitivity early in infection and more reliably detect vaccine-induced antibody (Eschbaumer et al. 2009; Oura et al. 2009). Two kits (by ID VET and Prionics) are commercially available in Germany.[1] However, their bias for multimeric antibody molecules such as immunoglobulin (Ig) M may have a negative impact on sensitivity during the transition from IgM to IgG in the development of the immune response (Eschbaumer et al. 2011a).

Beyond the performance of a group-specific assay, serotyping usually requires labour-intensive neutralization tests against a panel of reference viruses (Hamblin 2004). A serotype-specific antibody ELISA for BTV-8 has recently been implemented, but data on its performance are not yet available.

Regardless of the test format, the widespread use of inactivated whole virus vaccines in Europe (Zientara et al. 2010) interferes with serological surveillance. The commercially available VP7-based tests are unable to differentiate between infected and vaccinated animals (Mertens et al. 2009). Theoretically, inactivated vaccines should only elicit antibody responses to structural proteins. The discrimination potential of ELISAs based on non-structural proteins NS1 or NS3 has been demonstrated (Anderson et al. 1993; Barros et al. 2009) but is highly dependent on the purity of the vaccine. The carryover of non-structural proteins from the culture system used to produce the vaccine may result in antibodies to those proteins in vaccinated animals (Alpar et al. 2009), particularly after repeated vaccinations. Recent attempts at establishing a commercial NS1 ELISA were not successful. Vaccination with inactivated vaccines from different companies led to an increased number of unspecific results in the test, which, consequently, was not released by the manufacturer.

5.5 Epidemiology

5.5.1 Disease Transmission

When BTV serotype 8 (BTV-8) first appeared in Central Europe, no data on the putative vectors were available. Transmission of BTV had generally been supposed to be associated with *Culicoides imicola*, the most efficient and widespread vector in the Old World (Meiswinkel et al. 2007). Since this species is restricted to Africa and southern Europe, the occurrence of BT in more northern countries and its

[1] According to the list of products certified by the Friedrich-Loeffler-Institut pursuant to §17c of the Animal Diseases Act; see http://www.fli.bund.de/en/startseite/services/licensing-authority.html for the most recent version.

effective transmission by indigenous vectors was surprising, despite evidence from Italy suggesting that also *C. pulicaris* and midges of the *C. obsoletus* group can harbour BTV (Caracappa et al. 2003; Savini et al. 2004).

Meanwhile, entomological monitoring has shown that *C. imicola* is still not present in Germany and that members of the *C. obsoletus* and *C. pulicaris* complexes are relevant vectors for BT in Germany and Central Europe (Hoffmann et al. 2009c; Mehlhorn et al. 2009a). They are small midges of approximately 1–2 mm length which tend to be active in the evening, night and dawn. The length of time between ingestion of virus by a midge and its appearance in the saliva, the so-called extrinsic incubation period, is influenced by both temperature and salivary proteases of the vectors. At temperatures of 10–30°C, the extrinsic incubation period becomes progressively shorter with increasing temperatures since:

- Midges feed more frequently.
- Both the reproductive cycle of midges and virus replication in midges are faster.
- A greater proportion of the midge population becomes vector competent.
- Possibly more midge species become vector competent (Hateley 2009).

In Germany, transmission of BT is apparently interrupted or at least significantly reduced during the cold season (late autumn, winter and early spring) as a consequence of the low temperatures, which reduce the activity of *Culicoides* biting midges and BTV replication in the midgut of the biting midges. However, BTV might persist during the winter either in the vector population or in the host population (Wilson et al. 2008). Another possible way of overwintering is persistence in an as yet unknown wild ruminant population, e.g. red deer. In Belgium, relatively high levels of seroprevalence have been observed in red deer during 2007 and 2008. By contrast, the seroprevalence in roe deer was low, which was explained by the fact that red deer live in large groups, move more and therefore might be more exposed to insects (Linden et al. 2010). The results of the German wildlife monitoring on BT provide no evidence for the conclusion that a reservoir for BTV might have formed in wild ruminants in Central Europe. However, the relative importance of the remaining potential overwintering mechanisms remains unclear, too (Napp et al. 2011). As active midges were also found in the cold season, although in reduced numbers and mostly close to or within stables, a true midge-free period does not exist in Germany (Mehlhorn et al. 2009b).

Several investigation programmes were set up to share European BT outbreak data, e.g. in the BT51 group and as part of EPIZONE, an EU-funded Network of Excellence. Special attention has been paid to both mechanistic and stochastic predictive modelling which included a wide range of predictor variables to assess the spread of BTV (Faes et al. 2011; de Koeijer et al. 2011; Ducheyne et al. 2011; Willgert et al. 2011).

5.5.2 Introduction of Bluetongue Disease into Germany

In 2006, the first infections with bluetongue virus of serotype 8 (BTV-8), which was initially restricted to the sub-Saharan region, Asia and South America, invaded

Central Europe. On August 18th, BTV-8 was confirmed in the Netherlands (EU Rapid Press release Nr. IP/06/1112, ProMedMail, 20060828.2448) and Belgium (Toussaint et al. 2007). On 21 August 2006, the first outbreak was reported in the neighbouring district of Aachen on the German side of the border (OIE, immediate notification, Conraths et al. 2007). The virus hit an area with a high density of BT-naive animals, presence of suitable vectors (*Culicoides* spp.), climatic conditions favourable for virogenesis in local biting midges and for transmission. So far, the epidemiological situation in Germany has been dominated by BTV-8 (Conraths et al. 2009; Gethmann et al. 2010), but a few cases of infections with BTV-6 and a single case of BTV-1 in an imported animal were reported (Eschbaumer et al. 2010a).

In addition to entomological patterns, much research has been devoted to the determination of the most likely time and place of introduction of BTV-8 (Saegerman et al. 2010). The following hypotheses to explain the introduction of BTV-8 to Western Europe were taken into consideration:

- Legal or illegal import of viraemic susceptible animal species.
- Legal or illegal import of infected non-susceptible animal species, especially against the background that the World Equestrian Games took place in Aachen at the time when the first BT cases were noticed.
- Introduction via infected vectors, especially because midges are so light that they can drift by wind over hundreds of kilometres (Ducheyne et al. 2007). Alba et al. (2004) confirmed the possibility of introduction of infected midges to the Balearic Islands from Sardinia during the BT outbreak in the year 2000. It has also been proposed that BTV-infected biting midges might have been moved from northern Africa to Spain, Portugal or Italy by wind across the Strait of Gibraltar and the Mediterranean Sea in 2006 (Gloster et al. 2007; Hendrickx 2008). Another alternative is that infected vectors might have been directly introduced to Europe, e.g. by airplane with flowers imported from regions where BTV is enzootic. This hypothesis was also considered when BTV-6 occurred in the district of Grafschaft Bentheim (Lower Saxony) on the border to the Netherlands, although the respective virus closely resembled an isolate used in an attenuated live vaccine (Eschbaumer et al. 2010a).

Despite intensive epidemiological investigations, the source of the introduction has never been unambiguously identified.

5.5.3 Course of the 2006–2009 Bluetongue Epidemic in Germany

Until the end of 2006, a total of 890 cases/outbreaks were reported from the German federal states of North Rhine-Westphalia, Hesse, Rhineland-Palatinate and Lower Saxony to the German Animal Disease Notification System (TierseuchenNachrichtenSystem, TSN; accessed 23/11/2011). Between January and April 2007, when no transmission was expected, 185 further outbreaks were reported (Table 5.1). It is

Table 5.1 Number of reported outbreaks 2006/2007 and affected species

Year	Month	Cattle	Sheep	Goats	Wildlife	Total
2006	August	35	4	–	1	40
2006	September	64	37	–	–	101
2006	October	295	170	–	4	469
2006	November	137	95	–	5	237
2006	December	37	3	–	3	43
2007	January	80	–	–	1	81
2007	February	52	–	–	–	52
2007	March	19	1	–	–	20
2007	April	31	1	–	–	32
Total		750	311	0	14	1,075

likely that at least the vast majority of infected animals detected in winter 2006/2007 had contracted BT in summer or autumn 2006.

By analysing the outbreak data as reported to the TSN database and comparing them to the number of animals kept on the affected farms, it became apparent that at least 67,080 cattle, 9,825 sheep and 56 goats were present on premises affected by BTV-8 between August 2006 and April 2007 (Table 5.3). Of these animals, 1,529 cattle (2.28%) and 592 sheep (6.03%) were found infected. Eighty-four cattle and 222 sheep died. The case-fatality rate was much higher in sheep (37.5%) than in cattle (5.5%). These calculations are based on the assumption that all BT cases were reported. Since the infections caused only mild disease or remained even clinically inapparent in some animals, in particular cattle, it is likely that there was a substantial level of underreporting. As a consequence, the case-fatality rate in cattle might be slightly overestimated (Conraths et al. 2009).

Apparently, BTV-8 overwintered in the region and flared up again in 2007 to spread over most of western Germany during summer and autumn 2007. The first outbreak after the end of the cold season was confirmed in June 2007 in the district Oberbergischer Kreis, North Rhine-Westphalia, when a sentinel animal (cattle) sampled in May 2007 tested positive (Hoffmann et al. 2008).

The infection also re-emerged in other European countries that had been affected in 2006. BT spread rapidly through Belgium, the Netherlands, Germany, France and Luxembourg and reached the Czech Republic, Denmark, Italy, Spain, Switzerland and the United Kingdom.

Between May 2007 and April 2008, more than 22,600 cases/outbreaks were reported from Germany (Table 5.2; Conraths et al. 2009; Gethmann et al. 2010). Due to the enlargement of the affected territory in 2007, the exposed population of animals kept in farms with BT cases rose to at least 1,501,994 cattle, 505,934 sheep and 3,736 goats. The number of diseased animals on these farms amounted to 33,839 cattle, 32,158 sheep and 227 goats (Table 5.3). While mortality remained at relatively low levels as in 2006, the case-fatality rate rose to 10.8% in cattle and 41.5% in sheep.

To take potential underreporting of BT cases into account, we also determined the number of animals for which the owners received financial aid from the German

Table 5.2 Number of reported outbreaks 2007/2008 and affected species

Year	Month	Cattle	Sheep	Goats	Wildlife	Total
2007	May	–	–	–	–	0
2007	June	2	–	–	–	2
2007	July	10	8	–	–	18
2007	August	979	1,253	6	3	2,241
2007	September	5,142	4,792	60	21	10,015
2007	October	3,916	1,621	42	29	5,608
2007	November	1,755	96	6	14	1,871
2007	December	836	20	1	14	871
2008	January	854	11	2	5	872
2008	February	572	9	4	1	586
2008	March	428	5	1	3	437
2008	April	122	1	1	3	127
Total		14,616	7,816	123	93	22,648

Table 5.3 Exposed, diseased and dead animals and morbidity, mortality and case-fatality rate of BT infections for Germany in 2006 and 2007

Outbreak season	May 2006–April 2007			May 2007–April 2008			May 2008–April 2009		
Species	Cattle	Sheep	Goats	Cattle	Sheep	Goats	Cattle	Sheep	Goats
Premises	758	319	13	14,756	7,910	257	2,956	278	11
Animals kept on affected farms	67,080	9,825	56	1,501,994	505,934	3,736	425,959	62,915	141
Diseased animals	1,445	370	0	30,175	18,821	173	5,358	429	4
Dead animals	84	222	0	3,664	13,337	54	94	238	1
Morbidity (%)[a]	2.15	3.77	0.00	2.01	3.72	4.63	1.26	0.68	2.84
Mortality (%)[b]	0.13	2.26	0.00	0.24	2.64	1.45	0.02	0.38	0.71
Case fatality (%)[c]	5.49	37.50	0.00	10.83	41.47	23.79	1.72	35.68	20.00

[a]Number of diseased animals/number of animals in affected farms
[b]Number of dead animals/number of animals in affected farms
[c]Number of dead animals/number of infected animals

animal disease compensation funds (Tierseuchenkassen). In total, 10,240 cattle, 33,233 sheep and 102 goats were compensated for in 2007. This indicates an overall mortality of 0.08% in cattle and 1.36% in sheep. By focussing on the core region (North Rhine-Westphalia), where a prevalence of up to 100% can be assumed (by comparison to the situation in Belgium; Gethmann et al. unpublished), the mortality was 0.51% in cattle and 13.19% in sheep.

While several member states declared vector-free periods according to Commission Regulation (EC) No. 1266/2007 of 26 October 2007 during the cold season, i.e. defined as a period where the risk of virus transmission is deemed extremely low or negligible, thus allowing for a temporary lift of some trade restrictions, transmission was shown in Schleswig-Holstein in February 2008 (Hoffmann et al. 2008). This indicates that vector transmission on a low level may have played a role in the

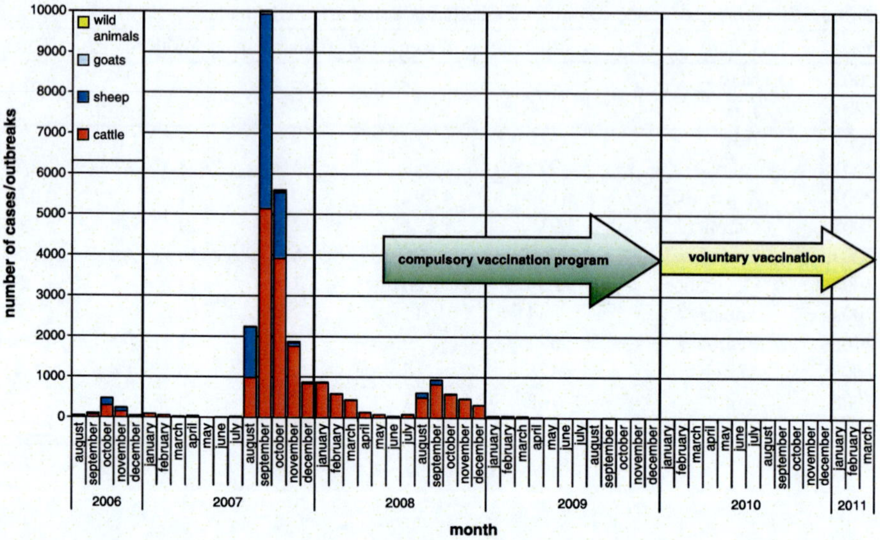

Fig. 5.7 Bluetongue disease, monthly incidence (*source*: TSN)

overwintering mechanism of BT in Germany and supported the view that declaration of a seasonally vector-free period was not appropriate for Germany.

In order to control BTV-8, to reduce the suffering of BT-infected animals and to mitigate the economic damage caused by the epizootic, it was decided to conduct a compulsory vaccination programme in Germany using inactivated vaccines which had not yet been licensed when the programme started (for details, see Chap. 5.6.2 Vaccination). The vaccination programme started in May and led to a massive decrease of new outbreaks in 2008 (Conraths et al. 2009) (Fig. 5.7, Table 5.4).

Between May and December 2008, a total of 3,083 new BTV-8 outbreaks plus 19 BTV-6 cases were reported (Table 5.4). They were mainly found in two regions in the north-west of Lower Saxony and Western parts of Baden-Württemberg. These cases can be explained by the relatively late onset of the immunization campaign because of initially limited supply of BTV-8 vaccines.

At the end of 2008, the genome of BTV serotype 6 (BTV-6) was detected in the district of Grafschaft Bentheim in BTV-1-vaccinated animals. In November and December, outbreaks in a total of 19 cattle farms were reported (Eschbaumer et al. 2010a). None of the animals showed clinical symptoms. Similar cases had previously been reported in the Netherlands (presentation at SCFCAH, section animal health and animal welfare, 08 December 2008, http://ec.europa.eu/food/committees/regulatory/scfcah/animal_health/presentations_en.htm). Despite comprehensive epidemiological investigations, the source of infection could not be identified. It cannot be excluded that animals were illegally vaccinated with an imported modified-live vaccine and that spread vaccine virus may have reassorted (Saegerman and Pastoret 2009). In 2009, no further cases of BTV-6 were detected despite intensive monitoring.

Table 5.4 Number of reported outbreaks 2008/2009

Year	Month	Cattle	Sheep	Goats	Wildlife	Total
2008	May	70	3	–	–	73
2008	June	33	1	–	–	34
2008	July	73	13	1	–	87
2008	August	492	118	–	1	611
2008	September	819	111	2	–	932
2008	October	567	20	1	1	589
2008	November	461	3	–	–	464
2008	December	307	3	–	2	312
2009	January	39	2	–	–	41
2009	February	44	–	–	1	45
2009	March	34	1	–	–	35
2009	April	12	–	–	–	12
Total		2,951	275	4	5	3,235

Since May 2009, only 12 outbreaks of BTV-8 in 9 cattle herds and 3 sheep flocks were reported, the last one occurred in November 2009 (TSN; accessed 23/11/2011).

5.6 Control Measures

After the first occurrence of BTV-8, Germany carried out the measures according to "Council Directive 2000/75/EC of 20 November 2000 laying down specific provisions for the control and eradication of bluetongue" in combination with national legislation (Verordnung zum Schutz gegen die Blauzungenkrankheit, Verordnung zum Schutz vor der Verschleppung der Blauzungenkrankheit). The measures focussed on (1) outbreak investigations in combination with monitoring and surveillance, (2) establishing restriction zones (e.g. movement restrictions) and (3) treating affected animals, farms, etc., with insecticides. In 2006, zones with a radius of at least 20 and 150 km were established around each outbreak farm (Fig. 5.8). In October 2007, measures regarding control, monitoring, surveillance and restrictions on movements of certain animals of susceptible species in relation to bluetongue were specified in Commission Regulation (EC) No 1266/2007 of 26 October 2007.

5.6.1 Movement Restrictions

Historically, BT has been regarded as an "exotic" disease in Europe, although sporadic incursions were observed on Cyprus in the first half of the twentieth century and in the mainland of Europe since the 1950s (Wilson and Mellor 2009). As a consequence, a protection zone with a radius of at least 100 km around

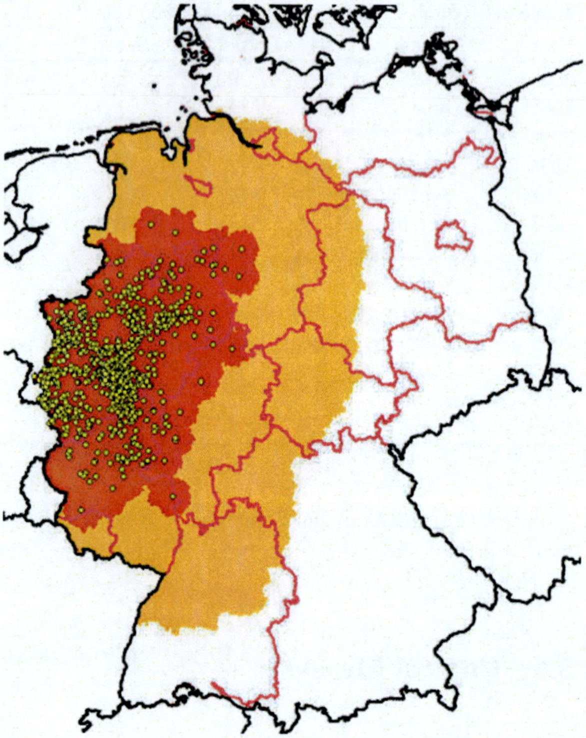

Fig. 5.8 Outbreaks and restriction zones in Germany by the end of April 2007

the infected holding and a surveillance zone with a depth of at least 50 km extending beyond the limits of the protection zone were provided for (Council Directive 2000/75/EC of 20 November 2000 laying down specific provisions for the control and eradication of bluetongue). An exit ban on animals and an epidemio-surveillance programme based on the monitoring of sentinel groups preferentially of bovine animals and of vector populations were imposed for both zones. While vaccination against BT can be allowed in the protection zone, animals must not be immunized against the disease in the surveillance zone. Trade within the same zone was allowed, but there were strict limitations for moving animals from one zone to another according to defined criteria (Commission Decision 2005/393/EC, replaced by Commission Regulation (EC) No 1266/2007), such as testing blood samples of the animals for BTV before movement, protecting them against vectors and treating them against insects prior to movement (Hateley 2009).

Since all control measures, including restrictions of animal transport, use of insecticides and indoor keeping of animals, had only limited effect during the BTV-8 epidemic that started in Belgium, Germany and the Netherlands in August 2006 (Mintiens et al. 2008b), vaccination of susceptible species with inactivated vaccines was included as the method of choice for BT control as soon as BTV-8-specific vaccines became available.

5.6.2 Vaccination

In general, vaccination is the only reliable means to protect animals from clinical bluetongue disease, while at the same time preventing the onward transmission of the virus. The two most common types of BTV vaccines are attenuated modified-live viruses and inactivated whole virus preparations with adjuvants. For modified-live vaccines, there is a delicate balance between achieving an acceptable reduction in virulence while at the same time maintaining the required level of immunogenicity (Alpar et al. 2009). In adverse circumstances, live vaccines can cause disease (Veronesi et al. 2005, 2010; Monaco et al. 2006), can be transmitted by vectors (Ferrari et al. 2005; Listes et al. 2009) and can exchange genetic information with field strains (Batten et al. 2008b; Maan et al. 2010). The repeated culture passages used for attenuation alter the tissue tropism of the virus; this can lead to teratogenic effects in pregnant animals (Kirkland and Hawkes 2004; Maclachlan et al. 2009). Hence, when BTV-8 was introduced to Europe in 2006, concerns over the safety of live vaccines prevented their use. The affected countries opted to wait until highly effective and safe inactivated vaccines became available in 2008 (Eschbaumer et al. 2009; Gethmann et al. 2009). A large-scale vaccination campaign all across Europe eventually brought the epizootic to a hold and it appears as if BTV-8 has been eradicated (Zientara et al. 2010). Since the end of mandatory vaccination in 2010, however, coverage is decreasing; even though the inactivated vaccines afford good long-term protection (Wäckerlin et al. 2010; Oura et al. 2012), replacement of stock will eventually return the animal population to its initial vulnerable state (Gethmann et al. 2010).

Infection with one serotype does not lead to cross-protective immunity, and neither does vaccination with a monovalent vaccine (Alpar et al. 2009; Bréard et al. 2011; Eschbaumer et al. 2011b). It has been shown experimentally, however, that sequential infection with several serotypes can give rise to neutralizing antibody against others (Jeggo et al. 1983). This suggests that broadly protective vaccines are possible, but none have been developed so far.

Another issue with currently available vaccines is the absence of a reliable strategy to differentiate infected from vaccinated animals (DIVA) (van Oirschot 1999; Mertens et al. 2009). Experimental vector vaccines (based on pox and herpes viruses) have shown some potential (Wade-Evans et al. 1996; Lobato et al. 1997; Boone et al. 2007; Calvo-Pinilla et al. 2009a; Franceschi et al. 2011). Since these vaccines only elicit an immune response to a subset of BTV proteins, the absent proteins can serve as negative markers in a DIVA strategy. The same goes for virus-like particles (Roy et al. 1994; Stewart et al. 2010), in vitro-assembled virus capsids without genome that do not evoke antibodies to non-structural proteins. Building on the reverse genetics system for BTV (Boyce et al. 2008), upcoming disabled infectious single-cycle vaccines (Matsuo et al. 2011) could combine the excellent immunogenicity of modified-live vaccines, the safety of inactivated vaccines and the DIVA capability of vector vaccines, if suitable companion tests are developed.

After the first introduction of BTV-8 to Belgium, Germany and the Netherlands in 2006, the massive spread of BTV-8 in 2007, reports about a large number of diseased and dead animals, and the failure of other control measures, the commission and the member states decided to carry out an harmonized vaccination programme to control BT (EU 2008: Bluetongue: Commission offers co-funding for vaccination campaign, Press release, IP/08/51, Brussels, 16 January 2008). Since the available vaccines against BTV-8 had not been registered at this time and safety and efficacy assessments were rudimentary, a large-scale safety study in combination with an efficacy study was conducted before a compulsory vaccination campaign involving the administration of millions of doses of largely untested BTV-8 vaccine was started in Germany. Participation in the study was prerequisite for a temporary emergency authorization to be granted by German law (Tierseuchengesetz §17c). Three monovalent inactivated BTV8 vaccines, precisely BLUEVAC® 8 (CZ Veterinaria), BTVPUR® AlSap 8 (Merial), and Zulvac® 8 Ovis or Bovis, respectively (Fort Dodge), were tested and proved to be safe and efficacious (Gethmann et al. 2009; Eschbaumer et al. 2009; Wäckerlin et al. 2010). For the basic immunization, administration of two doses was necessary in cattle, while a single dose was deemed sufficient in sheep and goats.

The first batches of the vaccines were delivered in May 2008. Until the end of 2008, about 20 million doses were administered to cattle and 2.6 million doses to sheep. In 2009, 13 million doses were applied to cattle and 2.1 million doses to sheep, so that the vaccination coverage was over 80%. In 2009, a further vaccine, Bovilis® BTV8 (Intervet), was introduced in the vaccination programme. By the end of 2009, the German federal states decided by majority vote to switch from a compulsory vaccination programme to a voluntary programme, resulting in a decrease of the administered vaccine doses. Only 5 million doses in cattle and 0.6 million doses in sheep were reported to the national animal database (HI Tier).

It has been pointed out, however, that low vaccination coverage or the introduction of other serotypes could result in further, potentially severe outbreaks in the future (Szmaragd et al. 2010).

5.6.2.1 Claims of Potential Adverse Reactions

Although the application of BTV-8 vaccines might induce moderate, short-term local inflammatory reactions at the injection site and a transient rise in body temperature shortly after booster vaccination, the vaccines proved to be well tolerated by both cattle and sheep (Bruckner et al. 2009; Gethmann et al. 2009). However, despite the proof of the safety of the vaccines, farmers, especially in south-eastern Germany and Switzerland, claimed a wide range of adverse reactions during the compulsory vaccination programme in 2008/2009, including reduction in milk yield, increase of somatic cell count in milk, mastitis or alterations of milk quality, reduced fertility and abortions. Officially, a total of 616 adverse reactions were reported in Germany to the Federal Agency for Vaccines and Biomedicines, thereof 547 in cattle (Hoffmann and Cußler 2009). In Switzerland, a total of 1,000 reports related to BTV-8 vaccination were received in 2009, the most frequently reported suspected adverse reactions being abortion, mastitis or alterations of milk quality (Müntener et al. 2010). However, in both countries evaluation of the data showed that plausible links between vaccination and the suspected adverse reactions could not be demonstrated (Probst et al. 2011; Tschuor et al. 2010). In any case, compared to the negative effect of BTV exposure on fertility, the possible side effect of vaccination seems to be rather small and therefore should not be an obstacle to vaccination (Nusinovicia et al. 2011).

5.6.3 Vector Evasion and Control

Vector evasion strategies and the use of repellents and insecticides alone are unlikely to lead to effective BT control (Mullens et al. 2001; EFSA 2007), although they may reduce vectorial capacity, i.e. reducing attack rates and the survival of adult midges (Mullens 1992). Vector evasion and control measures may thus be useful as auxiliary or mitigation measures which should be preferably applied in addition to vaccination against the relevant serotypes of BTV, the method of choice for the control of BT.

5.6.3.1 Vector Evasion Measures

It has been suggested that simple husbandry changes and practical midge control measures may help to diminish the risk of infection for susceptible animals, e.g. by housing livestock during times of maximum midge activity (from dusk to dawn) to reduce biting rates and thus transmission of BTV

Culicoides midges that carry BTV are believed to breed on animal dung and moist soil, either bare or covered in short grass. Identifying breeding grounds and breaking the breeding cycle may thus reduce the local midge population and hamper virus transmission.

It has been proposed that turning off taps, mending leaks and filling in or draining damp areas might also help to dry up breeding sites and that dung heaps or slurry pits should be covered or removed, and their perimeters regularly scraped to remove or destroy developing larvae of biting midges. Although it seems plausible that these measures might have an effect, published studies demonstrating their efficacy are lacking.

5.6.3.2 Use of Insecticides for Vector Control

Pyrethrum and synthetic pyrethroids are the most important compounds used against biting midges. They combine a repellent activity with toxic effects on insects. Due to the high efficiency of transmission of BTV from the biting midge to the vertebrate host, the repellent effect is particularly relevant for preventing BT infections as it may protect hosts from vector bites. However, an extremely high efficacy of the repellent seems to be required to achieve a significant level of protection against BT infections. Since the repellent activity of pyrethroids decreases much quicker than their toxic effect, it is difficult to take advantage of the repellent activity without applying the compounds repeatedly in short intervals. Frequent application may however increase the risk of side effects and lead to an unacceptable impact on non-target insects such as bees, beetles, etc.

Other potentially suitable compound groups include macrocyclic lactones, organophosphates, carbamates, chloronicotinyls (e.g. imidacloprid) and phenylpyrazoles (fipronil).

The efficacy of various pyrethroids against *Culicoides* spp. has been examined in several studies. Depending on the specific product and its formulation, deltamethrin, permethrin, cyfluthrin and cypermethrin protected animals for 3–5 weeks (Mehlhorn et al. 2008a, b; Liebisch and Liebisch 2008; Liebisch et al. 2008a, b; Papadopoulos et al. 2009; Schmahl et al. 2008, 2009a–c; Mullens et al. 2010). It is important to note that the treated animals were only protected if a sufficient concentration of the compound near the predilection sites of biting midges was warranted.

Pour-On and Spot-On Formulations of Repellents and Insecticides

Pour-on and spot-on formulations of repellents and insecticides have been successfully used against biting midges. It should be noted that pour-on and spot-on treatments with permethrin and deltamethrin lead to a dorsoventral gradient of the compound concentration, also after correct application of the product, with the consequence of a reduced insecticidal effectiveness in the bioassay (Mullens et al. 2000, 2001; Liebisch et al. 2008a). Moreover, a field study conducted in Brandenburg, Germany, showed that a regular pour-on treatment of bulls in intervals of 6 weeks had neither an effect on the total number of biting midges caught in a UV trap nor on the number of blood-fed midges (Bauer et al. 2009).

Nets

Fine mesh nettings and fabrics impregnated with insecticide, in particular pyrethroids, have been proposed to protect stables or to reduce contact of livestock with potentially infected midges (Braverman 1989; Carpenter et al. 2008) and were recently evaluated under field conditions (Bauer et al. 2009; Skrock et al. 2010; Skrock 2011). When Meiswinkel et al. (2000) gauzed all windows of a stable with a fine mesh screening and kept the doors closed, a 14-fold reduction in the number of *Culicoides* entering the stable was achieved. This approach may be useful as it reduces the biting rate. It is also a cheap measure that is easy to implement and requires only little maintenance.

Ear Tags

Holbrook (1986) treated cattle with one fenvalerate ear tag per animal and exposed adult *C. variipennis* in the laboratory to hair clippings recovered from the animals on several days post treatment. The insecticidal activity on the biting midges lasted throughout the 70-day test period, with decreased efficacy following rainfall and after the 49th test day. The duration of the repellent and toxic effect of ear tags may be shorter than that of pour-on or spot-on formulations and depend on the number of ear tags used per animal. For permethrin, a toxic and repellent activity of up to 7 days was observed if a single ear tag was applied, while the effects of two ear tags lasted for up to 19 days (Liebisch et al. 2008b). The repellent effect of cypermethrin on *C. sonorensis* lasted for 3–5 weeks (Reeves et al. 2010), while Liebisch and Liebisch (2008) determined an insecticidal effect of 14 (one ear tag) to 21 days (two ear tags).

Dipping

While this method has been successfully applied in many regions of the southern hemisphere including African and South American countries as well as Australia, farms in central and northern Europe are rarely equipped with dips, and the required formulations of pyrethroids or organophosphates are hardly available or their use has been suspended due to their environmental impact (pyrethroids) or discouraged because of their toxicity to users (EFSA 2007).

Systemic Use of Insecticides

The use of injectable macrocyclic lactones against biting midges yielded variable results (Standfast et al. 1985; Holbrook 1994; Holbrook and Mullens 1994). Since these compounds have no repellent, but only a toxic effect on insects, they may only act by reducing the population density of biting midges due to their effect on adult stages and on the larvae of dung-inhabiting *Culicoides*, thus reducing the vectorial capacity in a limited area surrounding the treated animals (EFSA 2007). A direct effect on BT transmission by reducing the biting rate of the relevant vectors cannot be expected.

5.7 Economic Impact

The financial impact of BTV-8 in the Netherlands including production losses, diagnosis, treatment and disease control amounted to 32 million Euros in 2006, 164–175 million Euros in 2007 (Velthuis et al. 2010) and about 41 million Euros in 2008 (Velthuis 2011).

Table 5.5 Economic losses in euros caused by the bluetongue disease epidemic in 2006–2009

Year	2006	2007	2008	2009	2010
Direct costs	2,878,082	57,738,386	8,201,700	22,789	0
Indirect costs	27,741,160	31,601,248	80,734,580	34,554,815	10,990,442
Control measures including vaccination	127,445	2,708,616	55,580,720	34,453,858	10,940,442
Trade restrictions, testing	27,124,515	23,131,575	24,533,595	0	0
Administrative costs	189,200	2,242,152	320,265	957	0
Monitoring/surveillance	3,218,905		No data	No data	No data
Public costs for research and development, consultancy	300,000	300,000	300,000	100,000	50,000
Total costs	30,619,243	89,339,634	88,936,281	34,577,604	10,990,442

The Animal Health Service of North Rhine-Westphalia, Germany, calculated mean production losses of 197 € per cow for a farm with 25 cows (http://www.landwirtschaftskammer.de/landwirtschaft/tiergesundheit/rgd/index.htm).

These data were used as reference for estimating the financial impact of BTV-8 for Germany as a whole. Outbreak data from the German animal disease notification system, information on payments made by the German animal disease compensation funds (Tierseuchenkassen) to BT-affected farms and administrative data were also included in the analysis. According to these calculations, the financial losses BTV-8 caused in Germany amounted to approximately 31 million Euros in 2006, 90 million Euros in 2007, 89 million Euros in 2008, 35 million Euros in 2009 and 11 million Euros in 2010, i.e. a total loss of 254 million Euros so far (Table 5.5). It has to be taken into account, however, that some parameters (e.g. morbidity, trade costs) could not be calculated exactly but had to be estimated.

Acknowledgements We thank the German animal disease compensation funds for data on compensation payments and veterinarians and farmers for their support. The excellent technical assistance of Kathrin Teske and Doris Kämer is gratefully acknowledged.

References

Afshar A (1994) Bluetongue: laboratory diagnosis. Comp Immunol Microbiol Infect Dis 17:221–242

Alba A, Casal J, Domingo M (2004) Possible introduction of bluetongue into the Balearic Islands, Spain, in 2000, via air streams. Vet Rec 155:460–461

Alpar HO, Bramwell V, Veronesi E, Darpel K, Pastoret PP, Mertens P (2009) Bluetongue virus vaccines past and present. In: Mellor P, Baylis M, Mertens P (eds) Bluetongue, Biology of animal infections. Elsevier, London

Anderson J, Mertens PP, Herniman KA (1993) A competitive ELISA for the detection of anti-tubule antibodies using a monoclonal antibody against bluetongue virus non-structural protein NS1. J Virol Methods 43:167–175

Anthony S, Jones H, Darpel KE, Elliott H, Maan S, Samuel A, Mellor PS, Mertens PP (2007) A duplex RT-PCR assay for detection of genome segment 7 (VP7 gene) from 24 BTV serotypes. J Virol Methods 141:188–197

Aradaib IE, Schore CE, Cullor JS, Osburn BI (1998) A nested PCR for detection of North American isolates of bluetongue virus based on NS1 genome sequence analysis of BTV-17. Vet Microbiol 59:99–108

Backx A, Heutink CG, van Rooij EMA, van Rijn PA (2007) Clinical signs of bluetongue virus serotype 8 infection in sheep and goats. Vet Rec 161:591–592

Bandyopadhyay SK, Kataria RS, Tiwari AK (1998) Detection of bluetongue virus genome segment 6 sequences by RT-PCR. Indian J Exp Biol 36:1034–1037

Barratt-Boyes SM, Maclachlan NJ (1994) Dynamics of viral spread in bluetongue virus infected calves. Vet Microbiol 40:361–371

Barratt-Boyes SM, Rossitto PV, Stott JL, Maclachlan NJ (1992) Flow cytometric analysis of in vitro bluetongue virus infection of bovine blood mononuclear cells. J Gene Virol 73:1953–1960

Barratt-Boyes SM, Rossitto PV, Taylor BC, Ellis JA, Maclachlan NJ (1995) Response of the regional lymph node to bluetongue virus infection in calves. Vet Immunol Immunopathol 45:73–84

Barros SC, Cruz B, Luis TM, Ramos F, Fagulha T, Duarte M et al (2009) A DIVA system based on the detection of antibodies to non-structural protein 3 (NS3) of bluetongue virus. Vet Microbiol 137:252–259

Batten CA, Bachanek-Bankowska K, Bin-Tarif A, Kgosana L, Swain AJ, Corteyn M et al (2008a) Bluetongue virus: European Community inter-laboratory comparison tests to evaluate ELISA and RT-PCR detection methods. Vet Microbiol 129:80–88

Batten CA, Maan S, Shaw AE, Maan NS, Mertens PP (2008b) A European field strain of bluetongue virus derived from two parental vaccine strains by genome segment reassortment. Virus Res 137:56–63

Bauer B, Jandowsky A, Schein E, Mehlitz D, Clausen PH (2009) An appraisal of current and new techniques intended to protect bulls against Culicoides and other haematophagous nematocera: the case of Schmergow, Brandenburg, Germany. Parasitol Res 105:359–365

Boone JD, Balasuriya UB, Karaca K, Audonnet JC, Yao J, He L et al (2007) Recombinant canarypox virus vaccine co-expressing genes encoding the VP2 and VP5 outer capsid proteins of bluetongue virus induces high level protection in sheep. Vaccine 25:672–678

Boyce M, Celma CC, Roy P (2008) Development of reverse genetics systems for bluetongue virus: recovery of infectious virus from synthetic RNA transcripts. J Virol 82:8339–8348

Braverman Y (1989) Control of biting midges *Culicoides* (Diptera:*Ceratopogonidae*), vectors of bluetongue and inducers of sweet itch: a review. Isr J Vet Med 45:124–129

Bréard E, Belbis G, Hamers C, Moulin V, Lilin T, Moreau F et al (2011) Evaluation of humoral response and protective efficacy of two inactivated vaccines against bluetongue virus after vaccination of goats. Vaccine 29:2495–2502

Brewer AW, Maclachlan NJ (1992) Ultrastructural characterization of the interaction of bluetongue virus with bovine erythrocytes in vitro. Vet Pathol 29:356–359

Bruckner L, Fricker R, Hug M, Hotz R, Muntwyler J, Iten C, Griot C (2009) Impfung gegen die Blauzungenkrankheit: Verträglichkeit und Immunantwort in der Praxis. Schweiz Arch Tierheilk 151:101–108

Calvo-Pinilla E, Rodríguez-Calvo T, Sevilla N, Ortego J (2009a) Heterologous prime boost vaccination with DNA and recombinant modified vaccinia virus Ankara protects IFNAR(-/-) mice against lethal bluetongue infection. Vaccine 28:437–445

Calvo-Pinilla E, Rodríguez-Calvo T, Anguita J, Sevilla N, Ortego J (2009b) Establishment of a bluetongue virus infection model in mice that are deficient in the alpha/beta interferon receptor. PLoS One 4:e5171

Caracappa S, Torina A, Guercio A, Vitale F, Calabro A, Purpari G, Ferrantelli V, Vitale M, Mellor PS (2003) Identification of a novel bluetongue virus vector species of Culicoides in Sicily. Vet Rec 153:71–74

Carpenter S, Mellor PS, Torr SJ (2008) Control techniques for *Culicoides* biting midges and their application in the U.K. and northwestern Palaearctic. Med Vet Entomol 22:175–187

Chaignat V, Nitzsche S, Scharrer S, Feyer D, Schwermer H, Thur B (2010) Milk concentration improves Bluetongue antibody detection by use of an indirect ELISA. Vet Microbiol 143:179–183

Clavijo A, Heckert RA, Dulac GC, Afshar A (2000) Isolation and identification of bluetongue virus. J Virol Methods 87:13–23

Conraths FJ, Kramer M, Freuling C, Hoffmann B, Staubach C, Gethmann J, Teifke J, Mettenleiter TC, Beer M (2007) Blauzungenkrankheit in Deutschland: Klinik, Diagnostik und Epidemiologie. Prakt Tierarzt 88:9

Conraths FJ, Gethmann JM, Staubach C, Mettenleiter TC, Beer M, Hoffmann B (2009) Epidemiology of bluetongue virus serotype 8, Germany. Emerg Infect Dis 15:433–435

Darpel KE, Batten CA, Veronesi E, Shaw AE, Anthony S, Bachanek-Bankowska K, Kgosana L, Bin-Tarif A, Carpenter S, Müller-Doblies UU, Takamatsu HH, Mellor PS, Mertens PRC, Oura CAL (2007) Clinical signs and pathology shown by British sheep and cattle infected with bluetongue virus serotype 8 derived from the 2006 outbreak in northern Europe. Vet Rec 161:253–261

Darpel K, Monaghan P, Anthony S, Takamatsu H, Mertens P (2009) Bluetongue virus in the mammalian host and the induced immune response. In: Mellor P, Baylis M, Mertens P (eds) Bluetongue. Elsevier, London

de Koeijer A, Boender G, Nodelijk G, Staubach C, Meroc E, Elbers A (2011) Quantitative analysis of transmission parameters for bluetongue virus serotype 8 in Western Europe in 2006. Vet Res 42:53

DeMaula CD, Jutila MA, Wilson DW, Maclachlan NJ (2001) Infection kinetics, prostacyclin release and cytokine-mediated modulation of the mechanism of cell death during bluetongue virus infection of cultured ovine and bovine pulmonary artery and lung microvascular endothelial cells. J Gen Virol 82:787–794

DeMaula CD, Leutenegger CM, Bonneau KR, Maclachlan NJ (2002a) The role of endothelial cell-derived inflammatory and vasoactive mediators in the pathogenesis of bluetongue. Virology 296:330–337

DeMaula CD, Leutenegger CM, Jutila MA, Maclachlan NJ (2002b) Bluetongue virus-induced activation of primary bovine lung microvascular endothelial cells. Vet Immunol Immunopathol 86:147–157

Drew CP, Gardner IA, Mayo CE, Matsuo E, Roy P, Maclachlan NJ (2010a) Bluetongue virus infection alters the impedance of monolayers of bovine endothelial cells as a result of cell death. Vet Immunol Immunopathol 136:108–115

Drew CP, Heller MC, Mayo C, Watson JL, Maclachlan NJ (2010b) Bluetongue virus infection activates bovine monocyte-derived macrophages and pulmonary artery endothelial cells. Vet Immunol Immunopathol 136:292–296

Ducheyne E, De Deken R, Bécu S, Codina B, Nomikou K, Mangana-Vougiaki O, Georgiev G, Purse BV, Hendickx G (2007) Quantifying the wind dispersal of Culicoides species in Greece and Bulgaria. Geospat Health 1:177–189

Ducheyne E, Lange M, Van der Stede Y, Meroc E, Durand B, Hendrickx G (2011) A stochastic predictive model for the natural spread of bluetongue. Prev Vet Med 99:48–59

EFSA (2007) Scientific Report of the Scientific Panel on Animal Health and Welfare on request from the Commission (EFSA-Q-2006-311) and EFSA Selfmandate (EFSA-Q-2007-063) on bluetongue. EFSA J 479:1–29; 480:1–20

Erasmus BJ (1990) Bluetongue virus. In: Horzinek MC (ed) Virus infections of vertebrates, Dinter Z, Morein B (eds) Virus infections of ruminants, vol 3. Elsevier, Amsterdam, pp 227–237

Eschbaumer M, Hoffmann B, König P, Teifke J, Gethmann J, Conraths FJ, Probst C, Mettenleiter TC, Beer M (2009) Efficacy of three inactivated vaccines against bluetongue virus serotype 8 in sheep. Vaccine 27:4169–4175

Eschbaumer M, Hoffmann B, Moss A, Savini G, Leone A, König P, Zemke J, Conraths F, Beer M (2010a) Emergence of bluetongue virus serotype 6 in Europe – German field data and experimental infection of cattle. Vet Microbiol 43:189–195

Eschbaumer M, Wäckerlin R, Rudolf M, Keller M, König P, Zemke J, Hoffmann B, Beer M (2010b) Infectious blood or culture-grown virus: a comparison of bluetongue virus challenge models. Vet Microbiol 146:150–154

Eschbaumer M, Schulz C, Wäckerlin R, Gauly M, Beer M, Hoffmann B (2011a) Limitations of sandwich ELISAs for bluetongue virus antibody detection. Vet Rec 168:643

Eschbaumer M, Wäckerlin R, Savini G, Zientara S, Sailleau C, Bréard E et al (2011b) Contamination in bluetongue virus challenge experiments. Vaccine 29:4299–4301

Faes C, Aerts M, Litière S, Méroc E, Van der Stede Y, Mintiens K (2011) Estimating herd prevalence on the basis of aggregate testing of animals. J R Stat Soc A 174:155–174

Ferrari G, De Liberato C, Scavia G, Lorenzetti R, Zini M, Farina F et al (2005) Active circulation of bluetongue vaccine virus serotype-2 among unvaccinated cattle in central Italy. Prev Vet Med 68:103–113

Franceschi V, Capocefalo A, Calvo-Pinilla E, Redaelli M, Mucignat-Caretta C, Mertens P et al (2011) Immunization of knock-out alpha/beta interferon receptor mice against lethal bluetongue infection with a BoHV-4-based vector expressing BTV-8 VP2 antigen. Vaccine 29:3074–3082

Gerdes GH (2004) A South African overview of the virus, vectors, surveillance and unique features of bluetongue. Vet Ital 40:39–42

Gethmann J, Hüttner K, Heyne H, Probst C, Ziller M, Beer M, Hoffmann B, Mettenleiter TC, Conraths FJ (2009) Comparative safety study of three inactivated BTV-8 vaccines in sheep and cattle under field conditions. Vaccine 27:4118–4126

Gethmann J, Hoffmann B, Probst C, Beer M, Conraths FJ, Mettenleiter TC (2010) Drei Jahre Blauzungenkrankheit Serotyp 8 in Deutschland – ein Überblick. Tierärztl Umschau 65:4–12

Gloster J, Mellor PS, Manning AJ, Webster HN, Hort MC (2007) Assessing the risk of windborne spread of bluetongue in the 2006 outbreak of disease in northern Europe. Vet Rec 160:54–56

Goldsmit L, Barzilai E (1985) Isolation and propagation of bluetongue virus in embryonating chicken eggs. Prog Clin Biol Res 178:307–318

Gould AR, Hyatt AD, Eaton BT, White JR, Hooper PT, Blacksell SD, LeBlanc-Smith PM (1989) Current techniques in rapid bluetongue virus diagnosis. Aust Vet J 66:450–454

Hamblin C (2004) Bluetongue virus antigen and antibody detection, and the application of laboratory diagnostic techniques. Vet Ital 40:538–545

Hateley G (2009) Bluetongue in northern Europe: the story so far. In Pract 31:202

Hemati B, Contreras V, Urien C, Bonneau M, Takamatsu HH, Mertens PP, Breard E, Sailleau C, Zientara S, Schwartz-Cornil I (2009) Bluetongue virus targets conventional dendritic cells in skin lymph. J Virol 83:8789–8799

Hendrickx G (2008) The spread of bluetongue in Europe. Small Ruminant Res 86:34–39

Higuchi R, Fockler C, Dollinger G, Watson R (1993) Kinetic PCR analysis: real-time monitoring of DNA amplification reactions. Biotechnology 11:1026–1030

Hoffmann A, Cußler K (2009) Impfkampagne zur Bekämpfung der Blauzungenkrankheit. D Tierärzteblatt 2:166–168

Hoffmann B, Sasserath M, Thalheim S, Bunzenthal C, Strebelow G, Beer M (2008) Bluetongue virus serotype 8 reemergence in Germany, 2007 and 2008. Emerg Infect Dis 14:1421–1423

Hoffmann B, Beer M, Reid SM, Mertens P, Oura CA, van Rijn PA, Slomka MJ, Banks J, Brown IH, Alexander DJ, King DP (2009a) A review of RT-PCR technologies used in veterinary virology and disease control: sensitive and specific diagnosis of five livestock diseases notifiable to the World Organisation for Animal Health. Vet Microbiol 139:1–23

Hoffmann B, Eschbaumer M, Beer M (2009b) Real-time quantitative reverse transcription-PCR assays specifically detecting bluetongue virus serotypes 1, 6, and 8. J Clin Microbiol 47:2992–2994

Hoffmann B, Bauer B, Bauer C, Bätza HJ, Beer M, Clausen PH, Geier M, Gethmann JM, Kiel E, Liebisch G, Liebisch A, Mehlhorn H, Schaub GA, Werner D, Conraths FJ (2009c) Monitoring of putative vectors of bluetongue virus serotype 8, Germany. Emerg Infect Dis 15:1481–1484

Hofmann MA, Renzullo S, Mader M, Chaignat V, Worwa G, Thuer B (2008) Genetic characterization of Toggenburg orbivirus, a new bluetongue virus, from goats, Switzerland. Emerg Infect Dis 14:1855–1861

Holbrook FR (1986) Exposure of *Culicoides variipennis* (Diptera: Ceratopogonidae) to hair clippings to evaluate insecticide-impregnated ear tags in cattle. J Econ Entomol 79:1127–1129

Holbrook FR (1994) Survival, fecundity, and egg fertility of Culicoides variipennis (Diptera: Ceratopogonidae) fed on calves inoculated with ivermectin. J Am Mosq Control Assoc 10:7–9

Holbrook FR, Mullens BA (1994) Effects of ivermectin on survival, fecundity, and egg fertility in Culicoides variipennis (Diptera:Ceratopogonidae). J Am Mosq Control Assoc 10:70–73

Hourrigan JL, Klingsporn AL (1975) Classification of ruminants, semen and ova for freedom from bluetongue virus. Aust Vet J 51:211–212

Howell PG (1960) A preliminary antigenic classification of strains of bluetongue virus. Onderstepoort J Vet Res 28:357–363

Hübschle OJB (1980) Bluetongue virus hemagglutination and its inhibition by specific sera. Arch Virol 64:133–140

Jeggo MH, Gumm ID, Taylor WP (1983) Clinical and serological response of sheep to serial challenge with different bluetongue virus types. Res Vet Sci 34:205–211

Jimenez-Clavero MA, Aguero M, San Miguel E, Mayoral T, Lopez MC, Ruano MJ, Romero E, Monaco F, Polci A, Savini G, Gomez-Tejedor C (2006) High throughput detection of bluetongue virus by a new real-time fluorogenic reverse transcription-polymerase chain reaction: application on clinical samples from current Mediterranean outbreaks. J Vet Diagn Invest 18:7–17

Jochim MM, Jones SC (1983) Identification of bluetongue and epizootic hemorrhagic disease viruses by immunofluorescence with monoclonal antibodies. Proc Am Assoc Vet Lab Diagn 26:277–286

Katz J, Alstad D, Gustafson G, Evermann J (1994) Diagnostic analysis of the prolonged bluetongue virus RNA presence found in the blood of naturally infected cattle and experimentally infected sheep. J Vet Diagn Invest 6:139–142

Kirkland PD, Hawkes RA (2004) A comparison of laboratory and 'wild' strains of bluetongue virus – is there a difference and does it matter? Vet Ital 40:448–455

Kramps JA, van Maanen K, Mars MH, Popma JK, van Rijn PA (2008) Validation of a commercial ELISA for the detection of bluetongue virus (BTV)-specific antibodies in individual milk samples of Dutch dairy cows. Vet Microbiol 130:80–87

Laman JD, van den Eertwegh AJ, Deen C, Vermeulen N, Boersma WJ, Claassen E (1991) Synthetic peptide conjugates with horseradish peroxidase and beta-galactosidase for use in epitope-specific immunocytochemistry and ELISA. J Immunol Methods 145:1–10

Liebisch G, Liebisch A (2008) [Efficacy of Flectron-eartags (cypermethrin) for control of midges (Culicoides) as the vectors of bluetongue virus in cattle: field studies and biossays]. Dtsch Tierarztl Wochenschr 115:220–230

Liebisch G, Krieger K, Heine S, Thienel S, Hinrichs P, Liebisch A (2008a) BAYOFLY Pour on R (Cyfluthrin) bei der Abwehr und Bekämpfung von Gnitzen (Ceratopogonidae: Culicoides) den Überträgern des Bluetongue Virus: Felduntersuchungen und Bioassay mit Milchkühen und Jungrindern. Prakt Tierarzt 89:411–426

Liebisch A, Liebisch G, Heine S, Thienel S, Hintichs P (2008b) Wirksamkeit von Auriplak Ohrclips (Permethrin) gegen Gnitzen (Culicoides) als Überträger des Bluetongue Virus bei Rindern. Prakt Tierarzt 89:128–141

Linden A, Grégoire F, Nahayo A, Hanrez D, Mousset B, Massart L, De Leeuw I, Vandemeulebroucke E, Vandenbussche F, De Clercq K (2010) Bluetongue in wild deer, Belgium, 2005–2008. Emerg Infect Dis 16:833–836

Listes E, Monaco F, Labrović A, Paladini C, Leone A, Di Gialleonardo L et al (2009) First evidence of bluetongue virus serotype 16 in Croatia. Vet Microbiol 138:92–97

Lobato ZI, Coupar BE, Gray CP, Lunt R, Andrew ME (1997) Antibody responses and protective immunity to recombinant vaccinia virus-expressed bluetongue virus antigens. Vet Immunol Immunopathol 59:293–309

Maan S, Maan NS, van Rijn PA, van Gennip RG, Sanders A, Wright IM et al (2010) Full genome characterisation of bluetongue virus serotype 6 from the Netherlands 2008 and comparison to other field and vaccine strains. PLoS One 5(4):e10323

Maan S, Maan NS, Nomikou K, Batten C, Antony F, Belaganahalli MN et al (2011a) Novel bluetongue virus serotype from Kuwait. Emerg Inf Dis 17:886–889

Maan S, Maan NS, Nomikou K, Veronesi E, Bachanek-Bankowska K, Belaganahalli MN, Attoui H, Mertens PP (2011b) Complete genome characterisation of a novel 26th bluetongue virus serotype from Kuwait. PLoS One 6:e26147

Maclachlan NJ (2004) Bluetongue: pathogenesis and duration of viraemia. Vet Ital 40:462–467

Maclachlan NJ (2011) Bluetongue: history, global epidemiology, and pathogenesis. Prev Vet Med 102:107–111

Maclachlan NJ, Barratt-Boyes SM, Brewer AW, Stott JL (1992) Bluetongue virus infection of cattle. In: Walton TE, Osburn BI (eds) Bluetongue, African horse sickness, and related orbiviruses. CRC Press, Boca Raton, pp 725–736

Maclachlan NJ, Jagels G, Rossitto PV, Moore PF, Heidner HW (1990) The pathogenesis of experimental bluetongue virus infection of calves. Vet Pathol 27:223–229

Maclachlan NJ, Drew CP, Darpel KE, Worwa G (2009) The pathology and pathogenesis of bluetongue. J Comp Pathol 141:1–16

Mahrt CR, Osburn BI (1986) Experimental bluetongue virus infection of sheep; effect of vaccination: pathologic, immunofluorescent, and ultrastructural studies. Am J Vet Res 47:1198–1203

Mars MH, van Maanen C, Vellema P, Kramps JA, van Rijn PA (2010) Evaluation of an indirect ELISA for detection of antibodies in bulk milk against bluetongue virus infections in the Netherlands. Vet Mcrobiol 46:209–214

Matsuo E, Celma CC, Boyce M, Viarouge C, Sailleau C, Dubois E et al (2011) Generation of replication-defective virus-based vaccines that confer full protection in sheep against virulent bluetongue virus challenge. J Virol 85:10213–10221

McColl KA, Gould AR (1991) Detection and characterization of bluetongue virus using the polymerase chain reaction. Virus Res 21:19–34

Mecham JO (1993) Detection of bluetongue virus from blood of infected sheep by use of an antigen-capture enzyme-linked immunosorbent assay after amplification of the virus in cell culture. Am J Vet Res 54:370–372

Mecham JO, Wilson WC (2004) Antigen capture competitive enzyme-linked immunosorbent assays using baculovirus-expressed antigens for diagnosis of bluetongue virus and epizootic hemorrhagic disease virus. J Clin Microbiol 42:518–523

Mehlhorn H, Walldorf V, Klimpel S, Jahn B, Jaeger F, Eschweiler J, Hoffmann B, Beer M (2007) First occurrence of *Culicoides obsoletus*-transmitted bluetongue virus epidemic in Central Europe. Parasitol Res 101:219–228

Mehlhorn H, Schmahl G, D'Haese J, Schumacher B (2008a) Butox 7.5 pour on: a deltamethrin treatment of sheep and cattle: pilot study of killing effects on *Culicoides* species (Ceratopogonidae). Parasitol Res 102:515–518

Mehlhorn H, Schmahl G, Schumacher B, D'Haese J, Walldorf V, Klimpel S (2008b) Effects of Bayofly on specimens of *Culicoides* species when incubated in hair taken from the feet of previously treated cattle and sheep. Parasitol Res 102:519–522

Mehlhorn H, Walldorf V, Klimpel S, Schmahl G, Al-Quraishy S, Walldorf U, Mehlhorn B, Bätza HJ (2009a) Entomological survey on vectors of transmission parameters for bluetongue virus serotype 8 in Western Europe in 2006. Vet Res 42:53

Mehlhorn H, Walldorf V, Klimpel S, Schaub G, Kiel E, Focke R, Liebisch G, Liebisch A, Werner D, Bauer C, Clausen H, Bauer B, Geier M, Hörbrand T, Bätza HJ, Conraths FJ, Hoffmann B, Beer M (2009b) Bluetongue disease in Germany (2007-2008): monitoring of entomological aspects. Parasitol Res 105:313–319

Meiswinkel R, Baylis M, Labuschagne K (2000) Stabling and the protection of horses from Culicoides bolitinos (Diptera: Ceratopogonidae), a recently identified vector of African horse sickness. Bull Entomol Res 90:509–515

Meiswinkel R, Van Rijn P, Leijs P, Goffredo M (2007) Potential new *Culicoides* vector of bluetongue virus in northern Europe. Vet Rec 161:564

Mellor PS, Wittmann EJ (2002) Bluetongue virus in the Mediterranean Basin 1998–2001. Vet J 164:20–37

Mertens PP, Maan NS, Prasad G, Samuel AR, Shaw AE, Potgieter AC, Anthony SJ, Maan S (2007) Design of primers and use of RT-PCR assays for typing European bluetongue virus isolates: differentiation of field and vaccine strains. J Gen Virol 88:2811–2823

Mertens PPC, Maan S, Batten C, Darpel KE, Shaw A, Maan NS et al (2009) Bluetongue virus diagnosis. In: Mellor P, Baylis M, Mertens P (eds) Bluetongue, Biology of animal infections. Elsevier, London

Mintiens K, Méroc E, Mellor PS, Staubach C, Gerbier G, Elbers AR et al (2008a) Possible routes of introduction of bluetongue virus serotype 8 into the epicentre of the 2006 epidemic in north-western Europe. Prev Vet Med 87:131–144

Mintiens K, Méroc E, Faes C, Abrahantes JC, Hendrickx G, Staubach C, Gerbier G, Elbers ARW, Aerts M, De Clercq K (2008b) Impact of human interventions on the spread of bluetongue virus serotype 8 during the 2006 epidemic in north-western Europe. Prev Vet Med 87:145–161

Monaco F, Cammà C, Serini S, Savini G (2006) Differentiation between field and vaccine strain of bluetongue virus serotype 16. Vet Microbiol 116:45–52

Mullens BA (1992) Integrated management of Culicoides variipennis: a problem of applied ecology. In: Walton TE, Osburn BI (eds) Bluetongue, African horse sickness and related orbiviruses. Proceedings of the second international symposium on bluetongue, Paris, 17–21 June 1991. CRC, Boca Raton, pp 193–196

Mullens BA, Velten RK, Gerry AC, Braverman Y, Endris RG (2000) Feeding and survival of Culicoides sonorensis on cattle treated with permethrin or pirimiphos-methyl. Med Vet Entomol 14:313–320

Mullens BA, Gerry AC, Velten RK (2001) Failure of a permethrin treatment regime to protect cattle against bluetongue virus. J Med Entomol 38:760–762

Mullens BA, Gerry AC, Monteys VS, Pinna M, González A (2010) Field studies on Culicoides (Diptera: Ceratopogonidae) activity and response to deltamethrin applications to sheep in northeastern Spain. J Med Entomol 47:106–110

Müntener CR, Bruckner L, Stürer A, Althaus FR, Caduff-Janosa P (2010) Vigilance der Tierarzneimittel: Gemeldete unerwünschte Wirkungen im Jahr 2009. Schweiz Arch Tierheilk 152:575–583

Napp S, Gubbins S, Calistri P, Allepuz A, Alba A, García-Bocanegra I, Giovannini A, Casal J (2011) Quantitative assessment of the probability of bluetongue virus overwintering by horizontal transmission: application to Germany. Vet Res 42:4

Nunamaker RA, Ellis JA, Wigington JG, Maclachlan NJ (1992) The detection of intracellular bluetongue virus particles within ovine erythrocytes. Comp Biochem Physiol 101A:471–476

Nusinovicia S, Seegersa H, Joly A, Beaudeaua F, Fourichon C (2011) A side effect of decreased fertility associated with vaccination against bluetongue virus serotype 8 in Holstein dairy cows. Prev Vet Med 101:42–50

Orru G, Santis PD, Solinas F, Savini G, Piras V, Caporale V (2004) Differentiation of Italian field and South African vaccine strains of bluetongue virus serotype 2 using real-time PCR. J Virol Methods 122:37–43

Orru G, Ferrando ML, Meloni M, Liciardi M, Savini G, De Santis P (2006) Rapid detection and quantification of bluetongue virus (BTV) using a molecular beacon fluorescent probe assay. J Virol Methods 137:34–42

Oura CA, Wood JL, Sanders AJ, Bin-Tarif A, Henstock M, Edwards L et al (2009) Seroconversion, neutralising antibodies and protection in bluetongue serotype 8 vaccinated sheep. Vaccine 27:7326–7330

Oura CA, Edwards L, Batten CA (2012) Evaluation of the humoral immune response in adult dairy cattle three years after vaccination with a bluetongue serotype 8 inactivated vaccine. Vaccine 30(2):112–115. doi:10.1016/j.vaccine.2011.10.100

Palacios G, Cowled C, Bussetti AV, Savji N, Weir R, Wick I, Travassos da Rosa A, Calisher CH, Tesh RB, Boyle D, Lipkin WI (2011) Rapid molecular strategy for orbivirus detection and characterization. J Clin Microbiol 49:2314–2317

Papadopoulos E, Bartram D, Carpenter S, Mellor P, Wall R (2009) Efficacy of alphacypermethrin applied to cattle and sheep against the biting midge Culicoides nubeculosus. Vet Parasitol 163:110–114

Parsonson IM (1992) Overview of bluetongue virus infection of sheep. In: Walton TE, Osburn BI (eds) Bluetongue, African horse sickness and related orbiviruses. Proceedings of the second international symposium on bluetongue, Paris, 17–21 June 1991. CRC, Boca Raton, pp 713–724

Pierce CM, Balasuriya UB, Maclachlan NJ (1998) Phylogenetic analysis of the S10 gene of field and laboratory strains of bluetongue virus from the United States. Virus Res 55:15–27

Pini A (1976) Study on the pathogenesis of bluetongue: replication of the virus in the organs of infected sheep. Onderstepoort J Vet Res 43:159–164

Probst C, Gethmann JM, Höreth-Böntgen D, Cussler K, Conraths FJ (2011) Lack of evidence for claims of farmers in south-eastern Germany regarding adverse reactions caused by BTV-8 vaccines. Berl Münch Tierärztl Wschr 124:282–287

Reeves WK, Lloyd JE, Stobart R, Stith C, Miller MM, Bennett KE, Johnson G. (2010) Control of *Culicoides sonorensis* (Diptera: Ceratopogonidae) blood feeding on sheep with long-lasting repellent pesticides. J Am Mosq Control Assoc 26:302–305

Ritter D, Roy P (1988) Genetic relationships of bluetongue virus serotypes isolates from different parts of the world. Virus Res 11:33–47

Roy P, Bishop DH, LeBlois H, Erasmus BJ (1994) Long-lasting protection of sheep against bluetongue challenge after vaccination with virus-like particles: evidence for homologous and partial heterologous protection. Vaccine 12:805–811

Ruckerbauer GM, Gray DP, Girard A, Bannister GL, Boulanger P (1967) Studies on bluetongue and detection of the virus in infected materials by immunofluorescence. Can J Comp Med 31:175–180

Saegerman C, Pastoret PP (2009) Bluetongue in northern Europe: appearance of new serotypes within an enzootic ground. Bull Off Int Epiz 1:35–39

Saegerman C, Mellor P, Uyttenhoef A, Hanon J-B, Kirschvink N et al (2010) The most likely time and place of introduction of BTV8 into Belgian ruminants. PLoS One 5:e9405. doi:10.1371/journal.pone.0009405

Savini G, Goffredo M, Monacol F, Di Gennaro A, de Santis P, Meiswinkel R, Caporale V (2004) The isolation of bluetongue virus from field populations of the Obsoletus Complex in central Italy. Vet Ital 40:286–291

Schmahl G, Walldorf V, Klimpel S, Al-Quraishy S, Mehlhorn H (2008) Efficacy of Oxyfly on Culicoides species – the vectors of bluetongue virus – and other insects. Parasitol Res 103:1101–1103

Schmahl G, Klimpel S, Walldorf V, Al-Quraishy S, Schumacher B, Jatzlau A, Mehlhorn H (2009a) Pilot study on deltamethrin treatment (Butox 7.5, Versatrine) of cattle and sheep against midges (*Culicoides* species, Ceratopogonidae). Parasitol Res 104:809–813

Schmahl G, Klimpel S, Walldorf V, Schumacher B, Jatzlau A, Al-Quraishy S, Mehlhorn H (2009b) Effects of permethrin (Flypor) and fenvalerate (Acadrex60, Arkofly) on *Culicoides* species-the vector of bluetongue virus. Parasitol Res 104:815–820

Schmahl G, Mehlhorn H, Abdel-Ghaffar F, Al-Rasheid K, Schumacher B, Jatzlau A, Pohle H (2009c) Does rain reduce the efficacy of Butox 7.5 pour on (deltamethrin) against biting midges (*Culicoides* specimens)? Parasitol Res 105:1763–1765

Shaw A, Monaghan P, Alpar HO, Anthony S, Darpel KE, Batten CA, Carpenter S, Jones H, Oura CAL, King DP, Elliott H, Mellor PS, Mertens PPC (2007) Development and initial evaluation of a realtime RT-PCR assay to detect bluetongue virus genome segment 1. J Virol Methods 145:115–126

Shone DK, Haig DA, McKercher DG (1956) The use of tissue culture propagated bluetongue virus for complement fixation studies on sheep sera. Onderstepoort J Vet Res 27:179–182

Skrock O (2011) Zur Wirksamkeit unterschiedlicher Netzprototypen zum Schutz von Milchvieh und Kälbern vor Überträgern (Gnitzen, Culicoides spp.) der Blauzungenkrankheit und anderen Lästlingsinsekten. Dissertation, Freie Universität Berlin

Skrock O, Manti B, Bauer B, Rohrmann KMA, Geericke N, Mehlitz D, Peters K-J, Frenzel K, Westerkamp A, Clausen P-H (2010) The effectiveness of insecticide-treated nets for the protection of cattle from biting midges and insect pests on dairy cattle. In: Joint meeting of the German Societies for Parasitology and Protozoology at Düsseldorf University, 17–20 Mar 2010. Düsseldorf University Press, Düsseldorf, p 179, ISBN:978-3-940671-37-0

Standfast HA, Muller MJ, Wilson DD (1985) Mortality of *Culicoides brevitarsis* fed on cattle treated with ivermectin. Prog Clin Biol Res 178:611–616

Stewart M, Bhatia Y, Athmaran TN, Noad R, Gastaldi C, Dubois E et al (2010) Validation of a novel approach for the rapid production of immunogenic virus-like particles for bluetongue virus. Vaccine 28:3047–3054

Szmaragd C, Wilson AJ, Carpenter S, Wood JLN, Mellor PS, Gubbins S (2010) The spread of bluetongue virus serotype 8 in Great Britain and its control by vaccination. PLoS One 5:e9353. doi:10.1371/journal.pone.0009353

Tabachnick WJ (1996) *Culicoides variipennis* and bluetongue-virus epidemiology in the United States. Annu Rev Entomol 41:23–43

Toussaint JF, Sailleau C, Mast J, Houdart P, Czaplicki G, Demeestere L, VandenBussche F, van Dessel W, Goris N, Bréard E, Bounaadja L, Thiry E, Zientara S, de Clercq K (2007) Bluetongue in Belgium, 2006. Emerg Infect Dis 13:614–616

Tschuor AC, Kaufmann T, Strabel D, Hässig M (2010) Abklärung von Aborten und anderen tiergesundheitlichen Problemen beim Rind im Zusammenhang mit der Blauzungenvirus-Impfung 2009. Schweiz Arch Tierheilk 152:501–506

Umeshappa CS, Singh KP, Nanjundappa RH, Pandey AB (2010) Apoptosis and immuno-suppression in sheep infected with bluetongue virus serotype-23. Vet Microbiol 144:310–318

Unger RE, Chuang RY, Chuang LF, Doi RH, Osburn BI (1988) Comparison of dot-blot and Northern blot hybridization in the determination of genetic relatedness of United States bluetongue virus serotypes. J Virol Methods 22:273–282

van der Walt NT (1980) A hemagglutination and hemagglutination inhibition test for bluetongue virus. Onderstepoort J Vet Res 47:113–117

van Oirschot JT (1999) Diva vaccines that reduce virus transmission. J Biotechnol 73:195–205

Vanbinst T, Vandenbussche F, Vandemeulebroucke E, De Leeuw I, Deblauwe I, De Deken G, Madder M, Haubruge E, Losson B, De Clercq K (2009) Bluetongue virus detection by real-time RT-PCR in Culicoides captured during the 2006 epizootic in Belgium and development of an internal control. Transbound Emerg Dis 56:170–177

Vandemeulebroucke E, De Clercq K, Van der Stede Y, Vandenbussche F (2010) A proposed validation method for automated nucleic acid extraction and RT-qPCR analysis: an example using bluetongue virus. J Virol Methods 165:76–82

Vandenbussche F, Vandemeulebroucke E, De Clercq K (2010) Simultaneous detection of bluetongue virus RNA, internal control GAPDH mRNA, and external control synthetic RNA by multiplex real-time PCR. Methods Mol Biol 630:97–108

Velthuis AG, Saatkamp HW, Mourits MC, de Koeijer AA, Elbers AR (2010) Financial consequences of the Dutch bluetongue serotype 8 epidemics of 2006 and 2007. Prev Vet Med 93:294–304

Velthuis AG, Mourits MC, Saatkamp HW, de Koeijer AA, Elbers AR (2011) Financial evaluation of different vaccination strategies for controlling the bluetongue virus serotype 8 epidemic in The Netherlands in 2008. PLoS One 6:e19612

Veronesi E, Hamblin C, Mellor PS (2005) Live attenuated bluetongue vaccine viruses in Dorset Poll sheep, before and after passage in vector midges (Diptera: *Ceratopogonidae*). Vaccine 23:5509–5516

Veronesi E, Darpel KE, Hamblin C, Carpenter S, Takamatsu HH, Anthony SJ et al (2010) Viraemia and clinical disease in Dorset Poll sheep following vaccination with live attenuated bluetongue virus vaccines serotypes 16 and 4. Vaccine 28:1397–1403

Wäckerlin R, Eschbaumer M, König P, Hoffmann B, Beer M (2010) Evaluation of humoral response and protective efficacy of three inactivated vaccines against bluetongue virus serotype 8 one year after vaccination of sheep and cattle. Vaccine 28:4348–4355

Wade-Evans AM, Mertens PPC, Bostock CJ (1990) Development of the polymerase chain reaction for the detection of bluetongue virus in tissue samples. J Virol Methods 30:15–24

Wade-Evans AM, Romero CH, Mellor P, Takamatsu H, Anderson J, Thevasagayam J et al (1996) Expression of the major core structural protein (VP7) of bluetongue virus, by a recombinant capripox virus, provides partial protection of sheep against a virulent heterotypic bluetongue virus challenge. Virology 220:227–231

Willgert K, Schroedle B, Schwermer H (2011) Spatial analysis of bluetongue cases and vaccination of Swiss cattle in 2008 and 2009. Geospat Health 5:227–237

Wilson WC, Chase CCL (1993) Nested and multiplex polymerase chain reactions for the identification of bluetongue virus infection in the biting midge, *Culicoides variipennis*. J Virol Methods 45:39–47

Wilson AJ, Mellor PS (2009) Bluetongue in Europe: past, present and future. Phil Trans R Soc B 364:2669–2681

Wilson WC, Stallknecht DE, Mecham JO (2004) Field–deployable realtime polymerase chain reaction detection of bluetongue and epizootic haemorrhagic disease viral ribonucleic acid. Vet Ital 40:587–593

Wilson A, Darpel K, Mellor PS (2008) Where does bluetongue virus sleep in the winter? PLoS Biol 6:e210

Wittwer CT, Herrmann MG, Moss AA, Rasmussen RP (1997) Continuous fluorescence monitoring of rapid cycle DNA amplification. Biotechniques 22(1):130–131

Yang J, Hua Q, Chen H, Lv J, Chen B, Ruan Z (2010a) A rapid assay for detecting antibody against bluetongue virus with a latex agglutination test using recombinant VP7 antigen. J Vet Diagn Invest 22:242–244

Yang J, Hua Q, Chen H, Lv J, Qin Z, Jin M et al (2010b) Development and evaluation of an immunochromatographic strip for the detection of serum antibodies against bluetongue virus. J Virol Methods 163:68–73

Zientara S, Bréard E, Sailleau C (2004) Bluetongue diagnosis by reverse transcriptase-polymerase chain reaction. Vet Ital 40:531–537

Zientara S, Maclachlan NJ, Calistri P, Sanchez-Vizcaino JM, Savini G (2010) Bluetongue vaccination in Europe. Expert Rev Vaccines 9:989–991

Chapter 6
Impact of Insecticide-Treated Nets on Insects of Medical and Veterinary Relevance

Burkhard Bauer, Dieter Mehlitz, and Peter-Henning Clausen

Abstract Global trade as much as climate change are favouring incursion and subsequent establishment of previously unknown vectors and/or the advent of exotic diseases. *Aedes albopictus* serves as an example for an insect vector now occurring in much of Europe. It is the highest ranking species listed on the "World's Worst Invasive Alien Species" and is transmitting a whole range of viral pathogens, for instance, Chikungunya, an exotic viral disease, which was recently detected in northern Italy. The emergence of bluetongue (serotype BTV-8) in much of north-western Europe in 2006/2007 serves as an example of a previously unknown vector-borne disease in Europe. It is now acknowledged that there is a need for effective and well-targeted vector control methods since protective vaccines will often not be available after a new disease has been diagnosed. This review deals with the development and subsequent assessment of a new and effective vector control method protecting livestock in sub-Saharan Africa as well as in Europe by means of insecticide-treated nets (ITN).

Keywords Climate change • Global trade • Insecticide treated net • Targeted vector control • Vector-borne diseases

6.1 Introduction

There is commonly accepted evidence that climate change associated with factors such as global mobility, environmental degradation, population growth and urbanisation have the potential to induce substantial changes in the occurrence and distribution of existing, emerging and re-emerging diseases and their vectors

B. Bauer • D. Mehlitz • P.-H. Clausen (✉)
Institute for Parasitology and Tropical Veterinary Medicine, Freie Universität Berlin, Königsweg 67, 14163 Berlin, Germany
e-mail: clausen.ph@vetmed.fu-berlin.de

H. Mehlhorn (ed.), *Arthropods as Vectors of Emerging Diseases*,
Parasitology Research Monographs 3, DOI 10.1007/978-3-642-28842-5_6,

(OIE 2007; IPCC 2007; WHO 2010). Climate is understood as part of several interconnected ecological and social factors (temperature, humidity, precipitation, soil moisture and rising sea levels) undergoing rapid change, thereby affecting animal and human disease distribution as well as severity and frequency through the presence or absence of potential vectors as well as their vectorial capacity (Kovats et al. 2003). The general trend of intensification and industrialisation of livestock production will continue, increasing the likelihood of emerging disease complexes, including zoonoses (OIE 2010).

Infectious vector-borne diseases (VBDs) are, among others, transmitted by arthropods (frequently insects) between human and animal hosts, thus causing a significant fraction of the global infectious disease burden in humans; nearly half of the world population is infected with at least one pathogen (CIESIN 2007; Lemon et al. 2008). In the case of malaria, 3.3 billion people are worldwide at risk, resulting in 250 million cases with nearly 1 million deaths per year. VBDs are considered as neglected diseases, profoundly restricting the socioeconomic status and development, particularly in poor countries with highest rates of infection, many of which are located in the tropics and subtropics (WHO 2007). A larger part of the emerging/re-emerging diseases is transmitted between humans and their livestock (61%). These zoonotic mostly VBDs are posing an increasing public health problem (WHO 2006; ILRI 2009). VBDs of livestock can result in severe economic losses; even the threat of infections will limit trade as a cause of OIE regulations.

Vector control strategies with a sufficiently high coverage are aiming at successfully reducing/interrupting disease transmission. This is presently addressed by employing new tools with minimal impact on the ecosystem and the environment, e.g. use of insecticide-treated bed nets to combat malaria, control of dengue with insecticide-treated curtains or netting material to cover water reservoirs, indoor insecticide application controlling Chagas disease and malaria and insecticide-treated traps and targets for the control of human and animal African trypanosomoses. The advent of pyrethroids, a particularly effective class of insecticides with low mammalian toxicity, has helped to replace other active ingredients with excellent insecticidal properties but with the disadvantage of posing eventual health hazards during application or due to their pervasive eco-toxicological effects. During the past decade, it became evident that global trade and a gradually changing climate have led to a re-emergence or propagation of hitherto largely unknown VBDs into new areas. Simultaneously, and as a consequence of the widespread, largely uncontrolled use of pyrethroids there have been reports about an increasing resistance of target insects against this particular class of insecticides (Kristensen and Jespersen 2007; Kristensen et al. 2001; Jandowsky et al. 2010).

Considering the likelihood of future VBD outbreaks and the scarcity of new active ingredients, there is an urgent need to have effective, targeted and environmentally non-hazardous vector control techniques as a first-line defence at hand before an effective vaccine will be developed. This was shown during the outbreak of bluetongue disease (BT) in north-western Europe in 2006, a midge-borne viral disease of sheep and cattle. Only little was known about the biological and ecological properties as well as effective techniques to control Palaearctic *Culicoides* species, the putative vectors of this viral disease (Carpenter et al. 2009; Hoffmann et al. 2009). The ensuing losses due to morbidity and mortality of BT virus (BTV) affecting livestock—in earlier work, FAO (2007) and Lemon et al. (2008) estimated annual losses of approximately US $3 billion resulting from morbidity and mortality of animals, trade embargoes and vaccination costs—may just give an example of future trends in emerging diseases. This plausible scenario warrants a shift in our approach with particular emphasis on integrated pest management (IPM) practices. While there is agreement that the mandatory vaccination programme of a large proportion of all ruminants against the causative BTV-8 resulted in a significant decline of new cases within 1 year, it has to be acknowledged that almost all of the ruminant population of north-western Europe suffered from considerable losses before the arrival of this vaccine.

The recent advent of a new vector control method, insecticide-treated nets (ITN), may offer an additional and promising tool for dealing with different vector populations. ITN were successfully used under varying ecological conditions in sub-Saharan Africa to control a range of disease vectors and/or nuisance flies (Bauer et al. 2001, 2006a, 2011; Kamau et al. 2009; Maia et al. 2010; Lassane et al. 2011). The promising results of the initial trials inspired further studies in moderate climates (Bauer et al. 2006b, 2009; Calvete et al. 2010; Skrock 2011). This review presents and discusses the potential and limitations of the ITN technology to control vectors of veterinary relevance in tropical and moderate climate zones.

6.2 Studies in Sub-Saharan Africa

The principle of using insecticide-treated nets for the protection of cattle was first tried in western Kenya (Bauer et al. 2001, 2006a). It is based on physiology and behaviour of many target insects (tsetse and horse flies as well as nuisance flies and mosquitoes), which are frequently attacking their hosts at altitudes below 1 m, thereby colliding with the net which is mostly invisible to the insects' eyes. Small-scale farmers keep more than 80% of the three million dairy cows in Kenya. Many of these cows are confined to zero-grazing units, thus reducing the exposure of the valuable cows to tick-borne diseases. However, in tsetse-infested areas like western Kenya, the tsetse flies invade the units and feed on these animals, thereby transmitting trypanosomes. Black multi-filament polyester netting of 75 denier (Vestergaard Frandsen A/S, Disease Control Textiles) was soaked in a 0.6% solution containing a residual pyrethroid insecticide, beta-cyfluthrin (125 g/l of active ingredient). Bands of the treated 150-cm-high mosquito netting were attached with wire to the poles of the zero-grazing units, completely surrounding the pens. The trial started in September 2001 and lasted until April 2002. The experimental trial group comprised 187 animals from 57 units that were permanently or partially protected. The control group consisted of 162 animals in 42 unprotected units. The mean hazard rate for trypanosome infections was significantly lower in protected cows, with a value of 0.007 as opposed to 0.02 for the control animals. Mean packed cell volumes (PCV) were significantly higher in protected cattle (29.7%) than in unprotected ones (27.6%). Farmers with protected animals also reported fewer nuisance flies and mosquitoes in their compounds.

The encouraging results of the study conducted in western Kenya induced Kamau et al. (2009) to use deltamethrin-treated polyethylene nets against tsetse and trypanosomosis for the protection of confined goats in zero-grazing units in Kwale District, Southeastern Coast of Kenya. The net was treated with coconut oil mixed at a ratio of 1:9 with hot water before treatment with 1% deltamethrin. The goats were confined in zero-grazing units surrounded by a 1.5-m treated and untreated polyethylene net. Tsetse fly densities were monitored by using Nzi traps (Mihok 2002), and the prevalence of trypanosome infections in goats was assessed with the Buffy Coat Technique. Trypanosome infections in unprotected, free-range goats increased from 38.9 to 61.1%. At the same time, the infection rates in goats protected with an untreated or a treated net decreased from 87.5% within the herd to 12.5% over the first 2 months and to nil in subsequent months. It was also observed that the protection with an untreated net resulted in an increase of fly numbers within the homesteads, whereas protection with the treated net decreased the fly densities.

As a result of the participating farmers' observations in the first study (Bauer et al. 2001, 2006a), it was tried to assess the effects of protecting cattle with insecticide-treated mosquito fences against nuisance and biting flies (Muscinae, Stomoxyinae and *Tabanidae*) in a follow-up study by Maia et al. (2010). The study took place at the end of the rainy season in the forest region of Kumasi, Ghana, during 6 weeks of October and November in 2005. Four similar sites at comparable distance from a water course were chosen on which four identical pens were built (Fig. 6.1). One pen served as negative control, with no animals and no netting; the second pen was protected with a black, untreated net containing two zebus; the third pen served as unprotected control with two zebus; and the last pen, which also contained two zebus, was protected with the identical (as for the second pen) but treated net (100 mg of deltamethrin/m^2, 150 denier multi-filament polyester fibre with a mesh width of 1 × 2 mm). The incorporated UV protection factor against sunlight ensured a longer persistency of the insecticide. Nuisance fly densities were weekly monitored using two mono-conical Vavoua traps (Laveissiére & Grébaut 1990) outside each pen at distances of 20–30 m. For the first 3 weeks, mean outside catches were highest around the second and the third pen with, respectively, 9.0 and 8.0 insects per trap per day compared with catches outside the negative control pen and the fourth pen (treated net protection) with 1.8 and 3.3 insects. This trend was corroborated during the fourth and the following weeks when catches increased sharply around pens 2 and 3 with, respectively, 155.7 and 172.8 insects during week 4 and following, while outside pens 1 and 4 significantly fewer insects (11.8 and 7.3) were caught. Pictures of selected body regions showed significantly fewer

Fig. 6.1 Insecticide-treated polyester fences protect cattle on the Boadi Cattle Farm in Kumasi, Ghana, 2005. The 1-m-high net consisted of black multi-filament polyester with a mesh width of 1 × 2 mm, pre-treated with deltamethrin (100 mg/m^2) and a UV protector

attacking insects inside pen 4, leading to significant nuisance reduction. Feed uptake and resting was undisturbed, contrasting with relentless disturbance of animals in pens 2 and 3.

Further confirmation of these effects was sought in another field study of peri-urban areas in Burkina Faso and Mali (Lassane et al. 2011) where improved livestock husbandry systems are maintained for dairy production. As in the previous study in Kumasi, a pre-treated multi-filament polyester net was attached to existing poles of the units containing cattle. Unprotected units served as controls. Densities of nuisance insects outside each pen were evaluated with mono-conical Vavoua traps. Mosquitoes were monitored with battery-driven BG traps (BioGents®sentinel traps; Rose et al. 2006). Distinct differences were recorded at the onset of the rainy season with significant increases of nuisance insects and mosquitoes in the vicinity of unprotected units, whereas the numbers of insects collected in the vicinity of protected units only showed insignificant increases.

Another study was conducted in two villages in the Eastern Region of Ghana where the national veterinary services had identified tsetse-transmitted trypanosomosis as the leading cause for sudden pig mortalities during April 2007 (Bauer et al. 2011).

Two villages were selected—one serving as control with 14 pigsties and one experimental village where 28 pigsties were protected with the netting material which had also been used during the previous trials in West Africa. The material was attached to surrounding timber poles and planks (Fig. 6.2). Bi-monthly monitoring of tsetse densities with ten geo-referenced bi-conical traps (Challier et al. 1977) per village showed a reduction of more than 90% in the protected village within

Fig. 6.2 Insecticide-treated polyester fences protect pigsties in the Eastern Region of Ghana, 2007. The 1-m-high net consisted of black multi-filament polyester with a mesh width of 1 × 2 mm, pre-treated with deltamethrin (100 mg/m^2) and a UV protector

2 months. Further reductions exceeding 95% were recorded during subsequent months. The tsetse population in the control village remained unaffected, only displaying seasonal variations.

The initial trypanosome prevalence, as assessed by thin blood smears, amounted to 76% and 72% of protected and control animals, respectively, and decreased to 16% in protected as opposed to 84% in control pigs 3 months after intervention. After 6 months, 8% of the protected pigs were infected contrasting with 60% in the control group. The results are a further confirmation of the efficacy of ITN, enabling farmers to keep their livestock alive in an area deemed unsuitable for livestock production because of high tsetse densities and the ensuing trypanosome risk. Bio-assays with susceptible *Musca domestica* showed a persistency of the ITN exceeding 8 months under tropical climate conditions.

Kagbadouno et al. (2011) reported in another tsetse control operation on Loos islands in Guinea-Conakry more than 90% reduction of the initial population following the use of ITN.

6.3 Studies in Europe

Since the principles of insect behaviour (horse and nuisance flies in this case) also apply for these insects in moderate climate, a study was undertaken in north-western Brandenburg to assess the efficacy of ITN protecting horses against attacks from biting and nuisance insects during the grazing period of summer 2004 (Bauer et al. 2006b). The ITN corresponded to the material previously used in Africa. The net was attached to the outside of already existing fences at a height of about 100 cm above ground. Three groups of horses that were kept on spatially separated pasture were retained for the trial. The first group consisted of mares and foals, serving as unprotected control. The second group was kept on a paddock that was fully protected. The third group, consisting of stallions, was partially protected, where only 13.4% of the total perimeter had been fenced. Fly densities were monitored with four Nzi traps outside each pasture. Fly catches showed a distinct reduction: when compared with the control area, the reduction was 67% outside the paddock (complete protection) and 57% in the vicinity of the enclosure for the stallions (partial protection). Recording of the attack rates of individual horses by flies with a digital camera on five different anatomical body regions (eye, neck, back, lateral and lower chest) revealed significant differences. Compared with the control, the attack rates of individual horses were reduced by 97% in the paddock and by 96%, respectively, of the horses kept on the partially protected pasture.

Following the outbreak of bluetongue (BTV-8) in many parts of north-western Europe and in Germany, efforts were undertaken to curb the spread of the disease, particularly in farms with valuable livestock, as on a stud bull farm in Schmergow, Brandenburg, Germany. Due to the abundance of *Culicoides* species, the putative BTV vectors, several control methods were applied, aiming at a reduction of the target insect populations. Insecticide-impregnated ear tags and regular treatments at 6 weeks intervals of all bulls with deltamethrin pour on were expected to achieve the intended reduction of biting midges. Since the trial results in Ghana (Maia et al. 2010) indicated an impact of the 1-m-high ITN on *Culicoides* spp., it was for the first time tried to control midges by using an incomplete ring of a 1.80-m-high fence protecting about 80% of the premises (Bauer et al. 2009). The net corresponded to the already described material. After an initial but only temporary reduction from more than 7,000 to below 500 midges—two BioGents® sentinel traps (Rose et al. 2006) had been used for measuring midge densities—the catches returned to their pre-intervention level. The transitory reduction could have been attributed to an initial protective effect of the fence or—what could not be excluded—to unfavourable weather conditions impairing the midges' activity for 2 weeks after deployment of ITN. It was also shown from microscopic examinations of midges and other haematophagous mosquitoes (*Aedes* and *Anopheles* spp.) that despite regular treatments of the bulls at 6 weeks intervals with a pour on formulation of deltamethrin and the simultaneous use of insecticide-impregnated ear tags, a high percentage of the target insects succeeded to feed. Between 10 and 35% of *Culicoides* spp. and more than 50% of *Aedes* and *Anopheles* spp., respectively, had engorged on the treated bulls. These observations illustrate the need for further R&D to identify more efficient control methods for midges and mosquitoes. It is also obvious that the modern structures of today's farms do not offer shelter against midges; hence, the recommended housing of valuable livestock will not prevent the midges from entering and feeding and, eventually, transmitting BTV (Clausen et al. 2009). Indoor survival of both *Culicoides* spp. as well as *Aedes* and *Anopheles* spp. cannot be excluded, transmission of BTV—albeit at a lower level—may continue during winter.

Fig. 6.3 Insecticide-treated polyester fences protect a stud bull farm in Schmergow, Brandenburg, Germany, 2009 (Rinderproduktion Berlin-Brandenburg (RBB)). The 3-m-high net consisted of black multi-filament polyester with a mesh width of 1.5 × 1.7 mm, pre-treated with micro-encapsulated deltamethrin (90 mg/m^2) and a UV protector

Building on the expertise acquired during the studies on the stud bull farm in Schmergow, a new approach was adopted, using—after further R&D—a different netting material in 2009. The fence consisted of black multi-filament polyester with a mesh width of 1.5 × 1.7 mm, pre-treated with micro-encapsulated deltamethrin (90 mg/m^2) and a UV protector. This new ITN prototype was used to protect—at a height of 3 m—the stud bull farm in Schmergow (Fig. 6.3) and also to protect with an incomplete ring (80%)—at 2 m height—zero-grazed cattle on a dairy farm in Lögow, Brandenburg. The incomplete ring allowed for daily operation procedures on the farm. A second unit served as a control. It was shown for the first time that the use of ITN resulted in significant reductions of *Culicoides* spp. on both farms during the trials. Structure of the net and mode of attachment ensured that the fence remained largely in place despite frequent gales during autumn 2009 and spring 2010. Chemical analyses of the amount of active ingredient (AI) allowed monitoring of an eventual, time-dependent decrease of deltamethrin. The loss of not more than 40–50% of the active ingredient after 1 year corroborates the results of the bio-assays. When exposed in a test box® (Patent No. 08803541.5 – 1260 PCT/EP 2008061570) for 10 s, susceptible *M. domestica* were still paralysed after more than 400 days of exposure of the netting material to the prevailing climate. Both laboratory-reared *C. nubeculosus* as well as wild-caught *C. obsoletus* were killed when alighting on or crossing the net even after short contacts for a few seconds (Manti et al. 2010). Catches with three BioGents® sentinel traps each and two

Glue-Fly Ribbon Traps per pen showed significant reductions of *Culicoides* spp. and nuisance flies, particularly *Stomoxys calcitrans* (Skrock 2011). In comparison to the control building, midges were reduced by 41.9% ($p = 0.0391$), and nuisance flies, by 77.2% ($p = 0.0039$).

Automatic recordings of defensive movements/dairy cows showed significantly fewer reactions in the protected pen due to the reduction of *S. calcitrans*. In turn, this resulted in a significant increase (1.2 kg/cow/day) of milk yield (Peters 2011). These observations are in line with previous work by Campbell et al. (1987) and Gerry et al. (2007), where it was found that five *S. calcitrans*/leg constitute an abundance value (= *economic injury threshold*) entailing unacceptable economic and/or health costs. In another paper, Mullens et al. (2006) suggested to evaluate fly-repelling behaviour, postulating that an average of 10 or more tail flicks/min would be equivalent to 5 stable flies/leg.

Environmental studies of beneficial arthropods (indicator species) showed that, for instance, bees (hymenoptera) or hovering flies (*Syrphidae*) were not or only negligibly affected.

Eco-toxicity of ITN was analysed in samples of milk, meat, faeces, soil and groundwater (no threshold value), indicating that these values were not attained (Frenzel 2011).

Calvete et al. (2010) studied in Spain the efficacy of surrounding yearling ewe pens with an untreated canvas barrier compared to a cypermethrin-treated canvas barrier to reduce the entry of *Culicoides* spp. and *C. imicola* in particular in the southwest of Spain. Analyses were based on comparisons of *Culicoides* catches in traps in pens with and without barriers, and in traps located outside pens. Although there was no clear reduction in the abundance of *Culicoides* other than *C. imicola* in pens with either barrier, the *C. imicola* presence was markedly reduced by the insecticide-treated barrier compared with the untreated barrier; the latter did not reduce the abundance of this species in pens. The results suggest that the use of insecticide-treated barriers may reduce contact between livestock and *C. imicola* in open areas or sheds. More research was deemed necessary to assess the degree of protection as a function of barrier height, *C. imicola* abundance and the size of the area to be protected.

6.4 Discussion and Recommendations

Integrated vector management (IVM) or integrated pest management (IPM) emphasise multi-disciplinary approaches where a proper understanding of the human–animal interface is considered a prerequisite for developing, implementing and validating strategic control methods and their impact evaluation over time (WHO 2008). Implementation of IPM/IVM takes into account synergies and antagonisms, achieving vector control in special settings (WHO 2007). A paradigm shift from a vertical single-disease to an inter-programmatic and inter-sectoral approach in the prevention and control of VBDs is required (WHO 2007, 2008).

Technical innovations of vector control should focus on their adaptability for the end users, enabling them to raise agricultural productivity while adjusting the technology to particular habitats (DFID 2008).

6.4.1 Economic Impact of Nuisance and Disease Transmission

It is estimated that attacks from *S. calcitrans* alone cause an overall annual economic loss to North American livestock producers of nearly $1 billion (Taylor and Berkebile 2006). Losses of comparable magnitude are here assumed for the industrialised livestock sector in all OECD countries. House flies (*M. domestica*) are a serious pest worldwide; apart from nuisance, they can transmit more than 60 known pathogens potentially causing zoonotic diseases. One of the major vectors of trachoma is *M. sorbens*, which is frequently associated with livestock keeping in sub-Saharan Africa. ITN may offer a particularly attractive solution in much of Africa and probably elsewhere due to the demographic changes: rapid urbanisation and scarcity of grazing resources are forcing farmers to settle and to intensify their animal husbandry management system. Living in proximity to their livestock will lead to an increase of synanthropic insects followed by the re-emergence of zoonotic diseases. African animal trypanosomosis has been and still is one of the major obstacles for the livestock sector, causing about $1 billion losses in cattle production per year and approximately five times more losses due to lost opportunities for development (FAO 2007). Human African trypanosomosis is also transmitted by tsetse flies from an animal reservoir (*Suidae* and *Bovidae*) to humans. West Nile and Rift Valley fever are examples for VBDs with increasing relevance, which may be greatly reduced through a strategic use of ITN. Up to now, farmers have been using pyrethroids as spray or pour on formulations. While it is acknowledged that the treatment of livestock with insecticides did have considerable success, the development of resistance against pyrethroids as a consequence of their largely uncontrolled use cannot be ignored. Affordability and thereby sustainability are increasingly questionable in resource poor farming systems of sub-Saharan Africa. An innovative control method such as ITN may just provide the much required attractive and affordable alternative.

6.4.2 Insecticide Resistance

Insecticide resistance (IR) shortens the effective life of any active ingredient. It will result in higher product costs: increase of active ingredient and frequency of use lead to higher costs and greater environmental impact (WHO 2008). While the preventive power of vector control in reducing VBD incidence is well acknowledged, there is also agreement that efforts focusing on a better understanding of the host–vector interface and/or research on designing innovative vector control strategies, the identification and development of more efficient/alternative molecules for vector control, have been paid less attention. The use of insecticides is frequently above its socially defined optimum, i.e. the benefits do not outweigh the costs (WHO 2008). Over-reliance on the most effective molecules, i.e. pyrethroids, has led to widespread resistance development in target insects, for instance, *M. domestica* (Jandowsky et al. 2010), and there are observations (Bauer unpublished) that other insect target species, e.g. *Stomoxys calcitrans*, *Drosophila repleta* and *Alphitobius diaperinus*, have also become tolerant or outright resistant (the two latter species) against pyrethroids. New EU regulations, the *E*uropean *R*egulation, *E*valuation, *A*uthorization and restriction of *Ch*emical Substances (REACH, EC 1907/2006), may further exacerbate the problem due to the disappearance of various active ingredients that are still effective (carbamates, organophosphates) but deemed too toxic during their handling. Still left are—apart from pyrethroids—neonicotinoid or spinosad baits and larvicides interrupting chitin biosynthesis. WHO in their Position Statement (WHO 2008) claim that "Insecticide resistance constitutes a significant threat for the effectiveness of pesticides that may be needed in urgent situations and a strong argument for IPM/IVM." It is also acknowledged that all pest and vector organisms will eventually develop IR if current patterns of insecticide use are continued. A holistic cross-cutting approach is required where the application of insecticides only constitutes the last resort. Since the protection with of ITN is considered as a targeted approach rather than the widespread and indiscriminate application of insecticides, it is likely to delay resistance development. However, in the end, target insects may as well adapt to this approach. For instance, insecticide-treated bednets greatly helped to reduce the malaria incidence in much of tropical Africa, but there are observations of behavioural changes and a rising resistance of *Anopheles* spp. to pyrethroids rendering bednets less effective. Considering our vulnerability in case of further adaptation of insect pests, there is a need for R&D to identify not only new and powerful active ingredients but also more effective and targeted approaches for vector control.

6.5 Conclusions

ITN have proven to protect livestock against tsetse-transmitted trypanosomosis. Nuisance insects (muscids and stable flies) were also effectively reduced—thus entailing a considerable increase of benefits. Population pressure of insect pests is distinctly higher in tropical areas of higher precipitation and temperatures. The use of ITN is likely to entail externalities such as the control of insects of public health relevance. Veterinary public health (VPH) is defined as "The contributions to the physical, mental and social well-being of humans through an understanding and application of veterinary science" (WHO/FAO/OIE definition 1999). It is also stated that "Human health, animal husbandry and animal health are closely connected and VPH is a fundamental part of public health whereby human health and well-being are the main objectives." Removing a constraint under these circumstances will therefore lead to higher gains.

In temperate climates, ITN significantly reduced *Culicoides* spp., which are not only vectors of bluetongue but also of African horse sickness. Preliminary observations in sub-Saharan countries as well as Europe point to distinct reductions of various mosquito species, warranting further research on a topic with potentially far reaching implications.

References

Bauer B, Oloo FP, Gitau D (2001) Management of tsetse-transmitted trypanosomosis in small-scale dairy farms in western Kenya: a private initiative for the good of the individual, but with external benefits for the neighbours and others. In: 26th Meeting of the International Scientific Council for Trypanosomiasis Research and Control (ISCTRC), Ouagadougou, abstract no 607

Bauer B, Gitau D, Oloo FP, Karanja SM (2006a) Evaluation of a preliminary trial to protect zero-grazed dairy cattle with insecticide-treated mosquito netting in western Kenya. Trop Anim Health Prod 38:29–34

Bauer B, Blank J, Heile C, Schein E, Clausen P-H (2006b) Einfacher Schutz von Pferden gegen Weidefliegen mit einem insektizidbehandelten Netzzaun im nördlichen Brandenburg. Berl Munch Tierarztl Wochenschr 119:421–424

Bauer B, Jandowsky A, Schein E, Mehlitz D, Clausen P-H (2009) An appraisal of current and new techniques intended to protect bulls against *Culicoides* and other haematophagous nematocera: the case of Schmergow, Brandenburg, Germany. Parasitol Res 105:359–365

Bauer B, Holzgrefe B, Mahama CI, Baumann MPO, Mehlitz D, Clausen P-H (2011) Managing tsetse transmitted trypanosomosis by insecticide treated nets – an affordable and sustainable method for resource poor pig farmers in Ghana. PLoS Negl Trop Dis 5(10):e1343

Calvete C, Estrada R, Miranda MA, Del Rio R, Borrás D, Beldron FJ, Martinez A, Calvo AJ, Lucientes J (2010) Protection of livestock against bluetongue virus vector *Culicoides imicola* using insecticide-treated netting in open areas. Med Vet Entomol 24:169–175

Campbell JB, Berry IL, Boxler DJ, Davis RL, Clanton DC, Deutscher GH (1987) Effects of stable flies (Diptera: Muscidae) on weight gain and feed efficiency of feedlot cattle. J Econ Entomol 80:117–119

Carpenter S, Wilson A, Mellor PS (2009) *Culicoides* and the emergence of bluetongue virus in northern Europe. Trends Microbiol 17(4):172–178

Challier A, Eyraud M, Lafaye A, Laveissière C (1977) Amélioration du rendement du piège biconique pour glossines (Diptera, Glossinidae): par I'emploi d'un cône inférieur bleu. Chiers ORSTOM séries Entomologie médicale et Parasitologie 15:283–286

CIESIN (Center for International Earth Science Information Network) (2007) Changes in the incidence of vector-borne diseases attributable to climate change. http://www.ciesin.columbia.edu/TG/HH/veclev2.html

Clausen P-H, Stephan A, Bartsch S, Jandowsky A, Hoffmann-Köhler P, Schein E, Mehlitz D, Bauer B (2009) Seasonal dynamics of biting midges (Diptera: Ceratopogonidae, *Culicoides* spp.) on dairy farms of Central Germany during the 2007/2008 epidemic of bluetongue. Parasitol Res 105:381–386

DFID (Department for International Development) (2008) DFID Research Strategy 2008–2013. Working Paper Series Sustainable Agriculture, April 2008

FAO (Food and Agricultural Organisation) (2007) Animal health disease cards: Bluetongue. http://www.fao.org/ag/againfo/subjects/en/health/diseasescards/bluetounge.html

Frenzel K (2011) Schlussbericht vom Teilprojekt 3 des BMELV Verbundvorhabens: Insektizid-behandelte Netze zur Bekämpfung von tiermedizinisch bedeutenden Vektorenseuchen. Tiergesundheitsdienst Bayern e.V.

Gerry AC, Peterson NG, Mullens BA (2007) Predicting and controlling stable flies on California dairies. University of California, Division of Agriculture and Natural Resources, Oakland, Publication No 8258

Hoffmann B, Bauer B, Bauer C, Bätza H-J, Beer M, Clausen P-H, Geier M, Gethmann J-M, Kiel E, Liebisch G, Liebisch A, Mehlhorn H, Schaub GA, Werner D, Conraths FJ (2009) Monitoring of putative vectors of bluetongue virus serotype 8, Germany. Emerg Infect Dis 15(9):1481–1484

ILRI (International Livestock Research Institute) (2009) New threats, new thinking at the animal-human disease interface. "Top Stories" ILRI Website, 5 June 2009

IPCC (Intergovernmental Panel on Climate Change) (2007) Working Group II: impacts, adaptation and vulnerability. Cambridge University Press, Cambridge

Jandowsky A, Clausen P-H, Schein E, Bauer B (2010) Vorkommen und Verbreitung von Insektizidresistenzen bei Fliegen (*Musca domestica* L.) in Milchviehbetrieben Brandenburgs. Prakt Tierarzt 91:590–598

Kagbadouno MS, Camara M, Bouyer J, Courtin F, Morifaso O, Schofield CJ, Solano P (2011) Tsetse control in Loos islands, Guinea. Parasit Vectors 4:18

Kamau SK, Ndegwa PN, Oyieke FA, Mukono JN (2009) Effects of confinement of goats in zero-grazing units and use of polyethylene nets against tsetse and trypanosomosis. In: Proceedings of the twenty-ninth meeting of the International Scientific Council for Trypanosomiasis Research and Control (ISCTRC), Luanda, Angola 2007, Publication No 124, pp 201–211

Kristensen M, Jespersen JB (2007) Susceptibility to thiamethoxam of *Musca domestica*. Pest Manag Sci. doi:10.1002/ps

Kovats S, Ebi KL, Menne B (2003) Methods of assessing human health vulnerability and public health adaption to climate change. In: Health and Global Environment Change Series No 1. World Health Organization, Copenhagen. 111 p

Kristensen M, Spencer AG, Jespersen JB (2001) The status and development of insecticide resistance in Danish populations of the house fly *Musca domestica* L. Pest Manag Sci 57:82–89

Lassane P, Sow M, Sidibe A, Diall O, Sidibe I, Belem AMG, Mahama C, Abavana M, Clausen P-H, Bauer B (2011) Management and control of vector-borne diseases in improved livestock production systems of peri-urban areas in Burkina Faso and Mali. In: Proceedings of the thirtieth meeting of the International Scientific Council for Trypanosomiasis Research and Control (ISCTRC), Kampala, Uganda 2009. ISBN:978-9966-7456-3-7, Publication No 125, pp 449–460

Laveissière C, Grébaut P (1990) Recherches sur les pièges à glossines (Diptera Glossinidae). Mise au point d'un modèle économique: Le piège ‹‹Vavoua ››. Trop Med Parasitol 41:165–192

Lemon SM, Sparling PF, Hamburg MA, Relman DV, Choffnes ER, Mack A (Rapporteurs) (2008) Vector-borne diseases: understanding the environment, human health, and ecological connections. Workshop Summary (Forum on Microbial Threats). National Academic of Science, USA. ISBN:978-0-309-10897-3, 350 p. http://www.nap.edu/catalo/11950.html

Maia M, Clausen P-H, Mehlitz D, Garms R, Bauer B (2010) Protection of confined cattle against biting and nuisance flies (Muscidae: Diptera) with insecticide-treated nets in the Ghanaian forest zone at Kumasi. Parasitol Res 106:1307–1313

Manti B, Bauer B, Mehlitz D, Westerkamp A, Clausen P-H (2010) Untersuchungen zur Empfindlichkeit von Gnitzen (*Culicoides* spp.) gegenüber pyrethroidbehandelten Netzen – Etablierung eines Testverfahrens zur Wirksamkeitsprüfung von Netzproben. Tagung der DVG-Fachgruppe Parasitologie und parasitäre Krankheiten, 7–9 Juli 2010 in München, Tierärztliche Praxis Großtiere 4, 2010, Abstract V 40

Mihok S (2002) The development of a multipurpose trap (the Nzi) for tsetse and other biting flies. Bull Entomol Res 92(5):389–403

Mullens BA, Lii K, MaoY MJA, Peterson NG, Szijj CE (2006) Behavioural responses of dairy cattle to *Stomoxys calcitrans* in an open field environment. Med Vet Entomol 20:122–137

OIE (Organisation Mondiale de la Santé Animale) (2007) Report of the *Ad Hoc* Group on climate change and vector-borne disease surveillance, Paris, 20–22 Nov 2007

OIE (Organisation Mondiale de la Santé Animale) (2010) 20th Conference of the OIE Regional Commission for the Americas, Montevideo, Uruguay, 16–19 Nov 2010

Peters K-J (2011) Untersuchungen zur Auswirkung von Insektizid-behandelten Netzen in offenen Stallsystemen der Milchviehhaltung auf das Verhalten und die Leistungsbereitschaft von Kälbern und Milchkühen als Beitrag zur Verbesserung der Tiergesundheit. BMELV Verbundprojekt, Schlussbericht, Teilprojekt 4, Humboldt Universität zu Berlin

Rose A, Kröckel U, Bergbauer R, Geier M, Eiras ÁE (2006) The BG-Sentinel, a novel mosquito trap for research and surveillance. Mitteilungen der Deutschen Gesellschaft für allgemeine und angewandte Entomologie 15:345–348

Skrock O (2011) Zur Wirksamkeit unterschiedlicher Netzprototypen zum Schutz von Milchvieh und Kälbern vor Überträgern (Gnitzen, *Culicoides* spp.) der Blauzungen krankheit und anderen Lästlingsinsekten. Dissertation, Freie Universität Berlin

Taylor DB, Berkebile D (2006) Comparative efficiency of six stable fly (Diptera: Muscidae) traps. J Econ Entomol 99:1415–1419

WHO (World Health Organisation) (2006) Neglected tropical diseases. http://www.who.int./zoonoses/diseases/en

WHO (World Health Organisation) (2007) Global plan to combat neglected tropical diseases, 2008–2015. 50 p

WHO (World Health Organisation) (2008) WHO position statement on integrated vector management. WHO/HTM/NTD/VEM/2008.2. 4 p

WHO (World Health Organisation) (2010) Climate change impacts on vector-borne diseases. *TDRnews*, May 2010

Chapter 7
The Changing Distribution Patterns of Ticks (Ixodida) in Europe in Relation to Emerging Tick-Borne Diseases

T. N. Petney, J. Skuballa, S. Muders, M. Pfäffle, C. Zetlmeisl, and R. Oehme

Abstract Ixodid ticks are the most important vectors of human pathogens and are significant vectors of animal pathogens in Europe. Evidence is accumulating that several tick species have extended their distributions, related at least in part to climate and habitat changes. With increasing anthropogenic modification of the environment, these distributional modifications are likely to continue, with the likelihood that tick-borne diseases will spread to new areas. We discuss those factors which are involved in the changing distributions of ixodid ticks and provide a list of possible invading species given potential changes in tick habitat.

Keywords Climate change • Distribution • Invading species • Ixodidae • Vector-borne disease

7.1 Introduction

The ecological cycles of vector-borne zoonotic pathogens differ depending on the species and the hosts and host density, the species and density of arthropods which can act as vectors, and the transmission dynamics between the two. These parameters all influence the likelihood of transmission to humans. This ecology is naturally dynamic, not only between seasons, but also between years depending on changes in host and vector populations and environmental and climatic conditions (Randolph et al. 2002). Superimposed on these natural changes are major anthropogenic

T.N. Petney (✉) • J. Skuballa • S. Muders • M. Pfäffle • C. Zetlmeisl
Department of Ecology and Parasitology, Zoology Institute, Karlsruhe Institute of Technology, Kornblumenstrasse 13, 76131 Karlsruhe, Germany
e-mail: petney@kit.edu

R. Oehme
Landesgesundheitsamt Baden-Württemberg, Ref. 93, Allgemeine Hygiene und Infektionsschutz, Nordbahnhofsstr. 135, 70191 Stuttgart, Germany

H. Mehlhorn (ed.), *Arthropods as Vectors of Emerging Diseases*,
Parasitology Research Monographs 3, DOI 10.1007/978-3-642-28842-5_7,

influences, such as land-use change, urbanization, human mobility, the introduction of invasive species, and global warming (Gilbert 2010; Reisen 2010). Unlike short-term natural variation, these changes are often long term and directional, such as the continuing trend to a warmer climate, or long-term and more or less abrupt, such as deforestation or reforestation (Foley et al. 2005; Rounsevell et al. 2006).

Studies on a wide variety of vector-borne disease systems have already established the likelihood that such dynamic environmental changes can have a strong influence on local and regional disease ecology (Patz et al. 2008). For example, malaria caused by *Plasmodium falciparum* is sensitive to rainfall and temperature changes, which can strongly influence the development and population dynamics of its mosquito vectors (Pascual et al. 2006). In addition, many mosquito species capable of transmitting malaria are highly sensitive to changes in land use (Yasuoka and Levins 2007; Imbahale et al. 2011), for example, deforestation in the northeast of Thailand has practically led to the eradication of malaria in this area (Petney et al. 2009).

Anthropogenic environmental changes are then an integral part of ecosystem change and, therefore, of changes in epidemiological cycles. Europe, for example, has been subject to the massive influence of humans on the environment from prehistoric times until now (Kalis et al. 2003; Schmidtchen and Bork 2003; Stoate et al. 2009). Intervention in the natural balance of the landscape started to increase from the end of the Stone Age owing to the growing density of the human population and incipient countrified landscape management (Jäger 1994; Rolle and Adroschko 1999; Ellenberg 2010). For instance, as early as 5500–1800 BC in the northern part and in the central and southern loess regions of Germany there were numerous peasants who cultivated the landscape, raised livestock, and let their cattle, pigs, sheep, and later also horses graze freely in the surrounding forests (Behre 2008; Ellenberg 2010). Such extensive grazing leads to the formation and maintenance of sites (e.g., neglected grassland) which serve as habitats for thermophilic species. During the seventh century AD, the first large-scale deforestation started, aimed at gaining arable farmland. A second phase of deforestation occurred because of the massive growth of the population in the thirteenth century (Heidel 2000). During the Neolithic the use of grassland and woodland led to the formation of heathlands, this formation reaching its maximum in the first half of the nineteenth century (Hallerbach 1994; Heidel 2000). The large open areas of heathland and farmland that developed by this means provide many native steppe inhabitants, such as the European brown hare (*Lepus europaeus*) and the wild rabbit (*Oryctolagus cuniculus*), with an optimal habitat, although today the subdivision of farmland into small lots has been widely replaced by large agricultural parcels (Schröpfer and Nyenhuis 1982; Pegel 1986; Kowarik 2010). The anthropogenic formation of such landscape is linked to a change in microclimatic factors, which can diverge strongly from the macroclimate and which create opportunities for microhabitat-specific arthropods to invade new environments (Ellenberg 2010).

In general, the emergence of a new disease in a certain area follows a set sequence involving introduction, establishment, and dispersal (Taraschewski 2006). In the case of vector-borne diseases, the requirements for each phase are complex, depending on the presence or absence of suitable reservoir hosts and vectors. In some cases native species are capable of taking over the role of both of them. A recent review dealing with the introduction of West Nile virus into North America highlighted three important factors leading to the rapid success of this introduced pathogen: (1) the abundance of hosts and vectors in human-modified environments, (2) adaptation to local vectors, and (3) focused feeding leading to new and unexpected hosts (Kilpatrick 2011).

If the natural vector was introduced previously to or is introduced concurrently with the pathogen, the number of vector individuals introduced must be suitable for the establishment of a viable, reproducing population. The likelihood that such a population will develop in a new area is dependent on the conditions found in the off-host environment, on the presence of suitable hosts, on propagule pressure (the number of individuals introduced per introduction event and the number of such events), and on the source population of the vector. The habitat suitability in a specific area can change, for example, owing to global warming, in which case the natural dispersal of hosts and vectors into new habitats can occur (Fuente et al. 2004; Kowarik 2010).

Introduced plants can influence host species in complex ways, based not only on their suitability in terms of microclimatic conditions, but also on their acceptance by hosts. In Europe, mast production of the North American red oak (*Quercus rubra*) is greater than that of the native European oak (*Quercus robur*). As rodent populations are potentially regulated by such production (Ostfeld et al. 2006), the introduction of this species could have influenced the spread of rodent-associated viruses such as hantavirus (Tersago et al. 2008) and the populations of vector species such as ticks (Randolph 2001).

Ticks are the most important vectors of human and animal pathogens in the temperate Northern Hemisphere (Randolph 2001). In Europe, these pathogens include a wide variety of zoonotic viruses, bacteria, and protozoa which cause significant morbidity but only limited mortality (Süss et al. 2004; Süss and Schrader 2004). Although our database on changes in the density and distributions of tick populations is increasing continually, there is little predictive information on which species and diseases are likely to, or may, invade central European countries.

Our aims are (1) to consider the factors likely to influence the emergence of a new tick-borne disease in an area, (2) to consider what changes in the distribution patterns of ticks have been observed in central Europe over the last few decades, how these changes may be related to tick attributes and dynamic climatic and environmental factors, and how this may influence the emergence of tick-borne diseases, and (3) to speculate on potential future introductions into central Europe.

7.2 Tick Ecology

At a local level, the distribution of ticks and their hosts can be influenced by a variety of factors. Tick survival and rate of development are both dependent on temperature and humidity. Low and high temperatures and saturation deficits preclude successful hatching. The microclimate is dependent on a variety of biotic and abiotic factors, such as the amount of vegetation present (shade and leaf litter provide lower temperatures but higher humidity) and the type of soil (water retention and potential interstitial refuges) (Merler et al. 1996; Schwarz et al. 2009). These factors are components of the habitat occupied by ticks. Hosts also have specific habitat preferences (see below), adding to the importance of habitat generally. Thus, each habitat must be considered separately in studies involving tick dynamics.

In one of the most comprehensive study on *Ixodes ricinus* to date, Estrada-Pena (2001) compared the abundance of ticks in 18 different habitats, defined by the vegetation present, over 3 years in Spain. He found that this species was absent from open, grassy habitats and hillsides as well as from young pine forest monocultures, and preferred sites with substantial secondary plant growth, in particular forests with oak species and fragmented forests with many ecotones. Estrada-Pena (2001) showed that 50% of the variation in tick abundance could be accounted for by temperature and vegetation characteristics of the habitat.

The dynamics of tick populations will follow the ecological changes occurring in a certain area depending on the habitat and host preferences of the different species. For the most important European vector species, *I. ricinus*, broadly assessed climatic conditions have a distinct influence on the activity patterns of all life history stages (Perret et al. 2004). For example, there has been a recent shift in the distribution of *I. ricinus* associated with higher average temperatures towards higher latitudes (Lindgren et al. 2000), as well as to higher altitudes (Daniel et al. 2003). Dautel et al. (2008) also suggested that the mild winter of 2006–2007 was responsible for the almost continuous yearly activity of *I. ricinus* in Berlin. Should such mild winters become more frequent, as predicted by climate models (Intergovernmental Panel on Climate Change data), the temporal window of exposure of the human population to tick-borne pathogens would increase. Not only *I. ricinus* is affected; *Dermacentor reticulatus* is not only increasing its area of distribution and population density, but is also moving out of its apparently preferred riparian habitats into drier, more forested areas (Dautel et al. 2006; Bullova et al. 2009).

Randolph (2001) suggested that the increases in both tick-borne encephalitis and Lyme borreliosis in central and northern Europe, as well as decreases in southern Europe, over the previous two decades may be due in part to changes in climate. She also pointed out that this problem has been increased by human-based land-use changes (e.g., habitat fragmentation), human behavior (e.g., outdoor recreation), and manipulation of host populations, which has allowed *I. ricinus* populations to increase in density. In addition, urban environments can differ from rural settings in the composition of the pathogen community in tick vectors. Fingerle et al. (2008), for example, showed unusually high prevalences of *Borrelia spielmanii* in the English Garden in Munich.

The predicted increases in temperature for Europe suggest that tick development will be accelerated during the coming decades if the predictions are correct. At the same time, tick survival is dependent on the humidity of the habitat occupied, with low humidity being unfavorable for survival and very high humidity potentially conducive to fungal infections (Stafford and Allan 2010). Thus, climatic variation, either natural or with an anthropogenic origin, will influence how quickly the ticks develop, how long they survive, their activity period, and their transmission potential (Dautel et al 2008). To complicate this situation, the survivorship of ticks may depend on infection with a pathogen. *I. ricinus* infected with *Borrelia burgdorferi* sensu lato survived longer under various thermohygrometric conditions than uninfected individuals (Herrmann and Gern 2010).

I. ricinus does not occupy a climatically homogeneous area. Estrada-Pena et al. (2006) showed that Europe contains nine groups of this species occupying climatically and ecologically significantly different areas. The differences in ecology are likely to play a major role in the dynamics of the host-tick pathogen system in these areas. During periods of glaciation, species would have been restricted to climatically suitable refuges and then would have moved to reoccupy their natural range as the climatic and host conditions became suitable again (Kowarik 2010).

Human-based land-use change has also undoubtedly influenced tick and tick-borne disease ecology from the earliest times. Deforestation to obtain agricultural land is likely to have reduced the population densities of *I. ricinus*, just as reforestation in the recent past will have increased the populations (Liebisch and Rahman 1976a, b). In the case of ticks, climate change has also been invoked as an agent causing expansion of the range of *I. ricinus* and *I. persulcatus* further to the north in Europe, in both cases accompanied by increasing prevalences of tick-borne encephalitis in the newly invaded areas (Gray et al. 2009; Tokarevich et al. 2011). Such change is consistent with the pattern found in many other species (Chen et al. 2011).

Habitat fragmentation, for example, via roads, may also be of significance, for example, by increasing host mortality or preventing or reducing the dispersal of non-avian hosts to new areas (Haemig et al. 2008; Petney et al. unpublished data). How this influences tick distribution is unknown, although preliminary data from the Hardtwald in northern Baden-Württemberg suggest that such fragmentation limits the dispersal of *D. reticulatus* (Petney et al. unpublished data).

7.3 Emergence of Tick-Borne Disease

Ticks generally have a very limited radius of movement (Crooks and Randolph 2006), so dispersal across geographical barriers must be accomplished via some additional transport mechanism. Europe is largely buffered from the natural movement of ticks from sub-Saharan Africa by the Sahara, although there is some potential for introductions via migratory birds (Waldenström et al. 2007; Elfving et al. 2010) or stock or companion animals (Menn et al. 2010). However, North African ticks are largely of Palearctic origin, and movement from this area across the Mediterranean is probably much more likely (Arthur 1965). Similarly, natural movement from the New World is blocked by the North Atlantic Ocean. Movement from Asia, however, is not blocked by any large-scale geographical barriers, and European species are found both in East Asia and in the Middle East (Pomerantzev 1950).

The potential for a particular species of tick to exist within an area is dependent on the macroclimate, which influences vegetation and the presence of host species, on microhabitat conditions present in relation to the tick's physiological requirements, and on the suitability of potential host species (Oliver 1989; Petney et al. 2011). These requirements differ widely between different tick species, with some being highly habitat and host specific, whereas others can be found in a wide variety of habitats and in or on a wide variety of hosts (Petney et al. 2011, 2012).

With increasing globalization, the movement of ticks through human activity has become increasingly possible. For example, transport times, which are significant in relation to the attachment times of ticks, have been vastly reduced over the last two centuries (Beierkuhnlein 2007; Kowarik 2010). There are numerous publications on the introduction of ticks to new countries via host animals (Keirans and Durden 2001; Burridge 2011). The most common example in central and northern Europe is the introduction of the brown dog tick, *Rhipicephalus sanguineus*, which is relatively host specific for dogs (Walker et al. 1999), with this host. *R. sanguineus* is the most widely distributed tick today, occurring throughout the world in temperate and tropical countries (Walker et al. 1999). In warmer areas it lives within the general environment; however, in the colder areas of Europe it is restricted to kennels or houses where the dogs live. It is introduced with those hosts that picked up the ticks in areas where the dogs' owners were on vacation, for example, in countries around the Mediterranean. Although it is not usually a direct problem for humans, *R. sanguineus* can multiply within this environment and reach high population densities (Kimmig et al. 2010). It is also a potential transmitter of *Rickettsia conorii*, the pathogen causing Mediterranean spotted fever, to humans (Kimmig et al. 2010). Central Europe and Norway also saw the introduction of the North American winter tick, *Dermacentor albipictus*, via the transport of horses (Lillehaug et al. 2002; Liebisch et al. 2006).

The emergence of tick-borne diseases within an area is therefore dependent on the environment as well as the hosts, tick species, and pathogens involved. Assuming that the environment is appropriate, one needs to know if suitable hosts for both the tick species and the pathogen are present. This is critical as the successful introduction of vertebrate species to an area is vastly less common than the introduction of arthropods (Pimentel et al. 2005). In this case, ticks with high host specificity are in general less likely to be successfully introduced unless that host is very common, as in the case of *R. sanguineus* and dogs, and *Rhipicephalus* (*Boophilus*) *microplus* and cattle (Cutullé et al. 2009). It is also necessary for these hosts to be sufficiently common to allow infestation and transmission rates that are high enough to maintain a reasonably stable enzootic cycle (Smith 1983).

For the further expansion and establishment of tick populations over wider areas, the radius of host movement is of critical importance. If a migratory species or one which moves across large distances, such as many bird species, is involved, then the transport of ticks to new areas is likely (Elfving et al. 2010). At the tick level, for an invading species not only is host specificity of importance but so is the number of hosts required in the life cycle (one-, two-, or three-host tick) as this determines the attachment (or contact) time with the host and therefore the likely distance over which it can be transported. Analogous with infectious disease ecology, the infective dose is important, as, in the case of ticks, is its life history stage. If a single gravid female is introduced, it would be capable of starting a population, whereas several nymphs or larvae would have to be introduced for nonparthenogenetic species or populations. In addition, the likelihood of a nymph or larva successfully becoming an adult and finding a mate would depend on the number introduced and the likelihood of finding a host and successfully engorging and molting (Hayes and Barry 2008).

At the level of the pathogen, host and vector specificity are also of importance; the rapid spread of West Nile virus throughout most of the USA was directly dependent on the presence of suitable avian reservoir hosts over winter and on the presence of native vector species (Kilpatrick 2011). Coupled with this are the usual factors defined by epidemiological models, such as how readily transmission occurs, how long the infection persists in the host or the tick, and if the host builds an immunity to the pathogen (Anderson and May 1991).

7.4 The Case of Central Europe

The most important central European tick species acting as vectors are listed in Table 7.1. Slovenia has been excluded from the central European list as it has a Mediterranean climate and coastline. Of the species present, *I. ricinus* and *D. reticulatus* are known to be extending their distributional limits (Gilbert 2010). Information on other species is too limited to provide significant evidence of range extension.

On the basis of the criterion that a successful introduction is likely to be a non-nest-dwelling generalist species with relatively wide environmental tolerance or a more specialized species with avian or domestic animals as hosts, a list of the most likely candidates for invading ixodid tick species can be constructed (Table 7.2). In most cases, such introductions will become more likely with warmer, drier climates such as those in southern European and North Africa, although species which currently live where there are continental climatic conditions may also benefit. Eleven species (57.9%) occupy arid, semiarid, or steppe environments, whereas *Haemaphysalis inermis*, *Haemaphysalis parva*, *Hyalomma lusitanicum*, *Hyalomma marginatum*, *Ixodes gibbosus*, *Ixodes redikorzevi*, *Rhipicephalus turanicus*, and *Rhipicephalus* (*Boophilus*) *annulatus* are all found in less harsh Mediterranean climates. With the exception of *I. redikorzevi*, which has been most commonly reported from rodents and other small mammals, and *R.* (*B.*) *annulatus*, which is predominantly a cattle tick, all of these species have a wide host range which frequently includes stock and companion animals (Table 7.2). Of particular importance are those species associated with such hosts as well as those which infest

Table 7.1 Common central European ixodid tick species known to transmit pathogens to humans and animals

Species	Range	Pathogens	Range expansion	Range contraction	Direction	Higher altitude
Dermacentor marginatus	Europe	OHFV, spotted fever group rickettsiae (e.g., *Rickettsia slovaca*, *R. raoultii*), *Francisella tularensis*, *Borrelia burgdorferi* s.l. (e.g., *B. lusitaniae*)	–	–	–	–
Dermacentor reticulatus	Central and eastern Europe	OHFV, spotted fever group rickettsiae (e.g., *R. slovaca*, *R. raoultii*, *R. helvetica*), *Coxiella burnetii*, *F. tularensis*, *Anaplasma phagocytophilum*, *Babesia canis*	Yes	No	East	
Haemaphysalis concinna	Europe	TBEV (*B. burgdorferi* s.l.), spotted fever group rickettsiae (e.g., *R. hulinii*), *F. tularensis*	–	–	–	–
Haemaphysalis inermis[a]	Eastern and southern Europe	High vector capacity for *R. helvetica* in Hungary				
Haemaphysalis punctata	Central and eastern Europe	TBEV, CCHFV	–	–	–	–
Ixodes ricinus	Europe	TBEV, *B. burgdorferi* s.l. and *Borrelia miyamotoi*, *A. phagocytophilum*, *Ehrlichia* spp. (e.g., *Ehrlichia canis*), *Bartonella* spp., spotted fever group rickettsiae (e.g., *R. helvetica*, *R. massiliae*, *R. monacensis*), *C. burnetii*, *F. tularensis*, *Babesia* spp. (e.g., *B. divergens*, *B. microti*)	Yes	No	North	Yes
Rhipicephalus sanguineus	Europe (introduced to central Europe)	*Ehrlichia* spp. (e.g., *E. canis*) spotted fever group rickettsiae (e.g., *R. conorii*, *R. siberica*, *R. rickettsi*, *R. massiliae*), *Babesia* spp. (e.g., *B. canis*, *B. gibsoni*)	No	No	–	No

Modified after Estrada-Pena and Jongejan (1999) and Petney et al. (2012)

OHFV Omsk hemorrhagic fever virus, *TBEV* tick-borne encephalitis virus, *CCHFV* Crimean-Congo hemorrhagic fever virus

[a]See Table 7.2, where *H. inermis* is also considered as it does not occur in the northern or western parts of central Europe

Table 7.2 Ixodid tick species with a significant vector capacity not currently occurring in central Europe (Germany) but with the potential to be transported to Germany

Species	Hosts required	Area of origin	Habitat	Generalist/specialist (at least one life history stage)	Bird hosts	Stock animals	Dogs/cats	Humans	Major diseases isolated from, associated with, or known to be transmitted by
Haemaphysalis inermis (central European but not in Germany)	3	EE, SE	Deciduous forests at various altitudes	G	Yes	Yes	Yes	No	Paralysis in roe deer by heavy infestations, spotted fever group rickettsiae (e.g., *R. helvetica*, *R. aeschlimannii*)
Haemaphysalis parva	3	EE, SE	Steppe and lower montane forests; urban environments	G	No	Yes	Yes	Yes	*F. tularensis*, *Babesia ovis*
Haemaphysalis sulcata	3	EE, NA, SE	Lowland and montane semideserts and steppes	G	Yes	Yes	Yes	No	Bhanja virus, *Ehrlichia canis*, *Rickettsia* spp., *Anaplasma* spp., *Babesia* spp., and *Theileria* spp.
Hyalomma anatolicum	2/3	NA, SE	Semidesert, steppe and savanna stock grazing areas	S	No	Yes	No	Yes	CCHFV, Thogoto virus, *Theileria annulata*, *T. mutans*, *Babesia equi*, *B. caballi*
Hyalomma asiaticum	3	EE	Desert areas	G	Yes	Yes	Yes	No	CCHFV, *C. burnetii*, spotted fever group rickettsiae (e.g., *R. sibirica mongolitimonae*, *Theileria* spp., *Anaplasma* spp.
Hyalomma excavatum	3	NA, SE	Semidesert, steppe and savanna nonagricultural habitats	S	Yes	Yes	Yes	Yes	CCHFV, *C. burnetii*, spotted fever group rickettsiae (e.g., *R. sibirica*), *Theileria* spp. (e.g., *T. annulata*,

(continued)

Table 7.2 (continued)

Species	Hosts required	Area of origin	Habitat	Generalist/ specialist (at least one life history stage)	Bird hosts	Stock animals	Dogs/ cats	Humans	Major diseases isolated from, associated with, or known to be transmitted by
									T. parva), *Babesia* spp. (e.g., *B. equi*, *B. ovis*)
Hyalomma impeltatum	3	NA	Semidesert, steppe and savanna environments	G	Yes	Yes	No	Rarely	CCHFV, Dugbe virus, Wanowrie virus
Hyalomma lusitanicum	3	NA, SE	Forested areas with natural grass patches	G	No	Yes	Yes	Yes	Spotted fever group rickettsiae (e.g., *Rickettsia conorii*), *T. annulata*, *T. equi*
Hyalomma marginatum	2	EE, NA, SE	Mediterranean environments	G	Yes	Yes	Yes	Yes	CCHFV, OHFV, Astrakhan virus, Dhori and other arboviruses, spotted fever group rickettsiae (e.g., *R. aeschlimanni*, *R. conorii*), *F. tularensis*, *C. burnetii*, *Babesia* spp. (e.g., *B. equi*, *B. caballi*), *T. annulata*
Hyalomma rifipes	2	SE, NA	Savanna environments	G	Yes	Yes	Yes	Yes	CCHFV, Tete, Dugbe, Jos, and Bhanja viruses, *R. conori*
Hyalomma scupense	1	EE, NA, SE	Areas with high moisture levels in arid areas	S	No	Yes	No	No	CCHFV, *C. burnetii*, *Theileria* spp. (e.g., *T. annulata*)
Ixodes gibbosus	3	SE	Deciduous forests and their ecotones with fields and grazing areas	G	Yes	Yes	No	No	Unknown

Ixodes redikorzevi	3	EE	Mediterranean environments; urban environments	S	Yes	No	Yes	Yes	Tick toxicosis, *A. phagocytophilum*
Rhipicephalus bursa	2	EE, NA, SE	Grassy areas, as well as arid shrub, steppe, and forest biomes	S	No	Yes	Yes	Yes	CCHFV, spotted fever group rickettsiae (e.g., *R. massiliae*), *A. marginale*, *A. ovis*, *Babesia* spp. (e.g., *B. bigemina*, *B. bovis*, *B. caballi*, *B. equi*, *B. motasi*, *B. ovis*, *Theileria* spp. (e.g., *T. ovis*)
Rhipicephalus pumilio	3	EE	Wide ranging from desert to humid habitats	G	No	Yes	Yes	Yes	CCHFV, spotted fever group rickettsiae (e.g., *R. conorii caspiensis*), *F. tularensis*, *C. burnetii*
Rhipicephalus pusillus	3	SE	Nest-dwelling; semiarid land and shrubland	S	No	No	Yes	No	Spotted fever group rickettsiae (e.g., *R. conorii*, *R. slovaca*)
Rhipicephalus rossicus	3	EE	Steppe and mountain steppe	G	No	Yes	Yes	Yes	CCHFV, *F. tularensis*, *C. burnetii*
Rhipicephalus turanicus[a]	3	NA, SE	Widely differing habitats; urban environments	G	Yes	Yes	Yes	Yes	Spotted fever group rickettsiae (e.g., *Rickettsia massiliae*), *C. burnetii*, *Babesia* spp. (e.g., *B. canis*, *B. trautmannii*)
Rhipicephalus (Boophilus) annulatus	1	EE, NA, SE	Mediterranean environments	S	No	Yes	Rarely	No	CCHFV, *Babesia bigemina*

Distribution and host information from Pomerantzev (1950), Balashov (1997), Manilla (1998), Walker et al. (1999), Wilamowski et al. (1999), Földvári et al. (2007), Kolonin (2009), and Guglielmone et al. (2010)

EE eastern Europe, *NA* North Africa, *SE* southern Europe (including Mediterranean Asia), *G* generalist, *S* specialist

[a]There is some taxonomic confusion in the *R. sanguineus*/*R. turanicus* species groups and these are currently undergoing systematic molecular analysis

birds, particularly migratory species, as these hosts provide a likely method of cross-border movement.

Taking all 19 species into account, 17 (89.5%) are found on stock (cattle, sheep, goats), 14 (73.7%) on companion animals, 10 (42.6%) on birds, and 11 (57.9%) have been reported from humans. *H. marginatum* and *Hyalomma rufipes* are known to reach central Europe irregularly, most likely via migratory birds (Siuda and Dutkiewicz 1979). The species belonging to the genera *Hyalomma* and *Rhipicephalus* are particularly significant vectors of both veterinary and medically important diseases. *H. marginatum*, for example, has been implicated in the introduction of Crimean-Congo hemorrhagic fever into Turkey in the late 1990s, with the first clinical cases being reported in 2002 (Estrada-Pena et al. 2010). Interestingly, neither single climate nor single vegetation variables could account for the spread of the disease; however, high levels of habitat fragmentation and connectivity were strongly associated with disease prevalence (Estrada-Pena et al. 2010).

One substantial problem in recognizing such emerging infectious diseases involves the continual advances being made in diagnostic medicine, as well as in our knowledge of the pathogens present in ticks, which may or may not have had a continual presence but have been unrecognized in the past, as is likely to be the case with tick-borne encephalitis (Süss 2008) or with several *Rickettsia* species (Walker et al. 2008; Dobler and Wölfel 2009). Another problem with tick studies is the lack of taxonomic expertise in several central European countries, including Germany, making the rapid and accurate identification of invading species difficult (Petney et al. 2012).

References

Anderson RM, May RM (1991) Infectious diseases of humans: dynamics and control. Oxford University Press, Oxford

Arthur DR (1965) Ticks of the genus *Ixodes* in Africa. Athlone, London

Balashov YS (1997) Distribution of ixodid ticks (Acarina, Ixodidae) over landscapes within their ranges. Entomol Rev 77:625–637

Behre KE (2008) Landschaftsgeschichte Norddeutschlands: Umwelt und Siedlung von der Steinzeit bis zur Gegenwart. Wachholtz, Neumünster

Beierkuhnlein C (2007) Biogeographie: Die räumliche Organisation des Lebens in einer sich verändernden Welt. Ulmer, Stuttgart

Bullova E, Lukan M, Satnko M, Petko B (2009) Spatial distribution of *Dermacentor reticulatus* tick in Slovakia in the beginning of the 21st century. Vet Parasitol 165:357–360

Burridge MJ (2011) Non-native and invasive ticks: threats to human and animal health in the United States. Florida University Press, Gainesville

Chen IC, Hill JK, Ohlemüller R, Roy DB, Thomas CD (2011) Rapid range shifts of species associated with high levels of climate warming. Science 333:1024–1026

Crooks E, Randolph SE (2006) Walking by *Ixodes ricinus* ticks: intrinsic and extrinsic factors determine the attraction of moisture or host odour. J Exp Biol 209:2083–2142

Cutullé C, Jonsson NN, Seddon J (2009) Population structure of Australian isolates of the cattle tick *Rhipicephalus (Boophilus) microplus*. Vet Parasitol 161:283–291

Daniel M, Danielova V, Kriz B, Nozicka J (2003) Shift of the tick *Ixodes ricinus* and tick-borne encephalitis to higher altitudes in Central Europe. Eur J Clin Microbiol Infect Dis 22:327–328

Dautel H, Dippel C, Oehme R, Hartelt K, Schettler E (2006) Evidence for an increased geographical distribution of *Dermacentor reticulatus* in Germany and detection of *Rickettsia* sp. RpA4. Int J Med Microbiol 296:149–156

Dautel H, Dippel C, Kämmer D, Werkhausen A, Kahl O (2008) Winter activity of *Ixodes ricinus* in a Berlin forest. Int J Med Microbiol 298:50–54

Dobler G, Wölfel R (2009) Fleckfieber und andere Rickettsiosen. Alte und neu auftretende Infektionen in Deutschland. Deutsches Ärzteblatt 106:348–354

Elfving K, Olsen B, Bergström S, Waldenström J, Lundkvist A, Sjöstedt A, Mejlon H, Nilsson K (2010) Dissemination of spotted fever rickettsia agents in Europe by migrating birds. PLoS One 5:e8572

Ellenberg H (2010) Vegetation Mitteleuropas mit den Alpen: In ökologischer, dynamischer und historischer Sicht. UTB, Stuttgart

Estrada-Pena A (2001) Distribution, abundance, and habitat preferences of *Ixodes ricinus* (Acari: Ixodidae) in northern Spain. J Med Entomol 38:361–370

Estrada-Pena A, Jongejan F (1999) Ticks feeding on humans: a review of records on human-biting Ixodoidea with special reference to pathogen transmission. Exp Appl Acarol 23:685–715

Estrada-Pena A, Venzal JM, Sanchez Acedo C (2006) The tick *Ixodes ricinus*: distribution and climate preferences in the western Palaearctic. Med Vet Entomol 20:189–197

Estrada-Pena A, Vatansever Z, Gargili A, Ergönul Ö (2010) The trend towards habitat fragmentation is the key factor driving the spread of Crimean-Congo haemorrhagic fever. Epidemiol Infect 138:1194–1203

Fingerle V, Schulte-Spechtel UC, Ruzic-Sabljic E, Leonhard S, Hofmann H, Weber K, Pfister K, Strle F, Wilske B (2008) Epidemiological aspects and molecular characterization of *Borrelia burgdorferi* s.l. from southern Germany with special respect to the new species *Borrelia spielmanii* sp. nov. Int J Med Microbiol 298:279–290

Földvári G, Márialigeti M, Solymosi N, Lukács Z, Majoros G, Kósa JP, Farkas R (2007) Hard ticks infesting dogs in Hungary and their infection with *Babesia* and *Borrelia* species. Parasitol Res 101:S25–S34

Foley JA, Defries R, Asner GP et al (2005) Global consequences of land use. Science 309:570–574

Fuente J, Naranjo V, Ruiz-Fons F, Vicente J, Estrada-Pena A, Almazán C, Kocan KM, Martín MP, Gortázar C (2004) Prevalence of tick-borne pathogens in ixodid ticks (Acari: Ixodidae) collected from European wild boar (*Sus scrofa*) and Iberian red deer (*Cervus elaphus hispanicus*) in central Spain. Eur J Wildl Res 50:187–196

Gilbert L (2010) Altitudinal patterns of tick and host abundance: a potential role for climate change in regulating tick-borne diseases? Oecologia 162:217–225

Gray JS, Dautel H, Estrada-Pena A, Kahl O, Lindgren E (2009) Effects of climate change on ticks and tick-borne diseases in Europe. Interdiscip Perspect Infect Dis. doi:10.1155/2009/593232

Guglielmone AA, Robbins RG, Apanaskevich DA, Petney TN, Estrada-Pena A, Horak IG, Shao R, Barker SC (2010) The Argasidae, Ixodidae and Nuttalliellidae (Acari: Ixodida) of the world: a list of valid species names. Zootaxa 2528:1–28

Haemig PD, Wladenström J, Olsen B (2008) Roadside ecology and epidemiology of tick-borne diseases. Scand J Inf Dis 40:853–858

Hallerbach J (ed) 1994 Die Wahner Heide. Landschafts- und Nutzungsgeschichte. Naturschutz. Studie im Auftrag Bündnis 90/Die Grünen, Köln

Hayes KR, Barry SC (2008) Are there any consistent predictors of invasion success? Biol Invasions 10:483–506

Heidel A (2000) Der mittelalterliche Mittelgebirgsraum-Waldwirtschaft und Landschaftshaushalt. GRIN, München

Herrmann C, Gern L (2010) Survival of *Ixodes ricinus* (Acari: Ixodidae) under challenging conditions of temperature and humidity is influenced by *Borrelia burgdorferi* sensu lato infection. J Med Entmol 47:1196–1204

Imbahale SS, Paaijmans KP, Mukabana WR, Van Lammeren R, Githeko AK, Takken W (2011) A longitudinal study on *Anopheles* mosquito larval abundance in distinct geographical and environmental settings in western Kenya. Malar J 10:81

Jäger H (1994) Einführung in die Umweltgeschichte. Wissenschaftliche Buchgesellschaft, Darmstadt

Kalis AJ, Merkt J, Wunderlich J (2003) Environmental changes during the Holocene climatic optimum in Central Europe – human impact and natural causes. Q Sci Rev 22:33–79

Keirans JE, Durden LA (2001) Invasion: exotic ticks (Acari: Argasidae, Ixodidae) imported into the United States. A review and new records. J Med Entomol 38:850–861

Kilpatrick AM (2011) Globalization, land use, and the invasion of West Nile virus. Science 334:323–327

Kimmig P, Pluta S, Naucke T (2010) Infektionsgefahr aus dem Süden. Globale Erwärmung. Biol Unserer Zeit 40:21–28

Kolonin GV (2009) Fauna of ixodid ticks of the world. http://www.kolonin.org/1.html. Assessed 15 Nov 2011

Kowarik I (2010) Biologische Invasionen: Neophyten und Neozoen in Mitteleuropa. Ulmer, Stuttgart

Liebisch A, Rahman MS (1976a) Zum Vorkommen und zur vektoriellen Bedeutung der Zecken *Dermacentor marginatus* (Sulzer, 1776) und *Dermacentor reticulatus* (Fabricius, 1794) in Deutschland. Tropenmed Parasitol 27:393–404

Liebisch A, Rahman MS (1976b) Zum Vorkommen und zur Ökologie einiger human- und veterinärmedizinisch wichtiger Zeckenarten (Ixodidae) in Deutschland. Z Angew Entomol 82:29–37

Liebisch G, Paufler S, Liebisch A (2006) Ein aktueller Fallbericht über die Gefahr der Verbreitung von Zecken durch die Aktivitäten des Menschen: Import der Zeckenart *Dermacentor albipictus* aus den USA mit einem Pferd nach Deutschland. Prakt Tierarzt 87:874–882

Lillehaug A, Mehl R, Gjerde B (2002) Importation of *Dermacentor albipictus* into Europe. Vet Rec 151:94–95

Lindgren E, Tälleklint L, Polfeldt T (2000) Impact of climatic change on the northern latitude limit and population density of the disease-transmitting European tick *Ixodes ricinus*. Environ Health Perspect 108:119–123

Manilla G (1998) Fauna d'Italia 36: Acari-Ixodida. Calderini, Bologna

Merler S, Furlanello C, Chemini C, Nicolini G (1996) Classification tree methods for analysis of mesoscale distribution of ixodes ricinus (Acari: Ixodidae) in Trentino, Italian Alps. J Med Entomol 33:888–893

Menn B, Lorentz S, Naucke TJ (2010) Imported and travelling dogs as carriers of canine vector-borne pathogens in Germany. Parasit Vectors 3:34

Oliver JH Jr (1989) Biology and systematics of ticks (Acari: Ixodidae). Annu Rev Ecol Syst 20:397–430

Ostfeld RS, Canham CD, Oggenfuss K, Winchcombe RJ, Keesing F (2006) Climate, deer, rodents, and acorns as determinants of variation in Lyme-disease risk. PLoS Biol 4:1058–1068

Pascual M, Ahumada JA, Chaves LF, Rodó X, Bouma M (2006) Malaria resurgence in East African highlands: temperature trends revisited. Proc Natl Acad Sci USA 103:5829–5834

Patz JA, Olson SH, Uejio CK, Gibbs HK (2008) Disease emergence from global climate and land use change. Med Clin North Am 92:1473–1491

Pegel M (1986) Der Feldhase (*Lepus europaeus* Pallas) im Beziehungsgefüge seiner Um- und Mitweltfaktoren. Schriftenreihe des Arbeitskreis Wildbiologie und Jagdwissenschaft an der Justus-Liebig-Universität Gießen, vol 16. F. Enke, Stuttgart

Perret J-L, Rais O, Gern L (2004) Influence of climate on the proportion of *Ixodes ricinus* nymphs and adults in a tick population. J Med Entomol 41:361–365

Petney TN, Sithithaworn P, Satrawaha R, Grundy-Warr C, Andrews RH, Wang YC, Feng CC (2009) Potential for the re-emergence of malaria in northeast Thailand. Emerg Inf Dis 15:1330–1331

Petney TN, Robbins RG, Guglielmone AA, Apanaskevich DA, Estrada-Peña A, Horak IG, Shao R (2011) A look at the world of ticks. Parasitol Res Monogr 2:283–296

Petney TN, Pfäffle M, Skuballa J (2012) An annotated checklist of the ticks of Germany. Syst Appl Acarol (May)

Pimentel D, Zuniga R, Morrison D (2005) Update on the environmental and economic costs associated with alien-invasive species in the United States. Ecol Econ 52:273–288

Pomerantzev BI (1950) Fauna of the USSR, Arachnida, vol. IV, no. 2. Academy of Sciences USSR, Moscow

Randolph SE (2001) The shifting landscape of tick-borne zoonoses: tick-borne encephalitis and Lyme borreliosis in Europe. Philos Trans R Soc Lond B 356:1045–1056

Randolph SE, Green RM, Hoodless AN, Peacey MF (2002) An empirical quantitative framework for the seasonal population dynamics of the tick *Ixodes ricinus*. Int J Parasitol 32:979–989

Reisen WK (2010) Landscape epidemiology of vector-borne diseases. Annu Rev Entomol 55:461–483

Rolle R, Adroschko FM (1999) Frühe Nutzung pflanzlicher Ressourcen. LIT, Berlin

Rounsevell MDA, Reginster I, Araújo MB et al (2006) A coherent set of future land use change scenarios for Europe. Agric Ecosyst Environ 114:57–68

Schmidtchen G, Bork HR (2003) Changing human impact during the period of agriculture in Central Europe: the case study Biesdorfer Kehlen, Brandenburg, Germany. In: Lang A, Hennrich K, Dikau R (eds) Lecture notes in earth sciences, vol 101. pp 183–200. Springer, Heidelberg

Schröpfer R, Nyenhuis H (1982) Die Bedeutung der Landschaftsstruktur für die Populationsdichte des Feldhasen (*Lepus europaeus* Pallas 1778). Z Jagdwissenschaft 28:213–231

Schwarz A, Maier WA, Kistemann T, Kampen H (2009) Analysis of the distribution of the tick *Ixodes ricinus* L. (Acari: Ixodidae) in a nature reserve of western Germany using Geographic Information Systems. Int J Hyg Environ Health 212:87–96

Siuda K, Dutkiewicz J (1979) *Hyalomma marginatum* Koch, 1844 (Acarina: Ixodidae) in Poland – an example of transport of southern tick by migratory birds. Wiad Parazytol 25:333–338

Smith RD (1983) *Babesia bovis*: computer simulation of the relationship between the tick vector, parasite, and bovine host. Exp Parasitol 56:27–40

Stafford KC, Allan SA (2010) Field applications of entomopathogenic fungi *Beauveria bassiana* and *Metarhizium anisopliae* F52 (Hypocreales: Clavicipitaceae) for the control of *Ixodes scapularis* (Acari: Ixodidae). J Med Entomol 47:1107–1115

Stoate C, Baldi A, Beja P, Boatman ND, Herzon I, Van Doorn A, De Snoo GR, Rakosy L, Ramwell C (2009) Ecological impacts of early 21st century agricultural change in Europe – a review. J Environ Manag 91:22–46

Süss J (2008) Tick-borne encephalitis in Europe and beyond–the epidemiological situation as of 2007. Euro surveillance: bulletin européen sur les maladies transmissibles. Eur Commun Dis Bull 13:270–277

Süss J, Schrader C (2004) Durch Zecken übertragene humanpathogene und bisher als apathogen geltende Mikroorganismen in Europa. Teil 1: Zecken und Viren. Bundesgesundheitsbl Gesundheitsforsch Gesundheitsschutz 47:392–404

Süss J, Fingerle V, Hunfeld KP, Schrader C, Wilske B (2004) Durch Zecken übertragene humanpathogene und bisher als apathogen geltende Mikroorganismen in Europa. Teil 2: Bakterien, Parasiten und Mischinfektionen. Bundesgesundheitsbl Gesundheitsforsch Gesundheitsschutz 47: 470–486

Taraschewski H (2006) Host and parasites as aliens. J Helminthol 80:99–192

Tersago R, Verhagen A, Servais P, Heyman P, Ducoffre G, Leirs H (2008) Hantavirus disease (nephropathia epidemica) in Belgium: effects of tree seed production and climate. Epidemiol Infect 137:250–256

Tokarevich NK, Tronin AA, Blinova OV, Buzinov RV, Boltenkov VP, Yurasova ED, Nurse J (2011) The impact of climate change on the expansion of *Ixodes persulcatus* habitat and the incidence of tick-borne encephalitis in the north of European Russia. Glob Health Action. doi:10.3402./gha.v410.8448

Waldenström J, Lundkvist A, Falk KI, Garpmo U, Bergström S, Lindegren G, Sjöstedt A, Mejlon H, Fransson T, Haemig PD (2007) Migrating birds and tickborne encephalitis virus. Emerg Infect Dis 13:1215–1218

Walker JB, Keirans JE, Horak IG (1999) The genus *Rhipicephalus* (Acari, Ixodidae). A guide to the brown ticks of the world. Cambridge University Press, Cambridge

Walker DH, Paddock CD, Dumler JS (2008) Emerging and re-emerging tick-transmitted rickettsial and ehrlichial infections. Med Clin North Am 92:1345–1361

Wilamowski A, Bromley-Schnur HJ, Ioffe-Uspensky I, Uspensky I (1999) Ticks (Ixodoidea) in Israeli towns. In: Proceedings of the 3rd international conference on urban pests, Czech University of Agriculture, Prague, pp 477–483

Yasuoka J, Levins R (2007) Impact of deforestation and agricultural development on anopheline ecology and malaria epidemiology. Am J Trop Med Hyg 76:450–460

Chapter 8
The Huge Risks Due to *Hyalomma* Ticks

Mohammed A. Bakheit, Abdalla A. Latif, Zati Vatansever, Ulrike Seitzer, and Jabbar Ahmed

Abstract Traditional tick taxonomy based on morphological characters is recently challenged by data generated from DNA analysis and several revisions in tick families have been proposed accordingly. Thus, names of some tick genera and their taxonomic positions have changed, species moved from one rank to another, while other names were invalidated. In this chapter, we update the genus *Hyalomma* species names as compiled from recent re-descriptions of species and tick reviews up to year 2011. *Hyalomma* species are known vectors of large numbers of parasites and pathogens transmitted to humans and livestock in different parts of the world making these ticks the economically most important ixodids.

Keywords Disease transmission • Genus *Hyalomma* • Species identification • Tick genera

8.1 Introduction to World Tick List

In the last 10 years, tick taxonomy and systematics passed through serious and contentious changes. Traditionally, tick taxonomy and phylogenetics were mainly based on morphological and biological characteristics. In the recent years, tick

M.A. Bakheit • U. Seitzer • J. Ahmed (✉)
Veterinary Infection-Biology and Immunology, Research Center Borstel, Leibniz Center for Medicine and Biosciences, Parkallee 22, 23845 Borstel, Germany
e-mail: jahmed@fz-borstel.de

A.A. Latif
Parasites, Vectors and Vector-borne Diseases, Onderstepoort Veterinary Institute, and Faculty of Veterinary Science, University of Pretoria, Private Bag X5, Onderstepoort 0110, South Africa

Z. Vatansever
Department of Parasitology, Kafkas University, 36100 Kars, Turkey
e-mail: zativet@gmail.com

H. Mehlhorn (ed.), *Arthropods as Vectors of Emerging Diseases*,
Parasitology Research Monographs 3, DOI 10.1007/978-3-642-28842-5_8,

Table 8.1 List of tick families and genera reported by several authors

Tick families and genera	Hoogstraal (1985)	Horak et al. (2002)	Barker and Murrell (2004)	Nava et al. (2009)	Guglielmone et al. (2010)	Mans (2011)
Nuttalliella	1	1	1	Omitted	1	1
Argasidae						
Argas	56	57	57	60	61	60
Antricola	8	Omitted	Omitted	17	17	17
Nothoaspis	1	Omitted	Omitted	1	1	1
Ornithodoros	100	38	37	106	112	45
Otobius	2	3	3	Omitted	2	2
Carios	Omitted	87	88	Omitted	Omitted	Omitted
Alectorobius	Omitted	Omitted	Omitted	Omitted	Omitted	66
Ixodidae						
*Amblyomma**	102	129	142	129	130	130
Anomalohimalaya	3	3	Omitted	Omitted	3	3
Aponoma	24	Omitted	Omitted	Omitted	Omitted	Omitted
*Boophilus**	5	Subgenus	Subgenus	Omitted	Omitted	Omitted
Cosmioma	1	1	1	1	1	1
Dermacentor	30	33	36	33	34	34
Haemaphysalis	155	168	166	164	166	166
Hyalomma	30	30	25	25	27	27
Ixodes	217	243	249	242	243	243
Margaropus	3	3	3	3	3	3
Nosoma	1	1	1	1	2	2
Rhipicentor	2	2	2	2	2	2
*Rhipicephalus**	70	80	79	81	82	82
*Bothriocroton**	Omitted	5	5	7	7	7

**Bothriocroton*: Created to include seven species of *Aponoma*. The other 17 *Aponoma* species are included in the genus *Amblyomma*
**Amblyomma*: 17 of *Aponoma* species are included here
**Boophilus* genus: Became a subgenus of *Rhipicephalus* and later the subgenus was omitted
Mans (2011): Followed Guglielmone et al. (2010) for hard ticks list. As his specialization focuses on argasids, he resurrected the earlier genus name *Alectorobius* with 66 species, retained 45 *Ornithodoros* and removed one species of *Argas*

systematics employing genetic markers have extensively been used and incorporated to compile phylogenetic trees. Thus, names of some tick genera and their taxonomic positions have changed, species moved from one rank to another, while other names were invalidated.

A summary of more than five recent reviews on genus valid names are shown in Table 8.1. The genera and also species names appearing in the different reviews disagree with one another to a greater or lesser extent. The authors of the most recent review on list of valid tick names (Guglielmone et al. 2010), who have traceable record in tick taxonomy/systematics, stated the following: 'The authors of this review disagree on the systematic status of several tick genera and share a concern about the validity of species names. They have adopted, but not necessary endorsed the genus level classification of Argasidae of Hoogstraal (1985)'.

They also stated that 'There is widespread disagreement concerning the genera in the family Argasidae and could not agree on the valid names' to appear in their abstract as they did for the Ixodidae. They also stated that *Aponoma*, *Anocentor* and *Boophilus* are still considered valid genera by some authors. The genus *Hyalomma*, likewise, has received an extensive revision during the last 10 years, specifically by Apanaskevich and Horak in series of more than ten publications from 2002 to 2010.

8.2 The Genus *Hyalomma*: Updated Species Names

Hoogstraal (1956) stated the following: Persons attempting to identify field collected material of *Hyalomma* should recognize that a certain proportion of specimens in many series will defy final determination of species. These had best be called '*Hyalomma* species'. He also referred to the heterogeneity of individuals within a species as it may have originated from the 'genetic instability in many specimens'. Since that time until to date, the 'criteria for identification to species level is in a chaotic state'. Apanaskevich and Horak, in series of publications from 2002 to 2010, attempted to review the taxonomy of the genus using the first author's expertise in the taxonomic status of ticks belonging to the genus *Hyalomma*. Moreover, he had access to the largest tick collection in the world, the United State National Tick Collection, Georgia Southern University. The following review is an attempt to update the species names up to the year 2011.

1. *Hyalomma aegyptium* Linnaeus, 1758.
   This is the type species of the genus *Hyalomma*, as discussed in Filippova, 1984.
2. *Hyalomma albiparmatum* Schulze, 1919.
3. *Hyalomma anatolicum* Koch, 1844.
   This species was classified as a subspecies of *H. anatolicum* Koch, 1844. It was listed as a subspecies of *H. anatolicum* by Camicas et al. (1998) and Horak et al. (2002). Recently, it is validated to species status by Apanaskevich and Horak (2005).
4. *Hyalomma arabica* Pegram, Hoogstraal and Wassef, 1982.
5. *Hyalomma asiaticum* Schulze and Schlottke, 1930.
   Apanaskevich and Horak (2010) treated the two previously subspecies *H. asiaticum kozlovi* and *H. asiaticum caucasicum* as junior synonyms of *H. asiaticum* and removed. The two subspecies were considered valid by Camicas et al. (1998) and Horak et al. (2002).
6. *Hyalomma brevipunctata* Sharif, 1928.
7. *Hyalomma dromedarii* Koch, 1844.
8. *Hyalomma excavatum* Koch, 1844.
   This species was previously classified as a subspecies of *H. anatolicum*. It was recognized as a subspecies of *H. anatolicum* by Camicas et al. (1998) and

Horak et al. (2002). It is omitted in the valid list by Barker and Murrell (2004). It is re-described and given species status by Apanaskevich and Horak (2005).

9. *Hyalomma franchinii* Tonelli-Rondelli, 1932.
10. *Hyalomma glabrum* Delpy, 1949.
    This name was re-instated recently and considered a synonym of *H. marginatum turanicum* by Apanaskevich and Horak (2006). It is a South African species found in the Karoo. The previous subspecies was omitted in the Horak et al. (2002) list but recognized by Guglielmone et al. (2009).
11. *Hyalomma hussaini* Sharif, 1928.
12. *Hyalomma hystricis* Dhanda and Raja, 1974.
13. *Hyalomma impeltatum* Schulze and Schlottke, 1930.
    Hyalomma erythraeum Tonelli-Rondelli, 1932 was recognized as a valid species and is included as a valid name in Barker and Murrell (2004). *Hyalomma erythraeum* is now shown to be a synonym of *H. impeltatum* by Apanaskevich and Horak (2009).
14. *Hyalomma impressum* Koch, 1844.
15. *Hyalomma isaaci* Sharif, 1928.
    This species was previously classified as a subspecies of *H. marginatum* listed by Camicas et al. (1998), Horak et al. (2002) and Guglielmone et al. (2009). It was omitted by Barker and Murrell (2004). It is now confirmed as a valid species by Apanaskevich and Horak (2008a).
16. *Hyalomma kumari* Sharif, 1928.
17. *Hyalomma lusitanicum* Koch, 1844 18.
18. *Hyalomma marginatum* Koch, 1844.
    This species was considered as a subspecies of *H. marginatum* Koch, 1844. It is now raised to a species level by Apanaskevich and Horak (2008a).
19. *Hyalomma nitidum* Schulze, 1919.
    Apanaskevich and Horak (2008b) considered *H. nitidum* a valid species.
20. *Hyalomma punt* Hoogstraal, Kaiser and Pedersen, 1969.
21. *Hyalomma rhipicephaloides* Neumann, 1901.
22. *Hyalomma rufipes* Koch, 1844.
    This species was considered as a subspecies of *H. marginatum* Koch, 1844. It was listed as such by Camicas et al. (1998) and by Horak et al. (2002). It is now raised to a species level by Apanaskevich and Horak (2008a).
23. *Hyalomma schulzei* Olenev, 1931.
24. *Hyalomma scupense* Schulze, 1918; Apanaskevich et al. (2010).
    This species was considered a subspecies of *H. detritum* Schulze, 1919. Because this is an economically important tick, Guglielmone et al. (2009) proposed referring to it as *H. scupense* (= *H. detritum*) in order to avoid confusion. *H. detritum*, as a valid tick species was synonymised with *H. scupense* by Apanaskevich et al. (2010).
25. *Hyalomma somalicum* Tonelli-Rondelli, 1935.
    This taxon was recently resurrected by Apanaskevich and Horak (2009) for *H. erythraeum*, Kaiser and Hoogstraal, 1968 (see *H. impeltatum*).
26. *Hyalomma truncatum* Koch, 1844.

27. *Hyalomma turanicum* Pomerantzev, 1946. This species was previously classified as a subspecies of *H. marginatum*. Listed by Horak et al. (2002) and Barker and Murrell (2004) as subspecies and was omitted in the list by Camicas et al. (1998). It is confirmed as a valid species by Apanaskevich and Horak (2008a).

8.3 Morphological Features of Males of *Hyalomma* Species of Economic and Ecological Importance (Hoogstraal 1956; Walker et al. 2003; Estrada-Pena et al. 2004)

8.3.1 Hyalomma scupense *(=* H. d. detritum, H. d. scupense*)*

Morphology of male tick (Fig. 8.1): Male ticks: Scutum is smooth, shiny and plain-coloured. Punctations are very few, large, shallow and scattered. Lateral lines (marginal grooves) are distinct up to a third of the scutum and extends beyond the mid-length of the scutum (may be obscured with indistinct punctations). The median groove is distinct and paramedian grooves are broad and distinct, with lines of rough surface in them. Scutum in the posterior region is formed into four raised ridges. Central festoon (parma) is pale coloured. Legs are not ringed (without bands). Subanal shields are small. There is extensive individual and geographical variability in the morphological characters of *H. scupense* Apanaskevich et al. (2010).

Hosts: Domestic cattle and horses are the most common hosts, but sheep, goats and camels may also be infested. All stages of development feed on the same host species. Adults attach on the inner thighs, udder, scrotum and perineum of cattle.

Distribution: The tick is present in Europe, Middle East, Asia, along the Mediterranean coast of Africa to Algeria and Morocco in the west. It also occurs in north-central Sudan but is not a common tick.

Life cycle: *Hyalomma scupense* (=*H. detritum*, Guglielmone et al. 2009) has a two-host life cycle that takes a year to complete. All tick stages feed on livestock. The adults are present during summer, and the larvae and nymphs in autumn. The detached engorged nymphs enter a winter diapause and moult to adults the following summer. Infestation is often associated with barns, stables and sheds, and livestock become infested when they are housed in these structures. The former *H. d. scupense* is biologically different. It is a one-host species.

8.3.2 Hyalomma rufipes

Morphology (Fig. 8.2): Male ticks: Large, robust, shiny black tick. Scutum is densely, uniformly covered with punctations obscuring lateral lines (grooves).

Fig. 8.1 *Hyalomma scupense*: male dorsal view and male ventral shields

Fig. 8.2 *Hyalomma rufipes*: male dorsal view and male ventral shields

Posterior margin of scutum is evenly rounded with little differentiation. Legs are distinctly annulated.

Hosts: Adults parasitize particularly the larger species of domestic and wild ungulates. They attach in the hairless perianal region and on the lower perineum and genitalia. The immature stages feed on hares as well as on ground-frequenting birds.

Life cycle: *Hyalomma rufipes* has a two-host life cycle that takes a year to complete. The adults are most numerous during the early part of the wet season and the immature stages during the dry season.

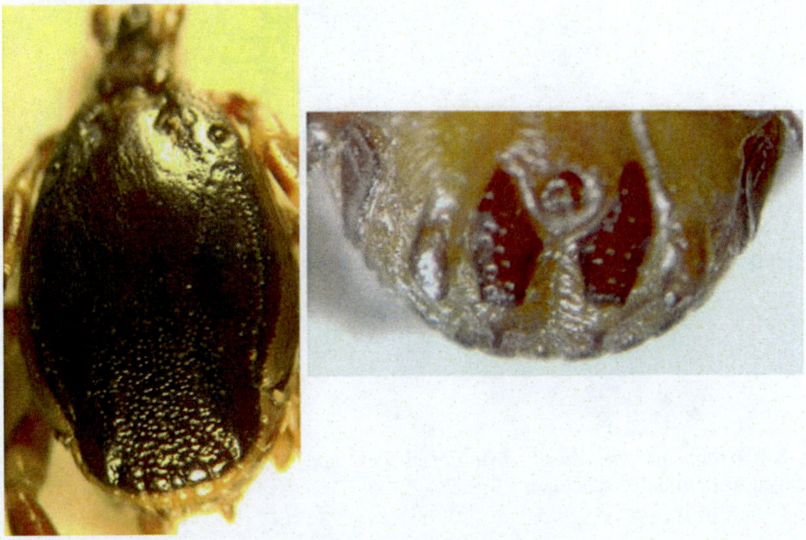

Fig. 8.3 *Hyalomma truncatum*: male dorsal view and male ventral shields

Distribution: This tick is widely distributed in most of the sub-Saharan African countries and southern Africa. Parasitism of birds by the immature stages undoubtedly contributes to the extensive distribution of this species.

8.3.3 *Hyalomma truncatum*

Morphology (Fig. 8.3): Male ticks: Lateral (marginal) lines are distinct grooves extending to the eyes. Scutum is dark, smooth and shiny and distinctly concave posteriorly (caudally). This region is densely punctated (contiguous punctations). There is in no parma.

Hosts: The preferred hosts of the adults are large domestic and wild herbivores. The ticks attach around the anus, on the lower perineum and on the legs, including around the feet. The immature stages feed on hares and on rodents, particularly gerbils.

Distribution: This tick is adapted to dry climates and is widely distributed all over Africa, occupying dry Savannah and high rainfall areas.

Biology and life cycle: *Hyalomma truncatum* has a two- or three-host life cycle, which normally takes a year to complete. Adults are present in the largest numbers in the late wet summer months and the immature stages in the dry autumn to spring months.

Fig. 8.4 *Hyalomma dromedarii*: male dorsal view and male ventral shields; top right fully engorged, bottom unfed specimen

8.3.4 *Hyalomma dromedarii*

Morphology (Fig. 8.4): Male ticks: Large, 5–7 mm. Scutum wide, smooth shiny and plain colour. Punctations are irregular, few, large and shallow, and rarely small ones occur. Marginal lines (lateral grooves) are very short limited to the posterior third of scutum and deep. Medians and paramedian grooves are deep, narrow with rough surface. The median groove extends from the distinct parma to scutal mid-length. The groove is surrounded on either side by four raised ridges. The subanal shields are prominent, always situated well exterior of the axis of adanal shields.

Hosts: The preferred hosts are camels, but cattle, sheep, goats and horses may also be infested. Adults attach on the inner thighs, udder and scrotum of camels. Larvae and nymphs feed on small burrowing animals and on hares, but the nymphs may also infest the same animals as the adults.

Distribution: This tick is common wherever camels occur in the Far, Middle and Near East. It is also present in Mauritania in West Africa and in Morocco, Algeria, Tunisia and Libya in North Africa and is well adapted to an arid and even desert environment. In North eastern and East Africa, it occurs in Sudan, Eritrea and northern, eastern and southern Ethiopia, northern Kenya and north-eastern Uganda.

Biology and life cycle: *Hyalomma dromedarii* has a two- or a three-host life cycle. The larvae may feed and moult to nymphs on small mammals or hares, and the adults feed on large herbivores. The larvae may feed on small mammal hosts, drop off and moult to nymphs, which can then either attach to other small mammals or feed on the same large animals as the adults. The life cycle appears to be continuous throughout the year.

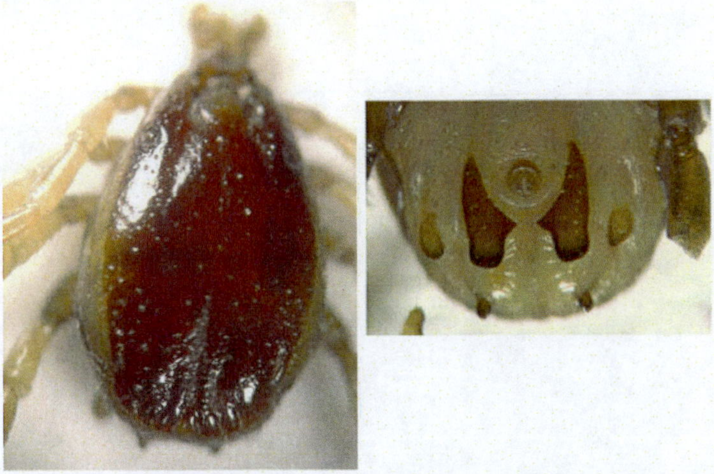

Fig. 8.5 *Hyalomma anatolicum*: male dorsal view and male ventral shields

8.3.5 *Hyalomma anatolicum*

Morphology (Fig. 8.5): Male ticks: Small ticks are referred to as 'small *Hyalomma*'; frail scutum is usually convex, has smooth shiny appearance and is pale in colour. Punctations are small and medium size, shallow, few in numbers and unevenly distributed. Median grooves long and narrow. Scutum in posterior (caudal) area is depressed between two indistinct lateral ridges. Large punctations are found here; surface is rough. Central festoon (parma) is pale coloured, single festoon on each side of the central festoon is of normal shape, separate and not fused (distinctive feature identifying this species from *H excavatum*). Legs are yellow coloured.

Female ticks: Knob-like genital aperture as a distinct bulge protruding from ventral surface; circular outline is characteristic.

Geographical distribution: *H. anatolicum* ranges from Somalia, northern parts of central Sudan (Southern limit), throughout North Africa, the Middle East, Iran, Indian sub-continent, China, Southern Russia and Turkey, up to southern Europe. This range covers three continents and different climates including desert, semi-desert, Mediterranean and subarctic.

Biology and life cycle: *H. anatolicum* has a great diversity of diapause mechanisms which regulate its seasonal activity and developmental rhythms in cold climates. Morphogenetic diapause has been found in the nymphal and the engorged female stages of the tick in Palaearctic regions. In North Africa, *H. anatolicum* adults also undergo behavioural diapause coinciding with shorter and less cold winter season. However, where both seasonal and circadian changes in environment are minimal, as in Sudan, all stages of the tick are active throughout

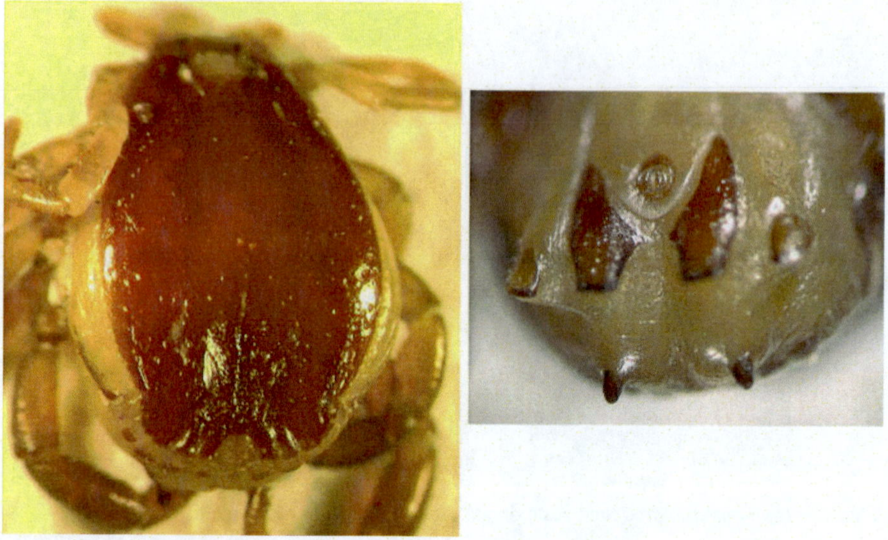

Fig. 8.6 *Hyalomma excavatum*: male dorsal view and male ventral shields

the year. All stages multiply and become a serious pest of livestock in stables and within zero-gazing farming system.

Hosts: *H. anatolicum* had a mixed two-host and three-host feeding pattern when fed on different host animals. When fed on rabbits, 54–98% of the larvae dropped as engorged nymphs, whereas on guinea pigs and Mongolian gerbils, greater percentage of larvae fed to engorgement and dropped. When *H. anatolicum* feeds on the normal hosts, e.g. cattle, horses, sheep and goats, a strict three-host cycle is observed.

8.3.6 *Hyalomma excavatum*

Morphology (Fig. 8.6): Male ticks: Larger ticks strong, shiny dark-brown scutum, with large few and small punctations widely scattered over entire surface, dense in the characteristically posterior (caudal) depression. Lateral lines (marginal grooves), limited to posterior 1/3 but deep and distinct. Posteromedian grooves (paramedian) are indistinct or absent. Scutum in posterior area is concave (depression) between two distinct lateral ridges. These ridges have steep internal margins. Punctations are concentrated here. Central festoon (parma) is pale coloured. The single festoon on each side of the central festoon extends to the anterior; their festoons fuse to form an arch (distinctive feature identifying this species

Fig. 8.7 *Hyalomma albiparmatum*: male dorsal view and male ventral shields

from *H. anatolicum*). Legs are brown in colour and have distinct pale bands at segments. Ventral shields (plates): adanal strong and broad. Subanals: larger than in *H. anatolicum*, distinct and dark.

Hosts, life cycle and biology: This is a two-host tick with the immature stages feed on smaller animals and the adults feed on livestock. It coexists with *H. anatolicum* but usually occupies the marginal areas. Where *H. anatolicum* and *H. excavatum* are sympatric, the former becomes more numerous and uniformly distributed than the latter. This co-existence of the two species has brought some confusion in the naming of *H. anatolicum* group.

8.3.7 *Hyalomma albiparmatum*

Morphology (Fig. 8.7): Male ticks: Lateral lines (marginal grooves) are distinct, long and deep and continue to the eye at lines of punctations. Scutum is dark, smooth and shiny with few punctations except posteriorly where it is densely punctate. Posterior region of the scutum is rectangular and concave (similar to *H. truncatum*). Central festoons (parma) is large whitish or brownish in colour (distinctive feature).

Hosts: The adults feed on domestic animals, mainly cattle, goats and sheep, while the immature stages feed on small animals.

Distribution: The tick is reported from the southern Kenya and northern Tanzania.

Fig. 8.8 *Hyalomma marginatum*: male dorsal view and male ventral shields

8.3.8 *Hyalomma marginatum*

Morphology (Fig. 8.8): Scutum is dark brown/black and is narrowly elongated, irregularly punctate, dense and large in the distal scapular areas and smaller and shallow and less dense centrally (different from *Hyalomma rufipes*), and in the posterior area (caudal), it is not strikingly differentiated and posterior margin is bluntly round. Marginal lines (lateral grooves) are long, reaching the eyes, but may be obscured anteriorly. The posteromedian groove reaches the scutal mid-length, being wider and narrow anteriorly. Paramedian grooves are wide near the festoon areas and taper anteriorly. Narrow heavily punctate ridge lies between the paramedian grooves and the lateral grooves. Legs may be entirely reddish (different from *Hyalomma turanicum*; bright enamelling).

Hosts, life cycle and biology: The tick undergoes a two-host life cycle with adults infesting cattle and other livestock and the immature stages feed small mammals.

Distribution: The tick is reported from North Africa (Morocco, Algeria and Tunisia) and southern Europe, Turkey and recently reported from Oman and the distribution extends to India.

8.3.9 *Hyalomma impressum*

Morphology (Fig. 8.9): Recent re-description of adults and larvae was provided by Apanaskevich and Horak (2007). Male ticks: Scutum is regularly densely covered

Fig. 8.9 *Hyalomma impressum*: male dorsal view and male ventral shields

by deep, large punctations obscuring the lateral grooves (marginal lines). The posterior (caudal) region of the scutum is distinctively narrower than the anterior part to form a rectangular outline (distinctive feature). Female ticks: The genital apron is broadly triangular in outline, with anterior bulging ridge and a deeply abruptly depressed narrower posterior button.

Distribution: *H. impressum* is mainly a West African tick, and it extends its range eastward into western Sudan.

Hosts: The chief hosts of the adult stage are cattle, but it was also reported from cattle and camels.

Biology and life cycle: Not studied.

8.3.10 Hyalomma impeltatum

Morphology (Fig. 8.10): Male ticks: Scutum is dark brown to black, covered by medium-size punctations (few large and shallow). Punctations are larger, deeper and denser on the posterior margin of scutum. Lateral lines (grooves) punctations are clear, extending anteriorly at least to mid-length of scutum and may be obscured by punctations. Subanal shields are large, exterior in position in fed specimen; subanal plates are always situated well exterior of the axis of adanal shields, usually borne on an udder-like swelling extending posterior beyond the body margin. Unfed specimen: subanal shields close to the ventral integument may appear in-line with adanal shields due to the unique tilting of subanal shields in medially directed position.

Distribution: The range covers Iran, Middle East, the Mediterranean, Sudan, Eritrea, Somalia, Northern Kenya, northern Tanzania, Chad and West African countries.

Hosts and life cycle: All large domestic animals are hosts of the adult stage, particularly cattle (high infestations were recorded) and camels. The immature

Fig. 8.10 *Hyalomma impeltatum*: male dorsal view and male ventral shields

Fig. 8.11 *Hyalomma aegyptium*: male dorsal view

stages feed on small animals like rodents, hares and ground birds. Under laboratory conditions, the tick had a three-host life cycle.

8.3.11 Hyalomma aegyptium

Morphology (Fig. 8.11): Male ticks: Subanal shields are present. Coxa I is simple (not deeply divided) with two wide branches. Scutum is smooth, shiny with few

Table 8.2 Viral diseases transmitted by *Hyalomma*

Viral disease	*Hyalomma* species	Host	Distribution
Crimean-Congo Haemorrhagic Fever	*H. marginatum, H. anatolicum, H. impeltatum, H. scupense*	Humans	Africa, Asia
Bhanja virus infection	*H. marginatum, H. detritum, H. dromedarii, H. truncatum and H. asiaticum*	Cattle sheep and goats	Asia, Africa and southern Europe
Thogoto, Dhori and Batken viral infections	*Hyalomma marginatum, H. dromedarii, H. marginatum*	Humans, guinea pigs, sheep?	Africa, Asia, Europe

large punctations. Festoons unfused. No lateral lines (grooves) or caudal (posterior) depressions.

Hosts: The tick feeds on tortoise in Mediterranean area.

8.4 Diseases Transmitted by *Hyalomma* Ticks

Hyalomma species transmit a number of viral, bacterial and parasitic diseases in different parts of the world, making these ticks the economically most important ixodids. The ticks may also act as reservoir of many viral pathogens. Besides disease transmission, certain species of *Hyalomma*, e.g. *H. truncatum*, contain toxin in their saliva that causes sweating sickness, an acute dermatitis, in cattle, particularly calves. The attachment of adult ticks to the inter-digital clefts and fetlocks of lambs results in lameness.

8.4.1 Viral Diseases Transmitted by Hyalomma *Ticks*

Table 8.2 illustrates some viruses whose transmission has been associated with *Hyalomma* ticks. In fact, several studies have shown the isolation of pathogenic viruses from *Hyalomma* ticks. There is, however, no proof that these ticks act as efficient biological vectors for these viruses. For example, Al-Khalifa et al. (2007) described the isolation of Sindbis, originally a mosquito-transmitted disease (Sane et al. 2011), Chick Ross and Kadam viruses from *H. dromedarii*, *H. anatolicum*, *H. impeltatum* and *H. schulzei* in Saudi Arabia. An earlier report demonstrated the isolation of Wanowrie, Thogoto and Dhori viruses from *Hyalomma* ticks infesting camels in Egypt (Williams et al. 1973). More recent data associating *Hyalomma* ticks to these viruses are scarce or not available.

Experimental transmission of some major zoonotic viruses infecting domestic livestock has been attempted. Linthicum et al. (1989) experimentally transmitted Rift Valley fever virus to *H. marginatum* and found that the infection peaks during developmental stages of the tick. However, transmissibility of the virus to Naive sheep has not been reported. On the other hand, Formosinho and Santos-Silva (2006) have transmitted West Nile fever virus to *H. marginatum* and showed that infected ticks in the stage of nymph and adult were able to transmit the infection to uninfected hosts, demonstrating that this tick could be involved in the natural circulation of West Nile fever in Portugal.

This chapter deals with those viral infections, which are studied in more details. In recent years, few viral infections caused a threat to livestock and public health. Of these, Crimean-Congo Haemorrhagic Fever (CCHF) is one of the most important.

8.4.1.1 Crimean-Congo Haemorrhagic Fever

CCHF is considered to be the most widespread tick-borne viral disease of humans. It has been recorded from more than 30 countries (Hoogstraal 1979; Ergönül 2006; Whitehouse 2004). The causative agent is a single-stranded RNA virus, which belongs to the *Nairovirus* genus in the Bunyaviridae family (Whitehouse 2004). CCHF causes viraemia without any clinical signs in its animal hosts and induces specific antibodies in many wild and domestic vertebrates such as cattle, sheep, goats, horses, camels, ostriches, hedgehogs, hares and some rodents and carnivores (Hoogstraal 1979; Nalça and Whitehouṣe 2007). Humans acquire the infection via tick bites, crushing the infected ticks or contact with viraemic animals' blood or tissues. Nosocomial infections due to contact with infected patients are also common (Hoogstraal 1979; Whitehouse 2004).

Ticks have been shown to transmit the virus by both trans-stadial and trans-ovarian methods (Hoogstraal 1979; Zeller et al. 1994; Turell 2007). The virus has been isolated from about 30 tick species; however, vector competence has been demonstrated only for *Amblyomma variegatum, H. marginatum, H. rufipes, H. anatolicum, H. asiaticum, H. truncatum, H. impeltatum, Dermacentor marginatus, Rhipicephalus evertsi* and *R. rossicus*. Among these, *Hyalomma* species are strictly associated with the global distribution of the disease (Hoogstraal 1979; Turell 2007; Camicas et al. 1994; Dohm et al. 1996; Logan et al. 1989; Korsunova and Petrova-Pointkovskaya 1949). Moreover, different species of *Hyalomma* are incriminated with transmission of CCHF in various geographical locations. Thus, *H. marginatum* is responsible for CCHF outbreaks in Turkey, Balkans, Crimea and Southern Russia. In Turkey, 16.43% of host seeking *H. marginatum* ticks were found to be infected with CCHF virus (Vatansever et al. 2010). *H. anatolicum* has been shown to be the vector in Iran, Pakistan, Turkmenistan and Tadjikistan. *H. asiaticum* transmits the disease in Central Asia and China, whereas *H. rufipes* in Africa.

H. marginatum is associated with wildlife and is adapted to steppe climate. It is a two-host tick whose immature (larvae and nymphs) stages feed on small mammals (hare, hedgehog) and ground-frequenting birds (rooks, partridges). Fed larvae moult on the host and become nymphs. Engorged nymphs moult on the ground to adults in about 4–20 days. The adults of *H. marginatum* actively seek/wait hosts horizontally on the ground. They mostly prefer artiodactyls (cattle, sheep/goats, horses, wild boars) but also aggressively attack humans. When attached to a host, they feed for about 9–14 days before females engorge and drop to lay eggs. The whole life cycle of *H. marginatum* takes about a year (Emelianova 2006; Pomerantsev 1950; Berezin 1971; Petrova-Piontkovskaya 1947; Ouhelli 1994). The ability of *H. marginatum* to overwinter as unfed adult is one of the main factors which allow the transmission of CCHF virus from year to year (Hoogstraal 1979; Berezin 1971).

CCHF epidemics in Balkans, Crimea and Southern Federal Districts of Russia have always been associated with ecological changes contributing to the maintenance and introduction of *H. marginatum* tick to regions where it was not known before. For example, the first cases of CCHF in Turkey were diagnosed in 2002 and until the end September 2010 a total of 5,318 laboratory-confirmed cases were reported from 2,200 rural settlements, representing a huge public health challenge. The disease risk is strongly associated with presence of *H. marginatum* and landscape fragmentation (Vatansever et al. 2007; Estrada-Pena et al. 2007). However, the empirical observations on habitat regeneration and its influences on wildlife and tick abundance still need to be determined.

CCHF is one of the most important emerging tick-borne infections, which is influenced by climatic and ecological changes. Research is needed regarding the biology of ticks, particularly *H. marginatum* taking into account the role of wild life in maintaining the ticks and consequently the circulation of the virus.

8.4.1.2 Bhanja Virus Infection

Bhanja virus is a member of the family Bunyaviridae. The virus is present in southern and central Asia, Africa and southern Europe. Bhanja virus was first isolated from *Haemphysalis intermedia* but has also been detected in a number of *Hyalomma* species including *H. marginatum*, *H. detritum*, *H. dromedarii*, *H. truncatum* and *H. asiaticum* (Hubálek 2009). Although it does not seem to be highly pathogenic, Bhanja virus has been shown to cause fever and CNS symptoms in young ruminant animals (lambs, kids and calves). Moreover, febrile illness caused by this virus has also been observed in humans (cited in Hubálek 2009). The virus is found in southern and Central Asia, Africa and southern and parts of central Europe.

8.4.1.3 Orthomyxovirus Virus Infection (Thogoto, Dhori and Batken Virus)

The family of Orthomyxoviridae comprises in addition to the three genera of influenza viruses the fourth genus of *Thogotovirus*. The latter include the arboviruses Thogoto, Dhori and Batken reported to be transmitted by ticks (Frese et al. 1997). Batken virus was obtained by Lvov et al. (1974) from *H. marginatum* ticks that were collected from sheep in Kyrgyzstan and has also been reported to be present in *H. marginatum* (Labuda and Nuttall 2004). Dhori virus has been reported in *H. dromedarii* and *H. marginatum* found in India, eastern Russia, Egypt and southern Portugal. Among other ticks, Thogoto virus has been isolated from *Hyalomma* species in Nigeria and Egypt (cited in Labuda and Nuttall 2004).

8.4.2 Bacterial Diseases Transmitted by Hyalomma *Ticks*

As in the case for viruses, several reports confirmed the presence of bacterial pathogens, including rickettsia, in *Hyalomma* ticks; however, because of the number of ticks investigated, these data need further corroboration before conclusions can be made regarding the significance of *Hyalomma* for bacterial pathogen transmission. For instance, disease agents such as *Anaplasma phagocytophilum* have been detected in *Hyalomma* ticks; however, there is no proof of the ability of the ticks to biologically transmit these organisms.

A list of the bacterial pathogens transmitted by *Hyalomma* ticks is given in Table 8.3. Of these, the Q fever causative agent *Coxiella burnetii* and *Anaplasma marginale*, which causes bovine anaplasmosis, might be of high relevance regarding human and animal *Hyalomma*-borne bacterial infections.

8.4.2.1 Rickettsiosis

Rickettsiae have also been reported to be transmitted by *Hyalomma* and other ticks. These are intracellular coccobacilli bacteria causing vasculitis, fever, hepatosplenomegaly, neurological signs, heart failure, renal insufficiency and bleeding. With the exception of epidemic typhus, which is food-borne disease, other rickettsioses are zoonoses and transmitted to human by arthropods such as fleas, lice and ticks. Among the *Hyalomma* species transmitting rickettsiosis are *H. marginatum* and *H. truncatum*, *H. rufipes* and *H. scupense* (see Table 8.3). Psaroulaki et al. (2005) reported on the identification of '*Rickettsia mongolotimonae*' (proposed name) detected simultaneously in a patient and in an *H. excavatum* tick, sampled from the patient in Greece. Polymerase chain reaction followed by restriction fragment-length polymorphism analysis on samples of haemolymph-positive ticks showed a rickettsia similar to that found in *H. marginatum* ticks from Morocco and Portugal to be present in *H. rufipes* (Beati et al. 1995). Keysary et al. (2011) brought molecular evidence for the presence of spotted fever group rickettsiae in ticks collected from roe deer, addax, red foxes and wild boars in Israel, and they could detect *R. aeschlimannii* in *H. marginatum* and *H. detritum*. Another recent publication reported the presence of *R. aeschlimannii* in *H. marginatum* and *R. sibirica* mongolitimonae in *H. excavatum* ticks (Chochlakis et al. 2011). In Senegal, *R. aeschlimannii* has been reported to be found in *H. rufipes* and *H. truncatum* and *R. sibirica mongolitimonae* in *H. truncatum* (Mediannikov et al. 2010).

Boutonneuse fever, which is also known as Mediterranean spotted fever, is caused by *R. conorii*. The disease was first found in Tunisia then reported from

Table 8.3 Bacterial diseases transmitted by *Hyalomma*

Bacterial/rickettsial species	*Hyalomma* species	Host	Distribution
Anaplasma marginale	*H. rufipes*	Cattle	Most tropical and subtropical countries and in some more temperate regions
Rickettsia conorii	*H. rufipes*, *H. truncatum*	Human	Mediterranean region, Africa, Asia
Anaplasma phagocytophilum	*H. lusitanicum*	Human, cattle	America, Asia, Europe
Rickettsia aeschlimannii	*H. marginatum*, *H. scupense* *H. truncatum*	Human	Africa, Europe
R. sibirica mongolitimonae	*Hyalomma asiaticum*, *H. excavatum*, *H. truncatum*	Human	Africa, Europe
Coxiella burnetii	*H. scupense*, *H. aegyptium*	Human	Worldwide
Borrelia turcica	*Hyalomma aegyptium*	Spur-thighed tortoise (*Testudo graeca*), Horsfield's tortoise (*Testudo horsfieldii*)	Turkey

many countries. *R. conorii* is mainly transmitted by *Rhipicephalus sanguineus*; however, many *Hyalomma* species were also reported to transmit the disease. Among these are *H. rufipes* and *H. truncatum* (Walker et al. 2003).

8.4.2.2 Anaplasmosis

Anaplasmosis is an infectious, non-contagious disease of humans or ruminants, including cattle, bison, sheep, goats, deer and elk. It is a widespread disease that is transmitted by at least five different ticks (Brayton et al. 2009). Bovine anaplasmosis is mainly caused by two *Anaplasma* species: *A. marginale* and *A. centrale*. Anaplasmosis due to *A. marginale* can be fatal; however, calves below 9 months old are usually immune. The disease has an incubation period of 4–6 weeks and affected animals develop anaemia and jaundice. They also become emaciated and abortion may occur. *A. centrale* infection is usually mild. Both pathogens induce mild infections in wild ruminants. This pathogen is transmitted by a number of ticks including *H. rufipes* (Bakirc et al. 2011).

8.4.2.3 Q Fever

Coxiella burnetii is the causative agent of Q fever. The pathogen is an obligatory intracellular Gram-negative bacterium with a high zoonotic potential and a wide distribution in mammals (Weisburg et al. 1989; Hirai and To 1998; Marrie and Raoult 1997; Reusse 1960). It has been shown that ticks function as vectors and distribute in transmitting *Coxiella burnetii* among wild animals using both trans-stadial as well as trans-ovarian transmission paths. Moreover, ticks can transmit the pathogen through biting via saliva and through tick faeces contaminated with the bacterium. In an experiment by Siroký et al. (2010), it could be shown that the tortoise tick *H. aegyptium* acquires *C. burnetii* infection from infected mammalian hosts and transmits the pathogen trans-stadially, whereby adult ticks were more efficient in transmitting the pathogen than larvae.

H. aegyptium has also been shown to be associated with the transmission of *Borrelia turcica* and DNA of this pathogen was detected in ticks infesting *Testudo graeca* or *Testudo horsfieldii Testudo* tortoise (Guner et al. 2003, 2004).

8.4.3 Parasitic Diseases Transmitted by Hyalomma *Ticks*

Among the diseases transmitted by the ixodid tick *Hyalomma*, theileriosis and babesiosis caused by members of the apicomplexan parasite genera *Theileria* and *Babesia*, respectively, are probably the most economically important. These diseases infect a wide number of wild and domestic animals worldwide. The disease, due to either of the organisms *Theileria* or *Babesia*, is also generally collectively known as piroplasmosis, referring to fact that these parasites form a ‘piroplasm’ stage parasitizing erythrocytes of their vertebrate hosts. *Theileria* and *Babesia* are phylogentically related, and both fall under the apicomplexan order Piroplasmorida Wenyon 1926 (according to Levine 1988). Infection of the animal with either of the two parasites is initiated by the inoculation of the sporozoite stage from infective ticks during a blood meal. Sporozoites enter leucocytes in case of *Theileria* or erythrocytes in case of *Babesia*. Thus, a major criterion to distinguish *Theileria* from *Babesia* is the presence, in the life cycle of *Theileria*, of an exo-erythrocytic schizont stage that arises through the development of the sporozoite inside the leucocytes. Schizonts normally parasitize lymphocytes and, in some *Theileria* species such as *T. annulata* and *T. lestoquardi*, also macrophages of their host animals, leading to cellular transformation, uncontrolled multiplication and immortalization. The schizont stage is responsible for the pathogenicity of the so-called lympho-proliferative *Theileria* species, in which schizogony occurs in multiple cycles; schizont-infected cells metastasize and invade different tissues and organs of the animal, causing severe damage and destruction. Schizonts also give rise to merozoites, which invade erythrocytes. In non-lympho-proliferative *Theileria* such as *T. mutans*, schizont-infected cells do not excessively multiply and hence do not play a major role in the pathogenesis of the disease. Instead, the severity of infection

Table 8.4 Parasitic diseases transmitted by *Hyalomma*

Parasite species	Hyalomma species	Host	Distribution
Theileria annulata	*H. anatolicum, H. dromedarii, H. rufipes, H. scupense.* Experimentally: *H. impeltatum, H. marginatum*	Cattle	Northern Africa, Southern Europe, Asia,
Theileria lestoquardi	*H. anatolicum*	Sheep, goats	Northern Africa, Asia
Theileria equi	*H. anatolicum, H. scupense*	Horses, donkeys	Africa, America, Asia, Australia, Europe
B. occultans	*H. rufipes*	Cattle	Africa, Asia?
B. caballi	*H. anatolicum, H. marginatum, H. truncatum*	Horses, donkeys	Africa, America, Asia, Australia, Europe

is brought about by the continued multiplication of merozoites invading new erythrocytes leading to the destruction, anaemia and jaundice. Comprehensive reviews on the life cycle of *Theileria* were given by Mehlhorn and Schein (1984, 1993) and *Babesia* by Uilenberg (2006). In *Babesia*, the pathogenicity is similarly due to the multiplication of merozoites and the invasion of new erythrocytes.

Of all *Theileria* species infecting livestock, the three lympho-proliferative species *T. parva* and *T. annulata* of cattle and *T. lestoquardi* of sheep and goats are the most economically important (Table 8.4). Two of these, namely *T. annulata* and *T. lestoquardi*, are exclusively transmitted by *Hyalomma* ticks, making the genus the most important *Theileria* transmitter. Besides, *Hyalomma* ticks are incriminated in the transmission of the only *Theileria* species infecting equine, *T. equi*. As long as *Babesia* is concerned, several species are also transmitted by *Hyalomma* including *B. beliceri* and *B. occultans* of cattle and *B. caballi* of equids.

8.4.3.1 Tropical Theileriosis

Tropical theileriosis, also known as tropical piroplasmosis and Mediterranean fever is a disease of cattle caused by *T. annulata* and transmitted by several *Hyalomma* species. The disease is widespread, occurring in southern Europe and extending to southern Russia, central Asia, the Middle East and India. In Africa, it occurs in the northern Littoral, Sudan and Eritrea (Uilenberg 1981) and has been reported in Mauritania (Jacquiet et al. 1990). Tropical theileriosis is of high economic importance and an overall number of 250 million cattle are estimated to be at risk. Natural host of *T. annulata* is the Asian water buffalo (*Bubalus bubulis*), in which it generally causes mild infections, though sever theileriosis due to *T. annulata* in water buffalo has been recently noted (Mahmmod et al. 2011). The parasite is highly pathogenic to cattle especially taurine breeds. Exotic wild animals such as the American bison (*Bison bison*) and the Tibetan Yak (*Bos grunniens*) were also occasionally found to be susceptible (Robinson 1982).

The transmission of *T. annulata* by *Hyalomma* is only trans-stadial (Uilenberg 1981), meaning that ticks become infected at the larval or nymphal developmental stage and infect a susceptible host at nymphal or adult stage. The main field vectors of *T. annulata* are *H. anatolicum*, *H. scupense* (= *H. detritum*) and *H. asiaticum* (Uilenberg 1981). In countries such as Mauritania, where neither *H. scupense* (= *H. detritum*) nor *H. anatolicum* occurs, *H. dromedarii* was found to be the main field vector (Jacquiet et al. 1990). Other *Hyalomma* species can transmit the disease experimentally, for instance, *H. impeltatum*, *H. truncatum* and *H. rufipes*. However, they are not considered efficient field vectors since their immature stages do not feed on cattle (Mustafa et al. 1983, Jongejan et al. 1983). *H. scupense* (= *H. detritum scupense*), the one-host tick types, transmitted *T. annulata* experimentally (Uilenberg 1981). Successful parasite transmission depends on several factors including the level of tick infestation and the tick infection rate. In Sudan, where *H. anatolicum* is the main field vector of *T. annulata*, natural infection rates in *H. anatolicum* with *Theileria* could reach as high as 86% (Walker et al. 1983).

As in the case with lympho-proliferative *Theileria*, the pathogenicity of *T. annulata* is attributed to the ability of the schizont-infected cells to multiply uncontrollably and metastasize (Ahmed et al. 1999). The course of infection is generally influenced by the dose of the parasite, the virulence of the strain and the innate resistance or susceptibility of the host (Pipano et al. 1974; Preston et al. 1992; Bakheit and Latif 2002; Glass and Jensen 2007; Ahmed et al. 2008; Seitzer and Ahmed 2008). The incubation period, between tick attachment and onset of fever, varies between 5 and 31 days (Gill et al. 1977). The infection is usually manifested by the swelling of the lymph nodes, draining the site of the tick bite followed by generalized lymphadenopathy (Uilenberg 1981; Boulter and Hall 1999). Symptoms associated with febrile infectious disease are usually manifested and death can occur within 3 weeks after the onset of clinical signs. These include anorexia, rapid loss of weight and condition, listlessness, lacrimation, mucous discharge nostrils, diarrhoea, dyspnea and abortion. In peracute cases, the animal may die as early as 3 days after the onset of clinical illness. In local cattle in endemic areas, the disease is usually mild and may pass unnoticed. Recovered animals usually become carriers and recovery is usually followed by solid immunity to both the homologous and heterologous isolates (Preston et al. 1999; Ahmed and Mehlhorn 1999).

8.4.3.2 Malignant Theileriosis of Sheep and Goats

This is a severe form of theileriosis that infects small ruminants caused by *T. lestoquardi*. The parasite resembles *T. annulata* in many aspects and has been speculated that both parasites have descended from a direct common ancestor (Katzer et al. 1998). Similar to *T. annulata*, the main field vector for *T. lestoquardi* is *H. anatolicum* and it has been associated with outbreaks of theileriosis in sheep and goats (Hooshmand-Rad and Hawa 1973; Tageldin et al. 1992; Latif et al. 1994;

Taha and Elhussein 2010; Taha et al. 2011). Infectivity of *H. anatolicum* with *T. lestoquardi* sporozoites has been demonstrated to peak between 2 and 4 days of tick feeding (Kirvar et al. 1998). Although the disease shares the same vector with tropical theileriosis, it seems to be less distributed. While it poses threats to animal production in countries like Sudan, where prevalence rate of up to 23.4% has been recorded (Salih et al. 2003), it has not been reported in countries such as Turkey and Israel where tropical theileriosis is well established.

Strains of *T. lestoquardi* are usually highly pathogenic to sheep. Heavy losses have been recorded even in indigenous sheep breeds (Hooshmand-Rad 1974; Tageldin et al. 1992) and mortalities vary but could reach nearly 100% during outbreaks (Latif et al. 1994; El Ghali and El Hussein 1995; Friedhoff 1997). Generally, goats show significant resistance to the disease compared to sheep; however, an outbreak involving mortality of over 70% in goats has recently been reported (Taha et al. 2011). Symptoms of malignant theileriosis are generally similar to those of tropical theileriosis and animals that survive the infection are immune to challenge.

8.4.3.3 Equine Theileriosis

Equine theileriosis is one of the major constraints for horse trade (Friedhoff et al. 1990). The disease is caused by *T. equi* (Mehlhorn and Schein 1998), and it is the only species known to infect equids. It is transmitted by two species of *Hyalomma*, namely *H. excavatum* and *H. plumbeum*. Other ixodid ticks reported to transmit *T. equi* include *Dermacentor* and *Rhipicephalus*, e.g. in South Africa the most important vector is *R. evertsi*. Equine theileriosis is distributed worldwide throughout most of the tropical and subtropical regions. *T. equi* is a non-lympho-proliferative species, and pathogenicity is related to the destruction of erythrocytes, leading to acute haemolytic anaemia in susceptible horses, abortion in pregnant mares and the occurrence of infections of the foetus in utero. This is usually followed by a chronic carrier state, which results in decreased performance of racehorses (Hailat et al. 1997).

8.4.3.4 Bovine Babesiosis

Hyalomma ticks are also incriminated in the transmission of *Babesia* species infecting cattle. *B. occultans* was recently described in South Africa (Gray and De Vos 1981) and was found to be transmitted to bovines by *H. rufipes*. Piroplasms formed by this species were morphologically similar to bovine *Babesia* spp. other than *B. divergens*, and merozoites in tick haemolymph morphologically resemble those of *B. bigemina*, though significantly larger. The species is highly infective to *H. rufipes* and the transmission is trans-ovarian. *B. occultans* is characterized by its low pathogenicity compared to the major species infecting bovine, *B. bigemina*, *B. divergens*, *B. bovis* and *B. major*.

B. beliceri is another species described in Russia and transmitted by *Hyalomma* ticks. More recently, another *Babesia* species of cattle similar to *B. occultans* and *B. beliceri* was described in China, which also has *Hyalomma* as vectors (Luo et al. 2002). Nonetheless, it is still pending to verify the identity of *B. beliceri* and the Chinese *Babesia* species as they could well be the same species as *B. occultans* (Uilenberg 2006).

References

Ahmed JS, Mehlhorn H (1999) the cellular basis of the immunity to and immunopathogenesis of tropical theileriosis. Parasitol Res 85(7):539–549

Ahmed JS, Schnittger L, Mehlhorn H (1999) Theileria schizonts induce fundamental alterations in their host cells. Parasitol Res 85(7):527–538

Ahmed JS, Glass EJ, Salih DA, Seitzer U (2008) Innate immunity to tropical theileriosis. Innate Immun 14(1):5–12

Al-Khalifa MS, Diab FM, Khalil GM (2007) Man-threatening viruses isolated from ticks in Saudi Arabia. Saudi Med J 28:1864–1867, Review

Apanaskevich DA, Horak IG (2005) The genus *Hyalomma* Koch, 1844. 11. Taxonomic status of *H.* (*Euhyalomma*) *anatolicum* Koch, 1844 and *H.* (*E.*) *excavatum* Koch, 1844 (Acari: Ixodidae) with redescription of all stages. Acarina 13:181–197

Apanaskevich DA, Horak IG (2006) The genus *Hyalomma* Koch, 1844. 1. Reinstatement of *Halomma* (*Euhyalomma*) *glabrum* Delpy, 1949 (Acari: Ixodidae) as a valid species with a redescription of the adults, the first description of its immature stages and notes on its biology. Onderstepoort J Vet 73:1–12

Apanaskevich DA, Horak IG (2007) The genus *Hyalomma* Koch, 1844. 111. Redescription of the adults and larvae of *H.* (*Euhyalomma*) *impressum* Koch, 1844 ((Acari: Ixodidae) with a first description of its nymph and notes on its biology. Folia Parasit 54:51–58

Apanaskevich DA, Horak IG (2008a) The genus *Hyalomma* Koch, 1844.V. Re-evaluation of the taxonomic rank of taxa comprising the *H.* (*Euhyalomma*) *marginatum* Koch complex of species (Acari: Ixodidae) with redescription of all parasitic stages and notes on biology. Int J Acarol 34:13–42

Apanaskevich DA, Horak IG (2008b) The genus *Hyalomma*. VI. Systematics of *H.* (*Euhyalomma*) *truncatum* and the closely related species, *H.* (*E.*) *albiparmatum* and *H.* (*E.*) *nitidum* (Acari: Ixodidae). Exp Appl Acarol 44:115–136

Apanaskevich DA, Horak IG (2009) The genus *Hyalomma* Koch, 1844. IX. Redescription of all parasitic stages of *H.* (*Euhyalomma*) *impeltatum* Schulze and Schlottke, 1930 and *H.* (*E.*) *somalicum* Tonelli Rondelli, 1935 (Acari: Ixodidae). Syst Parasitol 73:199–218

Apanaskevich DA, Horak IG (2010) The genus *Hyalomma* Koch, 1844. X1. Redescription of all parasitic stages of *H.* (*Euhyalomma*) *asiaticum* (Acari: Ixodidae) and notes on its biology. Exp Appl Acarol 52:207–220

Apanaskevich DA, Flippova NA, Horak IG (2010) The genus *Hyalomma* Koch, 1844. X. Redescription of all parasitic stages of *H.* (*Euhyalomma*) *scupense* Schulze, 1919 (= *H. detritum* Schulze) (Acari: Ixodidae) and notes on its biology. Folia Parasit 57:69–78

Bakheit MA, Latif AA (2002) The innate resistance of Kenana cattle to tropical theileriosis (*Theileria annulata* infection) in the Sudan. Ann N Y Acad Sci 969:159–163

Bakirc S, Sarali H, Aydin L, Latif A, Eren H, Karagenc T (2011) *Hyalomma rufipes* (Koch, 1844) infesting cattle in the West Aegean region of Turkey. Turk J Vet Anim Sci 35:359–363

Barker SC, Murrell A (2004) Systematics and evolution of ticks with a list of valid genus and species names. Parasitology 129:S15–S36

Beati L, Kelly PJ, Matthewman LA, Mason PR, Raoult D (1995) Prevalence of rickettsia-like organisms and spotted fever group rickettsiae in ticks (Acari: Ixodidae) from Zimbabwe. J Med Entomol 32:787–792

Berezin VV (1971). Investigation of the ecology of arboviruses in river deltas of the Caspian and Azov sea basins. (Avtoref. Diss. Soisk. Uchen. Step. Dokt. Biol. Nauk). Inst. Polio. Virusn. Entsefalitov, Akad. Med. Nauk SSSR, Moscow p 37 (In Russian). (In English, NAMRU3-T1160)

Boulter N, Hall R (1999) Immunity and vaccine development in the bovine theilerioses. Adv Parasitol 44:41–97

Brayton KA, Dark MJ, Palmer GH (2009) Anaplasma. In: Kole C, Nene VM (eds) Genome mapping in animal-associated microbes. Springer, Heidelberg, pp 85–116

Camicas JL, Cornet JP, Gonzalez JP, Wilson ML, Adam F, Zeller HG (1994) Crimean-Congo hemorrhagic fever in Senegal. Latest data on the ecology of the CCHF virus. Bull Soc Pathol Exot 87:11–16

Camicas JL, Hervy JP, Adam F, Morel PC (1998) Les tiques du monde. Nomenclature, stades décrits, hôtes, répartition (Acarida, Ixodida). Orstom, Paris

Chochlakis D, Ioannou I, Sandalakis V, Dimitriou T, Kassinis N, Papadopoulos B, Tselentis Y, Psaroulaki A (2011) Spotted fever group rickettsiae in ticks in cyprus. Microb Ecol 63:314–323

Dohm DJ, Logan TM, Linthicum KJ, Rossi CA, Turell MJ (1996) Transmission of Crimean-Congo hemorrhagic fever virus by *Hyalomma impeltatum* (Acari:Ixodidae) after experimental infection. J Med Entomol 33:848–851

El Ghali AA, El Hussein AM (1995) Diseases of livestock in Ed-Damer province: A two-years retrospective study. Sudan J Vet Sci Anim Hus 34:37–45

Emelianova IN (2006). *Hyalomma* Koch, 1844 (ACARI: IXODIDAE) ticks of Central Precaucasia and surrounding territories. (Distribution, ecology, role in the natural foci of Crimean-Congo haemorrhagic fever). MSc thesis in Biology. Stavropol: State University of Stavropol, p 146

Ergönül Ö (2006) Crimean-Congo haemorrhagic fever. Lancet Infect Dis 6:203–314

Estrada-Pena A, Bouattour A, Camicas JL, Walker AR (2004) Ticks of domestic animals in the Mediterranean Region. Bioscience Reports, 42 Commission Drive, Edinburgh EH10 5QR, Scotland, p 131

Estrada-Pena A, Vatansever Z, Gargili A, Aktas M, Uzun R, Ergönül Ö, Jongejan F (2007) Modeling the spatial distribution of Crimean-Congo hemorrhagic fever outbreaks in Turkey. Vector-Borne Zoonot 7:667–678

Formosinho P, Santos-Silva MM (2006) Experimental infection of *Hyalomma marginatum* ticks with West Nile virus. Acta Virol 50:175–180

Frese M, Weeber M, Weber F, Speth V, Haller O (1997) Mx1 sensitivity: Batken virus is an orthomyxovirus closely related to Dhori virus. J Gen Virol 78(Pt 10):2453–2458

Friedhoff KT (1997) Tick-borne diseases of sheep and goats caused by *Babesia*, *Theileria* or *Anaplasma* spp. Parassitologia 39:99–109

Friedhoff KT, Tenter AM, Muller I (1990) Haemoparasites of equines: impact on international trade of horses. Rev Sci Tech 9:1187–1194

Gill BS, Bhattacharyulu Y, Kaur D (1977) Symptoms and pathology of experimental bovine tropical theileriosis (*Theileria annulata* infection). Ann Parasitol Hum Comp 52:597–608

Glass EJ, Jensen K (2007) Resistance and susceptibility to a protozoan parasite of cattle-gene expression differences in macrophages from different breeds of cattle. Vet Immunol Immunopathol 120:20–30

Gray JS, De Vos AJ (1981) Studies on a bovine *Babesia* transmitted by *Hyalomma marginatum rufipes* Koch. Onderstepoort J Vet Res 48:215–223

Guglielmone AA, Robbins RG, Apanaskevich DA, Petney TN, Estrada-Peña A, Horak IG (2009) Comments on controversial tick (Acari: Ixodida) species names and species described or resurrected from 2003 to 2008. Exp Appl Acarol 48:311–327

Guglielmone AA, Robbins RG, Apanaskevich DA, Petney TN, Estrada-Pena A, Horak IG, Shao R, Barker SC (2010) The Argasids, Ixodidae and nuttalliellidae (Acari: Ixodida) of the world: a list of valid species names. Zootaxa 2528:1–28

Guner ES, Hashimoto N, Kadosaka T, Imai Y, Masuzawa T (2003) A novel, fast-growing *Borrelia* sp. isolated from the hard tick *Hyalomma aegyptium* in Turkey. Microbiology 149:2539–2544

Guner ES, Watanabe M, Hashimoto N, Kadosaka T, Kawamura Y, Ezaki T et al (2004) *Borrelia turcica* sp. nov., isolated from the hard tick *Hyalomma aegyptium* in Turkey. Int J Syst Evol Microbiol 54:1649–1652

Hailat NQ, Lafi SQ, al-Darraji AM, al-Ani FK (1997) Equine babesiosis associated with strenuous exercise: clinical and pathological studies in Jordan. Vet Parasitol 69(1–2):1–8

Hirai K, To H (1998) Advances in the understanding of *Coxiella burnetii* infection in Japan. J Vet Med Sci 60:781–790

Hoogstraal H (1956) African Ixodoidea, Volume 1, Ticks of the Sudan. Research Report NM 005 050.29.07 of the Vaval Medical Research Unit 3, Cairo. [Identification and biology of many of the important species of African ticks] p 1101

Hoogstraal H (1979) The epidemiology of tick-borne Crimean-Congo hemorrhagic fever in Asia, Europe, and Africa. J Med Entomol 15:307–417

Hoogstraal H (1985) Argasids and Nuttallielid ticks as parasites and vectors. Adv Parasit 24:135–238

Hooshmand-Rad P (1974) Blood protozoan diseases of ruminants. Bull Office Int des épizooties 81:779–792

Hooshmand-Rad P, Hawa NJ (1973) Transmission of *Theileria hirci* in sheep by *Hyalomma anatolicum anatolicum*. Trop Anim Health Pro 5:103

Horak IG, Camicas JL, Keirans JE (2002) The Argasidae, Iodidae and Nuttalliellidae (Acari: Ixodida): a valid list of tick names. Exp Appl Acarol 28:27–54

Hubálek Z (2009) Biogeography of tick-borne bhanja virus (bunyaviridae) in Europe. Interdiscip Perspect Infect Dis 2009:372691

Jacquiet P, Dia ML, Perie NM, Jongejan F, Uilenberg G, Morel PC (1990) Presence of *Theileria annulata* in Mauritania. Rev Elev Med Vet Pays Trop 43(4):489–490

Jongejan F, Morzaria SP, Mustaffa UE, Latif AA (1983) Infection rates of *Theileria annulata* in the salivary glands of the tick *Hyalomma marginatum rufipes*. Vet Parasitol 13:121–126

Katzer F, McKellar S, Kirvar E, Shiels B (1998) Phylogenetic analysis of *Theileria* and *Babesia equi* in relation to the establishment of parasite populations within novel host species and the development of diagnostic tests. Mol Biochem Parasitol 95:33–44

Keysary A, Eremeeva ME, Leitner M, Din AB, Wikswo ME, Mumcuoglu KY, Inbar M, Wallach AD, Shanas U, King R, Waner T (2011) Spotted Fever group rickettsiae in ticks collected from wild animals in Israel. Am J Trop Med Hyg 85:919–923

Kirvar E, Wilkie G, Katzer F, Brown CGD (1998) *Theileria lestoquardi*—maturation and quantification in *Hyalomma anatolicum anatolicum* ticks. Parasitology 117:255–263

Korsunova OS, Petrova-Pointkovskaya SP (1949) A virus isolated from *Hyalomma marginatmn marginatum* Koch, ticks. Zool Zh 28:186–187, Translation 793 (T793), Medical Zoology Department, US NAMRU, Cairo, Egypt

Labuda M, Nuttall PA (2004) Tick-borne viruses. Parasitology 129(Suppl):S221–S245, Review

Latif AA, Abdulla HM, Hassan SM, Zubeir N, Morzria SP, Osman AM, Mustafa UE (1994) Theileriosis of sheep in the Sudan. In: Tropical theileriosis in the Sudan. In: Proceedings of a Workshop Held at the Sudan Veterinary Association Residence, Khartoum, Sudan, 4–5 May, 1994. Ed. A.M. Atelmanan and S.M. Kheir pp 66–72

Levine ND (1988) The protozoan phylum Apicomplexa, vol II. CRC, Boca Raton, p 154

Linthicum KJ, Logan TM, Bailey CL, Dohm DJ, Moulton JR (1989) Transstadial and horizontal transmission of Rift Valley fever virus in *H. truncatum*. Am J Trop Med Hyg 41(4):491–496

Logan TM, Linthicum KJ, Bailey CL, Watts DM, Moulton JR (1989) Experimental transmission of Crimean-Congo hemorrhagic fever virus by *Hyalomma truncatum* Koch. Am J Trop Med Hyg 40:207–212

Luo J, Yin H, Guan G, Zhang Q, Lu W (2002) Description of a new Babesia sp. infective for cattle in China. Parasitol Res 88(suppl 1):S13–S15

Lvov DK, Karas FR, Tsyrkin YM, Vargina SG, Timofeev EM, Osipova NZ, Veselovskaya OV, Grebenyuk YI, Gromashevski VL, Fomina KB (1974) Batken virus, a new arbovirus isolated from ticks and mosquitoes in Kirghiz S.S.R. Arch Gesamte Virusforsch 44:70–73

Mahmmod YS, Elbalkemy FA, Klaas IC, Elmekkawy MF, Monazie AM (2011) Clinical and haematological study on water buffaloes (*Bubalus bubalis*) and crossbred cattle naturally infected with *Theileria annulata* in Sharkia province, Egypt. Ticks Tick Borne Dis 2:168–171

Mans BJ (2011) Evolution of vertebrate hemostatic and inflammatory control mechanisms in blood-feeding arthropods. J Innate Immun 3:41–51

Marrie TJ, Raoult D (1997) Q fever-a review and issues for the next century. Int J Antimicrob Agents 8:145–161

Mediannikov O, Diatta G, Fenollar F, Sokhna C, Trape JF, Raoult D (2010) Tick-borne rickettsioses, neglected emerging diseases in rural Senegal. PLoS Negl Trop Dis 4(9):e821

Mehlhorn H, Schein E (1984) The piroplasms: life cycle and sexual stages. Adv Parasitol 23:37–103

Mehlhorn H, Schein E (1993) The piroplasms: "A Long Story in Short" or "Robert Koch has seen it". Eur J Protistol 29:279–293

Mehlhorn H, Schein E (1998) Re-description of *Babesia equi* Laveran, 1901 as *Theileria equi* Mehlhorn, Schein 1998. Parasitol Res 84:467–475

Mustafa UE, Jongejan F, Morzaria SB (1983) Note on the transmission of *Theileria annulata* by *Hyalomma* ticks in the Sudan. Vet Quart 5:112–113

Nalça A, Whitehouse CA (2007) Crimean-Congo hemorrhagic fever virus infection among animals. In: Ergönül Ö, Whitehouse CA (eds) Crimean-Congo hemorrhagic fever a global perspective, 1st edn. Springer, Dordrecht, pp 155–165

Nava S, Guglielmone AA, Mangold AJ (2009) An overview of systematics and evolution of ticks. Front Biosci 14:2857–2877

Ouhelli H (1994) Comparative development of *Hyalomma marginatum* (Koch, 1844), *H. detritum* (Schulze, 1919), H. *anatolicum excavatum* (Koch, 1844), *H. lusitanicum* (Koch, 1844) and *H. dromedarii* (Koch, 1844) under laboratory conditions. Acta Parasitologica 39:153–157

Petrova-Piontkovskaya SP (1947) Biological and ecological data on the tick *Hyalomma marginatum* Koch in the northwestern Crimean hemorrhagic fever focus. Nov Med 5:21–24, Translation 864 (T864), Medical Zoology Department, US NAMRU, Cairo, Egypt

Pipano E, Wisman Y, Benado A (1974) The virulence of four local strains of *Theileria annulata*. Refu Vet 31(2):59–63

Pomerantsev BI (1950) [Ixodid ticks (Ixodidae)]. Fauna of the USSR, Paukoobraznyye. new series 41, 4(2):224

Preston PM, Brown CG, Bell-Sakyi L, Richardson W, Sanderson A (1992) Tropical theileriosis in *Bos Taurus* and *Bos tarus* x *Bos indicus* calves. Response to infection with graded doses of sporozoites. Res Vet Sci 53:230–243

Preston PM, Hall FR, Glass EJ, Campbell JD, Darghouth MA, Ahmed JS, Shiels BR, Spooner RL, Jongejan F, Brown CG (1999) Innate and adaptive immune responses co-operate to protect cattle against *Theileria annulata*. Parasitol Today 15:268–274

Psaroulaki A, Germanakis A, Gikas A, Scoulica E, Tselentis Y (2005) Simultaneous detection of "*Rickettsia mongolotimonae*" in a patient and in a tick in Greece. J Clin Microbiol 43:3558–3559

Reusse U (1960) Die Bedeutung des Q-Fiebers als Zoonose. Z Tropenmed Parasitol 11:223–262

Robinson PM (1982) *Theileria annulata* and its transmission—a review. Trop Anim Health Prod 14:3–12

Salih DA, El Hussein AM, Hayat M, Taha KM (2003) Survey of *Theileria lestoquardi* antibodies among Sudanese sheep. Vet Parasitol 111:361–367

Sane J, Kurkela S, Levanov L, Nikkari S, Vaheri A, Vapalahti O (2011) Development and evaluation of a real-time RT-PCR assay for Sindbis virus detection. J Virol Methods 179:185–188

Seitzer U, Ahmed J (2008) Tropical theileriosis: cytotoxic T lymphocyte response to vaccination. Vaccine 26(Suppl 6):G24–G28

Siroký P, Kubelová M, Modrý D, Erhart J, Literák I, Spitalská E, Kocianová E (2010) Tortoise tick *Hyalomma aegyptium* as long term carrier of Q fever agent *Coxiella burnetii*-evidence from experimental infection. Parasitol Res 107:1515–1520

Tageldin MH, Zakia AM, Nagwa EG, ElSawi SAS (1992) An outbreak of theileriosis in sheep in Sudan. Trop Anim Health Prod 24:15–16

Taha KM, Elhussein AM (2010) Experimental transmission of *Theileria lestoquardi* by developmental stages of *Hyalomma anatolicum* ticks. Parasitol Res 107:1009–1012

Taha KM, Salih DA, Ahmed BM, Enan KA, Ali AM, Elhussein AM (2011) First confirmed report of outbreak of malignant ovine theileriosis among goats in Sudan. Parasitol Res 109:1525–1527

Turell M (2007) Role of ticks in the transmission of Crimean-Congo hemorrhagic fever virus. In: Ergönül Ö, Whitehouse CA (eds) Crimean-Congo hemorrhagic fever a global perspective, 1st edn. Springer, Dordrecht, pp 143–154

Uilenberg G (1981) *Theileria* infections other than East Cost Fever. In: Ristic M, McIntyre I (eds) Diseases of cattle in the tropics. Martinus Nijhoff Publishers, The Hague, pp 411–427

Uilenberg G (2006) *Babesia*—a historical overview. Vet Parasitol 138:3–10

Vatansever Z, Uzun R, Estrada-Pena A, Ergönül Ö (2007) Crimean-Congo hemorrhagic fever in Turkey. In: Ergönül Ö, Whitehouse CA (eds) Crimean-Congo hemorrhagic fever A global perspective, 1st edn. Springer, Dordrecht, pp 59–74

Vatansever Z, Midilli K, Ergin S, Ozdarendeli A, Aktaş M, Gargili A (2010) The prevalence of Crimean-Congo haemorrhagic fever virus in the host seeking ticks in an endemic area of Turkey. INTERNATIONAL CONFERENCE EDEN, Emerging Vector-borne Diseases in a Changing European Environment. Le Corum, Montpellier, France

Walker AR, Latif AA, Morzaria SP, Jongajan F (1983) Natural infection rates of *Hyalomma anatolicum anatolicum* with *Theileria* in the Sudan. Res Vet Sci 35:87–90

Walker AR, Bouattour A, Camicas JL, Estrada-Pena A, Horak IG, Latif AA, Pegram RG, Preston PM (2003) Ticks of domestic animals in Africa. Bioscience Reports, 42 Commission Drive, Edinburgh EH10 5QR, Scotland, UK p 221

Weisburg WG, Dobson ME, Samuel JE, Dasch GA, Mallavia LP, Baca O, Mandelco L, Sechrest JE, Weiss E, Woese CR (1989) Phylogenetic diversity of the Rickettsiae. J Bact 171:4202–4206

Whitehouse CA (2004) Crimean-Congo hemorrhagic fever. Antiviral Res 64:145–160

Williams RE, Hoogstraal H, Casals J, Kaiser MN, Moussa MIJ, Entomol M (1973) Isolation of Wanowrie Thogoto, and Dhori viruses from Hyalomma ticks infesting camels in Egypt. J Med Entomol 10:143–146

Zeller HG, Cornet JP, Camicas JL (1994) Experimental transmission of Crimean-Congo hemorrhagic fever virus by west African wild ground-feeding birds to *Hyalomma marginatum rufipes* ticks. Am J Trop Med Hyg 50:676–681

Chapter 9
Flies as Vectors of Microorganisms Potentially Inducing Severe Diseases in Humans and Animals

Falk Gestmann, Maike Förster, Heinz Mehlhorn, Kai Sievert, Sabine Messler, Nicole Neuhausen, Sabine Petersdorf, and Klaus Pfeffer

Abstract This chapter reviews published and recent results on the detection of microorganisms on the surface or in the intestine of a variety of different fly species that may often occur in huge numbers in and around humans at rural and at urban sites. It is surely alarming when severe agents of diseases such as enterohemorrhagic *Escherichia coli*, enteropathogenic *E. coli*, enterotoxic *E. coli*, enteroaggregative *E. coli*, *Klebsiella*, *Campylobacter*, *Providencia*, *Staphylococcus aureus*, *Streptococcus viridans*, and *Candida albicans* and *Mucor* and *Aspergillus* spp. are found on the bodies of flies or within their feces, which may contaminate food or wounds of humans and/or animals. The arising dangers have to be intensively considered and evaluated. Furthermore, fly control measures have to be strengthened and research to find long-lasting, nonpoisonous insecticides has to be increased. Otherwise flies will take the leadership as vectors of several emerging diseases, which then will have an enormous killing or at least disease-causing potential in animal production and in current megatowns and in those of the future.

Keywords Aspergillus • Bacteria • Campylobacter • Candida • Flies • Fungi • Klebsiella • Staphylococcus • Vectors of diseases

F. Gestmann • M. Förster • H. Mehlhorn (✉)
Department of Zoology and Parasitology, Heinrich Heine University, Universitätsstr. 1, 40225 Düsseldorf, Germany
e-mail: mehlhorn@uni-duesseldorf.de

K. Sievert
Novartis Animal Health Inc., WRO-1032.3.82, Schwarzwaldallee 215, 4058 Basel, Switzerland
e-mail: kai.sievert@novartis.com

S. Messler • N. Neuhausen • S. Petersdorf • K. Pfeffer
Medical Microbiology and Hospital Hygiene, Heinrich Heine University, Universitätsstr. 1, 40225,
Düsseldorf, Germany
e-mail: klaus.pfeffer@hhu.de

H. Mehlhorn (ed.), *Arthropods as Vectors of Emerging Diseases*,
Parasitology Research Monographs 3, DOI 10.1007/978-3-642-28842-5_9,

9.1 Introduction

The number of food-borne diseases caused by bacterial pathogens or by parasites is constantly increasing worldwide although high standards of hygiene have been reached in European countries and in North America. However, mass production of food, globalization of food transport (including transport of plants and meat), concentration of humans in overcrowded towns, and meal production in short times in restaurants, canteens, etc., have created more and more chances for viruses, bacteria, fungi, and parasites to enter the human or animal food chains at different steps. Thus, there are infections of humans and animals which lead to serious diseases. The sources of such diseases—often occurring suddenly locally and then becoming larger epidemics (such as the outbreak of enterohemorrhagic *Escherichia coli*, EHEC, O104:H4 in the summer 2011 in northern Germany, in France, and even in some countries far away; Bielaszewska et al. 2011)—are mostly not found. The reason is that the infection might be imported by a nonsymptomatic person or animal, may be carried by further hosts without symptoms, and starts its disease-causing career often weeks after its first occurrence in a certain environment. Another possibility is that an agent of disease has been harbored for a long time by a symptomless reservoir host that might be an animal which lives close to a human population. Agents of severe diseases may also be introduced into a human community or into a rearing facility for animals by arthropod vectors. Those belonging to the insect family of flies often feed on or breed in human and/or animal feces and thus have intense contact with agents of diseases; therefore, it is not astonishing that many cases of fly-transmitted diseases have been reported in the last 40 years (Greenberg 1971, 1973). Some of these diseases are listed in Tables 9.1 and 9.2. The pathogens also listed there have enormous potential to induce severe and wide-reaching outbreaks of food-borne diseases. The WHO (2002a, b) calculated that up to 30% of the total number of cases of intestinal disease each year are for people who fall ill from food-borne disease (Abbott 2002; Ako-nai et al. 1991; Alam and Zurek 2004; Altekrues et al. 1997; Aspöck 2010; Banjo et al. 2005; Barrow and Feltham 2004; Beutin 1990; Bush et al. 1997; Chinery 2002; De Jesus et al. 2004; Ekdahl et al. 2005; Fetene and Worku 2009; Förster et al. 2012; Garmendia et al. 2005; Getachew et al. 2007; Greenberg et al. 1963, 1970; Hald et al. 2008; Haupt and Haupt 1998; Holt et al. 2007; Karch 2005; Khan and Huq 1978; Kobayashi et al. 1999; Kollaritsch and Paulke-Korinek 2010; Löscher and Burchard 2010; Mehlhorn and Piekarski 2002; Mehlhorn et al. 2011; Mian and Jacal 2002; Neumeister et al. 2009; Nichols 2005; Ngern-Klun et al. 2007; RKI 2006; Sasaki et al. 2000; Schaefer 2002; Steinmöller et al. 2006; Umeche and Mandah 1989; Wagner 1989; Wright 1983; Yap et al. 2008; Zarrin et al. 2007; Zumpt 1949).

Thus, to obtain more recent data on the existing threat by microorganisms in industrial regions, the University of Düsseldorf initiated a long-term investigation (2006–2011) into the load of pathogens on flies close to stables/meadows used by chickens, pigs, cattle, horses, rabbits, and dogs and in addition checked the refinding rates after experimental exposure of flies to such agents of diseases (Förster et al. 2007, 2009; Förster 2009). In 2009, a Safeguard project

Table 9.1 Reports on fly-transmitted agents of human diseases in the literature (selection)

Agent of disease	Diseases	Type of the agents of disease	Authors
Polio	Poliomyelitis	Viruses	Greenberg (1965)
Shigella sp.	Dysentery	Bacteria	Levine and Levine (1991)
Vibrio cholerae	Cholera	Bacteria	Fotedar (2001)
EHEC O157:H7	Enteritis	Bacteria	Iwasa et al. (1999)
ETEC	Diarrhea	Bacteria	Jordi et al. (2000)
Campylobacter spp.	Diarrhea	Bacteria	Szalanski et al. (2004)
Helicobacter pylori	Enteritis	Bacteria	Grübel et al. (1997)
Salmonella spp.	Salmonellosis	Bacteria	Olsen and Hammack (2000)
Salmonella typhimurium	Typhus	Bacteria	Greenberg (1973)
Mycobacterium sp.	Tuberculosis	Bacteria	Fischer et al. (2001)
Staphylococcus aureus	Enteritis	Bacteria	Fotedar et al. (1992a, b)
Chlamydia trachomatis	Trachoma	Bacteria	Emerson et al. (2004)
Aspergillus, *Penicillium*	Multiorgan invasion	Fungi	Sales et al. (2002)
Cryptosporidium sp.	Diarrhea	Protozoa	Szostakowska et al. (2004)
Giardia lamblia	Giardiasis, diarrhea	Protozoa	Szostakowska et al. (2004)
Ancylostoma duodenale	Hookworm disease	Nematodes	Oyerinde (1976)

EHEC enterohemorrhagic *Escherichia coli*, *ETEC* enterotoxic *Escherichia coli*

(*Campylobacter*) was started by the governments of the Netherlands and North Rhine-Westphalia, which is the German state bordering the Netherlands. The aim of this project was to check the flies at selected downtown sites (with a huge number of restaurants) and recreation sites with respect to their load of agents of diseases in order to obtain basic data which may help to establish methods and better prophylaxis measures to interrupt transmission of agents of diseases. This chapter gives some of the results obtained with respect to bacteria and fungi which occurred on the surface of flies and/or in their intestine. Chapter 10 by Förster et al. (2012) deals with the transmission of different groups of parasites by various species of flies captured in the wild and reports from laboratory experiments showing the capacity of flies to transmit various parasites.

9.2 Methods and Targets

9.2.1 Fly Catching Close to Animal Houses

Flies were caught on warm and dry days (20–30°C) close to a dog meadow, poultry house, cattle barn, horse stable, pigpen, and rabbit cages in the surroundings of the small town of Dormagen near Düsseldorf in Germany. To avoid contamination by the tissues of typical insect catchers, the flies were first caught either with sterile vessels or with commercial sterile plastic bags and were immediately placed for 5–10 min in sterile petri dishes on blood agar. After this incubation, the flies were removed from the petri dishes, killed by CO_2, and identified under a dissecting

Table 9.2 Reports on fly-transmitted agents of animal diseases in the literature (selection)

Agent of disease	Diseases	Type of the agent of disease	Host	Authors
H5N1 and others	Bird flu	Virus	Birds/humans	
Newcastle virus	Newcastle disease	Virus	Birds	Bram et al. (2002)
Herpes virus	Aujeszky's disease	Virus	Pigs	Medveczky et al. (1988)
PRRSV	Respiratory syndrome	Virus	Pigs	Otake et al. (2004)
FMDV	Foot and mouth disease	Virus	Cattle, pigs	Greenberg (1973)
Bacillus anthracis	Anthrax	Bacteria	Cattle	Greenberg (1973)
Mycobacterium sp.	Tuberculosis	Bacteria	Cattle, pigs	Fischer et al. (2001)
Campylobacter jejuni	Diarrhea	Bacteria	Birds	Shane et al. (1985)
Streptococcus suis	Diarrhea	Bacteria	Pigs	Enright et al. (1987)
Corynebacterium pseudotuberculosum	Mastitis	Bacteria	Cattle	Yeruham et al. (1996)
Brucella species	Brucellosis	Bacteria	Cattle, pigs	Greenberg (1973)
Cryptosporidium species	Diarrhea	Protozoa	Birds, cattle, pigs	Graczyk et al. (1999)
Poultry tapeworm	Diarrhea	Cestodes	Birds	Abrams (1976)
Parafilaria species	Skin disease	Nematodes	Cattle	Bech-Nielsen et al. (1982)
Thelazia species	Eye worm disease	Nematodes	Cattle	O'Hara et al. (1989)

PRRSV Porcine reproductive and respiratory syndrome virus, *FMDV* foot-and-mouth disease virus

microscope. Later, larger numbers of flies were also caught with freshly washed butterfly nets. In these cases only the intestinal contents of these flies were examined.

9.2.2 Fly Catching at Recreation Sites

The catching sites at four selected recreation sites (often with crowds of people and where there were also numerous birds—geese, swans, ducks, and pigeons) were situated in larger towns or close by them. The following four sites were selected:

1. The Old Town of Düsseldorf (Germany), with its many restaurants. The catching sites were meadows and bushes in the surroundings.
2. A park in a suburb of Düsseldorf (at Castle Eller).
3. A bathing lake close to Düsseldorf, which is crossed by the river Düssel, which gives its name to Düsseldorf. In summer thousands of people bathe there.
4. The six-lakes-region close to the town of Duisburg (Germany), where the lakes are visited by huge numbers of people.

The flies were caught at intervals of 7–14 days (depending on the weather) with the help of butterfly nets, the opening of which had a diameter of 25 cm. The flies were transferred singly into sterile plastic vessels immediately after the catch.

The feces of the birds was collected in sterile, one-way plastic tubes and transported to the institutes for microbiological and/or parasitological investigations.

Fig. 9.1 Motile *Musca domestica* larvae on their meat substrate

9.2.3 Determination of Microorganisms

The blood agar plates onto which the freshly caught flies had been placed were incubated at 37°C for 24 h. After the incubation, the petri dishes were checked for growth of microorganisms. For identification, the microorganisms were isolated and subcultured on blood agar and selective agar plates, and were finally incubated at 37°C for another 24 h. For identification of fungi, the microorganisms were isolated and were subcultured on Saboraud's dextrose agar, and were then incubated at room temperature for 4 days. After the incubation period, the bacterial colonies were identified by standard bacteriological procedures (e.g., coagulase, oxidase test, Gram's reaction, API®, Prolex®, VITEK®, etc.). Isolated *E. coli* bacteria were identified for their pathogenicity by in-house real-time polymerase chain reaction, differentiating enterotoxic *E. coli*, enteropathogenic *E. coli* (EPEC), EHEC, and enteroaggregative *E. coli*.

9.3 The Flies

9.3.1 Flies Around the Stables

The flies that were found close to the stables differed considerably with respect to their species relations from those found around dog meadows (Figs. 9.1–9.5). These differences are surely based on their different needs for different suitable breeding sites. Tables 9.3 and 9.4 summarize the occurrence of the different fly species at the different catching sites.

Around the stables, *Musca domestica* was the absolutely dominant species. In all catches in 3 years it was always present in the same relations:

- 44% of the catches in the pigpens
- 30% of the catches in horse stables

Fig. 9.2 Adult *Musca domestica* fly

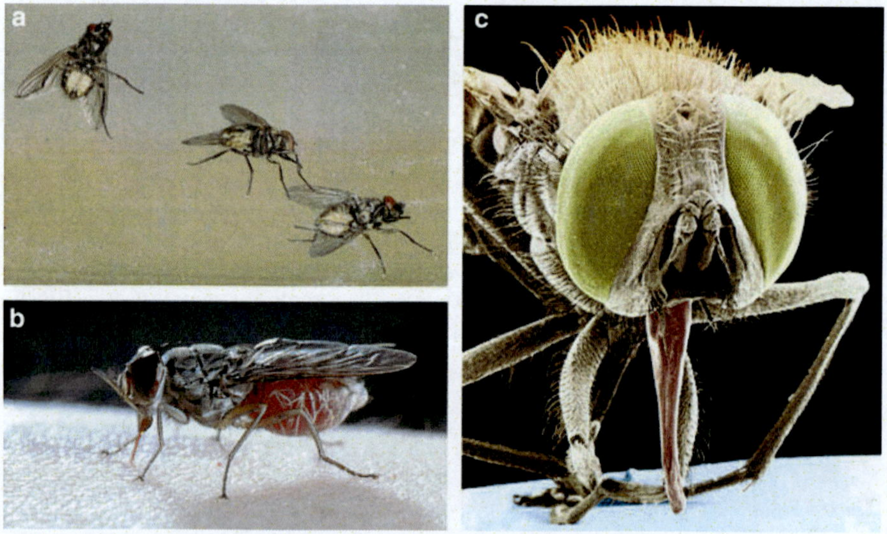

Fig. 9.3 *Stomoxys calcitrans*. (**a**, **b**) Macrophotographs of adults. (**c**) Scanning electron micrograph of the head with the piercing/sucking apparatus

- 17% of the catches in rabbit cages
- 9% of the catches in cattle barns

In several cases its occurrence in poultry houses reached approximately 90%, which coincided with several important and massive fly outbreaks which were investigated by our group:

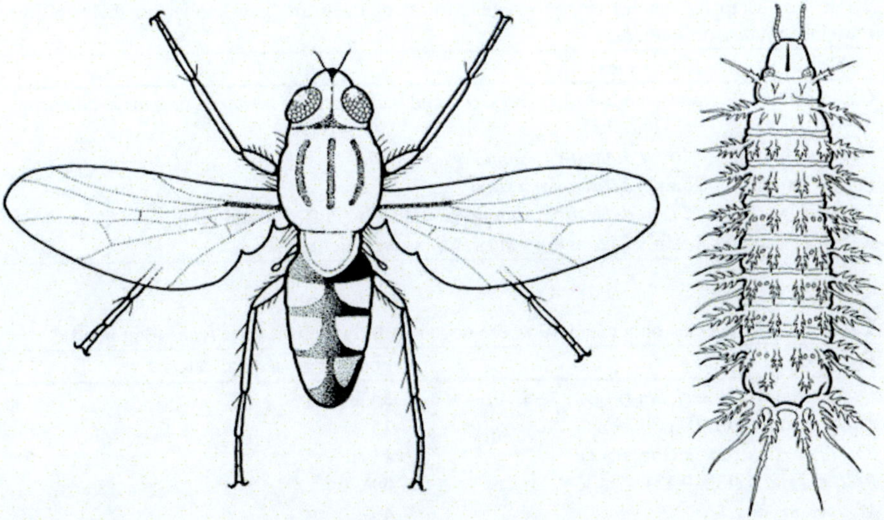

Fig. 9.4 *Fannia canicularis*, adult and larva

Fig. 9.5 *Fannia canicularis*, adult stage

(a) In a valley with large poultry houses and chicken feces stored in the open air
(b) After several field fertilizations with chicken feces close to villages
(c) In a small town with a compost plant, the units of which were opened nightly by feral pigs and in the early mornings by crows

In all these cases the private houses up to a distance of 1 km were filled with hundreds of *Musca domestica* adults, which were apparently attracted by the dark

Table 9.3 Main fly species close to stables (most of them are also known for their ability to introduce myiasis in humans)

Stable	Fly species
Cattle barns	*Musca domestica* (Figs. 9.1 and 9.2), *Polietes lardaria, Stomoxys calcitrans* (Fig. 9.3)
Horse stables	*Musca domestica, Stomoxys calcitrans*
Poultry houses	Mainly *Musca domestica*
Pigpens	*Musca domestica, Fannia canicularis* (Figs. 9.4 and 9.5)
Rabbit cages	Mainly *Musca domestica, Stomoxys calcitrans*

Table 9.4 Fly species on a meadow where dogs regularly deposit their feces (dog pond)

Fly species	Percentage of total catches
Lucilia caesar, L. sericata (Figs. 9.6–9.10)	54%
Muscina stabulans (Fig. 9.11)	15%
Sarcophaga carnaria (Fig. 9.12)	11%
Pollenia rudis (Fig. 9.13)	4%
Mydaea urbana	4%
Mydaea ancilla (Fig. 9.14)	2%
Muscina pabulorum (Fig. 9.15)	2%
Calliphora vomitoria (Fig.9.16)	2%
Fannia canicularis (Figs. 9.4 and 9.5)	2%
Calliphora vicina, Helina reversio (Fig. 9.17), *Phaonia errans* (Fig. 9.18), *Polietes lardaria*	Each species appeared as 1% of the catches

interior of the houses when arriving from the surrounding fields, from the compost plant, and/or from badly stored chicken feces.

Practically in all cases the sex ratio of the flies caught was nearly 50:50 independent of whether the flies were caught close to the stables (i.e., outdoors) or inside houses.

The bloodsucking species *Stomoxys calcitrans* was an exception: only 42% of the flies caught outside were females. This might be attributable to the behavior of the females to suck blood and then to lay their eggs inside stables.

Tables 9.5 and 9.6 provide information on the life cycles of several important fly species, which of course primarily attack farm animals, but, however, also molest humans by often numerous touchdowns. During these contacts, agents of diseases (see Tables 9.1 and 9.2) may be placed on humans and their food too. In these cases the flies could become vectors of zoonotic agents of diseases (see also Sect. 9.5 and Chap. 10).

Tables 9.5 and 9.6 in addition show the periods when the flies are most active at the in general moderate temperatures of central Europe. In warmer regions they are active for practically the whole year, so the potential transmission of agents of diseases may occur throughout the year. This makes it clear that only strict prophylaxis measures will protect animals and as a consequence also humans who live in the close neighborhood of animals.

Fig. 9.6 *Lucilia sericata*, eggs

Fig. 9.7 *Lucilia sericata*, larva

All flies caught in the surroundings of human dwellings and/or different animal houses or recreation sites are so-called synanthropic species (Sukopp and Wittig 1998). The definition of the term "synanthropy" characterizes plant or animal species which have adapted their life cycle to human housing areas. There they may reproduce in such high numbers that they do not need additions from outside to maintain their population at a constant level. In cases of an increase in food, many other individuals of this species might be attracted and this may lead to real plagues. Eusynanthropy and hemisynanthropy are subgroups of synanthropy. In the first case, the occurrence and

Fig. 9.8 *Lucilia sericata*, pupae

Fig. 9.9 *Lucilia sericata*, adult female

reproduction of a certain species is obligatory because of the co-occurrence with human houses. Members of the genus *Drosophila* are examples of flies in Europe. Examples of other insects are the bedbug (*Cimex lectularius*), the bread beetle (*Stegobium paniceum*), and imported cockroaches (*Blatta orientalis*, *Blattella germanica*, *Supella longipalpa*, *Periplaneta americana*), which can only exist in human houses, whereas forest cockroaches (e.g., *Ectobius silvestris*) enter human dwellings only occasionally. "Hemisynanthropy" (the facultative form) covers species which find an optimum amount of food and facilities for survival inside or around human dwellings, but

Fig. 9.10 *Lucilia caesar*, adult female and male

Fig. 9.11 *Muscina stabulans*, adult female

these species also survive easily outside human settlements. For example, *Lucilia caesar* is such an species as is the Eurasian swift (*Apus apus*), which likes to use human buildings as a starting point for its magnificent flights, although it may also start from natural rocks, etc.

With respect to these definitions of synanthropy, many flies are hemisynanthropic animals, which use the facilities of human settlements with all their garbage, superfluous food, and protected places for their—often immense—reproduction. They

Fig. 9.12 *Sarcophaga carnaria*, adult and pupa

Fig. 9.13 *Pollenia rudis*. These flies (6–12 mm) have *square silver dots* on their abdomen. They enter houses in autumn and winter

may, however, survive easily far away from human settlements, since dead animals, their carcasses, and their degrading feces offer practically unlimited food and propagation facilities. Therefore, even the best fly control measures cannot prevent the constant influx of flies from outside into human houses and into the stables of farmed animals. However, eusynanthropy and hemisynanthropy of most flies make it necessary for humans to control fly propagation by minimizing the breeding sites in and around human buildings of any kind. This can be done simply by reduction of garbage, removing remnant food, removing feces, changing straw, etc. In addition—especially in monocultures of many animals—chemical treatment of animals, of the stable walls, and/or of the straw might become necessary to prevent feeding farm animals from being disturbed or even to avoid infections with agents of diseases as listed in Tables 9.1 and 9.2.

Fig. 9.14 *Mydaea ancilla*, adult female. This species belongs to the family Muscidae

Fig. 9.15 *Muscina pabulorum*, adult female

Some species (e.g., *Calliphora vomitoria*, *Calliphora vicina*, *Fannia canicularis*, *Lucilia caesar*, *Musca domestica*, *Muscina stabulans*, and *Stomoxys calcitrans*) are eusynanthropics, which are very closely adapted to human settlements and farms, so they introduce higher risks for human and animal health.

9.3.2 Flies at Recreation Sites

To check whether the flies at urban sites pose problems similar to those which live in or around animal houses, a research project was initiated (by the governments of

Fig. 9.16 *Calliphora vomitoria*, adult stage and pupa

Fig. 9.17 *Helina reversio*, the head of which has a *frontal black line*. The *yellow-gray abdomen* is characterized by *two black dots*

Fig. 9.18 *Phaonia errans*. This 9–12-mm-sized fly has completely *yellow legs* and the tip of the abdomen of the male appears *reddish-yellow*

Table 9.5 Development data of flies of central Europe

Species	Size of adult (mm)	Eggs	Hatching of larvae	Larval development	Pupal rest	Life span of adults
Musca domestica	6–7	600–1,000 on feces (chicken, pig, horses, cattle) and other decaying organic matter	15°C: 50 h 20°C: 23 h 30°C: 10 h	15°C: 10–26 days 20°C: 8–10 days 30°C: 4–7 days	15°C: 18–21 days 20°C: 10–11 days 30°C: 4–5 days	60–70 days in stables
Musca autumnalis	5–7	600–900 on fresh feces	Temperature-dependent	4–7 generations per year	4–7 generations per year	Females of last generation hibernate
Fannia canicularis	4–6	Feces and putrescent material	25°C: 20–48 h	6 days	7–10 days	6–7 generations per year
Muscina stabulans	6–8	Eggs, larvae on chicken feces	Life cycle in summer about 2–3 weeks	Life cycle in summer about 2–3 weeks	Life cycle in summer about 2–3 weeks	4–5 generations per year
Stomoxys calcitrans	6–7	800 in groups of 25–50 in silage, in stables with urine and feces	1–2 days, temperature-dependent: 14 days up to months	6–8 days, temperature-dependent: 14 days up to months	6–8 days, temperature-dependent: 14 days up to months	Females live about 70–90 days
Haematobia irritans	4.5–5	In fresh cattle dung	Temperature-dependent: 24 days up to months, optimum 27–30°C	Temperature-dependent: 24 days up to months, optimum 27–30°C	Temperature-dependent: 24 days up to months, optimum 27–30°C	3–4 generations per year
Calliphora species	9–11 10–14	Eggs on feces with cadavers	Temperature-dependent	Temperature-dependent	10–40 days	1–2 months
Sarcophaga carnaria	10–19	Eggs on cadavers, agent of myiasis	Temperature-dependent	Temperature-dependent	10–40 days	1–2 months
Lucilia sericata, L. caesar	5–11	Eggs on feces, wounds, meat	24 h	4–7 days	1–2 weeks on the ground	1–2 months
Pollenia rudis	6–12	Eggs on feces, soil	Larvae parasitize in earthworms	4–8 days	1–2 weeks on carcasses	1–2 months

(continued)

Table 9.5 (continued)

Species	Size of adult (mm)	Eggs	Hatching of larvae	Larval development	Pupal rest	Life span of adults
Phaonia errans	9–12	Eggs are placed on wood powder or dry feces	Temperature-dependent	4–5 generations per year	10–40 days	1–2 months
Hydrotaea dentipes	6–7	Eggs on human and animal feces	Temperature dependent	4–6 generations per year	10–40 days	2–3 months
Oestrus ovis	8–15	Larvae are deposited at the nose or eyes	Immediately after laying	Larvae hibernate in the nose	2–4 weeks on the ground	4 weeks
Hypoderma bovis	13–15	600–800 glued on hair of cattle	4–7 days, then invading the skin	Inside the body until March	14–65 days on the ground	3–5 days (up to 28 days in the cold)
Hypoderma lineatum	11–13	5–20 eggs per hair, in a row	3–6 days, then invading the skin	Inside the body until March	23–28 days on the ground	3–5 days

Table 9.6 Periods of activity of selected flies that attack cattle and sheep in central Europe

Species	Breeding site	Jan	Feb	Mar	Apr	May	Jun	Jul	Aug	Sep	Oct	Nov	Dec
Musca domestica (housefly, typhoid fly)	Eggs and larva in and on animal feces, even in stables	$+^1$	$+^1$	$+^1$	$+^1$	+	+	+	+	+	+	$+^1$	$+^1$
Musca autumnalis (eye fly)	Eggs and larvae on animal feces outdoors, adults hibernate in buildings	0^2	0^2	0^2	+	+	+	+	+	+	+	0^2	0^2
Fannia canicularis (little housefly)	Eggs on feces of chicken and other animals, kitchen slops	0^2	0^2	0^2	0^2	+	+	+	+	+	+	0^2	0^2
Muscina stabulans (false stable fly)	Eggs, larvae often on chicken feces, predacious as well	$+^1$	$+^1$	$+^1$	$+^1$	+	+	+	+	+	+	$+^1$	$+^1$
Stomoxys calcitrans (stable fly; sucks blood every 3 days)	Oviposition on plant material in stables	+	+	+	+	+	+	+	+	+	+	+	+
Haematobia irritans (horn fly)	Eggs and larvae on fresh cow dung	–	–	–	–	+	+	+	+	+	–	–	–
Calliphora vicina, *C. vomitoria* (blowflies)	Eggs on feces and cadavers	–	–	+	+	+	+	+	+	+	+	+	–
Sarcophaga carnaria (flesh fly)	Eggs and larvae on meat, cadavers, earthworms	–	–	–	+	+	+	+	+	+	+	–	–
Lucilia sericata (gold fly)	Eggs in wounds, cadavers, feces	–	–	–	–	+	+	+	+	+	+	–	–
Lucilia caesar (carrion fly)	Eggs and larvae in feces, cadavers	–	–	–	–	+	+	+	+	+	+	–	–
Oestrus ovis (nose botfly)	Larvae are extruded into the nose of sheep and goats	Larvae in nose	Larvae in nose	Larvae in nose	Pupae on ground	+	+	+	+	+	+	+	Larvae in nose
Hypoderma bovis (large skin botfly)	600–800 eggs adhere to hair of cattle	Larvae in body	Larvae in body	Larvae in body	Pupae on ground	+	+	+	+	+	Larvae in body	Larvae in body	Larvae in body
Hypoderma lineatum (small botfly)	5–20 eggs per hair of cattle	Larvae in body	Larvae in body	Larvae in body	Larvae in body	+	+	+	Larvae in body	Larvae in body	Larvae in body	Larvae in body	Larvae in body

+, activity of adult flies in the wild and in stables; $+^1$, activity of fly stages in stables; 0^2, overwintering in buildings or stables; –, in winter rest

Germany and the Netherlands) that aimed to investigate the occurrence of flies, their load with potential agents of diseases, and the relations between flies and bird feces, which is a well-beloved target of flies and thus might be a source of the take up of agents of emerging diseases in urban communities. During this project, recreation sites were selected in the Netherlands and in Germany especially in and around the cities of Düsseldorf and Duisburg (see Sect. 9.2).

The following fly species were caught during their months of outdoor activity (in 2009–2011):

1. *Sarcophaga carnaria*, *Sarcophaga argyrostoma*: 48%
2. *Lucilia caesar*, *Lucilia cuprina*, *Lucilia sericata:* 31%
3. *Musca domestica*: 11%
4. *Calliphora vicina*, *Calliphora vomitoria*: 10%

Biting flies, i.e., bloodsuckers, were set free if they had been found occasionally in the butterfly catcher.

For the main species of flies, there was more variation at the recreation sites than close to the different stables, but there was much less variation than on the meadow with a lot of dog feces. This distribution is in correlation with the different breeding substrates that are preferred by the different species.

Among the fly genera caught, only specimens of the genus *Lucilia* occurred at each of the four recreation sites: i.e., inside the park at Castle Eller in Düsseldorf, where surely more dog feces occurred than in the so-called Old Town (downtown Düsseldorf), at Lake Unterbach (a bathing lake), where specimens of the genera *Calliphora*, *Lucilia*, and *Sarcophaga* were caught, and in the rather natural surroundings of the six lakes close to Duisburg, where only specimens of two genera were caught (*Sarcophaga*, *Lucilia*). The occurrence of *Musca domestica* at just one site, which is situated rather close to houses and is used as a playground, picnic site, etc., is very probably because that site has the best contact to human dwellings when compared with the three other catching sites.

9.4 The Pathogens in/on Flies Caught Close to Rural Stables

The flies have licking, punch-like protruding mouthparts which end at the so-called labellae (Fig. 9.19a, b). There are fine deepenings which distribute the saliva that is used to surround food upon contact. Then the mixture of saliva and food particles is sucked into the mouth. With the thus-enclosed food, agents of diseases are also sucked in. A further morphological peculiarity of the fly increases the capability to propagate agents of diseases: this is the formation of the structures at the terminal ends of the feet (Fig. 9.20). There are bunches of fine chitinous hair which will be covered by agents of diseases if the feet are placed onto a contaminated substrate.

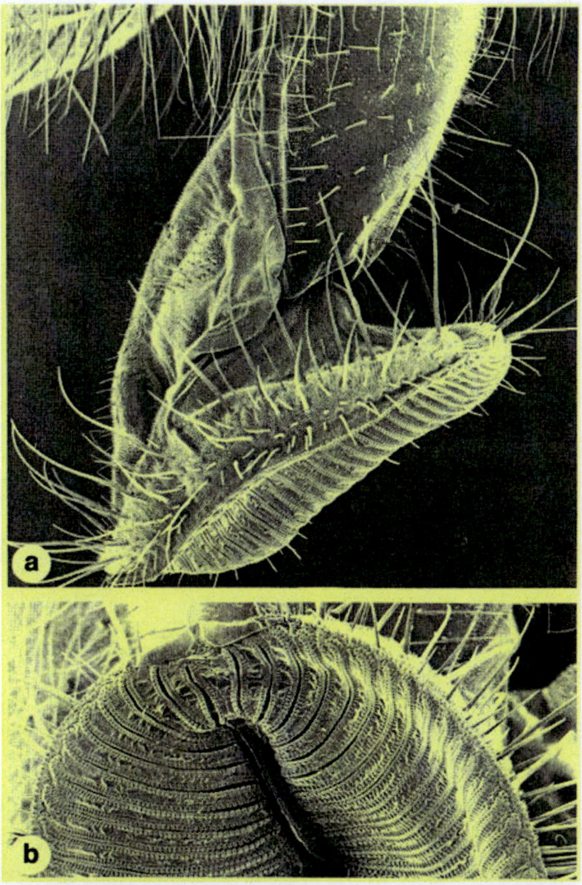

Fig. 9.19 Scanning electron micrograph of the mouthparts of the fly *Calliphora erythrocephala* (= *C. vicina*) showing the lateral aspect (**a**) and a surface view on the fine deepenings which spread the saliva (**b**) (Labellum)

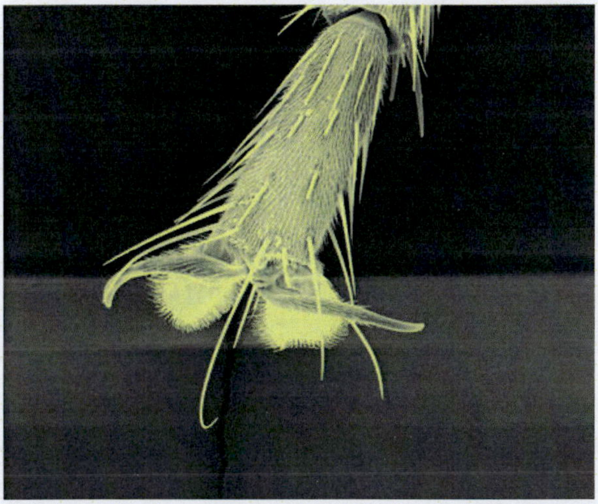

Fig. 9.20 Scanning electron micrograph of a tarsus of a fly showing numerous fine chitinous hairs that help attachment of the fly and that is filled with agents of diseases upon contact

Table 9.7 Bacteria found on flies on the exoskeleton (*Exo*) and in the intestines (*IN*) at the different catching sites

Bacteria	Dog meadow		Cattle barn		Horse stable		Pigpen	
	Exo	IN	Exo	IN	Exo	IN	Exo	IN
Acinetobacter iwoffi	–	–	–	–	–	–	+	–
Aerobe spore former	+++	+	+++	+	+++	++	+++	++
Campylobacter jejuni	–	+	–	–	–	–	–	+
Campylobacter sp.	–	–	–	–	–	–	+	+
Corynebacterium sp.	++	–	–	–	–	–	++	–
Enterobacter aerogenes	+++	–	–	–	–	–	–	–
Enterobacteriacea	+++	+++	+++	+++	+++	+++	+++	+++
Enterococcus faecium	+	–	–	–	–	–	–	–
Enterococcus sp.	+++	+++	+	+	+	+	+	+++
Escherichia coli	+++	+++	+	+	+++	+++	+++	+++
EAEC	+	–	–	–	–	–	–	–
EHEC	–	–	–	+	–	–	–	+
EPEC	+++	+++	–	+	–	–	–	++
ETEC	+++	–	–	–	–	–	++	+
Klebsiella ornitholytica	+	–	–	–	–	–	–	–
Klebsiella oxytoca	–	–	–	–	+	–	–	–
Klebsiella pneumoniae	+	–	–	–	–	–	–	–
Morganella morganii	++	–	–	–	+	–	+++	–
Nonfermenter	–	–	–	–	+	–	–	–
Pantoea sp.	–	–	–	–	–	–	+	–
Proteus mirabilis	+	–	–	–	+	–	+	–
Proteus sp.	+++	+++	+++	+	+++	+	+++	+++
Providencia rettgeri	++	–	–	–	–	–	++	–
Pseudomonas sp., oxidase-positive	+	–	–	–	–	–	–	–
Sphingomonas paucimobilis	–	–	–	–	+	–	–	–
Staphylococci, koagulase-negative	+++	–	+++	+	+++	+++	+++	+++
Staphylococcus aureus	+	–	+++	–	++	–	–	–
Streptococcus sp.	++	–	+	+	+++	–	++	–
Streptococcus viridans	+++	+++	+	–	+	+	+++	+++

EAEC, enteroaggregative *Escherichia coli*; EHEC, enterohemorrhagic *Escherichia coli*; EPEC, enteropathogenic *Escherichia coli*; ETEC, enterotoxic *Escherichia coli*; +, at least in one fly; ++, in several flies; +++, often

9.4.1 Bacteria

In total, 29 species of bacteria were detected in and on the flies caught in the present study around stables and/or on the dog meadow. These bacteria belonged to 18 different genera, which are listed in Tables 9.7 and 9.8. In and on the bodies of flies from the surroundings of the cattle barn, eight species of bacteria were detected. The feces of the flies from the dog meadow contained seven species of bacteria, whereas on their surface 20 bacterial species were found after cultivation. Flies caught in the surroundings of the horse stable contained six species of bacteria and their surfaces harbored 13 bacterial species. Whereas the surfaces of flies

Table 9.8 Examples of isolated bacteria found in/on flies from different catching sites

Fly species	Bacteria	Catching site			
		D	C	H	P
Calliphora vomitoria	Enterobacteriacea	x			
	Enterococcus sp.	x			
	Escherichia coli	x			
	Morganella morganii	x			
	Proteus mirabilis	x			
	Staphylococci, koagulase-negative	x			
Lucilia caesar	Aerobe spore former	x			
	Escherichia coli	x			
	ETEC	x			
	Proteus sp.	x			
Musca domestica	Aerobe spore former			x	x
	Corynebacterium				x
	Enterobacteriacea			x	x
	Escherichia coli		x	x	
	ETEC				x
	Morganella morganii				x
	Pantoea sp.				x
	Proteus sp.		x	x	x
	Providencia rettgeri				x
	Sphingomonas paucimobilis			x	
	Staphylococci, koagulase-negative		x	x	x
Polietes lardaria	Aerobe spore former	x	x		
	EPEC	x			
	Proteus sp.	x	x		
Sarcophaga carnaria	Aerobe spore former	x			
	Escherichia coli	x			
	EAEC	x			
	ETEC	x			
	Enterococcus faecium	x			
	Klebsiella ornitholytica	x			
	Klebsiella pneumoniae	x			
	Morganella morganii	x			
	Proteus sp.	x			
	Providencia rettgeri	x			
	Pseudomonas sp., oxidase-positive	x			
	Streptococcus sp.	x			

C cattle barn, *D* dog meadow, *H* horse stable, *P* pigpen, *ETEC* enterotoxic *Escherichia coli*, *EPEC* enteropathogenic *Escherichia coli*, EAEC enteroaggregative *Escherichia coli*

originating from the surroundings of pigpens were contaminated with up to 15 species of bacteria, the analysis of their intestines showed 11 bacterial species.

With respect to the different bacterial species, it was noted that some of the fly species were carrying many bacteria at the same time. For example, individuals of *Sarcophaga carnaria* harbored in total 12 species of bacteria, and the specimens of

Table 9.9 Important human diseases, the agents of which were found recently on or inside flies

Species	Disease, symptoms, general transmission
ETEC	Watery diarrhea for 1–9 days owing to attachment to the intestinal wall and release of LT and/or ST toxins; fever; no blood cells in the feces; infections need 10^8 pathogens; transmission via dirty food or drinking water
EPEC	Watery diarrhea with fever and vomiting; destruction of microvilli by attachment of the bacteria to the intestinal wall, especially in children below 2 years; diarrhea persistent
EAEC	Up to 33% of travelers may suffer from this diarrhea, which is due to large bacterial aggregates in the lumen of the intestine. These bacteria produce toxins (EAST 1), petechiae; chronic diarrhea and malabsorption
EHEC	Gastroenteritis with vomiting and diarrhea; infections are often due to fewer than 100 single pathogens; transmission via drinking water and contact with human or cattle feces; the most common serotype is O157:H7; the recent outbreak was due to O104:H4, which caused bloody diarrhea and in approximately 30% of cases hemorrhagic uremic syndrome (http://www.rki.de/cln_109/nn_205760/DE/Content/InfAZ/E/EHEC/EHEC-Abschlussbericht,templateId=raw,property=publicationFile.pdf/EHEC-Abschlussbericht.pdf), with kidney destruction. Bacteria produce Shiga toxins. In many cases diarrhea stops after 4–10 days
Campylobacter jejuni	This species represents 95% of the slimy, bloody diarrheas due to *Campylobacter* bacteria, 400–500 million cases per year, transmission via feces of birds, pigs, and cattle, milk, meat, dirty drinking water; 500 bacteria are sufficient to initiate infections; pathogens are found in the stool for 1–3 weeks; incubation period 12 h to 3 days
Staphylococcus aureus	Watery diarrhea, vomiting nausea; incubation period 1–18 days after eating contaminated food, wound infections
Klebsiella pneumoniae	In immunocompromised persons, pneumonia. *K. rhinoscleromatis* induces nose infections, leading to a bad smelling nose
Proteus mirabilis, P. vulgaris	They are gram-negative bacteria leading to intestinal diseases and also extraintestinal infections
Enterococcus	*Enterococcus faecalis* is a gram-positive bacterium which may act as a nosocomial agent of disease leading to sepsis, endocarditis, peritonitis, urethral infections, cholecystitis, etc.
Providencia rettgeri	These species are gram-negative bacteria which lead to urethral infections and may induce diarrheas, sepsis
Morganella	These gram-negative bacteria may lead to severe diarrheas, sepsis
Pseudomonas aeruginosa	This species is a gram-negative bacterium which as a nosocomial agent may induce infections of the urethral channels, wounds, brain, and breathing system; the leading symptom is the production of a greenish-blue sanies

ETEC enterotoxic *Escherichia coli*, *EPEC* enteropathogenic *Escherichia coli*, EAEC enteroaggregative *Escherichia coli*, *EHEC* enterohemorrhagic *Escherichia coli*

Musca domestica carried 11 species of bacteria (Table 9.8). Among these bacterial pathogens, several are of great concern, e.g., the different types of *E. coli* and the *Staphylococcus* and *Streptococcus* species (see Table 9.9).

Table 9.10 Fungi found on the surface (*Exo*) or in the intestine (*IN*) of flies at different catching sites

Fungi	Dog meadow		Cattle barn		Horse stable		Pigpen	
	Exo	IN	Exo	IN	Exo	IN	Exo	IN
Aspergillus fumigatus	–	–	+	–	+	–	+	–
Aspergillus sp.	+	–	++	–	+++	–	+++	+++
Candida albicans	+	–	–	–	–	–	–	–
Candida tropicalis	++	++	–	–	–	–	–	–
Candida sp.	+	++	++	+++	–	+	++	+++
Cladosporium sp.	–	–	–	–	–	+	–	–
Geotrichum sp.	++	+	–	++	–	+	+	++
Hyphomycetes	+	–	+++	–	+++	++	+	–
Mucor sp.	+++	++	+++	–	+++	–	+++	++
Rhizomucor sp.	+	–	+	–	–	–	–	–
Trichosporon sp.	–	++	–	+	–	–	–	–

–, not found; +, rare; ++, not rare; +++, more often, common

These findings of a considerable load of pathogens on flies are even more impressive when they are compared with the rather low number of species of bacteria found on agar cultures onto which flies that had been reared in the laboratory had been placed (Table 9.13).

It is especially noteworthy that in the present fly-catching project the very important human pathogenic bacteria such as the different pathogenic strains of *E. coli* and *Campylobacter* were found practically exclusively in or on flies which had been captured in the close surroundings of pigpens and/or from the dog meadow (with the exception of EHEC and EPEC, which were isolated from flies caught close to horse stables and cattle barns too).

This shows that these sites need special attention in terms of prophylactic care to avoid spreading of severe agents of diseases in the close neighborhood of humans.

On the other hand, it was somewhat astonishing that *Salmonella* or *Shigella* serovars were not detected any of the flies caught. This might be because either all farmed animals were free of these bacteria or the number of flies studied was too low to determine a low-grade infection rate of the farm animals, or both.

9.4.2 Fungi

On the surface of the flies caught close to the animal stables, 11 species of fungi were found belonging to eight genera. In the case of flies from the cattle barn, six species of fungi were found on their surface and three inside their feces. The flies from the dog meadow contained five species of fungi, whereas the flies from the surroundings of the horse stables contained in total six species of fungi in their feces and four species of fungi on their cuticle. The flies from the pigpens contained four species of fungi in their intestine and six species on their exoskeletons (Tables 9.10 and 9.11). The individual

Table 9.11 Important fungal diseases of humans the agents which have been found in flies

Fungi	Disease and symptoms
Aspergillus fumigatus	Aspergillosis: allergic pulmonary disease, aspergilloma
Candida sp.	Candidiasis: dermatitis, intertrigo, intestinal infections
Mucor sp.	Invasive myosis in the nose and brain
Trichosporon sp.	Invasive skin mycosis (white piedra)
Geotrichum sp.	Fungemia: stages in respiratory excretions
Cladosporium sp.	Skin mycosis, leading to dark scars

findings showed that fungi belonging to the genera *Aspergillus* and *Mucor* as well as several member of the Hyphomycetes were isolated rather often from the surfaces of the flies caught at all four catching sites. Furthermore, flies from the surroundings of the pigpens and from the dog meadow also contained either *Mucor* or *Aspergillus* specimens in their intestines. Fungi of the genera *Candida* and *Geotrichum* were found in flies from all catching sites. *Aspergillus fumigatus* was found exclusively at the surface of flies from all catching sites and never in the intestines. Fungi of the genus *Rhizomucor* also occurred exclusively at the surface of flies from the dog meadow and from the cattle barn. Exclusively intestinal findings were noted in the case of fungi of the genera *Cladosporidium* (in flies from horse stables) and *Trichosporon* (in flies from the dog meadow).

It is important, however, that the human pathogenic fungi *Candida albicans* and *Candida tropicalis* only occurred at the dog meadow catching site, which shows how important sanitation measures are around dog meadows inside and outside cities.

9.5 The Pathogens Found in/on Flies at Downtown Sites and at Recreation Sites

The agents of disease isolated from flies that had been caught at recreation sites had a less variable spectrum with respect to the different species of bacteria and fungi. The flies at the different catching sites contained the following species and genera:

1. Six-lakes region close to Duisburg:
   - *Lucilia*: EPEC, *Aspergillus* sp., *Candida* sp.
   - *Sarcophaga*: EPEC, *Aspergillus* sp., *Candida* sp.
2. Lake Unterbach (Düsseldorf):
   - *Sarcophaga*: EPEC, EHEC, *Candida*, *Aspergillus* sp.
   - *Lucilia*: EPEC, EHEC, *Candida*, *Aspergillus* sp.
   - *Calliphora:* EPEC, *Aspergillus*, *Candida* sp.

3. Park at Castle Eller (Düsseldorf)

   – *Musca domestica:* EPEC, *Candida*, *Aspergillus*
   – *Sarcophaga:* EPEC, *Candida*, *Aspergillus*
   – *Lucilia:* EPEC, *Candida*, *Aspergillus*
   – *Calliphora:* EPEC, *Candida*, *Aspergillus*

4. Downtown Düsseldorf

   – *Sarcophaga:* EPEC, *Candida*, *Candida albicans*, *Aspergillus*, *Mucor*
   – *Lucilia*: EPEC, *Candida*, *Aspergillus*
   – *Calliphora:* EPEC, *Candida*, *Aspergillus*

The total infection rate of flies was very high, whereby the feces showed especially high rates: 60% of the flies contained EPEC, *Candida* sp. was found in 70% of the files, and *Aspergillus* was found in 80% of the flies.

These results, which were obtained from probes until midsummer of 2011, show that both the number of fly species caught and the number of species that are agents of diseases determined were considerably lower than in rural regions close to stables. However, it would be a severe error to believe that there is no chance of an outbreak of a disease, since all important flies are present and can take up agents of diseases if they are present in open garbage, food, dog feces, or bird feces. Thus, attention is always needed to minimize the number of possible breeding sites of flies.

9.6 The Danger Index

The danger index (modified according to Mihály 1967) is of course an artificial collection of data which allow, however, a survey of the dangers that may be connected with the occurrence (at least in larger numbers) at certain places under some given conditions.

The danger index $D = (a + b + c + d) \times (e + f + g) \times m$ thus gives a relative grade of the risks that are induced for animal and human health if there are considerable amounts of flies close to humans and animals. The different variables *a*–*g* and *m* have the gradings 0, 0.5, or 1. Thus, the highest danger index is 12. The different variables (*a*–*g*, *m*) are based on the following definitions:

a, *b*, *c*: Grade 1 is given when both females and males visit human or animal feces for feeding and/or egg deposition.
d: Grade 1 is given if these species visit wounds and suck infectious "juices"!
e: Grade 1 is given if the species is eusynanthropic (i.e., mainly based in human surroundings), and grade 0.5 is given if the species is hemisynanthropic (i.e., may also exist in larger numbers in the wider world).
f: Grade 1 is given if the flies set down on milk or other human food.
g: Grade 1 is given if the flies also set down on fruits.
m: Grade 1 is given if the fly is not bigger than *Musca domestica*.

Table 9.12 Comparative danger indices (*D*) of several licking fly species and of *Stomoxys* (in the latter case the index is misleading): $D = (a + b + c + d) \times (e + f + g) \times m$

Fly species	*a*	*b*	*c*	*d*	*e*	*f*	*g*	*m*	*D*
Calliphora vicina	1	1	1	1	1	1	1	1	12
Calliphora vomitoria	1	1	1	1	1	1	1	1	12
Fannia canicularis	1	1	1	1	1	1	1	0.5	6
Helina reversio	1	1	1	0	0.5	0	0	0.5	0.75
Lucilia caesar	1	1	1	1	1	1	1	1	12
Musca domestica	1	1	1	1	1	1	1	1	12
Muscina pabulorum	1	1	1	0	0.5	1	0	1	4.5
Muscina stabulans	1	1	1	0	1	1	0	1	6
Mydaea ancilla	1	1	1	0	0.5	0	0	1	1.5
Mydaea urbana	1	1	1	0	0.5	0	0	1	1.5
Phaonia errans	1	1	1	0	0.5	0	0	1	1.5
Polietes lardaria	1	1	1	0	0.5	0	0	0.5	0.75
Pollenia angustigena	1	1	1	0	0.5	0	0	0.5	0.75
Sarcophaga carnaria	1	1	1	0	0.5	1	0	1	4.5
Sarcophaga agyrostoma	1	1	1	0	0.5	1	0	1	4.5
Stomoxys calcitrans	0	0	0	1	1	0	0	1	1

Although the danger index makes some sense in the case of licking flies, it cannot be applied to bloodsucking Brachycera such as *Stomoxys calcitrans* and/or tabanids. In these cases many of the parameters *a*–*g* do not fit for the bloodsuckers and would lead to a too low danger index, which does not reflect the possible mechanical transmission of important agents of diseases (such as poliomyelitis, several bacterioses, and even worm diseases) during bloodsucking (Mehlhorn 2008; Eckert et al. 2008). Table 9.12 summarizes the danger indices for several—often severe—agents of diseases. The inclusion of *Stomoxys calcitrans* under the conditions given here obviously makes no sense since its importance as a bloodsucker and mechanical vector of agents of diseases is much higher than grade 1.

Furthermore, species such as *Sarcophaga carnaria* which may enter tissues too are underestimated in this danger index system. On the other hand, species with a low danger index may also be successful vectors of severe agents if their body comes into contact with a source that contains a significantly high number of pathogens.

Thus, in all cases fly control in houses and animal stables must have high priority if the health of humans and animals could be endangered.

9.7 Experimental Approaches

9.7.1 Exposure of Flies to Selected Pathogens

Because it is known that flies are able to transmit many types of agents of diseases, the tests in the present series of studies were limited. Only one fly was placed in a

Table 9.13 Bacteria found on and in *Musca domestica* from laboratory rearing facilities

Bacteria	Exoskeleton	Intestine
Aerobe spore former	+	–
Enterococcus sp.	+	–
Klebsiella oxytoca	–	+
Proteus mirablis	–	+
Staphylococcus, koagulase-negative	+	–

+, present; –, absent

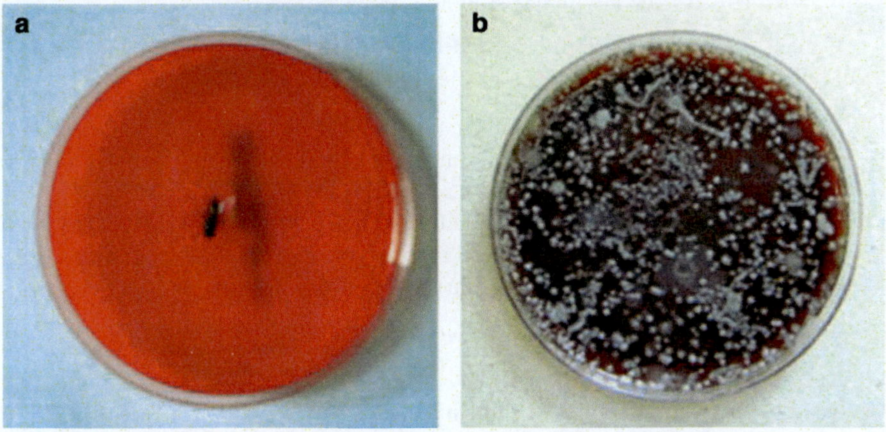

Fig. 9.21 Bacterial growing plate at the beginning (**a**) when a *Musca* fly was placed onto it (*left*) and 24 h later showing numerous bacterial cultures (**b**)

closed petri dish with a culture of bacteria for 5 min and then it was placed on a fresh culture medium for another 5 min. Colonies developed (Table 9.13) with aspects similar to those in the case when a fly caught in the wild was placed on a fresh bacterial growth medium (Fig. 9.21). Furthermore, cultured flies were placed on fresh blood agar, and this led to the occurrence of other bacteria. More and more intensive trials were done with parasites (see Chap. 10), and showed that there are many proven transmissions of parasitic stages.

9.8 Measures to Reduce the Infection Risks Due to Flies

Flies are always attracted if there are sufficient food and breeding sites. Owing to their short reproduction cycles—especially at high temperatures—mass outbreaks may often occur. Therefore, the following *management measures* will help:

1. Avoid open storage of animal feces.
2. Change the straw in stables often.

3. Cover or enclose any food (meat, milk, cheese, animal food) so that flies cannot deposit their eggs in it or have the chance to lick it.
4. Do not deposit food remnants on compost piles.
5. Always keep garbage containers closed.
6. Remove dog feces from the garden.
7. Place fine insect nets or fine gauze before windows/doors of stables and houses.
8. Place fly-catching systems (e.g., sticky bands in kitchens and storage rooms, which should be kept closed at all times prior to or after entering them).
9. Keep a mechanical fly flap with a transparent end in the kitchen, since otherwise the flies will see the attempted swat and fly away.

If there are stables close to human dwellings, the following *chemical measures* will help to reduce the occurrence of flies in stables and nearby houses:

1. Insecticides (deltamethrin, cyfluthrin, permethrin, cyhalothrin, fenvalerate, etc.) that are poured on the hair of cattle, sheep, etc., and will kill and repel *adult flies*.
2. Insecticides that are sprayed on the walls of stables and kill insects that try to rest there.
3. Insect growth regulators (e.g., diflubenzuron,, hexaflumuron, methopren, dicyclanil, cyromazine), which are sprayed on fly-larva-infested sites. These compounds will block the molt of larvae and this will kill them.

In addition, the following *physical methods* should be used:

1. Keep straw dry in stables.
2. Remove feces from stables at short intervals.
3. Thoroughly stack droppings so that temperatures occur inside (around 40°C) which block larval development.
4. Ensure there is a fresh climate in stables.
5. Install fly gauze in front of windows.
6. Install UV traps in stables.

Also, some *biological methods* are available to keep fly populations low in stables:

1. Introduce the commercially available fly *Ophyra anaescans* into pigpens. Its larvae are predators of other larvae. For example, a single *Ophyra* larva feeds on up to 20 *Musca domestica* larvae. The adult flies themselves remain mainly in low numbers in stables and do not molest humans. Furthermore, ichneumon wasps of the genera *Nasonia* and *Muscidifurax* are commercially available. The females introduce their eggs into fly larvae and then the wasp larvae feed on the fly larvae from inside.

9.9 Conclusions

In regions with low hygiene standards but also in regions with huge numbers of animal production sites or in urban communities with a dense population and thus needs for food imports, the danger of an increase in emerging food-borne diseases must be kept under constant observation. The outbreak of the *E. coli* epidemic due to the strain EHEC O104:H4 (Bielaszewska et al. 2011; Gault et al. 2011; Rasko et al. 2011) in the summer of 2011 in northern Germany, in France, and in several other countries (leading in many patients to hemolytic uremic syndrome) which destroyed kidneys and killed 43 people is a huge warning signal (http://www.rki.de/cln_109/nn_205760/DE/Content/InfAZ/E/EHEC/EHEC-Abschlussbericht,templateId=raw,property=publicationFile.pdf/EHEC-Abschlussbericht.pdf).

This awareness of lurking dangers is especially important, since until now arthropods as vectors of important agents of diseases have been rather often neglected, even though the transmission of bacteria and fungi by flies has been known for a long time. Even the possible transmission by flies of the rather aggressive strain EHEC O157:H7 was described by Iwasa et al. (1999). Thus, flies could have been the vector in the O104:H4 epidemic. The recent findings summarized in this chapter clearly show that there are large numbers of microorganisms around human settlements in regions which apparently feel safe and are proud of their high standards of hygiene. However, the number and variety of often severe agents of disease described mean that all fly-control measures must be strengthened and should initiate more research in this field. Otherwise fatal epidemics may occur in cities with several million inhabitants living close together.

Of course, bacteria and fungi found on the exoskeleton and in the feces of flies do not necessarily induce an infection. However, with respect to the quick reproduction of bacteria and fungi, large numbers of pathogens may easily occur on suitable substrates such as meat, milk, and cheese which can then become the starting organisms for an invasion of human or animal bodies. The invasion might be hard to stop once it has started.

Acknowledgments We gratefully acknowledge support of the studies that are the basis of this chapter by Novartis, Basel, and by the joint governmental project (the Netherlands/North Rhine-Westphalia) Safeguard, subproject *Campylobacter*. Furthermore, we are grateful to I. Schaefers and S. Walter, who typed the text and arranged the photographic plates, respectively.

References

Abbott SP (2002) Insects and other arthropods as agents of vector-dispersal in fungi. http://www.thermapure.com/pdf/AbbottInsectdispersal.pdf, pp 1–5

Abrams L (1976) Cestodosis in battery-housed laying hens. J S Afr Vet Assoc 47:171–3

Ako-nai AK, Oguniyi AD, Lamikanra A, Torimire SEA (1991) The characterization of clinical isolates of *Staphylococus aureus* in Ile-Ife, Nigeria. J Med Microbiol 34:109–112

Alam MJ, Zurek L (2004) Association of *Escherichia coli* O157:H7 with houseflies on a cattle farm. Appl Environ Microbiol 70:7578–7580

Altekrues SF, Cohen MT, Swerdlow DI (1997) Emerging foodborne diseases. Emerg Infect Dis 3:285–293

Aspöck H (ed.) (2010) Sick through arthropods, Denisia vol 30. Austrian Museal Publisher, Vienna

Banjo AD, Lawal OA, Adedujji OO (2005) Bacteria and fungi isolated from house fly (*Musca domestica* L.) larvae. Afr J Biotechnol 4:780–784

Barrow GI, Feltham RKA (2004) Cowan and Steel's manual for the identification of medical bacteria, 3rd edn. Cambridge University Press, Cambridge, 352 pp

Bech-Nielsen S, Bornstein S, Christensson D, Wallgren TB, Zakrisson G, Chirico J. (1982) http://www.ncbi.nlm.nih.gov/pubmed/7103186 Parafilaria bovicola (Tubangui 1934) in cattle: epizootiology-Vector studies and experimental transmission of Parafilaria bovicola to cattle. Am J Vet Res 43:948–54

Beutin L (1999) *Escherichia coli* as a pathogen in dogs and cats. Vet Res 30:285–298

Bielaszewska M, Mellmann A, Zhand W, Köck R, Fruth A, Bauwens A, Peters G, Karch H (2011) Characterization of the *E. coli* strain associated with an outbreak of haemolytic uremic syndrome in Germany. Lancet. doi:10.1016/S 1473-3099 (11) 70165-7

Bram RA, George JE, Reichar RE, Tabaciinic WJ (2002) http://www.ncbi.nlm.nih.gov/pubmed/12061432 Threat of foreign arthropod-borne pathogens to livestock in the United States. J Med Entomol 39:405–16

Bush O, Lafferty AD, Lotz JM, Shostak AW (1997) Parasitology meets ecology on its own terms: Margolis et al. revisited. J Parasitol 83:575–583

Chinery M (2002) Pareys Buch der Insekten, vol III. Parey, Berlin, 328 pp

De Jesus AJ, Olsen AR, Bryce JR, Whiting RC (2004) Quantitative contamination and transfer of *Escherichia coli* from food by houseflies, *Musca domestica* L. (Diptera: Muscidae). Int J Food Microbiol 93:259–262

Eckert J, Deplazes P, Friedhoff KT, Zahner H (2008) Parasitology for veterinarians. Enke, Stuttgart

Ekdahl K, Normann B, Andersson Y (2005) Could flies explain the elusive epidemiology of campylobacteriosis? BMC Infect Dis 5:1–11

Emerson PM, Lindsay SW, Alexander N, Bah M, Dibba SM, Faal HB, Lowe KO, McAdam KP, Ratcliffe AA, Walraven GE, Bailey RL (2004) Role of flies and provision of latrines in trachoma control: cluster-randomized controlled trial. Lancet 363:1093–1098

Enright MR, Alexander TJ, Clifton-Hadley FA (1987) http://www.ncbi.nlm.nih.gov/pubmed/3672850 Role of houseflies (Musca domestica) in the epidemiology of Streptococcus suis type 2. Vet Rec 121:132–3

Fetene T, Worku N (2009) Public health importance of non-biting cyclorrhaphan flies. Trans R Soc Trop Med Hyg 103:187–191

Fischer O, Mátlová L, Dvorská L, Svástová P, Bartl J, Melichárek I, Weston RT, Pavlik I (2001) Diptera as vectors of mycobacterial infections in cattle and pigs. Med Vet Entomol 15:208–211

Förster M (2009) Gefährliche Plagegeister. Rundschau Fleischhyg Lebensmittelüberwach 7:4–6

Förster M, Klimpel S, Mehlhorn H, Sievert K, Messler S, Pfeffer K (2007) Pilot study on synanthropic flies as vectors of pathogenic microorganisms. Parasitol Res 101:243–246

Förster M, Klimpel S, Sievert K (2009) The house fly (*Musca domestica*) as a potential vector of metazoan parasites caught in pig-pen in Germany. Vet Parasitol 160:163–167

Förster M, Gestmann F, Mehlhorn H, Sievert K, Messler S, Neuhausen N, Petersdorf S, Pfeffer K (2012) Flies as vectors of parasites potentially inducing severe diseases in humans and animals. In: Mehlhorn H (ed) Arthropods as vectors of emerging diseases, vol 3, Parasitology research monographs. Springer, Heidelberg

Fotedar R (2001) Vector potential of houseflies (Musca domestica) in the transmission of Vibrio cholera in India. Acta Trop 78:31–34

Fotedar R, Banerjee U, Samantray JC, Shirniwa S (1992a) Vector potential of hospital houseflies with special reference to *Klebsiella* species. Epidemiol Infect 109:143–147

Fotedar R, Banerjee U, Singh S, Shirniwa S, Verma AK (1992b) The housefly (*Musca domestica*) as carrier of pathogenic microorganisms in a hospital environment. J Hosp Infect 20:209–215

Garmendia J, Frankel G, Crepin VF (2005) Enteropathogenic and enterohemorrhagic *Escherichia coli* infections: translocation, translocation, translocation. Infect Immun 73:2573–2585

Gault G, Weill FX, Mariani-Kurkdjian P et al (2011) Outbreak of haemolytic uremic syndrome and bloody diarrhoea due to *E. coli* O104:H4, south-west France, June 2011. Euro Surveill 16 (26), pii=19905

Gestmann F, Förster M, Mehlhorn H, Messler S, Neuhausen N, Pfeffer K (2012) Flies as vectors of microorganisms potentially inducing severe diseases in humans and animals. In: Mehlhorn H (ed) Arthropod as vectors of emerging diseases, vol 3, Parasitology research monographs. Springer, Heidelberg

Getachew S, Gebre-Michael T, Erko B, Balkew M, Medhin G (2007) Non-biting cyclorrhaphan flies (Diptera) as carriers of intestinal human parasites in slum areas of Addis Ababa, Ethiopia. Acta Trop 103:186–194

Graczyk TK, Thompson RC, Fayer R, Adams P, Morgan UM, Lewis EJ (1999) http://www.ncbi.nlm.nih.gov/pubmed/10548284 Giardia duodenalis cysts of genotype A recovered from clams in the Chesapeake Bay subestuary, Rhode River. Am J Trop Med Hyg 61:526–9

Greenberg B (1965) Flies and diseases. Sci Am 213:92–99

Greenberg B (1971) Flies and disease, vol I. Princeton University Press, Princeton

Greenberg B (1973) Flies and disease, vol II. Princeton University Press, Princeton

Greenberg B, Kowalski JA, Klowden MJ (1970) Factors affecting the transmission of *Salmonella* by flies: natural resistance to colonization and bacterial interference. Infect Immun 2:800–809

Greenberg B, Verela G, Bornstein A, Hernandez H (1963) Salmonellae from flies in a Mexican slaughterhouse. Am J Hyg 77:177–183

Grübel P, Hoffman JS, Chong FK, Burstein NA, Mepani C, Cave DR (1997) Vector potential of houseflies (*Musca domestica*) for *Helicobacter pylori*. J Clin Microbiol 35:1300–1303

Hald B, Skovgard H, Pedersen K, Bunkenborg H (2008) Influx insects as vectors for *Campylobacter jejuni* and *Campylobacter coli* in Danish broiler houses. Poult Sci 87:1428–1434

Haupt I, Haupt A (1998) Fliegen und Mücken. Naturbuch, Augsburg

Holt PS, Geden CJ, Moore RW, Gast RK (2007) Isolation of *Salmonella enteritica* serovar enteritis from houseflies (*Musca domestica*) found in rooms containing *Salmonella* serovar enteritis-challenged hens. Appl Environ Microbiol 73:6030–6035

Iwasa M, Makino S, Asakura H, Kobori H, Kaljser B (1999) Detection of *Escherichia coli* O157:H7 from *Musca domestica* (Diptera: Muscidae) at a cattle farm in Japan. J Med Entomol 36:108–112

Karch H (2005) Enterohaemorrhagic *Escherichia coli* in human medicine. Int J Med Microbiol 295:405–418

Khan AR, Huq F (1978) Disease agents carried by flies in Dacca city. Bangladesh Med Res Counc Bull 4:86–93

Kobayashi M, Sasaki T, Saito N, Tamura K, Suzuki K, Watanabe H, Agui N (1999) House flies: not simple mechanical vectors of enterohemorrhagic *Escherichia coli* O157:H7. Am J Trop Med Hyg 61:625–629

Kollaritsch H, Paulke-Korinek M (2010) Durchfallerkrankungen. In: Löscher T, Burchard KT (eds) Reisekrankheiten. Thieme, Stuttgart

Levine OS, Levine MM (1991) House flies (*Musca domestica*) as mechanical vectors of shigellosis. Rev Infect Dis 13:688–696

Löscher T, Burchard GD (eds) (2010) Tropenmedizin in Klinik und Praxis. Thieme, Stuttgart

Medveczky I, Kovács L, Kovács F, Papp L (1988) http://www.ncbi.nlm.nih.gov/pubmed/2856539 The role of the housefly, Musca domestica, in the spread of Aujeszky's disease (pseudorabies). Med Vet Entomol 2:81–6

Mehlhorn B, Mehlhorn H, Walldorf V (2011) Schach den Blutsaugern und Schädlingen. Düsseldorf University Press, Düsseldorf

Mehlhorn H (ed) (2008) Encyclopedia of parasitology, 3rd edn. Springer, Heidelberg

Mehlhorn H, Piekarski G (2002) Grundriss der Parasitenkunde, 7th edn. Spectrum, Heidelberg

Mian LSH, Jacal JV (2002) Isolation of salmonellas from muscoid flies at commercial animal establishments in a Bernardino country, California. J Vector Ecol 27:82–85

Mihály F (1967) The danger-index of synathropic flies. Acta Zool Acad Sci Hung 13:373–377

Neumeister B, Geiss HK, Braun RW, Kimmig P (eds) (2009) Mikrobielle Diagnostik, 2nd edn. Stuttgart, Thieme

Nichols GL (2005) Fly transmission of *Campylobacter*. Emerg Infect Dis 11:361–364

Ngern-Klun R, Sukotason K, Mathanitikorn R, Vogtsberger RC, Sukontason KL (2007) Fine structure of *Chrysoma nigripes* (Diptera: Calliphoridae), a fly species of medical importance. Parasitol Res 100:993–1002

O'Hara JE, Kennedy MJ (1989) http://www.ncbi.nlm.nih.gov/pubmed/2795384 Prevalence and intensity of Thelazia spp. (Nematoda: Thelazioidea) in a Musca autumnalis (Diptera: Muscidae) population from Central Alberta. J Parasitol 75:803–6

Olsen AR, Hammack TS (2000) Isolation of *Salmonella* spp. From the housefly *Musca domestica* L., and the dump fly, *Hydrotaea aenescens* (Wiedemann) (Diptera: Muscidae), at cage-layer houses. J Food Prot 63:958–960

Otake S, Dee SA, Moon RD, Rossow KD, Trincado C, Pijoan C (2004) http://www.ncbi.nlm.nih.gov/pubmed/14756503 Studies on the carriage and transmission of porcine reproductive and respiratory syndrome virus by individual houseflies (Musca domestica). Vet Rec 154:80-5

Oyerinde JPO (1976) The role of the housefly (*Musca domestica*) in the dissemination of hookworm. Ann Trop Med Parasitol 70:455–462

Rasko DA, Webster DR, Sahl IW et al (2011) Origins of the *E. coli* strain causing on outbreak of haemolytic uremic syndrome in Germany. N Engl J Med. doi:10.105/NEJ Moa 11069LO

RKI (Robert-Koch Institut) (2006) Information. Epidemiol Bull 41:351–362

Sales Mde, da Costa GL, Bittencourt VR (2002) Isolation of fungi in *Musca domestica* Linnaeus, 1758 (Diptera: Muscidae) captured at two natural breeding grounds in the municipality of Serop dica, Rio de Janeiro, Brazil. Mem Inst Oswaldo Cruz 97:1107–1110

Sasaki T, Kobayashi M, Agui N (2000) Epidemiological potential of excretion and regurgitation by *Musca domestica* (Diptera: Muscidae) in the dissemination of *Escherichia coli* O157:H7 to food. J Med Entomol 37:945–949

Schaefer M (2002) Brohmer – Fauna von Deutschland, 21st edn. Quelle & Meyer, Wiebelsheim

Shane SM, Montrose, Harrington KS (1985) Transmission of Campylobacter jejuni by the housefly (Musca domestica). Avian Dis 19:384–391

Steinmüller N, Demma L, Bender JB, Eidson M, Angulo FJ (2006) Outbreaks of enteric disease associated with animal contact: not just a foodborne problem anymore. Emerg Infect Dis 43:1596–1602

Sukopp H, Wittig R (eds) (1998) Stadtökologie, 2nd edn. Stuttgart, Fischer

Szalanski AL, Owens CB, McKay T, Steelman CD (2004) Detection of *Campylobacter* and *Escherichia coli* O157:H7 from filth flies by polymerase chain reaction. Med Vet Entomol 18:241–246

Szostakowska B, Kruminis-Lozowska W, Racewicz M, Knight R, Tamang L, Myjak P, Graczyk TK (2004) *Cryptosporidium parvum* and *Giardia lamblia* recovered from flies on a cattle farm and in a landfill. Appl Environ Microbiol 70:3742–3744

Umeche N, Mandah LE (1989) *Musca domestica* as a carrier of intestinal helminths in Calabar, Nigeria. East Afr Med J 66:349–352

Wagner AB (1989) Bacterial food poisoning. Texas Agricultural Extension Service publication no. L-1540. Texas Agricultural Extension Service, College Station

WHO (2002a) Food safety and foodborne illness. Fact sheet no. 237. World Health Organization, Geneva

WHO (2002b) Foodborne diseases, emerging. Fact sheet no. 124. World Health Organization, Geneva

Wright EP (1983) The isolation of *Camplylobacter jejuni* from flies. J Hyg 91:223–226

Yap KL, Kalpana M, Lee HL (2008) Wings of the common housefly (*Musca domestica*): importance of mechanical transmission of *Vibrio cholera*. Trop Biomed 25:1–8

Yeruham I, Braverman Y, Shpigel NY, Chizov-Ginzburg A, Saran A, Winkler M (1996) http://www.ncbi.nlm.nih.gov/pubmed/8903139 Mastitis in dairy cattle caused by Corynebacterium pseudotuberculosis and the feasibility of transmission by houseflies. I. Vet Q 18:87–9

Zarrin M, Vazirianzadeh B, Shams SS, Zarei Mahmoudabadi A, Rahdar M (2007) Isolation of fungi from housefly (*Musca domestica*) in Ahvaz, Iran. Pak J Med Sci 23:917–919

Zumpt F (1949) Hausfliegen als Überträger von Seuchen, insbesondere der spiralen Kinderlähmung. Anz Schädl 11:161–163

Chapter 10
Flies as Vectors of Parasites Potentially Inducing Severe Diseases in Humans and Animals

Maike Förster, Falk Gestmann, Heinz Mehlhorn, Kai Sievert, Sabine Messler, Nicole Neuhausen, Sabine Petersdorf, and Klaus Pfeffer

Abstract Many fly species feed regularly on feces or their larvae develop inside it. Therefore, it is not astonishing that parasites in the form of durable cysts or eggs or even as fragile trophozoites may be found in fly feces or attached to their legs and bodies. These parasites may then be transported to humans and animals or to their food, from where the parasites may enter (via oral uptake) the inner organs and can induce an infection. These infections might be severe in the case of pathogenic parasites or if high numbers of parasites are transferred. Since many articles in the past reported—often only from occasional findings—that parasites might be transported by licking flies, their role as regular vectors of microorganisms and parasites is therefore eminent. The activity of flies as vectors of pathogens was mostly neglected especially in countries that are proud of their hygiene standards. However, in recent times of intensive globalization of goods and humans and of shrinking distances between rural and downtown centers, flies and their potential load of microorganisms and parasites must be constantly controlled in order to avoid outbreaks of highly transmissible agents of diseases (e.g., viruses, bacteria, parasites such as *Cryptosporidium* oocysts). This chapter reviews the results of former reports and elucidates the situation near some German mid-sized towns with their rural surroundings. By investigation of flies caught close to pigpens, cattle

M. Förster • F. Gestmann • H. Mehlhorn (✉)
Department of Zoology and Parasitology, Heinrich Heine University, Universitätsstr. 1, 40225 Düsseldorf, Germany
e-mail: mehlhorn@uni-duesseldorf.de

K. Sievert
Novartis Animal Health Inc., WRO-1032.3.82, Schwarzwaldallee 215, 4058 Basel, Switzerland
e-mail: kai.sievert@novartis.com

S. Messler • N. Neuhausen • S. Petersdorf • K. Pfeffer
Medical Microbiology and Hospital Hygiene, Heinrich Heine University, Universitätsstr. 1, 40225 Düsseldorf, Germany
e-mail: klaus.pfeffer@hhu.de

H. Mehlhorn (ed.), *Arthropods as Vectors of Emerging Diseases*,
Parasitology Research Monographs 3, DOI 10.1007/978-3-642-28842-5_10,

barns, poultry houses, rabbit cages, and horse stables as well those from dog meadows and/or from recreation sites in or close to downtown sites, it is shown that certain flies are specific for these different sites. In all cases it was noted that a considerable number of flies carried parasites besides numerous bacteria. Some of them carried specimens of parasitic species which have zoonotic importance. When animal feces from these places was investigated, it was found that the parasitic species were the same as those that had been determined before in/on flies. In additional experiments, where laboratory bred, parasite-free flies were placed onto food (milk) with eggs of *Trichuris suis* and *Ascaris suum*, it was proven that these flies took up large amounts of these two worm egg types. These results in addition to the finding of parasites on flies caught in the wild as well as the findings of existing publications make it clear that flies have an extremely high potential as vectors of agents of disease that "might lurk unexpectedly before the doors of our well-believed islands of safety and hygiene."

Keywords Bacteria • Fly-borne diseases • Fly species • Insects • Parasites • Protozoa • Transmission • Vectors • Viruses • Worms

10.1 Introduction

Flies fulfill important tasks in nature, since they help to degrade organic materials. Some species, however, have become parasites and enter body openings (e.g., nose, mouth, anus, genital vulvae, ears, eyes) or their larvae even bore into skin or penetrate into the intestinal tract (Mehlhorn 2008). In addition to these truly parasitic activities, adult flies molest humans and animals by numerous touchdowns on their skin, leading to restlessness, which—in the case of animals—disturbs regular food uptake and thus brings about considerable economic losses. The occurrence of flies, however, may have even much worse consequences than those described above. This is the case when flies are vectors of microorganisms and/or parasites, which they ingest when feeding on human or animal feces. Some of these pathogens may reproduce inside the flies, and then become distributed with the fly's feces onto the skin or on mucous layers of humans and/or animals or onto the food of both groups.

The danger of becoming sick due to fly-transmitted pathogens is often neglected in central Europe and in North America, where people feel safe owing to apparently high, governmentally confirmed hygiene standards. That this safety is far from being true was shown in numerous publications reviewed by Greenberg (1973), Fischer et al. (2001), Olsen and Hammack (2000), Bramley et al. (1985), Banjo et al. (2005), Bush et al. (1997), Busse and Aka (1997), Damriyasa et al. (2004), Dipeolu (1977, 1982), Haupt and Haupt (1988), Howard (2001), Nansen and Roepstorff (1998), Olsen et al. (1958), Roepstorff and Nansen (1994), Szostakowska et al. (2004), and Umeche and Mandah (1989) and in numerous diagnostic textbooks (Eckert et al. 2008; Mehlhorn et al. 1993, 1995, 2012; Mehlhorn 2012; Rommel et al.

2002). These data, which were more or less occasionally obtained during various activities in different biotopes, were strongly supported by recent field and laboratory studies which were executed over several years by our research groups at Heinrich Heine University in Düsseldorf (Förster et al. 2007, 2009a, b; Gestmann et al. 2012).

Any agent of disease needs a particular initial inoculation dose to establish an infection with symptoms of disease. Whereas about 100 single bacteria may be sufficient in the case of, e.g., enterohemorrhagic *Escherichia coli* infections (Löscher and Burchard 2010), other species need 10,000 bacteria or more to overcome the host's self-defense systems. This number is, however, easily reached in the case of bacteria owing to their huge reproduction rate in very short intervals if warmth and sufficient substrate are available. In general, these conditions exist when the bacteria are in human or animal feces. On the other hand, it is much more difficult for the in general larger parasites to become transmitted. The first condition is that the parasitic stage is very stable (e.g., as cystic stages such as oocysts and worm eggs), whereas tissue stages with their tiny limiting membranes have mostly only poor chances of surviving outside a body. A second condition for a successful transmission via flies is that the parasitic stages reach the mouth of a host and are thus swallowed. This can happen if a fly transports the parasites directly onto the mouth of a host or places them onto its food when licking or when defecating there.

Thus, compared with the recently shown rather easy pathways of transmission of bacteria and fungi (see Chap. 9 by Gestmann et al. 2012), it was suggested and expected that the transmission of parasites in cysts would be much more difficult. However, the field trials with flies caught close to animal stables and the experimental infestations described in this chapter show that transmission of parasites by licking flies also works extremely well. Thus, flies help to propagate not only bacteria (and surely additionally also viruses) and fungi but also parasites to such a considerable degree that control measures are urgently required. This chapter summarizes some findings of parasites on and in flies that were captured in the wild close to humans and reports findings from experiments where what happens when flies come in contact with substrates that contain various parasites was simulated.

10.2 Materials and Methods

Flies were caught close to pigpens, horse stables, cattle barns, rabbit cages, and a poultry house as well as on a dog meadow close to Dormagen near Düsseldorf (Gestmann et al. 2012). Similar fly catching was done at recreation sites close to Düsseldorf and Duisburg. The catching periods covered the fly-activity periods in 2007–2011. All procedures conducted for the catching of flies for the work reported in this chapter were the same as those described in the following for the flies obtained in or around the pigpen (Förster et al. 2009a, b).

The pigpen close to Dormagen contained approximately 50 pigs of different ages, from piglets to finishers. The pigs were kept in pairs of five to ten pigs in big boxes, which were separated by only half-high walls, allowing contact with other pigs.

The flies were caught individually with the help of glass boxes or with commercial plastic bags and were immediately transported to the laboratory. The flies were identified according to the keys of Greenberg (1971), Chinery (2002), Haupt and Haupt (1998), Lane and Crosskey (1993), Schaefer (2002), and Weidner and Sellenscho (2010).

Afterwards, a subsample of the flies were anesthetized at 4°C for 5–10 min and killed by decapitation. Heads (with mouth parts) and torsi (thorax and abdomen) of the flies were washed and kept separately in 1.5 ml sterile solution (0.9% NaCl) in microtubes in pairs of ten. Then the torsi were removed from the wash solution, washed with ethanol (70%), and air-dried. All microtubes were centrifuged (5 min, 800 rpm). The sediments from the centrifugation (0.5 ml or more) were investigated for protozoan and metazoan parasitic stages under a light microscope. Intestines of the flies were dissected, colored with one drop Lugol's iodine, and were also examined for parasites under a light microscope. Some flies from every catching of flies were examined individually using the same method. The isolated parasites were identified by descriptions given in the textbooks by Rommel et al. (2002), Mehlhorn et al. (1995), Mehlhorn (2012), and Eckert et al. (2008).

For evaluation of the parasites in the pig feces, mixed samples were taken twice from different boxes in the pigpen. The feces was then examined with the merthiolate–iodine–formaldehyde concentration (MIFC) method (Mehlhorn et al. 1995).

The possible transmission of nematode eggs by flies was tested experimentally in the laboratory. For these experiments we used 120 sterile-bred, freshly hatched *Musca domestica* flies from the fly-breeding site of the institute and exposed them to embryonated eggs of the pig-parasitic nematodes *Ascaris suum* and *Trichuris suis*. Experimental solutions contained 1.5 ml ultraheated milk (0.1% fat contents) and 840 *Ascaris suum* eggs or 200 *Trichuris suis* eggs in water and were placed into the fly cage. After 1 h all the fluid was sponged by the flies. The flies were examined for the eggs of both nematode species with the same method as was used in the case of wild flies (see above). The calculation of the prevalence, intensity, and mean intensity followed the recommendations of Bush et al. (1997).

10.3 What Are Parasites?

In the present paper chapter the parasite load of flies is investigated, whereas in Chap. 9 by Gestmann et al. (2012) the load of flies with bacteria and/or fungi was considered. Parasites are defined as animals that live inside other animals and humans (endoparasites) or on their surface constantly (stationary ectoparasites) or at least occasionally for a given period (temporary ectoparasites). These parasitic stages obtain all their food from their hosts and thus have the potential to act as agents of diseases or as their vectors (Mehlhorn 2008).

The range of the severity of diseases differs depending on the parasitic species and/or on the pathogenicity of a specific strain of the species as well as on the immune status of the host. In this chapter special attention is given to those parasites which have a zoonotic potential, since the stages of these species might be transferred by flies from animals (and/or their feces) even to those humans who have no direct contact with animals. Such infections might be especially dangerous since in the case of unspecific symptoms of disease parasites will not be considered immediately as reasons for the disease and thus will have the chance to propagate and reproduce undisturbed inside the new host. In this study, the groups of parasites described in the following sections were of importance:

10.3.1 Protozoa: Unicellular Parasites

Protozoa that might be transmitted by flies must have developed a cyst stage, the wall of which protects the parasite from desiccation outside the host and/or from aggressive bacteria when excreted in animal or human feces. Parasitic stages which are limited by just a single membrane (such as trophozoites of amoebae, *Giardia lamblia*, or coccidians) do not have the chance to survive on the mouthparts, on the exoskeleton, and/or in the intestine of flies. Thus, this study focused predominantly on the search for cystic stages of amoebae, *Giardia lamblia*, and coccidians such as cysts of the genera *Cryptosporidium*, *Toxoplasma*, *Sarcocystis*, *Eimeria*, *Isospora*, and *Balantidium* (Figs. 10.1–10.4).

10.3.2 Trematodes: Sucking Plathyhelminths

Trematodes have a complicated life cycle involving several final hosts. While they are in the intermediate hosts, asexual reproduction occurs, and the sexual stages are formed inside the final hosts. Depending on the organ parasitized, the final hosts (animals, humans) excrete worm eggs via feces, sputum, or urine. However, these eggs—being protected with either a thick or a thin wall—are not infectious if they are taken up immediately by humans or animals that have excreted them, but always need at least one intermediate host for further development. Therefore, these eggs of trematodes are unimportant in studies where flies are investigated as vectors of agents of diseases. Flies only help to propagate trematode eggs within a certain area, and are never involved in a direct human-to-human or animal-to-animal transmission of trematode species (Fig. 10.5).

10.3.3 Cestodes: Tapeworms

In general, like trematodes, most tapeworms need at least one intermediate host, within which the larva reaches a size that allows it to be transmitted (mostly by

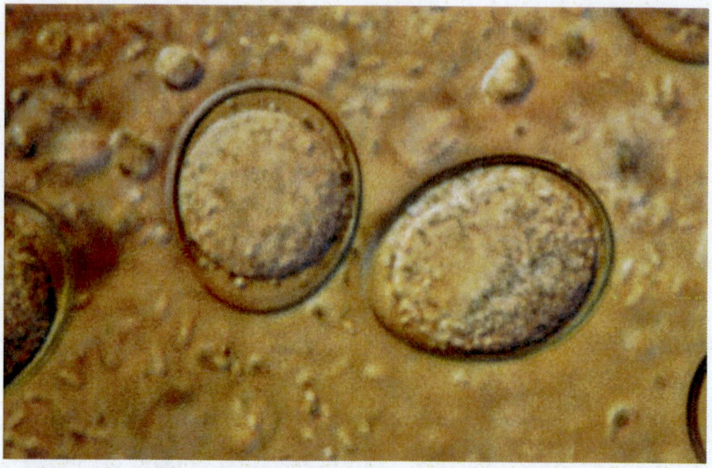

Fig. 10.1 Light micrograph of unsporulated oocysts of the type *Toxoplasma gondii* or Isospora from cat feces

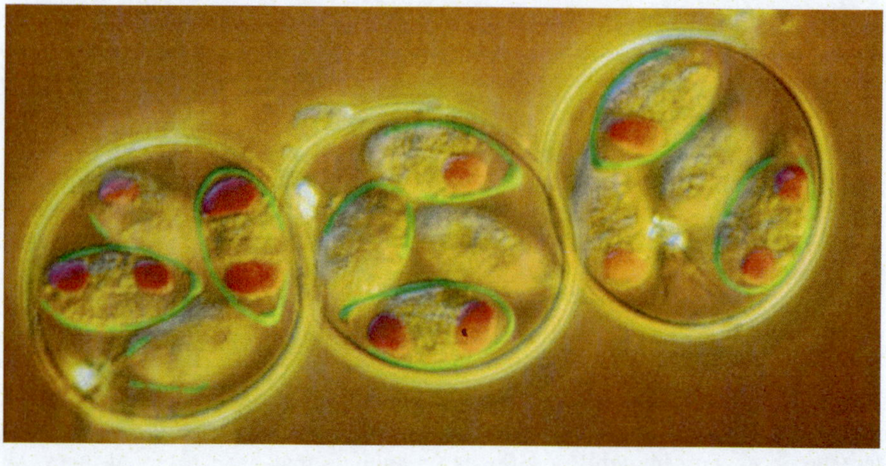

Fig. 10.2 Light micrograph of sporulated *Eimeria tenella* oocysts from a chicken containing four sporocysts each with two sporozoites

feeding on raw or undercooked meat) to a final host (such as humans or predator animals). However, in contrast to trematodes, there are important exceptions among the cestodes. For example, *Hymenolepis* species and also the human pig-derived tapeworm *Taenia solium* are able to infest the final host (humans) by means of excreted eggs (Fig. 10.6). In these cases, humans might become intermediate hosts (bearer of larvae) besides their normal activity as final hosts, where the adult tapeworm lives in the intestine and dispatches terminal proglottides that are filled with eggs. In those cases where the proglottides are disrupted inside the intestine or

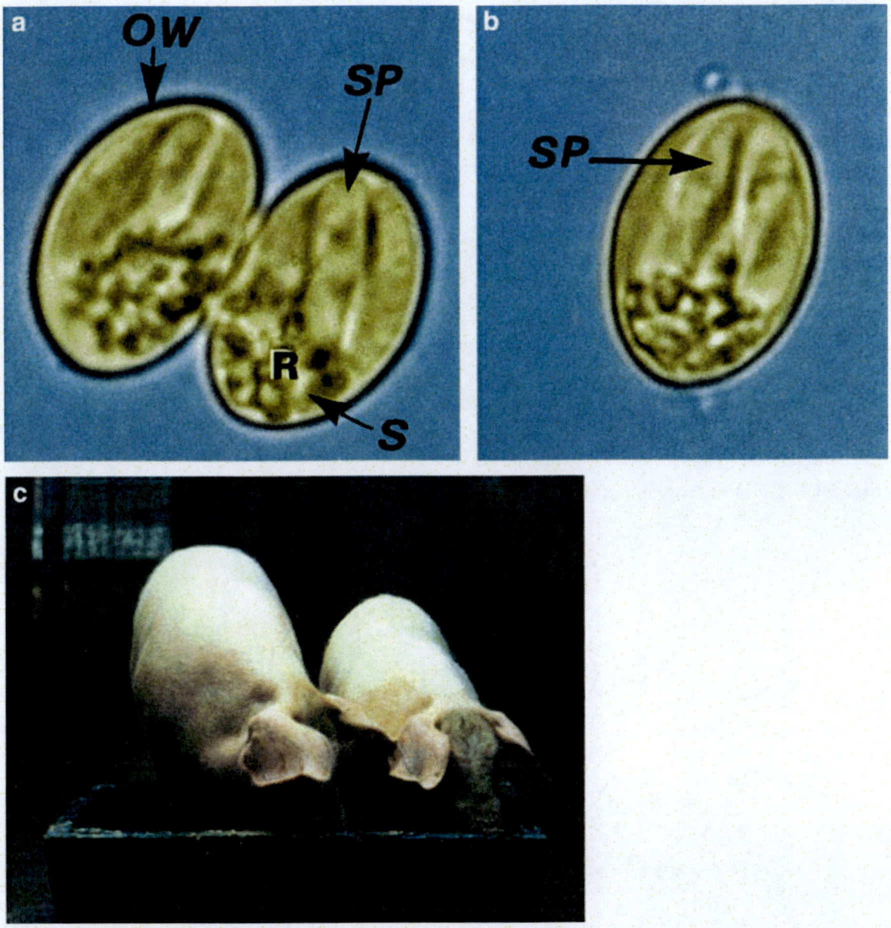

Fig. 10.3 (**a**) Light micrograph of an sporulated oocyst of *Sarcocystis suihominis* from human feces containing two sporocysts each with four sporozoites. (**b**) Single sporocyst. (**c**) Two pigs of the same age: one with *Sarcocystis* infection, the other without infection. *OW* oocyst wall, *R* residual body, *S* sporocyst, *SP* sporozoite

become digested as a consequence of an antitapeworm drug treatment, eggs may release the infectious larvae (oncosphaera or cysticercoid) inside the intestine (Figs. 10.7 and 10.8). In the case of *Hymenolepis* species, these larvae produce new adult tapeworms, whereas in the case of *Taenia solium*, a cysticercus may be formed, e.g., in human brain, or a large tissue cyst may develop when humans ingest *Echinococcus* eggs. If the cysticercus formation unfortunately occurs in the liver or in the brain, the severe symptoms of a hepatic cysticercosis or neurocysticercosis may occur (Mehlhorn 2008, 2012). Therefore, flies may transport eggs/ larvae of some tapeworms from human to human and/or animal to animal. Thus, deposition of human feces close to highway parking lots, camping grounds, etc.,

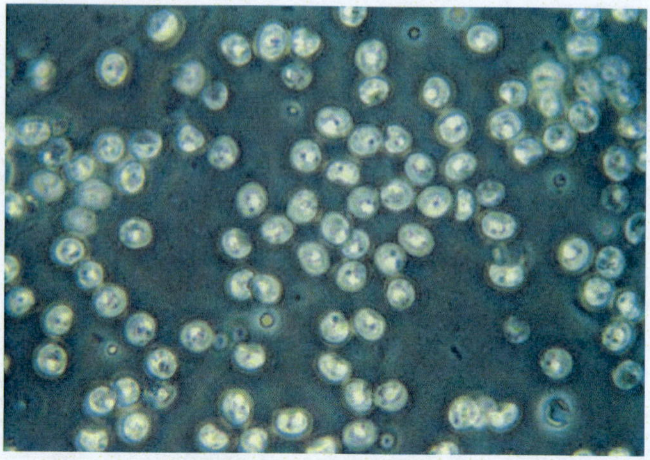

Fig. 10.4 Light micrograph of oocysts of *Cryptosporidium* sp. from feces of cattle

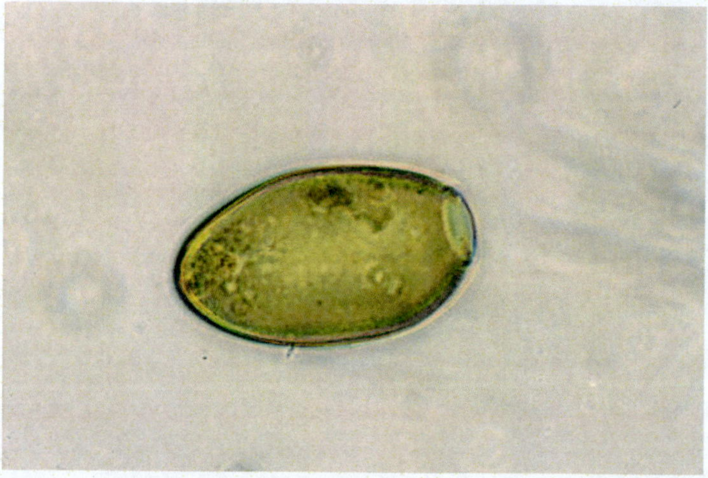

Fig. 10.5 Light micrograph of an egg of the trematode *Fasciola hepatica* (large liver fluke of ruminants)

should be strictly avoided, since feces principally attracts flies, which may become loaded there with tapeworm eggs as described above besides other pathogens.

10.3.4 Nematodes

Only rather few nematodes (roundworms) that infest land-living hosts have a complicated life cycle involving final and intermediate hosts. For example,

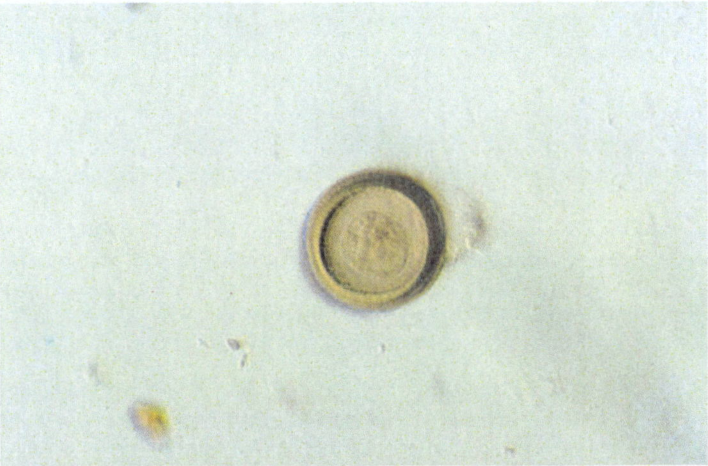

Fig. 10.6 Light micrograph of an egg of the human tapeworm *Taenia solium* showing its thick outer wall (embryophore)

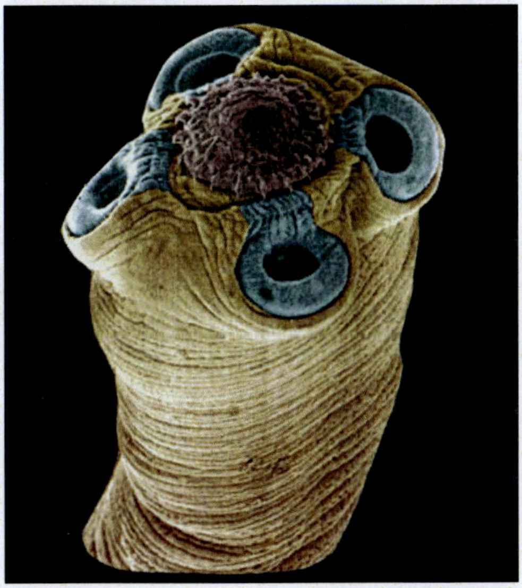

Fig. 10.7 Scanning electron micrograph of the scolex of a *Hymenolepis* tapeworm

Trichinella spiralis and the mosquito-transmitted filariae such as *Onchocerca volvulus* and *Wuchereria bancrofti* have obligatorily multihost life cycles and thus cannot be transmitted from final host to final host by licking flies, which owing to their peculiarities in transmission never come into contact with the transmissible stage of the nematode. However, a large number of nematodes infesting humans as well their pets and food animals have a direct life cycle

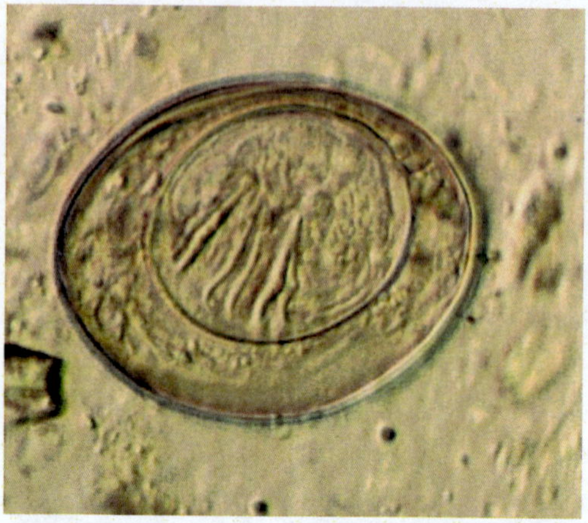

Fig. 10.8 Light micrograph of the egg of the tapeworm *Hymenolepis nana*

Fig. 10.9 Macrophotograph of adult *Toxocara canis* worms excreted after treatment

which ends with the fecal excretion of very resistant eggs that often have very thick shells (Figs. 10.9–10.11). After temperature-dependent development of a larva inside the eggshell, the egg (or the hatched larva) is ready for transmission, which occurs in general by ingestion of fecally contaminated food of humans or animals. This type of regular transmission of these eggs offers, of course, many possibilities for flies to become mechanical vectors of such eggs. Therefore, it is not astonishing that many nematodes with a worldwide distribution and infections occurring in many millions of hosts (e.g., *Ascaris*, *Ancylostoma*, *Trichuris*) profit from an additional pathway of transmission by licking flies (Anderson 2000; Bramley et al. 1985).

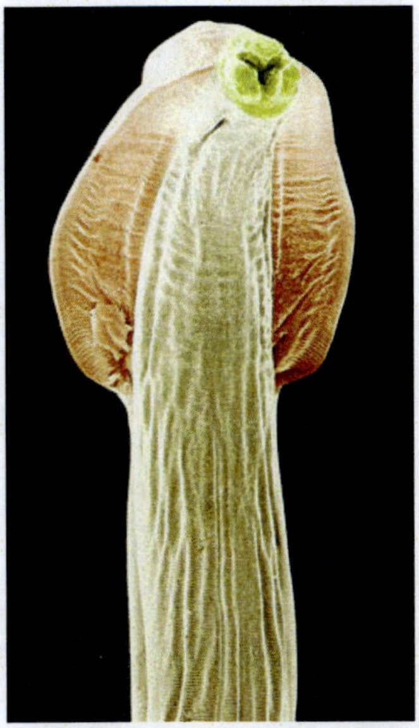

Fig. 10.10 Scanning electron micrograph of the anterior region of the nematode *Toxocara canis* showing lateral alae (broadenings at the anterior pole)

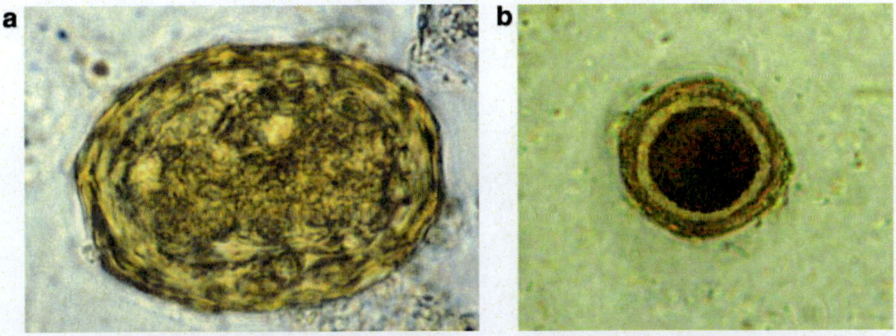

Fig. 10.11 (**a**) Light micrograph of a thick-shelled egg of the human roundworm *Ascaris lumbricoides*. (**b**) Egg of *Toxocara canis* found in flies caught at a dog meadow

10.3.5 Mites and Flies

If one looks in the wild especially when capturing insects, it is obvious that many insects are covered by numerous small stages of mites and ticks, which use beetles, dragonflies, butterflies, etc., as means of transportation (phoresy). Some of these "voyagers," however, even act as parasites, since they suck the hemolymph of their transporters. Thus, it is not astonishing that flies are also targets of such arthropods, which may enter flies when they are sitting on the soil or on animals. If the flies come into contact with larval stages of such arthropods, their transmission to other hosts and/or sites seems likely (Figs. 10.12 and 10.13).

Fig. 10.12 Light micrograph of different developmental stages of the bloodsucking mite *Dermanyssus gallinae*, some of which had sucked (*red*) and others had not yet sucked (*white*)

Fig. 10.13 Photograph of two adult *Lucilia* flies

10.4 Flies Close to Animal Stables and Downtown Recreation Sites

The main fly species that was found at all stables was *Musca domestica*, whereas close to the recreation sites *Musca domestica* was only ranked in fourth place, behind *Sarcophaga*, *Calliphora*, and *Lucilia* species. Only specimens of the genus *Calliphora* occurred in lower numbers (Gestmann et al. 2012). However, the highest species variation occurred at the dog meadow, where 12 species of flies were caught, headed by *Lucilia caesar*, *Muscina stabulans*, and *Sarcophaga carnaria*, all being attracted by the numerous portions of dog feces. Only single specimens of *Musca domestica* were found there, reaching less than 0.5% of the total catch (Gestmann et al. 2012). This distribution supported expectations that parasites on those flies caught at different sites would belong to different types of parasites. All species are known to infest humans and to lead to myiasis (Table 10.1).

10.5 Parasites Found on Flies Caught in the Wild

In total, 22 different parasites were found in or on flies caught close to stables or at the dog meadow. They belonged to the groups of protozoa (with six different coccidians, *Giardia lamblia*, Fig. 10.14, *Herpetomonas muscarum*, *Balantidium coli*), nematodes (with five defined and several undefined species), and three determined arachnid species in addition to several undetermined ones. In the following sections, more details on the flies caught and their parasitic load are given (Fig. 10.15).

10.5.1 Parasitic Load of Flies at Pigpens

The results of the examination of 593 flies caught close to a pigpen are given in Table 10.2.

In addition to the parasites listed in Table 10.2, numerous flagellate stages of *Herpetomonas muscarum* were found in 116 intestines of 593 flies. However, this species is parasitic inside flies and this finding therefore has no relevance for the role of flies as vectors of parasites. In another test done exclusively on 100 *Musca domestica* flies, similar results were obtained. It was shown that all parasites listed in Table 10.2 were seen with prevalences ranging between 0.30 and 5.5% depending on the parasite species. Worm eggs (e.g., from nematodes) were significantly more common than protozoans. For the pig feces studied with help of the

Table 10.1 Characteristics of the main fly species observed at recreation sites

Species	Size of adults (mm)	Feeding sites	Characteristics
Musca domestica	7–8	Larvae: omnivorous Adults: fruit, feces, dead bodies	Adults: black with gray powder Males: eyes without contact Females: posterior of abdomen yellowish
Sarcophaga carnaria	~14	Larvae: in earthworms, dead bodies Adults: fruit juices	Adults: light gray; eyes are red Female: deposit larvae instead eggs; enter houses
Lucilia (Phaenicia) sericata	5–7	Larvae: necrophagous, in wounds Adults: close to humans	Adults: greenish metallic; third segment 4 times longer than broad
Lucilia caesar	6–8	Larvae: necrophagous Adults: in the wild, plant and fecal juices	Adults: greenish metallic Males: possess a large, shiny hypopygium
Muscina stabulans	7–9	Larvae: saprophagous Adults: on fruits, walls, trees, markets	Adults: body *Musca*-like, yellowish legs Larvae: spiracle slits widely separated
Calliphora vicina (syn. *erythrocephala)*	8–10	Larvae: feces, wounds Adults: close to or in houses	Body bluish shiny Males: face yellowish brownish
Calliphora vomitoria	10–14	Larvae: feces, wounds Adults: rare in houses	Body bluish, head reddish, fly sound dense
Tachina fera	12–15	Larvae: in larvae of butterflies Adults: on flowers	Head yellow, thorax black, abdomen yellow
Fannia canicularis	4–6	Larvae: feces, decaying material Adults: toilets, dung	Adults are covered with grey hair, thorax brownish-grey with three brown stripes
Pollenia angustigena *Pollenia rudis*	5–12	Larvae: in earthworms Adults: overwinter in houses	Adults appear blackish with square dots on the abdomen

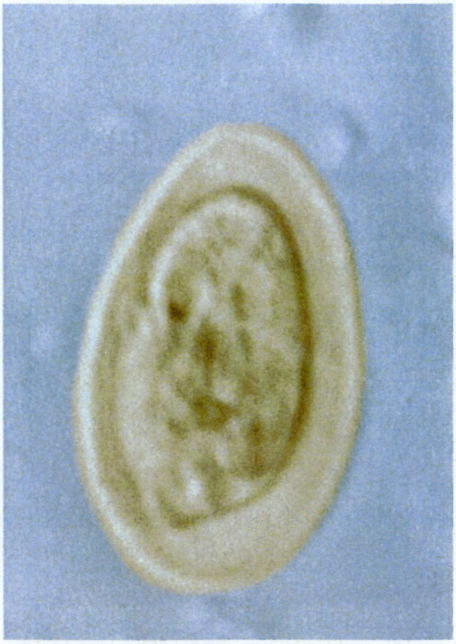

Fig. 10.14 Type of *Giardia* cysts seen in feces

MIFC method, a very similar range of parasites was found. This proves that the flies caught had intensive contact with the feces in the pigpen (Figs. 10.16 and 10.17).

10.5.2 Parasitic Load of Flies at Cattle Barns

The findings at the cattle barn are summarized in Table 10.3. The number of parasites observed seems low. However, considering that in the surroundings of the cattle hundreds of flies have several hundred touchdowns on their hosts in a few minutes, the infection rate might be estimated as very high.

10.5.3 Parasitic Load of Flies at Horse Stables

The parasites for 50 flies caught close to horses are summarized in Table 10.4.

The results of this series of investigations corresponded to the findings in other catches. Although again the number of observed parasitic stages seems low, it must be considered that propagation may occur on any of the numerous touchdowns or on any defecation of the flies, which molest the animals the whole day long for weeks. It is also noteworthy that *Parascaris equorum*, which is an important agent

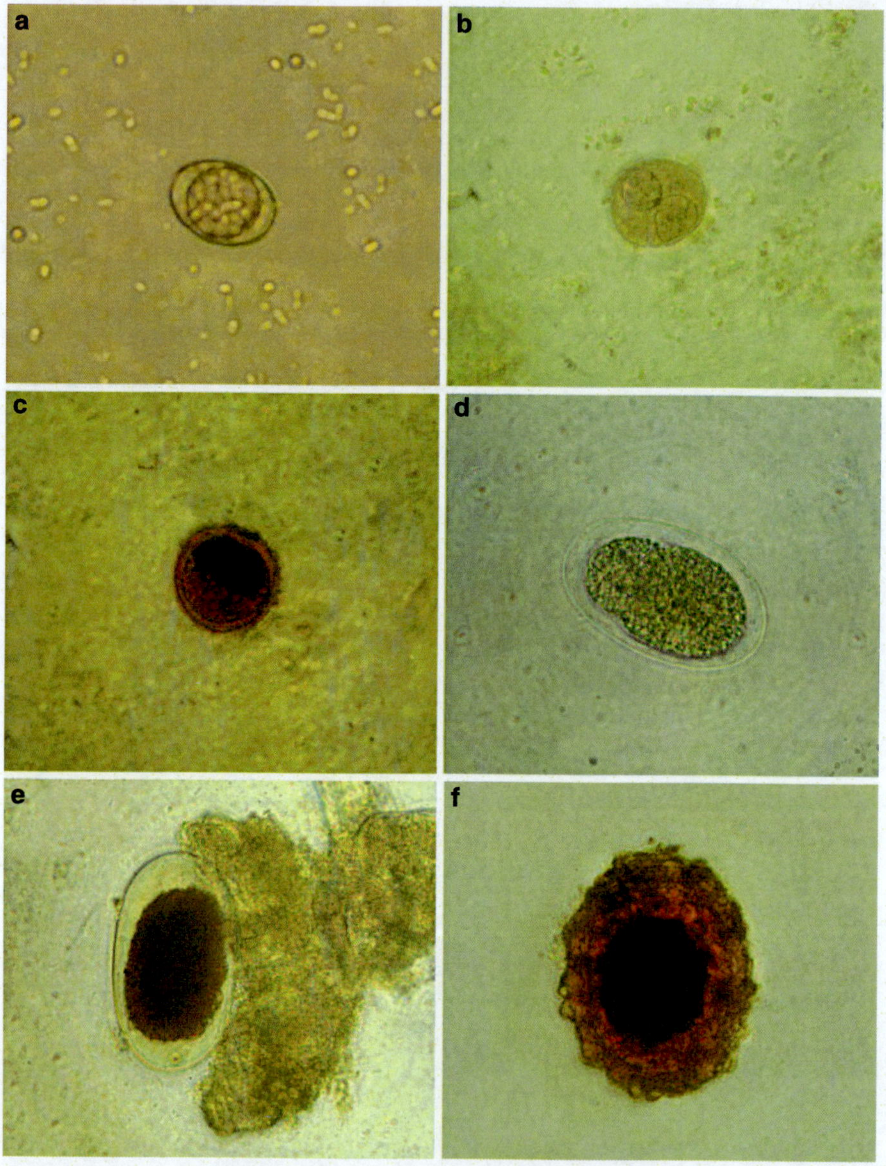

Fig. 10.15 Micrographs of parasites seen on or inside flies caught in pigpens. (**a**, **b**) Unsporulated and sporulated oocysts of *Isospora suis*. (**c**) *Balantidium coli* cyst. (**d–f**) Eggs of nematodes

of horse disease, was among the parasites detected on the body of flies as well as in their feces (Fig. 10.18).

Table 10.2 The results of the examination of 593 flies caught close to a pigpen

Parasite	Stage	Number of flies testing positive	Number of parasites detected	
			IN	EXO
Giardia lamblia	Trophozoite	2	100	–
	Cyst	2	–	8
Eimeria perminuta	Oocyst	2	8	–
Isospora suis	Oocyst	7	18	70
Balantidium coli	Cyst	13	16	–
Ascaris suum	Egg	12	4	16
Metastrongylus apri	Egg	8	–	74
	Larva	8	–	29
Strongyloides ransomi	Egg	4	72	7
	Larva	4	11	4
Strongyloidea	Egg	27	19	24
	Larva	27	21	17
Haematopinus suis	Nymph	1	–	1
Macrocheles muscae domestica	Adult	6	–	18
Histiostoma sp.	Nymph	2	–	2

EXO, on the surface; IN, in the intestine, in the feces

10.5.4 Parasitic Load of Flies at Rabbit Cages

The parasites observed on flies that were caught close to rabbit cages are listed in Table 10.5.

This list contains two *Eimeria* species which have high pathogenic effects on rabbits, which may even die during a massive infection with these organisms. In addition, the load with other parasites was comparably high.

10.5.5 Parasitic Load of Flies at Dog Meadows

The parasites that were found on flies close to a dog meadow are summarized in Table 10.6. Repeated catches always showed similar relationships. Thus, it can be concluded that this focal glimpse represents the general infection situation around the dog meadow. With respect to pathogenicity, there are three species (two *Cystoisospora* species and *Toxocara canis*) which are very dangerous especially for young dogs. *Toxocara canis* eggs have, in addition, a zoonotic potential, since the larvae may start a migration phase in humans (again especially in young hosts, i.e., children). If the eyes or the brain becomes parasitized by *Toxocara canis*, the damage may be severe.

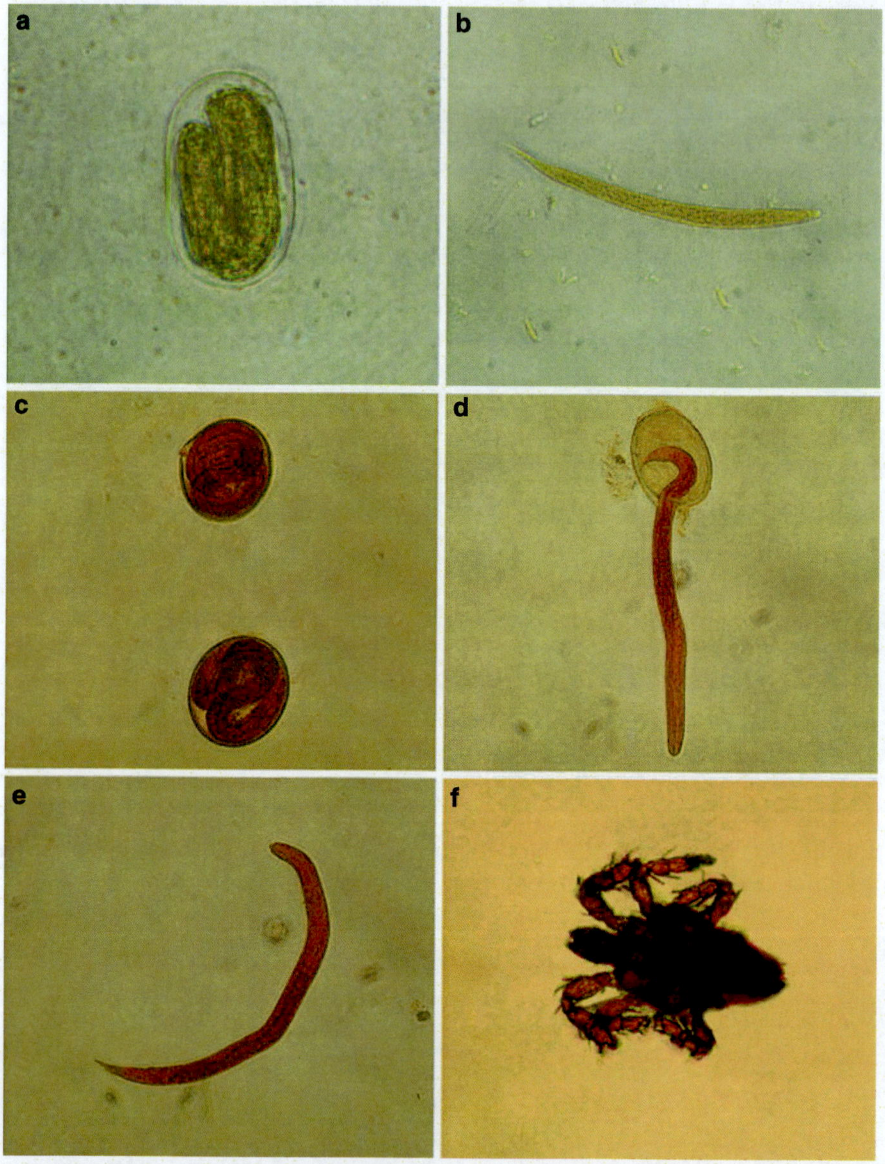

Fig. 10.16 Micrographs of parasites found in and on flies caught at a pigpen. (**a**, **b**) Egg and larva of *Strongyloides ransomi*. (**c**–**e**) Eggs and larvae of *Metastrongylus apri*. (**f**) Nymph of the bloodsucking pig louse *Haematopinus suis*

10.5.6 Parasitic Load of Flies at Downtown Recreation Sites

The rather few flies collected at recreation sites in the surroundings of cities also contained parasites. However, as well as the number, the variety of parasitic species

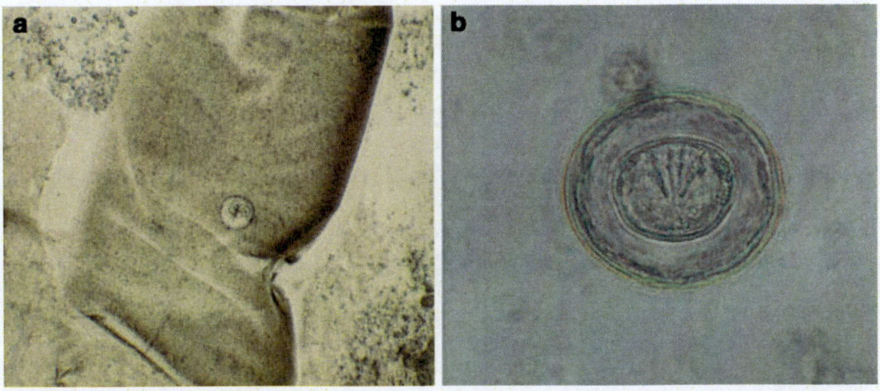

Fig. 10.17 Light micrographs of *Hymenolepis diminuta* in the fly's intestine (**a**) and in the washing solution from the surface (**b**)

was much lower than that of flies caught close to stables. Most common were the tiny, only 5 μm in diameter oocysts of species of the genus *Cryptosporidium*. Of course, the exact species could not be determined since no morphological criteria are available. However, many *Cryptosporidium* species are not very host specific, so many of them may introduce human infections especially in immunosuppressed, young, or very old people. Since such oocysts are found on flies as well as in their feces, this proves that the flies had touchdowns on feces and had also ingested portions of the feces. In addition to the cryptosporidians, eggs of mites were found on the surface of flies. These mites had apparently also been attracted by feces at the recreation sites investigated.

10.5.7 Parasitic Load of Bird Feces Collected at Recreation Sites

Since flies of practically all species visit at least temporarily feces of many sources, bird feces was also collected at the recreation sites and examined for its parasitic content. Since all sites investigated had large or small watering places, water birds were most common in the surroundings. The microscopic investigation of their collected feces with the help of the MIFC method revealed a broad spectrum of parasites. There were oocysts of species of the genera *Cryptosporidium*, *Isospora*, *Eimeria*, and *Sarcocystis*. These might have developed inside the birds' intestines and/or may have just been passengers after the birds had ingested them during feeding. The same is true for numerous eggs and larvae of nematodes also found in the feces. The finding of eggs of the tapeworm genus *Hymenolepis* completes the spectrum of potentially transmitted intestinal parasites.

Considering that the MIFC method is effective only for rather high infection rates, since only very small portions (i.e., 1 g) of collected feces were investigated,

Table 10.3 Parasites found on or in 251 *Musca domestica* flies at cattle barns

Parasite	Stage	Number of flies testing positive	Number of parasites detected	
			IN	EXO
Giardia lamblia	Trophozoite	1	10	–
Herpetomonas muscarum	Trophozoite	12	610	–
Toxocara vitulorum	Egg	2	2	2
Undetermined nematode	Larvae	1	–	1
Histiostoma sp.	Nymph	1	–	14
Macrocheles muscae domestica	Adult	1	–	4
Undetermined mite	Adult	1	–	1

EXO, on the surface; IN, in the intestine

Table 10.4 Parasite load on or in 50 flies caught close to a horse stable

Parasite	Stage	Number of flies testing positive	Number of parasites detected	
			IN	EXO
Parascaris equorum	Egg	3	2	2
Strongylid nematodes	Egg	6	3	3
	Larva	8	8	24
Undetermined mite	Egg	2	–	2
	Adult	1		1
Histiostoma sp.	Nymph	1	1	1
Macrocheles muscae domestica	Adult	4	–	8

EXO, on the surface; IN, in the intestine

it is clear that feces of any kind at recreation sites contains considerable numbers of parasites, which—depending on their host specificity—may harm humans.

Thus, the results obtained with respect to parasites are in good agreement with the findings of numerous bacteria and fungi at the same time (Gestmann et al. 2012).

10.6 Infections of Flies After Their Experimental Exposure to Parasites

To check the efficacy of flies in the uptake of agents of disease, repeated experiments were done and showed similar results, such as with the following test with 120 laboratory-bred *Musca domestica* flies of both sexes. The flies were offered a milk fluid (see Sect. 10.2) that contained eggs of the nematode species *Ascaris suum* and *Trichuris suis*. These species are rather common in normal pigpens and occur in even higher numbers in "biological" farms. The intestines of 13 flies (10.8%) contained eggs of both species, whereas nematode eggs of one or the other species were detected in 46 flies (38.3%). Eggs of *Ascaris suum* (840 eggs had been included in the food solution) were found in or on 31 flies (25.8%), and eggs of *Trichuris suis* (200 eggs had been placed in the food solution) were detected

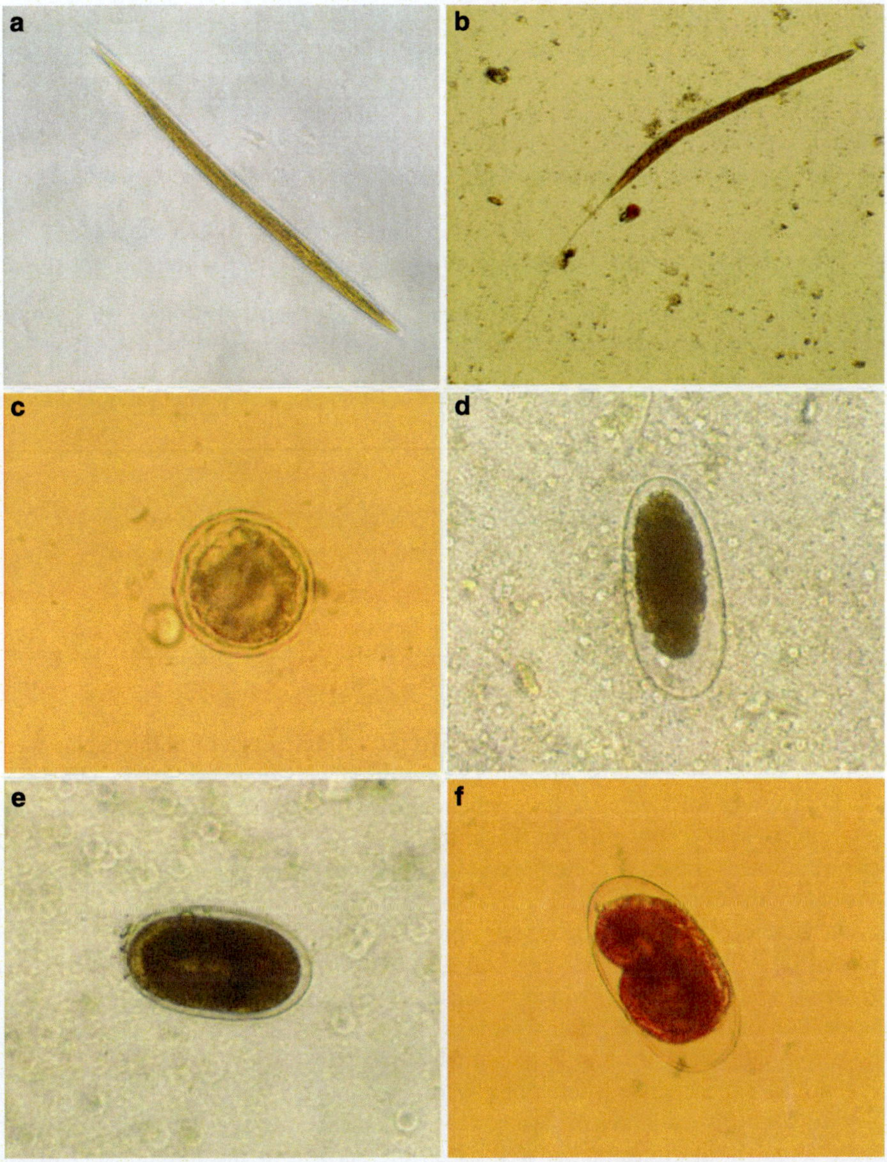

Fig. 10.18 Parasites found on or in flies close to a horse stable. (**a, b**) Larvae of horse strongylids. (**c**) Egg of *Parascaris equorum*. (**d–f**) Eggs of horse strongylids seen in fly feces

in the intestines of 41 flies (34.2%). The individual load per fly ranged from one to 11 eggs in the case of *Ascaris suum* and from one to six eggs in the case of *Trichuris suis*. In total, 77 of the original 840 eggs of *Ascaris suum* were refound in the intestine, as were 73 of the original 200 *Trichuris suis* eggs. In addition, 36 worm

Table 10.5 Parasites on or in 226 flies caught close to rabbit cages

Parasite	Stage	Number of flies tesing positive	Number of parasites detected	
			IN	EXO
Eimeria magna	Oocyst	1	6	–
Eimeria perforans	Oocyst	17	23	4
Strongylid nematodes	Larva	3	7	10
Undetermined mites	Egg	4	–	10
	Adult	5	–	5
Macrocheles muscae domestica	Adult	1	–	1

EXO, on the surface; IN, in the intestine

Table 10.6 Parasites on or in 248 flies caught at a dog meadow

Parasite	Stage	Number of flies testing positive	Number of parasites detected	
			IN	EXO
Cystoisospora burrowsi	Oocyst	3	110	4
Cystoisospora canis	Oocyst	4	2	2
Herpetomonas muscarum	Trophozoite	11	545	62
Toxocara canis	Egg	6	2	8
Undetermined nematodes	Larva	2	1	1
Histostoma sp. mite	Nymph	1	–	1
Undetermined mite	Adult	1	–	1

EXO, on the surface; IN, in the intestine

eggs were found on the surface of the fly body. Single eggs of both species were detected on the mouthparts, whereas 29 eggs of *Ascaris suum* occurred on the body as did five eggs of *Trichuris suis*, which apparently had less gluing capacity. This experiment shows how efficient flies might be in their capacity for transmission of parasitic stages if the latter are present in a reasonable number inside the substrates visited by the flies.

In another experiment, laboratory-bred specimens of *Musca domestica* were placed on a fluid containing eggs of the tapeworms *Hymenolepis diminuta* and *Hymenolepis microstoma*, which were cultivated in laboratory rats or mice. From inspection of the surface (after washing) and the intestinal contents, the infectious eggs were all found again in large numbers (Fig. 10.18).

Further experiments where flies were exposed to laboratory-bred larvae of *Angiostrongylus cantonensis* again showed positive results, since these larvae were found in the intestine as well as on the surface of the flies. Owing to the possible danger of an infection with *Angiostrongylus cantonensis*, only larvae 1 were used and not larvae 3 (which are infectious to humans and which are normally excreted by the intermediate host, i.e., snails) (Fig. 10.19).

Similar positive results for a potential fly transmission were obtained when different stages (eggs, larvae) of the bloodsucking so-called red chicken mite *Dermanyssus* were exposed to flies. The mites originated from a culture kept in the institute. Numerous eggs and larvae of the mites (Fig. 10.20) were recovered in and on flies exposed to such mite cultures. Furthermore, many of the flies caught

Fig. 10.19 Light micrograph of a larva 1 of *Angiostrongylus cantonensis* refound in flies after their exposure to a fluid with such larvae

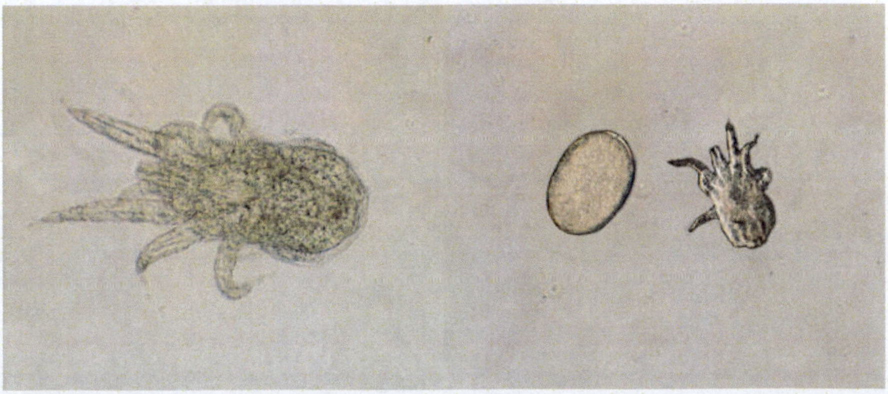

Fig. 10.20 Light micrographs of an egg and of larvae of the red chicken mite *Dermanyssus gallinae*

in the wild were covered with "traveler mites" (Figs. 10.21 and 10.22). This phenomenon shows how easy it is for pathogenic mite species to be translocated from the wild or stables to human dwellings.

The broad spectrum of parasites found on flies caught in the wild and their rates of infection with different parasites between 0.3 and 5.5% show the high potential of flies to act as vectors of parasites. This is even more understandable considering the individual infection rate (about 0.003%) of mosquitoes in highly endemic malaria regions. This latter very low infection rate of mosquitoes is, however, sufficient to infect practically all human inhabitants in many regions of the world.

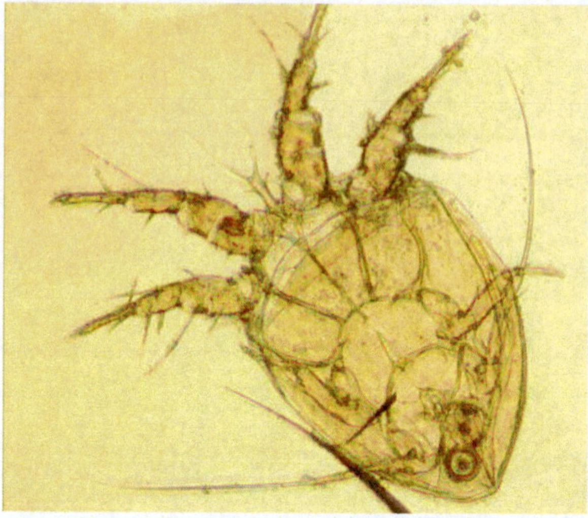

Fig. 10.21 Light micrograph of *Histiostoma* sp., a traveler mite on flies

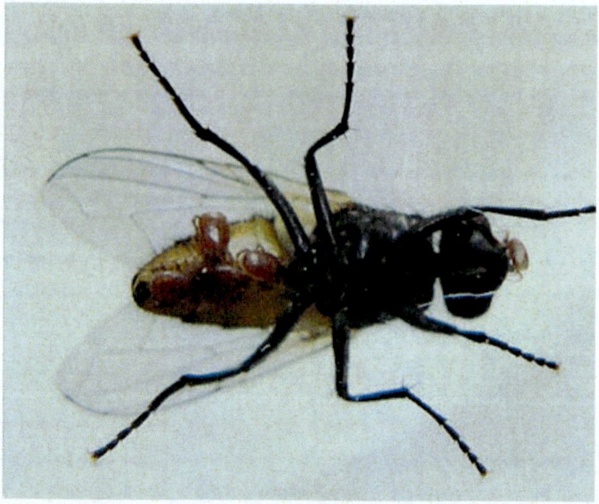

Fig. 10.22 Light micrograph of a fly with three migrating *Macrocheles* mites (*red dots*)

10.7 Conclusions

This review of fly catches close to various animal stables, close to dog meadows, and close to downtown recreation sites shows that there are site-specific flies depending on the available substrates (feces, etc.). It was found that the parasites which were detected in the feces of the farmed animals or in that of the dogs or the birds at recreation sites could also be detected on or in the bodies of the flies caught nearby.

These fly catches and the experimental exposure of flies to worm eggs show that flies regularly take up any available parasitic stage and are able to transport it to other places. Similar effects were discussed by Gestmann et al. (2012) in Chap. 9 with respect to transmission of microorganisms such as bacteria and fungi by flies. All these findings confirm the findings reported in numerous articles where it was shown that human pathogenic parasites such as *Cryptosporidium* species, *Giardia lamblia* cysts, eggs of *Ascaris lumbricoides*, *Hymenolepis nana*, *Hymenolepis diminuta*, *Taenia* species, and *Trichuris trichiura* might be transmitted by different licking flies (Szostakowska et al. 2004; Rao et al. 1971; Greenberg 1971, 1973; Dipeolu 1977, 1982; Umeche and Mandah 1989; Howard 2001). Besides the transportation of important dangerous agents of infectious diseases, there is also a spreading of pathogenic and nonpathogenic mites (Figs. 10.21 and 10.22).

The present examinations and those in the past prove that flies are especially dangerous because of their transmission of zoonotic parasites in regions where it is believed that there is a high standard of hygiene. In these countries, the transmission of parasites by flies is often neglected. It is believed that this type of parasite transmission only occurs in countries where human feces is found outside houses or in badly cleaned toilets. The finding of human parasites in the present study done close to large towns or at inner recreation sites shows that there is high potential for underestimated transmissions not only of microorganisms (Förster et al. 2012) but also of parasites. Species of the genera *Cryptosporidium*, *Giardia*, *Ascaris*, *Toxocara*, *Trichuris*, etc., have a high pathogenic potential for humans. Their finding, however, underlines that the flies may carry any agent of disease which is available in their substrates, where they lick their food, or wherein their larvae grow. In this context potential transmission has to be remembered even in (as the inhabitants believe) "clean" countries at the numerous places where human feces lies around (e.g., at camping sites, parking lots along highways—only very few are equipped with toilets—and also at construction sites in mid-downtown).

In urban surroundings, animal stables offer excellent breeding sites for flies, which then may become infected when feeding on the feces of the animals. Thus, there must be intense fly control around the stables by use of ecologic measures and additionally by use of insecticides. A particular problem is posed by dog feces especially in towns, since this offers breeding and feeding sites for flies and is a reservoir for reproducing microorganisms and also for zoonotic parasites such as *Toxocara canis* and *Giardia* strains (Mehlhorn 2008; Eckert et al. 2008). Thus, regular cleaning of such dog meadows is absolutely needed.

The infection rates of the flies caught in the wild and/or at downtown recreation sites seem low at first sight. However, when it is considered that each day each fly has hundreds of touchdowns on humans and animals and/or on feces, the chances of transmission of agents of diseases are extremely high. For example, the agents of one of the human malaria species (*Plasmodium falciparum*) that occur in only three of 1,000 female *Anopheles* mosquitoes in highly endemic regions may induce a high percentage of human infections. However, when the numerous bites per day that unprotected inhabitants suffer in certain regions of Africa or Asia are counted, there are so many infected mosquitoes that practically all people in these regions have been infected with malaria by the time they reach their tenth year of life. Similar rates can be reached in, on, or close to our big towns with respect to transmission of microorganisms by flies, since a single touchdown with transfer of a significant amount of agents of diseases might be the starting point for various diarrheas or other diseases with severe consequences. Therefore, we must maintain a constant watch on and control of flies as potential vectors of emerging diseases.

Acknowledgments We gratefully acknowledge support of the studies that are the basis of this chapter by Novartis, Basel, and by the joint governmental project (the Netherlands/North Rhine-Westphalia) Safeguard, subproject *Campylobacter*. Furthermore, we are grateful to I. Schaefers and S. Walter, who typed the text and arranged the photographic plates, respectively.

References

Anderson RC (2000) Nematode parasites of vertebrates: their development and transmission, 2nd edn. CAPI, New York

Banjo AD, Lawal OA, Adeduji OO (2005) Bacteria and fungi isolated from house fly (*Musca domestica* L.) larvae. Afr J Biotechnol 4:780–784

Bramley AJ, Hillerton JE, Higgs TM, Hogben EM (1985) The carriage of summer mastitis pathogens by muscoid flies. Br Vet J 141:618–627

Bush O, Lafferty AD, Lotz JM, Shostak AW (1997) Parasitology meets ecology on its own terms: Margolis et al. revisited. J Parasitol 83:575–583

Busse FW, Aka E (1997) Vorkommen von Endo- und Ektoparasiten bei Schweinen im Weser-Ems-Gebiet sowie Behandlungsstrategien. Prakt Tierarzt 78:686–691

Chinery M (2002) Pareys Buch der Insekten, 3rd edn. Parey Buchverlag, Berlin

Damriyasa IM, Failing K, Volmer R, Zahner H, Bauer C (2004) Prevalence, risk factors and economic importance of infestations with *Sarcoptes scabiei* and *Haematopinus suis* in sows of pig breeding farms in Hessen, Germany. Med Vet Entomol 18:361–367

Dipeolu OO (1977) Field and laboratory investigations into the role of the *Musca* species in the transmission of intestinal parasitic cysts and eggs in Nigeria. J Hyg Epidemiol Microbiol Immunol 21:209–214

Dipeolu OO (1982) Laboratory investigations onto the role of *Musca vicina* and *Musca domestica* in the transmission of parasitic helminth eggs and larvae. Int J Zoonoses 9:57–61

Eckert J, Friedhoff KT, Deplazes P, Zahner H (2008) Parasitologie für die Veterinärmedizin, 2nd edn. Enke, Stuttgart

Fischer G, Pospischil G, Nentwig G (2001) Flies as transmitters of tuberculosis. Med Vet Entomol 15:126–138

Förster M, Klimpel S, Mehlhorn H, Sievert K, Messler S, Pfeffer K (2007) Pilot study on synanthropic flies (e.g. *Musca, Sarcophaga, Calliphora, Fannia, Lucilia, Stomoxys*) as vectors of pathogenic microorganisms. Parasitol Res 101:243–246

Förster M, Klimpel S, Sievert K (2009a) The house fly (*Musca domestica*) as a potential vector of metazoan parasites caught in a pig-pen in Germany. Vet Parasitol 160:163–167

Förster M, Messler S, Pfeffer K, Sievert K (2009b) Synanthropic flies as potential transmitters of pathogens to animals and humans. Mitt Dtsch Ges Allg Angew Entomol 17:327–329

Gestmann F, Förster M, Mehlhorn H, Sievert K, Messler S, Neuhausen N, Petersdorf S, Pfeffer K (2012) Flies as vectors of microorganisms inducing severe diseases in humans and animals. In: Mehlhorn H (ed) Arthropods as vectors of emerging diseases, vol 3, Parasitology Research Monographs. Springer, Heidelberg

Greenberg B (1971) Flies and disease, vol I. Princeton University Press, Princeton

Greenberg B (1973) Flies and disease, vol. 2, Princeton University Press, Princeton

Haupt J, Haupt H (1998) Fliegen und Mücken. Naturbuch Verlag, Augsburg

Howard J (2001) Nuisance flies around a landfill: patterns of abundance and distributions. Waste Manag Res 19:308–313

Lane RP, Crosskey RW (1993) Medical parasites and arachnids. Chapman and Hall, London

Löscher T, Burchard GD (eds) (2010) Tropenmedizin in Klinik und Praxis. Thieme, Stuttgart

Mehlhorn H (ed) (2008) Encyclopedia of parasitology, 3rd edn. Springer, Berlin

Mehlhorn H (2012) Die Parasiten des Menschen und die von ihnen hervorgerufenen Erkrankungen. Spektrum, Heidelberg

Mehlhorn H, Düwel D, Raether W (1993) Diagnose und Therapie der Parasiten von Haus-, Nutz- und Heimtieren, 2nd edn. Fischer, Stuttgart

Mehlhorn H, Eichenlaub D, Löscher T, Peters W (1995) Diagnosis and therapy of human parasitosis, 2nd edn. Fischer, Stuttgart

Mehlhorn B, Mehlhorn H, Walldorf V (2012) Schach den Blutsaugern und Schädlingen. Düsseldorf University Press, Düsseldorf

Nansen P, Roepstorff A (1998) Worm infections in pigs. A general review with particular references to epidemiology and control. In: Proceedings of the 15th IPVS congress, Birmingham, UK

Olsen AR, Hammack TS (2000) Isolation of *Salmonella* spp. From the housefly *Musca domestica* L., and the dump fly, *Hydrotaea aenescens* (Wiedemann) (Diptera: Muscidae), at cage-layer houses. J Food Prot 63:958–960

Olsen LS, Kelly GW, Sen HG (1958) Longevity and egg-production of *Ascaris suum*. Trans Am Mircosc Soc 77:380–383

Rao CK, Krishnaswami AK, Gupta SR, Biswas H, Raghavan NGS (1971) Prevalence of amoebiasis and other intestinal parasitic infections in a selected community. Indian J Med Res 59:1365–1373

Roepstorff A, Nansen P (1994) Epidemiology and control of helminth infections in pigs under intensive and non-intensive production system. Vet Parasitol 54:69–85

Rommel M, Eckert J, Kutzer E, Körting W, Schnieder T (2002) Veterinärmedizinische Parasitologie, 6th edn. Parey, Berlin

Schaefer M (2002) Brohmer-Fauna von Deutschland, 21st edn. Quelle & Meyer, Wiebelsheim

Szostakowska B, Kruminis-Lozowska W, Racewicz M, Knight R, Tamang L, Myjak P, Graczyk TK (2004) *Cryptosporidium parvum* and *Giardia lamblia* recovered from flies on a cattle farm and in a landfill. Appl Environ Microbiol 70:3742–3744

Umeche N, Mandah LE (1989) *Musca domestica* as a carrier of intestinal helminths in Calabar, Nigeria. East Afr Med J 66:349–352

Weidner H, Sellenscho U (2010) Vorratsschädlinge und Hausungeziefer, 7th edn. Spektrum, Heidelberg

Chapter 11
Lice as Vectors of Bacterial Diseases

Günter A. Schaub, Astrid H. Kollien, and Carsten Balczun

Abstract Two species of lice are ectoparasites of humans, *Pediculus humanus* with the two subspecies *P. humanus capitis* (head louse), living in head hair, and *P. humanus humanus* (body or clothing louse), staying in clothes and approaching the skin only to feed, and *Pthirus pubis* (crab louse), colonising mainly the hairs of the genital region. Besides humans, only some monkeys are hosts of these obligate parasites that spend the entire life cycle on the host. The strong association with the host is reflected in morphological peculiarities: very short antennae, reduced eyes, and no wings. Since they are hemimetabolous insects, all postembryonic stages suck blood. Symbionts are located in mycetomes and are transmitted during egg development. Only *P. humanus humanus* is relevant as a vector of *Rickettsia prowazekii* (louse-borne epidemic typhus), *Bartonella quintana* (trench or 5-day fever), and *Borrelia recurrentis* (louse-borne relapsing fever). Since economic instability and wars cause large migrations of humans, louse infestations become more prevalent, increasing the risk of transmission of neglected diseases.

Keywords Hematophagy • lice control • pathogen • Pediculus • Pthirus

11.1 Introduction

The co-evolution of lice and humans started about 12 or 19 million years ago (summarised by Aspöck and Walochnik 2007; Habedank 2010). According to calculations of the molecular biology clock, the ancestors of hominids possessed ancestors of the extant lice species. During evolution of hominids, the thick body hairs were reduced and the body was covered in a layer of fine hair, but on the head

G.A. Schaub (✉) • A.H. Kollien • C. Balczun
Group Zoology/Parasitology, Ruhr-Universität Bochum, Universitätsstr. 150, 44801 Bochum, Germany
e-mail: guenter.schaub@rub.de

H. Mehlhorn (ed.), *Arthropods as Vectors of Emerging Diseases*, Parasitology Research Monographs 3, DOI 10.1007/978-3-642-28842-5_11,

and the genital regions the thick hairs remained. Several reasons have been suggested for this evolution; one of them is the defense reaction against ectoparasites, resulting in a removal of refuges and an increase in the chance of detecting the ectoparasites (Dean and Siva-Jothy 2012). By the change to fine body hairs, colonies of the ancestor lice remained in the two regions with thick hairs and developed to head and crab lice, about 13 million years ago (Habedank 2010). During migration into colder regions, about 70,000–40,000 years ago, hominids protected the body with skins, and clothing lice developed (Kittler et al. 2003; Habedank 2010). However, on the timescale of speciation, this is a relatively short period of separation. Thus, the classification as subspecies—*Pediculus humanus capitis* and *P. humanus humanus* (synonym *P. humanus corporis*)—or as species—*P. capitis* and *P. humanus*—depends on the respective author (see Sect. 11.2.2). The search for lice at archaeological sites is complicated by degradation of hairs, but a 10,000-year-old strand of human hair contained an egg of a head louse (Araújo et al. 2000).

11.2 Taxonomy

11.2.1 General Taxonomy

Groups of the order Phthiraptera offer an interesting example of the development to bloodsucking ectoparasites. The suborders Amblycera (3,000 species) and Ischnocera (300 species) are ectoparasites of birds and mammals, feeding on keratinous substances of feathers, hairs, and the upper layers of the skin, dermal secretion fluids, and in some species on blood (Dettner 1999; Habedank 2010). The head is broader than the thorax, and the chewing mouthparts are visible below the head. Only three species belong to the suborder Rhynchophthirina. They suck blood of elephants and African pigs (Habedank 2010). The head is prolonged to a long rostrum with mandibles at the distal end to open the skin (Dettner 1999). The suborder Anoplura (synonym Siphunculata) contains about 500 species, all bloodsuckers with a maximum length of 6 mm (Habedank 2010).

11.2.2 Taxonomy of Head and Body Lice

The debate whether or not body and head lice are species, subspecies, or races started about 100 years ago (Hase 1915; summarised by Habedank 2010). Both remain distinct when they occur on the same individual, a strong indicator to classify them as different species, *P. capitis* and *P. humanus* (Busvine 1978). However, on strongly infested humans, body lice also occupy the head hairs and head lice also occupy the body hairs, and in the laboratory intermediate forms can

Table 11.1 Characteristics of lice of humans

Characteristics	*Pediculus humanus capitis*	*Pediculus humanus humanus*	*Pthirus pubis*
Length[a] of eggs	0.5–0.8[m]	0.6–0.8[m]	<1[h]
Duration[b] of embryogenesis	8[g], 6–9[m]	7[g], 5–6[m]	5–8[g], 7–8[i], 5–8[m]
Duration[b] of larval development	7–10[f]	7–10[f], 8[g], 13[m]	13–17[i], 27[m]
Length[a] of males	2.4–2.6[g], 2.4–2.9[m], 2.0–3.0[n]	2–3[k], 2.7–3.8[m], 2.7–3.7[n]	1.3–1.6[m]
Length[a] of females	2.6–3.1[g], 2.6–4.0[m], 2.1–3.5[n]	3–4[k], 3.0–4.8[m], 2.7–4.4[n]	1.3–1.6[m]
Lifespan[b] of males		31[e], 29[m]	
Lifespan[b] of females	9[f], 31.9[l], 9–10 or 20–25[m], 36[n]	34[e], 9[f], 30–40[g], 30–60[m]	26[g], 30[m]
Number of eggs/day	6[f], 6.6[l], 4–9[m], 3–7[n]	9–10[e], 12[f], 5–14[g], 5–7(10)[m]	3[f], 1–3[m]
Number of eggs/lifespan	55[f], 210[l], 50–150[m]	270–300[e], 110[f], 300[g], 200–300[m]	30[g], 40–50[m], 26–50[n]
Generation time[b] (egg–egg)	21[g], 17[m], 17[n]	15[g], 22–23[m]	25[g]
Maximal starvation capacity[c] (incubation temperature[d])	55[f](23), 30[m](24)	85[f](23)	24–48[m], 26–42 (20)[n]
Temperature[d] preference	28–29[n]	31–33[g], 31–33[m]	
Lethal temperature[d] (incubation period, min)		50 (<30)[g], 90–100 (1)[g]	

[a]In millimeters
[b]In days
[c]In hours
[d]In degrees Celsius
[e]Data from Buxton (1947)
[f]Data from Ibarra (1993)
[g]Data from Peters (1999)
[h]Data from Burkhart et al. (2000)
[i]Data from Wenk and Renz (2003)
[k]Data from Service (2004)
[l]Data from Lang cited by Burgess (2004)
[m]Data from Mehlhorn and Mehlhorn (2009)
[n]Data from Habedank (2010)

be produced. Correlated with the distribution is the temperature preference, and according to the temperature in the microhabitat, body lice prefer higher temperatures (Table 11.1).

The differences induced the search for morphological criteria. If glued to body hairs, head lice use more glue to connect the egg with the hair (Peters 1999; Mehlhorn and Mehlhorn 2009). *P. humanus capitis* is slightly smaller than *P. humanus humanus* (Table 11.1), but the lengths of the body overlap (Busvine 1978). In addition, in *P. humanus capitis* antennae, tibiae, and the second pairs of legs are shorter than in *P. humanus humanus*, and the abdominal segmental borders are laterally more distinctly separated (Busvine 1978; Habedank 2010).

Some molecular biological investigations supported a conspecific development, others, using sequencing of microsatellite DNA, the classification as separate species, and others indicated two genetically distinct lines developing to body and head lice or only head lice (Leo and Barker 2002, 2005; Leo et al. 2002, 2005; Reed et al. 2004; summarised by Li et al. 2010). A large-scale investigation of mitochondrial DNA markers shows more similarities between local populations of head and body lice than between populations of both subspecies originating from different countries, i.e., body lice arose several times from head lice (Leo et al. 2002; Li et al. 2010). Therefore, we follow the classification of body and head lice as subspecies.

11.3 Distribution and Morphology of Human Lice

Owing to the strong co-evolution, human lice are cosmopolitans. Since all species are wingless, the transmission requires an intimate contact or the common use of fomites such as combs and head gear (Habedank 2010). On the head and in the clothes of Eskimos and dark-skinned people, black head and body lice are present, not the normal pale beige or grayish head and body lice (Raoult and Roux 1999; Peters 1999) (Fig. 11.1), and in Brazil all intermediate colour forms of body lice can be found owing to reproduction of black and normal-coloured lice (Kollien unpublished).

Head lice, *P. humanus capitis*, prefer the hairs on the neck and behind the ears, but in strong infestations they colonise all regions on the head and also eyelashes, eyebrows, and hairs in the armpit (Habedank 2010). They are mainly transmitted by head-to-head contact, and transmission is less likely by fomites such as combs and caps (Burgess 2004; Takano-Lee et al. 2005).

Body lice, *P. humanus humanus*, especially infest humans who rarely change and wash their clothes or possess only one set of clothes. Since lice react sensitively to high temperatures, body lice are more prevalent in regions with a temperate climate than in the tropics (Peters 1999). They prefer clothes, but in strong infestations they also colonise the hairs of the head (Habedank 2010). Body lice are adapted to the change to nighttime clothes and survive longer periods without a host than head lice. The morphology of both *Pediculus* ssp. is very similar (see Sect. 11.2.2), and the three pairs of legs are of similar lengths (Figs. 11.1 and 11.2a).

Crab lice, *Pthirus pubis*, are mainly transmitted during sexual contact; other modes of transmission seem to occur occasionally (Habedank 2010). The prevalence of crab lice might be reduced by the increasing shaving of hair in the genital region, but this species prefers the thickest hairs and thus also colonises the coarse hairs of the head and body, such as eyelashes, eyebrows, and hairs in the armpit (Peters 1999; Schaub 2008). The morphology of *Pthirus* differs strongly from that of *Pediculus*. In *Pthirus*, the thorax and abdomen are broader, and the legs stretch more sidewards, thus resembling the appearance of a crab (Habedank 2010). The first pair of legs are smaller than the second and third pairs (Fig. 11.2b).

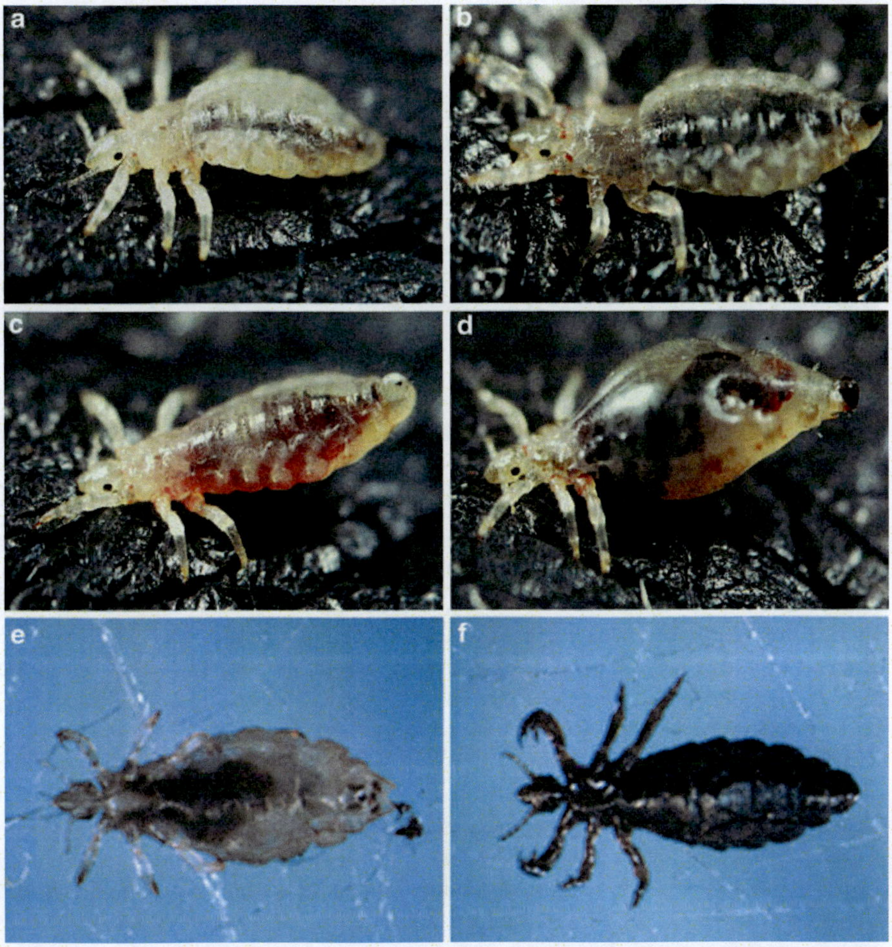

Fig. 11.1 Light micrographs of a female of *Pediculus humanus humanus* about 3.5 mm in length during bloodsucking on human skin (**a**–**d**). The skin was black-inked to enhance the contrast of the micrograph. Light micrographs of typical beige (**e**) and black (**f**) morphs of *P. humanus humanus* (photographs by A.H. Kollien)

Lice possess some morphological peculiarities: very short antennae (five segments), reduced eyes, no wings, and legs which have strong claws to cling onto the hairs of their hosts or fibers of the clothes (Schaub 2008). These curved claws at the end of the tarsus are opposite a thumb-like spine of the short tibia and thereby firmly grip hairs or fibers of clothes between them (Service 2004). In addition, the three segments of the thorax are fused, the end of a fusion tendency within the Phthiraptera (Dettner 1999). Also three segments of the abdomen are fused (Wenk and Renz 2003). The cuticle of lice is leathery and tough and not easy to crush. In both *Pediculus* ssp., females are slightly bigger than males. Females and males of *P. pubis* are of similar size and about 40% shorter than *Pediculus* ssp. (Table 11.1).

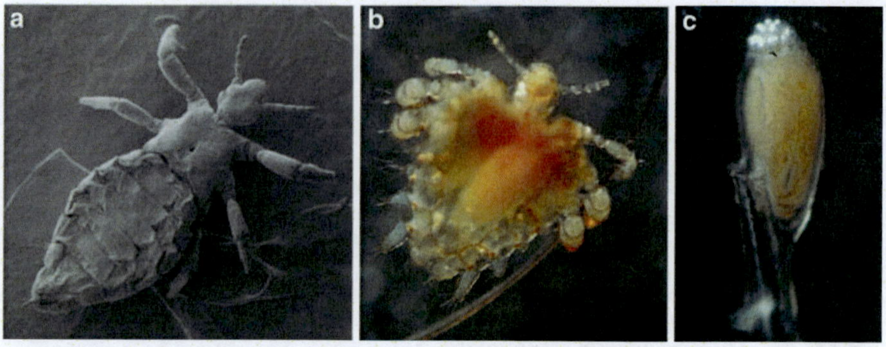

Fig. 11.2 Scanning electron micrograph of a male of *P. humanus humanus* (**a**). Light micrograph of an adult of *Pthirus pubis* gripping a human hair (**b**). Light micrograph of an egg of *P. pubis* glued to a human hair (**c**); the developing embryo is visible through the eggshell, and aeropyles for air and moisture supply of the embryo are located at the top of the egg (photographs by A.H. Kollien)

11.4 Biology of Human Lice

Lice spend their entire life cycle on the host and usually leave it only for transfer to a new host. However, if the host has a fever or dies, this induces a migration of the lice. All lice are highly host specific. Human lice will only infest some monkeys, and animal lice will not establish a population on humans (Habedank 2010). This has to be verified for the louse of gorillas, *Pthirus gorillae*, and the *Pediculus* sp. of monkeys, but the contact is too rare to be of importance. Transmission of head lice is supported by a behavioural peculiarity: After ingestion of blood, head lice go to the ends of the hairs, enabling an easier transfer. Some minutes later, they go back to the base of the hairs and remain hidden near the skin (Mehlhorn and Mehlhorn 2009). Of the three larval instars, females, and males of head lice, the first-instar larvae showed the lowest tendency to go to a new host, whereas adults were the most likely to disperse (Takano-Lee et al. 2005). Hungry body lice move towards the light (Hase 1915; Habedank 2010), and when engorged they hide close to the skin in the dark, especially between turnups at the seams of clothes, where they also prefer to lay eggs (Raoult and Roux 1999; Habedank 2010). Even exercise-induced increases of body temperature are sufficient to increase the activity of body lice (Habedank 2010).

Only once has a colony of body lice accepted a host other than humans (Culpepper 1944, 1946, 1948; Raoult and Roux 1999). These lice can be reared on rabbits or in vitro, and groups of both maintenance conditions show no strong differences in the development (Culpepper 1948; Habedank 2010). However, in attempts to maintain head lice in vivo and in vitro, development of membrane-fed older larvae was much slower and the fecundity of females and the hatching rate of their progeny were much lower (Takano-Lee et al. 2003a, b; summarised by Habedank 2010).

11.5 Development and Reproduction

The hemimetabolous development of lice includes the egg, three larval stages, and the adults. The eggs of the three human lice are similar in length, 0.5–0.8 mm (Table 11.1) (Burkhart et al. 2000; Habedank 2010). All are glued in a species- and subspecies-specific manner to the shafts of hairs, just above the skin, or fibers of clothes (Martini 1952). The glue is not dissolved by water or soap (Mehlhorn and Mehlhorn 2009). The cap-like operculum of the eggs points to the tip of the hair shafts or fibers of clothes and contains species-specifically arranged aeropyles for air and moisture supply of the embryo (Martini 1952; Burkhart et al. 2000). In *P. humanus capitis* and *P. humanus humanus*, the ventral side of the embryo is located at the hair, whereas the embryo of *P. pubis* is orientated laterally to the hair (Martini 1952). Eggs of the latter also possess a higher operculum and longer aeropyles than both *Pediculus* ssp. (Martini 1952; Peters 1999) (Fig. 11.2c). Like in all insects, the duration of embryonic development is temperature-dependent, lasting about 1 week under optimal conditions (Table 11.1) (Raoult and Roux 1999; Habedank 2010).

Like all terrestrial insects, the fully developed louse in the egg swallows air and forces the operculum off (Service 2004). After hatching, the cuticle of the first-instar larvae sclerotizes, and the larvae, which are 1-mm long, take the first blood meal (Raoult and Roux 1999; Habedank 2010). Usually they suck every 2–5 h, but body lice have to feed only at least once daily (Raoult and Roux 1999; Peters 1999; Habedank 2010). The duration of larval development is also temperature-dependent. Usually the molts to the second and third larval instars and to the adults occur 3, 5, and 10 days after egg hatching (Raoult and Roux 1999). Whereas head and crab lice live at relatively constant temperatures, the development of body lice is slowed if the daytime clothes are discarded at night.

Besides the smaller size, males of *Pediculus* can be distinguished from females by the rounded posterior end of the abdomen and the stylet-like penis in the sexual orifice (Fig. 11.2a), whereas at the end of the female abdomen a split appearance is caused by two large lobes that grip the hairs or fibers during egg laying (Ibarra 1993; Peters 1999; Service 2004). In addition, the dorsal side of the abdomen of males possesses dark transverse bands, and the forelegs are broader and bear a larger tibial thumb-like spine and tarsal claw than in females (Ibarra 1993). About 10 h after molting, adults copulate (Wenk and Renz 2003; Mehlhorn and Mehlhorn 2009). Whereas the female crab louse has a spermatheca to store the sperms of the male and presumably requires to be fertilised only once during its life, the two *Pediculus* ssp. lack this organ and have to be fertilised several times (Burgess et al. 1983; Raoult and Roux 1999). One day or 2 days after copulation, females start to lay eggs (Wenk and Renz 2003; Mehlhorn and Mehlhorn 2009; Haberkorn 2010). The number of eggs per day and the life spans of adults differ (Table 11.1), but body lice seem to live longer and produce more eggs than head lice, and both *Pediculus* ssp. lay more eggs than *P. pubis*. In body lice, the number of eggs per female is affected by the frequency of feeding (Gooding 1963). After membrane feeding using blood exclusively from women or men, there was no significant difference between the 24-h survivorships of the respective groups of head lice (Takano-Lee et al. 2005). The starvation capacity is species- and temperature-dependent. At 18 and 26°C, body lice survive on average for 35 and 24 days, respectively (Lang cited according to Burgess 2004). Near the optimum temperature, body lice survive longer than at lower and higher temperatures: at 6, 24, and 31°C all were dead after 6, 11, and 9 days, respectively (Mumcuoglu et al. 2006). However, even before movement ceases, they are unable to feed (Burgess 1995).

11.6 Haematophagy

The acquisition of blood as the sole food source causes many problems for ectoparasites. Temporary ectoparasites have to find a host, and all ectoparasites have to gain access to obtain the blood without being killed by defence reactions of the host. Finally, during and after blood ingestion, all are confronted with several defence systems of the host that have evolved against parasites, such as the complement system and antibodies, or the loss of blood via injuries, haemostasis (Andrade et al. 2005).

11.6.1 Biting and Blood Ingestion

Lice belong to the vessel-feeders, i.e., they ingest blood directly from a blood vessel (Dettner 1999). Their mouthparts consist of strongly modified labrum, labium, mandibles, and hypopharynx. A stylet is built from two half-pipes, the dorsal one originating from the hypopharynx and forming the blood channel and the ventral one originating from the labium. Between them or in the hypopharynx is the duct of the salivary glands. At the host, the labrum is pushed out of the head capsule in a hydraulic manner and contacts the skin (Peacock 1919). Then in alternating movements both parts of the stylet penetrate the skin. After a blood vessel has been pierced, blood is ingested. Pumping is performed by the cibarial pump and especially the pharyngeal pump. When blood ingestion has finished, muscles retract the mouthparts into the head capsule (Ramcke 1965; Dettner 1999; Wenk and Renz 2003).

11.6.2 Salivary Glands and Saliva

The two pairs of salivary glands, one pair reniform and the other bifurcated and tubular, lie in the anterior part of the thorax beside the pharynx (Peacock 1919). The saliva of lice is expected to store only limited amounts of protein (Mumcuoglu et al. 1996c). However, like in many vessel-feeders, the saliva of lice contains a vasodilator to increase the speed of blood ingestion (Jones 1998). Like all bloodsucking insects, lice have to overcome the normal haemostatic mechanisms of the host (Ribeiro 1987; Ribeiro and Francischetti 2003). The saliva of lice contains inhibitors of factor X and thrombin, thus preventing the synthesis of fibrin from fibrinogen. An inhibitor of thrombin also acts against one of the pathways of platelet aggregation; another one is inhibited by apyrases in the saliva (Mumcuoglu et al. 1996c). In contrast to triatomines (Meiser et al. 2010; Balczun et al. 2012), only a limited number of inhibitors have been identified, and in contrast to many dipteran bloodsuckers, lice possess no hyaluronidase, which is suggested to increase the permeability of tissues for other salivary components (Volfova et al. 2008).

11.7 Intestine, Blood Digestion, and Excretion

11.7.1 Intestinal Tract

The intestinal tract of lice is a relatively simple tube. The anterior midgut is a strongly distensible region and possesses two anterior caeca (Buxton 1947; Dettner 1999). It is followed by the narrow, less distensible posterior midgut. At the border of the hindgut, the four Malpighian tubules end. In the middle of the relatively long, narrow hindgut, six rectal papillae are located in a short, wide hindgut region (Buxton 1947; Dettner 1999).

11.7.2 Blood Digestion

Lice feed as permanent ectoparasites frequently, commonly every few hours, taking small blood meals. *P. humanus capitis* takes up only 0.19–0.3 times its unfed weight; females ingest on average 0.17 mg blood, males 0.07 mg, and nymphs 0.04 mg (Speare et al. 2006). Therefore, a heavy infection with head lice does not lead to clinically significant blood loss, even in iron-deficient people. If body lice are starved for 1 day, females and males ingest 0.9 and 0.3 mg blood, respectively (Martini 1952). In vitro fed females, males, and first-instar larvae suck 1.7, 0.7, and 0.08 mg blood, respectively (Habedank 2010). According to the frequent ingestion of blood, lice have the fastest digestion among haematophagous insects, about 4 h per blood meal, but the particular time for the digestion is always affected by several factors, such as ambient temperature, age of the insect, and mating status (Lehane 2005).

Like most haematophagous insects (Lehane 1991), lice use a range of alkaline digestive proteases. A leucine aminopeptidase of about 67–69 kDa has been partially characterised from the body louse (Ochanda et al. 2000). This enzyme has maximum activity at pH 8 and seems to be associated with the midgut epithelial cell membranes since homogenisation of midguts in the presence of detergent results in a tenfold higher aminopeptidase activity (Ochanda et al. 1998). If starved body lice are fed once in the laboratory, the leucine aminopeptidase is stimulated by a blood meal and reaches maximum activity 48 h after feeding. Under natural conditions, blood is ingested at short intervals and, therefore, the activity is supposed to be constantly stimulated and fairly constant (Ochanda et al. 1998). Partial purification of this enzyme indicates that only one aminopeptidase or at least a family of charged isomers of the same enzyme are present (Ochanda et al. 2000).

Trypsin-like enzyme activity appears in a laboratory experiment 24 h after feeding starved body lice once in the laboratory, i.e., much earlier than that of the aminopeptidases, and this activity disappears 48 h after feeding (Borovsky and Schlein 1988). Two trypsin-encoding genes (*Try1*, *Tyr2*) have been identified, and the *Tyr1* gene seems to be expressed with identical levels in unfed first-instar larvae and fed adults. In addition, this gene is constitutively expressed in adults 2–24 h after a blood meal (Kollien et al. 2004). The expression of *Tyr1*, *Tyr2*, and an also characterized chymotrypsin-encoding gene (*Chy1*) has been localised in the distensible anterior region of the midgut (stomach) and—to a low extent—in the narrow posterior region of the midgut using whole-mount in situ hybridisation (Waniek et al. 2005). *Tyr2* and *Chy1* are also constitutively expressed for up to 24 h after blood feeding, the *Tyr2* gene always being expressed at much lower levels that *Tyr1* and *Chy1*.

The consecutive activities of aminopeptidase and trypsin after a blood meal indicate a network of different proteases is responsible for the digestive processes in the midgut of lice. A first hint at a regulatory network was given by the cleavage of an in vitro synthesised trypsinogen from *P. humanus humanus* by chymotrypsin leading to an activated trypsin (Kollien et al. 2004). Sequencing of the transcriptome and the genome of *P. humanus humanus* revealed the first insights into a putative network of digestive proteases (Pedra et al. 2003; Pittendrigh et al. 2006; Kirkness et al. 2010), and sequences for aminopeptidases, carboxypeptidases, trypsins, and chymotrypsins are now available in GenBank and VectorBase (http://www.vectorbase.org/). In future, the open access to such data will promote research on the digestion of lice since understanding of digestion is important for elucidating the basic biology of these insects, particularly with regard to their control as vectors of pathogens.

11.7.3 Excretion and Defecation

Like all bloodsucking insects, lice ingest a food in which water is the major constituent, about 80% (Lehane 2005); therefore, the Malpighian tubules secrete a proportion of this water. Body lice starved for about 1 day ingest so much blood that they defecate blood even during ingestion (Fig. 11.1d). In these populations, leakage of blood into the haemocoel also occurs, resulting in a transient red colouration (Raoult and Roux 1999; Kollien unpublished). This leakage seems to have no obvious effect on the survival (Kollien unpublished). During the normal period of digestion, lice defecate extremely dry and powdery faeces containing only 2% water (Burgess 1995). Similarly to the faeces of triatomines, the faeces contains a large amount of ammonium acting as an attractant for other lice (Raoult and Roux 1999; Schaub 2009).

11.8 Symbionts and Antibacterial Compounds of Lice

Like many bloodsucking insects, lice require symbionts for a normal development (Buxton 1947; Puchta 1955; Buchner 1965). Whereas in some of these insects the symbionts are located in the lumen of the intestine, tsetse flies and lice possess special organs, mycetomes, which are not connected to the intestine (Buchner 1965). Already in the first-instar larva, the mycetome is visible as a lenticular organ on the intestine in the middle of the abdomen. In males, this remains the only symbiont-bearing region. In third-instar larvae which will develop into females, all or nearly all symbionts migrate via the haemolymph to ovarial ampullae (Puchta 1955; Eberle and McLean 1982; Wenk and Renz 2003). From these proximal cells of the oviduct, the symbionts infect the egg via the micropyle and develop an infection mycetome in the egg yolk. In the developing embryo the midgut possesses a protrusion which is colonised by the symbionts and strangulated, thus losing the connection to the intestine (Puchta 1955; Wenk and Renz 2003). Immediately after hatching, the symbionts multiply in the mycetome. Aposymbiotic lice can be obtained by careful centrifugation of eggs. These larvae require supplementation of the blood with yeast or vitamins of the vitamin B complex (Puchta 1955).

11.9 Peculiarities of Lice Genomes

The human body louse *P. humanus humanus* and the human head louse *P. humanus capitis* have the smallest known insect genomes, spanning 104–108 megabases (Mb) (Johnston et al. 2007; Kirkness et al. 2010; Pittendrigh et al. 2006). The recent sequencing of the genome of *P. humanus humanus* is a milestone in genomic research on hemimetabolous insects because the often large (2,000 Mb) to very large (up to 16,300 Mb) genomes of other hemimetabolous insect species have prevented large-scale genomic sequencing efforts until now. Fortunately, the louse genome retains a remarkably complete basal insect repertoire of 10,773 protein-encoding genes and 57 microRNAs (Kirkness et al. 2010), and hence the genome provides a model for genomes of other hemimetabolous insects as well as for holometabolous insects.

Compared with other insect genomes, the louse genome contains significantly fewer genes for environmental sensing and response, including odourant and gustatory receptors and detoxification enzymes (Kirkness et al. 2010; Lee et al. 2010). In particular, in comparison with *Drosophila melanogaster* and *Anopheles gambiae*, lice possess only approximately half of the genes encoding cytochrome P450 monooxygenase, glutathione *S*-transferase, esterase, and ATP-binding cassette transporter. The relatively small number of genes might indicate that certain gene families have been lost or at least contracted most probably owing to the simple life history and the obligate parasitism of a single host species (Lee et al. 2010).

The gene composition, gene order, and structure of the mitochondrial genome are remarkably stable across bilaterian animals. In multiple lineages of lice, this genomic stability has been lost. Instead of all mitochondrial genes being on a single chromosome, the 37 genes are located on minichromosomes, 18 in *P. humanus humanus*, each encoding only a few of the genes (Shao et al. 2009; Cameron et al. 2011). The advantages of a fragmented mitochondrial genome, if any, are currently unknown, and the coevolution with blood feeding in the sucking lice is not supported since fragmentation is also found in non-blood-feeding louse groups and is not found in the blood-feeding genus *Heterodoxus* (Cameron et al. 2011). Most probably, fragmentation is due to the apparent loss of the nuclear-encoded mitochondrial single-stranded binding protein (Kirkness et al. 2010), which is required for the mitochondrial genome replication in insects and mammals, and the loss of its function is lethal in late third larval instar/pupal stages of *D. melanogaster* (Maier et al. 2001).

11.10 Control of Lice

11.10.1 Diagnosis

After the first itching reactions on the head or the body, the living lice and/or eggs must be observed before treatment is initiated. These can be difficult to find at low infestation levels, which are normal (summarised by Raoult and Roux 1999; Habedank 2010). The use of specific lice combs can help (Burgess 2004; Habedank 2010). Usually an egg with a developing embryo is only visible by very careful inspection, but after hatching it is filled with air and is thus white (Ibarra 1993). In addition, because of the growth of the hairs during the embryonic development, the empty eggs, named nits, are further away from the skin and are better visible. Since the nit remains glued to the hair, the presence of nits does not indicate a failure of treatment.

11.10.2 Treatment

The easiest and often practiced treatment of head lice is shaving of hair and for body lice a change of clothes and storage of old clothes at higher temperatures (at least 70°C for 1 h). Shaving of head hair is done for boys and men, but can rarely be used for girls and women (Habedank 2010).

Insecticides are commonly used, but none of them kill all developing embryos, which are protected by the eggshell. Therefore, treatments have to be repeated, optimally after the hatching of all larvae and before the end of larval development and the egg laying by females. This is also necessary when plant extracts or oils, e.g., of the seeds of the neem tree, are used (Mehlhorn and Mehlhorn 2009). In head lice, the second treatment should be performed 8–10 days after the first one (Habedank 2010). Increasing resistances against different compounds, including the most widely used pyrethroids, complicate the treatment (Raoult and Roux 1999; Burgess 2004; Habedank 2010).

11.10.3 New Strategies for Lice Control

Considering the increasing resistance of lice to insecticides, the development of new strategies for lice control is a high priority. Immunisation of rabbits with midgut extracts of body lice and subsequent feeding on these animals showed a significantly higher mortality rate of lice compared with a control group (Ben-Yakir et al. 1994). In addition, in lice fed on immunised rabbits, the proportion of lice with ruptured guts was significantly higher, the size of the ingested blood meal was lower, females laid fewer eggs, and fertility was reduced. In a similar approach, immunisation of rabbits with fecal extracts of lice caused similar deleterious effects on the lice population fed on these rabbits (Mumcuoglu et al. 1997). The immunogenic proteins of the lice extracts were localised on the microvilli of the midgut epithelial cells and were partially characterised (Mumcuoglu et al. 1996b; Ochanda et al. 1996). Lysis of lice midgut cells is supposed as a mechanism of the action against the lice (Mumcuoglu et al. 1996b). This theory was derived from results of experiments on ticks fed on vaccinated cattle in which leukocytes and/or other components of the host's blood seemed to leak into the haemocoel and to destroy other tissues (Agbede and Kemp 1986; Kemp et al. 1986; Lehane 1994; Willadsen and Billingsley 1996).

Different lice species share common antigens in their midguts and faeces and, therefore, material obtained from colonies of human body lice might be useful for immunisation of domestic animals against their specific lice (Mumcuoglu et al. 1996a). These efforts in producing resistance to lice are promising, but further investigations are needed to promote the development of effective vaccines. However, a vaccination against human lice is complicated by the critical attitude of many people to all vaccinations, often the same people who do not wish to use insecticides.

11.11 Medical Importance

11.11.1 Skin Reactions

The saliva of all bloodsucking insects induces itching reactions, which differ strongly between humans (Peters 1999). Different from other insects, the area surrounding the ingestion position of head and body lice changes colour from light red to bluish. Scratching is the first indication of head lice, but in the first infection this usually occurs about 3–4 weeks after the beginning of the infestation (Burgess 1995). In infestations of volunteers previously used to feed *P. humanus humanus*, immediate skin reactions developed more rapidly than in volunteers without previous contact with lice (Burgess et al. 1983). Scratching increases the itching reaction and can result in skin lesions and secondary infections, supported by the bacteria in the faeces. In strong, long-lasting infestations with head lice, a strong dermatosis develops and exudates of the skin glue the hairs and eggs together in a so-called plica polonica (German: Weichselzopf) (Gnanaraj et al. 2007; Habedank 2010). High rates of infestation with body lice and permanent scratching might result in a brown melanisation of the skin (Peters 1999). The itching reaction to crab lice is much less than for *Pediculus*. Since crab lice prefer to suck at the same location and since their saliva modifies the haemoglobin at the ingestion location, the skin there shows a bluish point (Peters 1999; Habedank 2010).

11.11.2 Vectors of Pathogens

Lice transmit three bacterial diseases to humans, but only body lice are important vectors (Raoult and Roux 1999; Fournier et al. 2002; Habedank 2010). Head and crab lice can experimentally transmit the pathogens (Robinson et al. 2003; Habedank 2010). One important pathogen is *Rickettsia prowazekii*, the etiologic agent of louse-borne epidemic typhus (typhus exanthemicus), which should not be mistaken for typhus abdominalis with *Salmonella typhi* as the etiologic agent (Wenk and Renz 2003). Louse-borne epidemic typhus is an anthroponosis, i.e., only humans are the reservoir. The symptoms are fever with headache and spots of haemorrhages on the body and extremities which develop about 4–6 days after the beginning of the fever (Wenk and Renz 2003). After ingestion of the Gram-negative pathogen with a blood meal, the rickettsiae invade the cells of the stomach wall of the lice and multiply enormously, thus distending the cells, which rupture about 4 days after infection (Raoult and Roux 1999; Service 2004). These injuries kill many lice 8–12 days after infection. The lethal effects are delayed in lice after ingestion of blood containing clinically attainable levels of the antibiotics doxycycline and rifampin since they, but not antibodies against *R. prowazekii*, inhibit the initial development of rickettsiae in lice (Boese et al. 1973). Since the gut wall is destroyed locally, haemoglobin induces a reddish colouring of the haemolymph ("red louse disease"). Thereby, the rickettsiae are present not only in the intestinal content and passed out with the faeces of the louse, but also in the haemocoel. The fever of humans induces a departure of body lice, increasing the risk of spreading the disease (Wenk and Renz 2003). Several modes of infection of humans occur: contact of skin lesions and mucous membranes with haemolymph of crushed lice or faeces and inhalation of powdered dry faeces (Service 2004). In the latter, the pathogens remain infective for up to 66 (100) days, but for a much shorter period under more humid conditions (Raoult and Roux 1999; Wenk and Renz 2003). Humans are usually infective within the first 10–14 days after infection, but in asymptomatic cases bacteria are present in the blood for many years (Wenk and Renz 2003; Service 2004). Also a recrudescence is possible (Brill–Zinsser disease) (Ibarra 1993; Raoult and Roux 1999; Service 2004). Broad-spectrum antibiotics bring about a quick recovery (Huys et al. 1973). However, before the development of these compounds, the disease caused many deaths, e.g., three million in the Russian revolution (Patterson 1993). Without a chance of treatment, about 30–50% of infected people died (Raoult and Roux 1999; Wenk and Renz 2003). Although the development of a vaccine seemed promising (Woodward 1986), efficient antibiotic treatment reduced the priority for this. However, typhus should be considered a serious threat since it has a severe epidemic potential and can dramatically reemerge, e.g., as in Burundi in 1997 (Bise and Coninx 1997; Raoult et al. 1997, 1998).

Bartonella (synonym *Rochalimaea) quintana* is the etiologic agent of trench or 5-day fever (Service 2004). This nonfatal disease was first recognised in eastern Europe during World War I among soldiers in the trenches. Meanwhile it also occurs in Africa, and in Western Europe and North America mainly in homeless people (Stein and Raoult 1995; Jackson and Spach 1996; Foucault et al. 2002; Service 2004; Brouqui and Raoult 2006). Since humans are the only hosts, trench fever is an anthroponosis (Raoult and Roux 1999). The bacteria attach to the cells of the intestine, but do not penetrate them. They multiply in the lumen of the intestine and 5–10 days after infection they are present in the faeces. The routes of infection by the dry faeces are identical to those of the louse-borne epidemic typhus. According to observations in a refugee camp in Burundi, the same louse can apparently be infected by two pathogens, *B. quintana* and *R. prowazekii* (Raoult et al. 1998). Therefore, in outbreaks of louse-borne epidemic typhus, trench fever should also be considered. Without treatment, e.g., with tetracycline or chloramphenicol, less than 1% of infected people die (Raoult and Roux 1999).

In louse-borne relapsing fever, also an anthroponosis, the etiologic agent is *Borrelia recurrentis* (Ibarra 1993; Service 2004). Within 1 day after ingestion of blood containing the spirochetes, the bacteria have passed through the stomach wall, and about 4 days after infection they slowly multiply in the haemolymph, reaching enormous numbers after an additional 6–8 days (Ibarra 1993; Service 2004). Apparently lice are not impaired by an infection (Raoult and Roux 1999). Since spirochetes are not present in the faeces, they can only be transmitted to humans by crushing the louse between the fingernails or cracking them between the teeth. Skin lesions or the mucous membranes in the mouth are the entry ports

(Service 2003). Without treatment, 10–40% of infected people die. However, relapsing fever can be successfully treated with chloramphenicol, penicillin, tetracycline, and erythromycin (Raoult and Roux 1999).

Infections of *R. prowazekii*, *B. quintana*, and *B. recurrentis* in lice are detectable with high sensitivity and specificity using polymerase chain reaction. Instead of collection of blood samples, the collection of lice and subsequent analysis of infection in a medical or biological laboratory is a convenient tool in epidemiological studies of louse-borne diseases (Roux and Raoult 1999).

11.12 Conclusions

Since economic instability and wars cause strong migrations of humans and a decline in hygiene conditions, louse infestations become more prevalent (Raoult and Roux 1999; Brouqui 2011). Also as a result of insecticide resistances, the rates of infestation with head lice have increased remarkably in the last 20 years, especially in schoolchildren (Peters 1999; Burgess 2004). In addition to a reduced communication in cases of infestations in a group of children, many parents show an aversion to the use of insecticides and to using them according to the recommendations a second time. Therefore, not only new compounds for insecticidal treatment have to be developed, but also our knowledge of fundamental louse biology and physiology has to be improved. Current misunderstanding and lack of knowledge result in a great waste of financial and time resources by professional and family caregivers during louse management.

Acknowledgements We are deeply grateful for cooperation with many colleagues—recognisable by co-authorships—and the funding of the investigations and travel by the German Academic Exchange Service (DAAD) and Ruhr-Universität Bochum. Especially we thank Dr. Júlio Vianna Barbosa, Instituto Oswaldo Cruz, Rio de Janeiro, Brazil, for his strong support during collection of lice from homeless people.

References

Agbede RIS, Kemp DH (1986) Immunization of cattle against *Boophilus microplus* using extracts derived from adult female ticks: histopathology of ticks feeding on vaccinated cattle. Int J Parasitol 16:35–41

Andrade BB, Teixeira CR, Barral A, Barral-Netto M (2005) Haematophagous arthropod saliva and host defense system: a tale of tear and blood. An Acad Bras Cienc 77:665–693

Araújo A, Ferreira LF, Guidon N, de Serra M, Freire N, Reinhard KJ, Dittmar K (2000) Ten thousand years of head lice infection. Parasitol Today 16:269

Aspöck H, Walochnik J (2007) Die Parasiten des Menschen aus der Sicht der Koevolution. Denisia 20:179–254

Balczun C, Meiser CK, Schaub GA (2012) Triatomines as vectors of American trypanosomiasis. In: Mehlhorn H (ed) Arthropods as vectors of emerging diseases, Parasitology research monographs, vol 3. Springer, Berlin

Ben-Yakir D, Mumcuoglu KY, Manor O, Ochanda J, Galun R (1994) Immunization of rabbits with a midgut extract of the human body louse *Pediculus humanus humanus*: the effect of induced resistance on the louse population. Med Vet Entomol 8:114–118

Bise G, Coninx R (1997) Epidemic typhus in a prison in Burundi. Trans R Soc Trop Med Hyg 91:133–134

Boese JL, Wisseman CL, Walsh WT, Fiset P (1973) Antibody and antibiotic action on *Rickettsia prowazeki* in body lice across the host-vector interface, with observations on strain virulence and retrieval mechanisms. Am J Epidemiol 98:262–282

Borovsky D, Schlein Y (1988) Quantitative determination of trypsinlike and chymotrypsinlike enzymes in insects. Arch Insect Biochem Physiol 8:249–260

Brouqui P (2011) Arthropod-borne diseases associated with political and social disorder. Annu Rev Entomol 56:357–374

Brouqui P, Raoult D (2006) Arthropod-borne diseases in homeless. Ann N Y Acad Sci 1078:223–235

Buchner P (1965) Endosymbiosis of animals with plant microorganisms. Interscience, New York

Burgess IF (1995) Human lice and their management. Adv Parasitol 36:271–342

Burgess IF (2004) Human lice and their control. Annu Rev Entomol 49:457–481

Burgess I, Maunder JW, Myint TT (1983) Maintenance of the crab louse, *Pthirus pubis*, in the laboratory and behavioural studies using volunteers. Commun Med 5:238–241

Burkhart CN, Gunning W, Burkhart CG (2000) Scanning electron microscopic examination of the egg of the pubic louse (Anoplura: *Pthirus pubis*). Int J Dermatol 39:201–202

Busvine JR (1978) Evidence from double infestations for the specific status of human head lice and body lice (Anoplura). Syst Entomol 3:1–8

Buxton PA (1947) The louse, 2nd edn. Edward Arnold, London

Cameron SL, Yoshizawa K, Mizukoshi A, Whiting MF, Johnson KP (2011) Mitochondrial genome deletions and minicircles are common in lice (Insecta: Phthiraptera). BMC Genomics 12:394

Culpepper GH (1944) The rearing and maintenance of a laboratory colony of the body louse. Am J Trop Med Hyg s1–24:327–329

Culpepper GH (1946) Factors influencing the rearing and maintenance of a laboratory colony of the body louse. J Econ Entomol 39:472–474

Culpepper GH (1948) Rearing and maintaining a laboratory colony of body lice on rabbits. Am J Trop Med Hyg 28:499–504

Dean I, Siva-Jothy MT (2012) Human fine body hair enhances ectoparasite detection. Biol Lett. doi:10.1098/rsbl.2011.0987

Dettner K (1999) Phthiraptera, Tierläuse. In: Dettner K, Peters W (eds) Lehrbuch der Entomologie. Fischer, Stuttgart, pp 804–807

Eberle MW, McLean DL (1982) Initiation and orientation of the symbiote migration in the human body louse *Pediculus humanus* L.. J Insect Physiol 28:417–422

Foucault C, Barrau K, Brouqui P, Raoult D (2002) *Bartonella quintana* bacteremia among homeless people. Clin Infect Dis 35:684–689

Fournier P-E, Ndihokubwayo J-B, Guidran J, Kelly PJ, Raoult D (2002) Human pathogens in body and head lice. Emerg Infect Dis 8:1515–1518

Gnanaraj P, Venugopal V, Pandurangan CN (2007) Plica polonica in association with pediculosis capitis and scabies – a case report. Int J Dermatol 46:151–152

Gooding RH (1963) Studies on the effect of frequency of feeding upon the biology of a rabbit-adapted strain of *Pediculus humanus*. Parasitology 49:516–521

Habedank B (2010) Läuse – Biologie, medizinische Bedeutung und Bekämpfung. In: Aspöck H (ed) Krank durch Arthropoden, Denisia, vol 30. Biologiezentrum/Oberöstereichische Landesmuseen, Linz, pp 655–672

Hase A (1915) Beiträge zur Biologie der Kleiderlaus. Flugschr Dtsch Ges Angew Entomol 1:1–95

Huys J, Freyens P, Kayihigi J, Van den Berghe G (1973) Treatment of epidemic typhus. A comparative study of chloramphenicol, trimethoprim-sulphamethoxazole and doxycycline. Trans R Soc Trop Med Hyg 67:718–721

Ibarra J (1993) Lice (Anoplura). In: Lane RP, Crosskey RW (eds) Medical insects and arachnids. Chapman and Hall, London, pp 517–528

Jackson LA, Spach DH (1996) Emergence of *Bartonella quintana* infection among homeless persons. Emerg Infect Dis 2:141–144

Johnston JS, Yoon KS, Strycharz JP, Pittendrigh BR, Clark JM (2007) Body lice and head lice (Anoplura: Pediculidae) have the smallest genomes of any hemimetabolous insect reported to date. J Med Entomol 44:1009–1012

Jones D (1998) The neglected saliva: medically important toxins in the saliva of human lice. Parasitology 116(Suppl):S73–S81

Kemp DH, Agbede RIS, Johnston LAY, Gough JM (1986) Immunization of cattle against *Boophilus microplus* using extracts derived from adult female ticks: feeding and survival of the parasite on vaccinated cattle. Int J Parasitol 16:115–120

Kirkness EF, Haas BJ, Sun W, Braig HR, Perotti MA, Clark JM, Lee SH, Robertson HM, Kennedy RC, Elhaik E, Gerlach D, Kriventseva EV, Elsik CG, Graur D, Hill CA, Veenstra JA, Walenz B, Tubio JMC, Ribeiro JM, Rozas J, Johnston JS, Reese JT, Popadic A, Tojo M, Raoult D, Reed DL, Tomoyasu Y, Kraus E, Mittapalli O, Margam VM, Li HM, Meyer JM, Johnson RM, Romero-Severson J, Vanzee JP, Alvarez-Ponce D, Vieira FG, Aguade M, Guirao-Rico S, Anzola JM, Yoon KS, Strycharz JP, Unger MF, Christley S, Lobo NF, Seufferheld MJ, Wang NK, Dasch GA, Struchiner CJ, Madey G, Hannick LI, Bidwell S, Joardar V, Caler E, Shao R, Barker SC, Cameron S, Bruggner RV, Regier A, Johnson J, Viswanathan L, Utterback TR, Sutton GG, Lawson D, Waterhouse RM, Venter JC, Strausberg RL, Berenbaum MR, Collins FH, Zdobnov EM, Pittendrigh BR (2010) Genome sequences of the human body louse and its primary endosymbiont provide insights into the permanent parasitic lifestyle. Proc Natl Acad Sci USA 107:12168–12173

Kittler R, Kayser M, Stoneking M (2003) Molecular evolution of *Pediculus humanus* and the origin of clothing. Curr Biol 13:1414–1417

Kollien AH, Waniek PJ, Pröls F, Habedank B, Schaub GA (2004) Cloning and characterization of a trypsin-encoding cDNA of the human body louse *Pediculus humanus*. Insect Mol Biol 13:9–18

Lee SH, Kang JS, Min JS, Yoon KS, Strycharz JP, Johnson R, Mittapalli O, Margam VM, Sun W, Li HM, Xie J, Wu J, Kirkness EF, Berenbaum MR, Pittendrigh BR, Clark JM (2010) Decreased detoxification genes and genome size make the human body louse an efficient model to study xenobiotic metabolism. Insect Mol Biol 19:599–615

Lehane MJ (1991) Managing the blood meal. In: Lehane MJ (ed) Biology of blood sucking insects. HarperCollins, London, pp 79–105

Lehane MJ (1994) Digestive enzymes, haemolysins and symbionts in the search for vaccines against blood-sucking insects. Int J Parasitol 24:27–32

Lehane MJ (2005) Managing the blood meal. In: Lehane MJ (ed) The biology of blood-sucking in insects, 2nd edn. Cambridge University Press, Cambridge, pp 84–115

Leo NP, Barker SC (2002) Intragenomic variation in ITS2 rDNA in the louse of humans, *Pediculus humanus*: ITS2 is not a suitable marker for population studies in this species. Insect Mol Biol 11:651–657

Leo NP, Barker SC (2005) Unravelling the evolution of the head lice and body lice of humans. Parasitol Res 98:44–47

Leo NP, Campbell NJH, Yang X, Mumcuoglu K, Barker SC (2002) Evidence from mitochondrial DNA that head lice and body lice of humans (Phthiraptera: Pediculidae) are conspecific. J Med Entomol 39:662–666

Leo NP, Hughes JM, Yang X, Poudel SKS, Brogdon WG, Barker SC (2005) The head and body lice of humans are genetically distinct (Insecta: Phthiraptera, Pediculidae): evidence from double infestations. Heredity 95:34–40

Li W, Ortiz G, Fournier P-E, Gimenez G, Reed DL, Pittendrigh B, Raoult D (2010) Genotyping of human lice suggests multiple emergences of body lice from local head louse populations. PLoS Negl Trop Dis 4(3):e641. doi:10.1371

Maier D, Farr CL, Poeck B, Alahari A, Vogel M, Fischer S, Kaguni LS, Schneuwly S (2001) Mitochondrial single-stranded DNA-binding protein is required for mitochondrial DNA replication and development in *Drosophila melanogaster*. Mol Biol Cell 12:821–830

Martini E (1952) Lehrbuch der Medizinischen Entomologie, 4th edn. Fischer, Jena

Mehlhorn B, Mehlhorn H (2009) Läusealarm. Düsseldorf University Press, Düsseldorf

Meiser CK, Piechura H, Werner T, Dittmeyer-Schäfer S, Meyer HE, Warscheid B, Schaub GA, Balczun C (2010) Kazal-type inhibitors in the stomach of *Panstrongylus megistus* (Triatominae, Reduviidae). Insect Biochem Mol Biol 40:345–353

Mumcuoglu KY, Ben-Yakir D, Gunzberg S, Ochanda JO, Galun R (1996a) Immunogenic proteins in the body and faecal material of the human body louse, *Pediculus humanus*, and their homology to antigens of other lice species. Med Vet Entomol 10:105–107

Mumcuoglu KY, Rahamim E, Ben-Yakir D, Ochanda JO, Galun R (1996b) Localization of immunogenic antigens on midgut of the human body louse *Pediculus humanus humanus* (Anoplura: Pediculidae). J Med Entomol 33:74–77

Mumcuoglu KY, Galun R, Kaminchik Y, Panet A, Levanon A (1996c) Antihemostatic activity in salivary glands of the human body louse, *Pediculus humanus humanus* (Anoplura: Pediculidae). J Insect Physiol 42:1083–1087

Mumcuoglu KY, Ben-Yakir D, Ochanda JO, Miller J, Galun R (1997) Immunization of rabbits with faecal extract of *Pediculus humanus*, the human body louse: effects on louse development and reproduction. Med Vet Entomol 11:315–318

Mumcuoglu KY, Friger M, Cohen R (2006) Use of temperature and water immersion to control the human body louse (Anoplura: Pediculidae). J Med Entomol 43:723–725

Ochanda JO, Mumcuoglu KY, Ben-Yakir D, Okuru JK, Oduol VO, Galun R (1996) Characterization of body louse midgut proteins recognized by resistant hosts. Med Vet Entomol 10:35–38

Ochanda JO, Oduor EAC, Galun R, Imbuga MO, Mumcuoglu KY (1998) Partial characterization and post-feeding activity of midgut aminopeptidase in the human body louse, *Pediculus humanus humanus*. Physiol Entomol 23:382–387

Ochanda JO, Oduor EAC, Galun R, Imbuga MO, Mumcuoglu KY (2000) Partial purification of the aminopeptidase from the midgut of the human body louse, *Pediculus humanus humanus*. Physiol Entomol 25:242–246

Patterson KD (1993) Typhus and its control in Russia, 1870–1940. Med Hist 37:361–381

Peacock AD (1919) The structure of the mouthparts and mechanism of feeding in *Pediculus humanus*. Parasitology 11:98–117

Pedra JHF, Brandt A, Li H-M, Westermann R, Romero-Severson J, Pollack RJ, Murdock LL, Pittendrigh BR (2003) Transcriptome identification of putative genes involved in protein catabolism and innate immune response in human body louse (Pediculicidae: *Pediculus humanus*). Insect Biochem Mol Biol 33:1135–1143

Peters W (1999) Läuse. In: Dettner K, Peters W (eds) Lehrbuch der Entomologie. Fischer, Stuttgart, pp 676–678

Pittendrigh BR, Clark JM, Johnston JS, Lee SH, Romero-Severson J, Dasch GA (2006) Sequencing of a new target genome: the *Pediculus humanus humanus* (Phthiraptera: Pediculidae) genome project. J Med Entomol 43:1103–1111

Puchta O (1955) Experimentelle Untersuchungen über die Bedeutung der Symbiose der Kleiderlaus *Pediculus vestimenti* Burm.. Z Parasitenkd 17:1–40

Ramcke J (1965) Kopf der Schweinelaus (*Haematopinus suis* L., Anoplura). Zool Jb Anat 82:547–663

Raoult D, Roux V (1999) The body louse as a vector of reemerging human diseases. Clin Infect Dis 29:888–911
Raoult D, Roux V, Ndihokubwayo JB, Bise G, Baudon D, Marte G, Birtles R (1997) Jail fever (epidemic typhus) outbreak in Burundi. Emerg Infect Dis 3:357–360
Raoult D, Ndihokubwayo JB, Tissot-Dupont H, Roux V, Faugere B, Abegbinni R, Birtles RJ (1998) Outbreak of epidemic typhus associated with trench fever in Burundi. Lancet 352:353–358
Reed DL, Smith VS, Hammond SL, Rogers AR, Clayton DH (2004) Genetic analysis of lice supports direct contact between modern and archaic humans. PLoS Biol 2(11):e340
Ribeiro JMC (1987) Role of saliva in blood-feeding arthropods. Annu Rev Entomol 32:463–478
Ribeiro JMC, Francischetti IMB (2003) Role of arthropod saliva in blood feeding: sialome and post-sialome perspectives. Annu Rev Entomol 48:73–88
Robinson D, Leo N, Prociv P, Barker SC (2003) Potential role of head lice, *Pediculus humanus capitis*, as vectors of *Rickettsia prowazekii*. Parasitol Res 90:209–211
Roux V, Raoult D (1999) Body lice as tools for diagnosis and surveillance of reemerging diseases. J Clin Microbiol 37:596–599
Schaub GA (2008) Lice. In: Mehlhorn H (ed) Encyclopedia of parasitology, vol 1, 3rd edn. Springer, Heidelberg, pp 715–719
Schaub GA (2009) Interactions of trypanosomatids and triatomines. Adv Insect Physiol 37:177–242
Service M (2004) Lice (Anoplura). In: Service M (ed) Medical entomology for students, 3rd edn. Cambridge University Press, Cambridge, pp 181–192
Shao R, Kirkness EF, Barker SC (2009) The single mitochondrial chromosome typical of animals has evolved into 18 minichromosomes in the human body louse, *Pediculus humanus*. Genome Res 19:904–912
Speare R, Canyon DV, Melrose W (2006) Quantification of blood intake of the head louse: *Pediculus humanus capitis*. Int J Dermatol 45:543–546
Stein A, Raoult D (1995) Return of trench fever. Lancet 345:450–451
Takano-Lee M, Yoon KS, Edman JD, Mullens BA, Clark JM (2003a) In vivo and in vitro rearing of *Pediculus humanus capitis* (Anoplura: Pediculidae). J Med Entomol 40:628–635
Takano-Lee M, Velten RK, Edman JD, Mullens BA, Clark JM (2003b) An automated feeding apparatus for in vitro maintenance of the human head louse, *Pediculus capitis* (Anoplura: Pediculidae). J Med Entomol 40:795–799
Takano-Lee M, Edman JD, Mullens BA, Clark JM (2005) Transmission potential of the human head louse, *Pediculus capitis* (Anoplura: Pediculidae). Int J Dermatol 44:811–816
Volfova V, Hostomska J, Cerny M, Votypka J, Volf P (2008) Hyaluronidase of bloodsucking insects and its enhancing effect on *Leishmania* infection in mice. PLoS Negl Trop Dis 2(9): e294.
Waniek PJ, Hendgen-Cotta UB, Stock P, Mayer C, Kollien AH, Schaub GA (2005) Serine proteinases of the human body louse (*Pediculus humanus*): sequence characterization and expression patterns. Parasitol Res 97:486–500
Wenk P, Renz A (2003) Rickettsiosen und Läuse. In: Wenk P, Renz A (eds) Parasitologie – Biologie der Humanparasiten. Thieme, Stuttgart, pp 209–215
Willadsen P, Billingsley PF (1996) Immune intervention against blood-feeding insects. In: Lehane MJ, Billingsley PF (eds) Biology of the insect midgut. Chapman & Hall, London, pp 323–344
Woodward TE (1986) Rickettsial vaccines with emphasis on epidemic typhus. Initial report of an old vaccine trial. S Afr Med J (Suppl):73–76

Chapter 12
Triatomines as Vectors of American Trypanosomiasis

Carsten Balczun, Christian K. Meiser, and Günter A. Schaub

Abstract Triatomine bugs are the biggest blood-sucking insects, with adults up to 41 mm long. They are well adapted to hematophagy, possessing fine mouthparts to ingest blood from the capillaries of the warm-blooded hosts. Hematophagy is supported by the salivary glands, a "chemical factory" producing hundreds of compounds, some of them acting as local anaesthetics and as inhibitors of the complement system and the blood coagulation cascade. The latter is more strongly inhibited by Kazal-type inhibitors from the stomach, in which the blood is stored essentially undigested. Digestion occurs in the small intestine. In contrast to most other insects, the intestinal pH is slightly acidic, and triatomines use cysteine and aspartate proteases, cathepsins B, L and D. The majority of triatomines occur on the American continent, especially in Latin America. About 12 of the 141 species are well adapted to houses and important vectors of the etiologic agent of Chagas disease, *Trypanosoma cruzi*. The development of this flagellate in the vector is strongly adapted to the different regions of the intestinal tract. The vector is only slightly, i.e. "subpathogenically" affected by the parasite, but an activation of the intestinal immune system is evident.

Keywords Chagas disease • Hematophagy • Latin America • Triatomines • *Trypanosoma cruzi*

12.1 Introduction

Trypanosomiasis is one of the "big six" of tropical diseases (Schaub and Wülker 1984). Under this heading, two diseases with different clinical symptoms, African and American trypanosomiases, were summed up, because the etiologic agents,

C. Balczun (✉) • C.K. Meiser • G.A. Schaub
Group Zoology/Parasitology, Ruhr-Universität Bochum, Universitätsstr. 150, 44801 Bochum, Germany
e-mail: carsten.balczun@rub.de; guenter.schaub@rub.de

H. Mehlhorn (ed.), *Arthropods as Vectors of Emerging Diseases*, Parasitology Research Monographs 3, DOI 10.1007/978-3-642-28842-5_12,

Trypanosoma brucei spp. and *Trypanosoma cruzi*, respectively, belong to the same genus. In American trypanosomiasis, more popularly known as Chagas disease, the etiologic agent is mainly transmitted by triatomines (Telleria and Tibayrenc 2010). Correlated with the distribution of these insects, Chagas disease is mainly restricted to Latin America, but increasingly introduced to the United States and Europe by infected immigrants (Coura and Viñas 2010).

The blood-sucking vectors are night active, stay in hiding places during the daytime and are thereby often found in houses in rural areas made of adobe bricks or wooden frames covered with mud (Schaub 2009). Of the 141 species of triatomines, only about 12 species are well adapted to houses, and most important vectors are *Triatoma infestans*, *Rhodnius prolixus*, *T. dimidiata*, *Panstrongylus megistus* and *T. brasiliensis* (Patterson and Guhl 2010). The first three species are distributed across several countries. Since only two compounds for chemotherapy are available, which often induce severe side effects in the patients (Apt 2010), control campaigns focus on the interruption of the transmission by the vectors. Besides education and house improvements, the campaigns mainly base on intense insecticide sprayings against the species colonizing the houses, especially *T. infestans*. The campaigns have strongly reduced the prevalence of Chagas disease within the last 20 years from about 30 million infected people in 1990 to about 10 million in 2010 (WHO 2008, 2012). However, eradication is impossible since Chagas disease is an anthropozoonosis, and all mammals in the domestic area and in the wood might serve as reservoir hosts, but many livestock species possess a low susceptibility (Jansen and Roque 2010). In addition, all species colonizing houses possess sylvatic populations, causing a permanent risk of a reinvasion after eradication of domestic populations. If a species, which has been transported far away from its origin by human migration and possesses there only domestic but no sylvatic populations, is eradicated in the houses by the insecticide campaigns, other indigenous species of triatomines invade these houses (Abad-Franch and Monteiro 2005). Resistances to insecticides also aggravate a long-term success of the control campaigns (Gorla et al. 2010). Since especially poor people are affected, American trypanosomiasis is classified as neglected disease (Telleria and Tibayrenc 2010), and the control does not have a high priority in the health administration.

The deficit in treatment and the vectors as main target emphasize the necessity for investigations on parasite–vector interaction. However, the diversity of triatomines and the etiologic agent aggravate the step from basic research to applications. This is also recognizable in the recent reviews on the interactions of *T. cruzi* and triatomines (e.g. Vallejo et al. 2009; Schaub 2009; Garcia et al. 2010; Walochnik and Aspöck 2010; Telleria and Tibayrenc 2010; Schaub et al. 2011). In contrast to these reviews, the present chapter focusses on the vector, its saliva and digestion.

12.2 The Vectors, Triatomines

12.2.1 Taxonomy

The subfamily Triatominae belongs to the family Reduviidae and thus to the order Hemiptera in which all species possess specific piercing-sucking mouthparts, and in the suborder Heteroptera, the winged species additionally have hemelytra, i.e. forewings with a basal sclerotization (Kettle 1984). Reduviidae are mainly predators on other invertebrates, and triatomines are believed to have developed to the blood-sucking habitat (Bargues et al. 2010). However, occasionally several Reduviidae suck blood, and some Triatominae are regularly predaceous. In addition to the food, Triatominae are defined by the straight three-segmented rostrum and its peculiarities (Bargues et al. 2010). They are the biggest blood-sucking insects (Schaub 2008), females and males of *Dipetalogaster maxima* possessing lengths of 41–42 mm and 33–35 mm, respectively (Lent and Wygodzinsky 1979).

Following the recent updates of the classifications of Triatominae, the 141 species are classified into 15 or 18 genera and five or six tribes—Alberproseniini, Bolboderini, Cavernicolini, Rhodniini, Triatomini and Linshcosteini, the latter also classified into the tribe Triatomini (Galvão et al. 2003; Schofield and Galvão 2009; Bargues et al. 2010). Only the Rhodniini appear to represent a monophyletic tribe. In addition to clear morphological characteristics, the habitat distinguishes them from other triatomines. They primarily prefer palm trees and glue the eggs to the leaves or feathers of birds (Zeledón and Rabinovich 1981).

12.2.2 Distribution and Important Species

The majority of triatomines occur on the American continent between the Great Lakes of North America and Argentina, i.e. from latitude 42°N to 46°S (Lent and Wygodzinsky 1979; Schofield 1994; Gorla et al. 1997). Species of the genus *Linshcosteus* possess a uniquely Indian distribution. According to phylogenetic analyses, they are closely related to the eight species forming the *Triatoma rubrofasciata* group (Bargues et al. 2010). These originate from the New World and have been distributed by the association with rats on sailing ships to many ports in tropical and subtropical regions (Haridass and Ananthakrishnan 1981; Gorla et al. 1997). Although the species outside of Latin America are not infected with *T. cruzi*, all species of triatomines are suggested to be susceptible for the flagellate, and in over half of the species natural or experimental infections were found (Bargues et al. 2010). According to the association to human dwellings, triatomines are classified into sylvatic, peridomestic and domestic species, but especially domestic species also colonize the other habitats.

Considering the five most important vectors of Chagas disease, *Triatoma infestans* is a medium-sized triatomine, females and males ranging 26–29 mm and 21–26 mm, respectively (Lent and Wygodzinsky 1979). This species is the most important vector of *T. cruzi*, reaching high population densities in houses. In 1991, it inhabited all countries south of the Amazon from Peru to Argentina, in total 6,278,081 km^2. In 2006, it had been eliminated from Chile, Uruguay and Brazil and be strongly reduced in other countries, thus occupying only 913,485 km^2. It remains in the Gran Chaco, a highland ranging from South Bolivia and West Paraguay to North Argentina (Schofield et al. 2006). There sylvatic populations exist, an indication for the origin of this species. The bugs are colonizing nests of rodents and guinea pigs, and during domestication of the latter as meat animals for human consumption, it colonized the houses. Human trading and migration distributed this bug strongly, but outside of the Gran Chaco, rarely sylvatic populations established (Schofield 1994).

R. prolixus is the smallest of the five species, females and males possessing lengths of 19.5–21.5 mm and 17.5–20.0 mm, respectively (Lent and Wygodzinsky 1979). This species transmitted *T. cruzi* in Venezuela, Colombia, and some countries of Central America, including Honduras and Nicaragua, but not those neighboured to Colombia. Since some species of this genus are particularly difficult to distinguish (Monteiro et al. 2010), records from Panama, Ecuador, Bolivia, French Guiana, Guiana, Suriname and Brazil are presumably due to confusion with other species (Gorla and Noireau 2010). For the discontinuity of the distribution area, two explanations are discussed, migratory birds or an accidental escape of insects from a laboratory in Central America (Gorla and Noireau 2010). Recently it was successfully eradicated from Central America (Hashimoto and Schofield 2012). This species served as model insect in the outstanding physiology investigations of Sir Wigglesworth (Wigglesworth 1972).

The other three important vectors belong to the tribe Triatomini. *Triatoma dimidiata* is one of the bigger triatomines, females and males developing to lengths of 24.5–35.0 mm and 24.5–32.0 mm, respectively (Lent and Wygodzinsky 1979). The *T. dimidiata* complex of species occurs in Mexico, all countries of Central America, Venezuela, Colombia and Ecuador (Gorla and Noireau 2010). The colour of different populations varies, resulting in the classification of subspecies which possess a different epidemiological importance. After successful campaigns against domestic *R. prolixus*, often peridomestic or sylvatic populations of *T. dimidiata* invade the houses. In the field, they occupy a great variety of habitats, such as hollow trees, palm trees, rock piles and nests of opossums and armadillos and other mammals (Gorla and Noireau 2010).

Panstrongylus megistus has a similar size as *T. infestans*, females and males possessing lengths of 26–34 mm and 29–38 mm, respectively (Lent and Wygodzinsky 1979). It is an important vector in Uruguay, Paraguay, Brazil and Argentina and can be found in palm tree crowns and often in terrestrial burrows, tree-root cavities or hollow trees colonized by opossums. In the digestive tract of this triatomine, Dr. Carlos Chagas first detected *Trypanosoma cruzi* (Chagas 1909). At that time, this triatomine was the main vector in Minas Gerais and the Mata Atlantica, from whence it presumably originated. It lived in the woods and adapted to the houses during the colonization of these regions (Forattini 1980; Schofield et al. 1999). Probably since 1930, it was progressively replaced by *T. infestans*, which became the main vector of *T. cruzi* by about 1980 (Pereira et al. 2006; Gorla and Noireau 2010). Since the Southern Cone Initiative eliminated the domestic populations of *T. infestans* in many of these countries (Schofield and Dias 1999), the sylvatic and peridomestic populations of *P. megistus* reinfested the domestic ecological niche, and today, *P. megistus* is considered the main vector of *T. cruzi* in the eastern states of Brazil (Barbosa et al. 2001; Villela et al. 2005).

Triatoma brasiliensis is of similar colouration, but smaller than *T. infestans*; females are 23.0–25.5 mm long and males 22–55 mm (Lent and Wygodzinsky 1979). This species is an important vector in the semiarid northeast region of Brazil and colonizes sylvatic, peridomestic and domestic habitats (Dias et al. 2000; Herrera et al. 2003). In that region, the elimination of *T. infestans* resulted in an increased participation of *T. brasiliensis* in domestic and peridomestic cycles (Costa et al. 2003).

12.2.3 Development and Reproduction

Triatomines are hemimetabolous insects. Thus, the developmental cycle includes eggs, five immature stages and the adults (Schaub 2008). Like in all hemimetabolous insects, immature stages and adults occupy the same habitat and feed on the same hosts. The immature stages, named nymph or larva, successively increase in size, and the future wing pads are visible after the first instar. All adults possess wings—except *Mepraia* sp.—and gender-specific genitalia. Adults prefer to walk, but strongly starved adults fly at optimal night temperatures (Schofield 1979).

Prior to copulation, the male approaches the female and induces an immobilization, then entering a dorsolateral position on the female (Lent and Wygodzinsky 1979). Copulation lasts 5–15 min, providing sufficient sperm for the whole life of the female. About 1–4 weeks later, oviposition begins which may continue for several months. Under optimal conditions—availability of hosts, temperature and humidity, a single female can produce 500–1,000 eggs. Eggs are ovoid and pearly white after oviposition. In many species, the colour changes to light and then dark pink due to the colour of the cuticle of the developing embryo which is visible through the egg shell. Embryogenesis is strictly temperature dependant and lasts about 10–30 days (Lent and Wygodzinsky 1979). The first instar presses off the operculum, hatches and immediately runs away. Before the first blood ingestion, the cuticle has to harden changing to a brown or black colour. Usually 2–3 days after hatching, the nymphs are ready to suck blood. One full engorgement of about 6–12 times their own body weight or several small blood meals are necessary for the development to the next instar. The increase in size up to the fifth instar is about fivefold, e.g. in *T. infestans*, the length increases from 4 to 6, 9, 12 and 19 mm (Stadler et al. 2011). The increase in size of the abdomen activates distension receptors. This induces a nervous stimulus resulting in the secretion of diuresis and moulting hormones (Noireau and Dujardin 2010). Depending on the instar and temperature, the nymphs moult 7–28 days after the engorgement. Thereafter, some instars can starve up to about 1 year (Schaub and Lösch 1989). The total developmental time of nymphs depends on the species and the availability of food. Feeding *R. prolixus* and *T. infestans* in the laboratory always 4 days after the moult, one generation can be completed within 3 or 4 months, respectively. In the field, often 1 year is required.

12.2.4 Hematophagy

Blood is an energy-rich food source, but the acquisition causes many problems for the blood-sucking ectoparasite. The host has to be found, and the blood has to be made available without attracting the attention of the host and thereby a mechanical defence reaction. During and after blood ingestion, the ectoparasite is confronted with the different variants of the biomechanical defence of the host, especially the complement system and the blood clotting (Andrade et al. 2005).

12.2.4.1 Biting and Blood Ingestion

The triatomines belong to the vessel feeders, i.e. they ingest blood directly from a blood vessel (Lavoipierre 1965). Their mouthparts consist of the labrum, labium, mandibles and maxillae. The visible part of the biting apparatus, the proboscis, is formed by the labrum and labium and protects the internal mandibles and maxillae. In the resting position, the proboscis is bent backwards and closely appressed to the ventral surface of the head and thorax, reaching the first pair of legs with the apex. The mandibles, the actual stylets, are serrated at the proximal ends, while the two maxillae are joined to a maxillary bundle that is placed between the mandibles. Left and right maxillae are moveable against each other in the longitudinal axis. Between left and right maxilla, the dorsal food canal and the ventral salivary canal are formed, possessing diameters of about 38.6 μm and 4.8 μm, respectively. The two maxillae are separated at the proximal end of the maxillary bundle (Wirtz 1987).

The bugs are able to detect the warmest regions on the skin of their warm-blooded hosts, regardless of the temperature of host and background and bite preferentially at these regions (Ferreira et al. 2007). Apparently, this strategy is very efficient for locating blood vessels within the wide temperature range of the different putative hosts as well as on different areas of the host body. At the host, the proboscis is swung forwards and downwards to a vertical or straight forward position and contacts the skin while small amounts of saliva are released (Amino et al. 2001). With a jerky movement of the head, the skin is punctured. The upper layers of skin are opened by the barbed mandibles, which are anchored there, while the maxillary bundle is delving deeper into the skin. According to measurements of the length of the mouthparts of *R. prolixus*, adults may penetrate the hosts' skin at least 2,100 μm with their maxillae and 750 μm with their mandibles (Wenk et al. 2010). The maxillary bundle is performing whipping movements in the skin until a suitable blood vessel is found, causing minor hematoma. Several times, the bugs probe to suck blood on a trial and error basis ("probing phase"), but no blood is ingested from the hematoma. If a suitable blood vessel is found, the right maxilla penetrates into the vessel, while the left maxilla attaches on the vessel wall (Lavoipierre et al. 1959; Wirtz 1987). The duration of the probing phase differs, independent of the triatomine or host species between only a few seconds and several minutes (Lavoipierre et al. 1959; Sant'Anna et al. 2001). The bug starts new attempts elsewhere, if no suitable vessel is found or the bug is disturbed (Lavoipierre et al. 1959; Sant'Anna et al. 2001). The feeding itself lasts—depending on host and triatomine species as well as the developmental stage—about 20–50 min (Soares et al. 2000). During feeding, the blood vessel is narrowing in the rhythm of the aspiration of the food pump (Lavoipierre et al. 1959; Wirtz 1987). At the same rhythm, saliva is delivered continuously into the blood vessel (Soares et al. 2006).

12.2.4.2 Salivary Glands and Saliva

The salivary glands of triatomines are located in the anterior part of the thorax beside the pharynx. They consist of a single epithelial layer that is covered by muscles and tracheae. The number of glands differs between Rhodniini and Triatomini. Rhodniini possess two paired salivary glands, the larger reddish main gland, which shows a weak division into an anterior and a posterior lobe, and the colourless supplemental gland (Baptist 1941; Meirelles et al. 2003). Triatomini possess three pairs of glands: the circular main gland D1, the kidney-shaped supplemental gland D2 and the big reservoir gland D3. In addition to their shape, the three glands differ also in their colouring: the D1 is slightly yellow, D2 milky white and D3 colourless clear. The luminal content of the three salivary glands is mixed during the release of saliva (Barth 1954; Lacombe 1999).

The saliva of triatomines is a complex protein mixture, which contains many pharmacologically active substances. According to investigations of the proteome of salivary glands of several species, the saliva of each species contains at least 100 different proteins (Costa et al. 2011; Assumpção et al. 2011; Meiser and Schaub unpublished). Moreover, analysis of cDNA libraries of *T. infestans*, *T. brasiliensis* and *D. maxima* indicated even several hundred to more than a thousand putative secreted proteins (Santos et al. 2007; Assumpção et al. 2008, 2011). Only few of these proteins have been investigated so far in detail. Different proteins, e.g. apyrases, have the same enzymatic action, but differ in their immunogenicity in the vertebrate host (Schwarz et al. 2009), thus excluding an interference of antibodies with all of these enzymes.

12.2.4.3 Effects of Saliva on Hosts Defence Mechanisms

The saliva of the triatomines *P. megistus*, *T. brasiliensis*, *R. prolixus* and *T. infestans* inhibits the classical and/or alternative pathway of the complement system (Cavalcante et al. 2003; Barros et al. 2009). In addition, the saliva of *T. infestans* possesses an anaesthetic function similar to the effect of many local anaesthetics (Dan et al. 1999). It also contains a sialidase with suggested anti-inflammatory function (Amino et al. 1998). A similar effect has a 39-kDa protein in the saliva of *R. prolixus*, which also inhibits platelet aggregation (Ribeiro and Garcia 1981b; Ribeiro 1982; Ribeiro and Walker 1994). A broad range of other inhibitors of primary hemostasis occur frequently in the saliva of triatomines. In the saliva of *R. prolixus,* the collagen-induced adhesion of platelets is inhibited by at least two proteins and the thromboxane A_2-mediated aggregation by another compound (Ribeiro and Garcia 1981b). The *R. prolixus* aggregation inhibitor-1 (RPAI-1) binds ADP and thus prevents platelet aggregation (Francischetti et al. 2000). Another inhibitor of collagen-induced activation of platelets is pallidipin from the saliva of *Triatoma pallidipennis* (Noeske-Jungblut et al. 1994). Also the saliva of *T. infestans* contains inhibitors of collagen-induced aggregation with high sequence similarities to pallidipin, named triplatin-1 and triplatin-2 (Morita et al. 2006). Furthermore, the primary hemostasis is inhibited by apyrases in saliva of all triatomines (Ribeiro et al. 1998).

Also the vasoconstriction of the injured vessel is affected. While the saliva of all species of Triatomini inhibits the vasoconstriction in the presence of the endothelium and without endothelium partly a vasoconstriction is induced, the saliva of Rhodniini induces vasodilation in both cases (Ribeiro et al. 1998). This effect is based, like in the saliva of the bedbug *Cimex lectularius*, on nitrophorins that occur within the Triatominae only in the saliva of Rhodniini and give the saliva a characteristic cherry red colour (Soares et al. 2000). Seven proteins of this group, with molecular masses between 19 and 21 kDa and with no sequence similarities to the nitrophorins of *Cimex lectularius*, occur in the saliva of *R. prolixus*. These heme proteins release nitric oxide at a slightly basic pH, i.e. in the blood vessel, and bind it at acidic pH. The vasodilatory effect of the saliva is therefore most probably based on the relaxing effect of nitric oxide on the smooth muscle of the vascular wall (Kaneko et al. 1999; Weichsel et al. 2000; Martí et al. 2008).

These compounds are also active in other systems. Nitrophorin 2 of *R. prolixus* inhibits the fibrin formation through interaction with the coagulation factor IX (Zhang et al. 1998; Isawa et al. 2000). Another inhibitor of the coagulation cascade is the 17-kDa triabin from the saliva of *T. pallidipennis*, which binds to the fibrin recognition site of thrombin (Noeske-Jungblut et al. 1995). Two triabin-like proteins in the saliva of *T. infestans* inhibit the reciprocal activation of factor XII and prekallikrein in the intrinsic activation of the coagulation cascade (Isawa et al. 2007). The inhibition of fibrin formation by the saliva of triatomines differs markedly between species. In the genus *Rhodnius*, homogenates of the salivary glands of *R. robustus* induce the longest prolongation of the clotting time after recalcification (approximately 70-fold per pair of glands) and those of *R. neglectus* the shortest (0.6-fold extension). Similar differences are evident in the tribe Triatomini: homogenates of the glands of *Triatoma tibiamaculata* extend the clotting time about 28-fold, while the glands of *Triatoma nitida* induce no significant prolongation (Ribeiro et al. 1998). In the detailed comparisons of the different ways of inhibition by the saliva of *T. infestans* and *P. megistus*, that of *T. infestans* prolonged the coagulation after intrinsic and extrinsic activation and that of *P. megistus* only after intrinsic activation (Pereira et al. 1996). Since these inhibitors seem to avoid the sticking or clogging of the mouthparts by fibrin during and immediately after a blood meal, it remains to be elucidated how the triatomines that lack such inhibitors avoid this reaction. Until recently, a fibrinolytic activity of the saliva was not shown for hematophagous insects and was excluded for *Glossina morsitans morsitans* (Parker and Mant 1979). Also triapsin, the first protease isolated from the saliva of *T. infestans*, does not degrade fibrin (Amino et al. 2001). In contrast, a very similar protease in the saliva of *P. megistus*, named PmSgP1, showed fibrinolytic activity. This compound originates from the salivary glands D2 (Meiser et al. 2010a; Meiser and Schaub unpublished).

12.2.4.4 Effects of Compounds in the Stomach on Hosts Defence Mechanisms

Inhibitors of the complement system and the serine proteases of the coagulation cascade are present not only in the saliva of triatomines but also in the anterior part of the digestive tract. The classical and the alternative activation pathways of the complement system are inhibited by the stomach contents of *R. prolixus*, *T. brasiliensis* and *T. infestans*, but interacting factors are unknown (Barros et al. 2009).

Competent inhibitors of the coagulation cascade have been found in the stomach of *D. maxima*, *T. infestans* and *T. brasiliensis* as well as in whole body homogenates of *R. prolixus* (van de Locht et al. 1995; Lange et al. 1999; Mende et al. 1999; Campos et al. 2002; Lovato et al. 2006; Araujo et al. 2007). Different Kazal-type inhibitors interfere with the blood coagulation. These inhibitors are expressed as multi-domain precursors. The deduced full-length amino acid sequences of the Kazal-type inhibitor from *D. maxima* possess six, those of *T. infestans* and *P. megistus* seven and the inhibitor of *T. brasiliensis* eight Kazal-like domains (Mende et al. 1999; Lovato et al. 2006; Araujo et al. 2007; Meiser et al. 2010b). In the content of the stomach, the precursors are processed post-translationally into the active inhibitors (Meiser et al. 2010b). Rhodniini, the inhibitor isolated from the stomach of *R. prolixus,* consists of two Kazal-like domains (Friedrich et al. 1993). In the other species, processing results in inhibitors that consist of a single or two Kazal-like domains. These processed inhibitors target different proteases in the coagulation cascade. From each species, two double domain inhibitors are targeting thrombin, the key enzyme of the coagulation cascade. Other targets are factors XII and X (Campos et al. 2004). In addition to the factors of the coagulation cascade, also other serine proteases, such as plasmin, trypsin, subtilisin or elastase, are inhibited by these compounds (Campos et al. 2002, 2004; Mende et al. 2004; Lovato et al. 2006; Meiser et al. 2010b).

12.2.4.5 Effects of Saliva and Stomach Contents on Feeding

In triatomines, the saliva rarely contains compounds interacting with the fibrin formation, and those reported seem to bind not to the active site of the serine proteases of the coagulation cascade but to exosites (Noeske-Jungblut et al. 1995; Sun et al. 1996). In a comparison of the anticoagulant activity of the saliva and stomach content of three different triatomines, *T. infestans*, *T. brasiliensis* and *R. prolixus*, the inhibition by the stomach content was significantly higher, varying from 1.6-fold higher for *R. prolixus* to 70-fold higher for *T. brasiliensis* (Paim et al. 2011). Saliva seems to be less important for blood ingestion since the feeding performance of salivary gland-rectomized *R. prolixus* is mainly affected in the initial probing phase, but the bugs ingest identical volumes of blood (Ribeiro and Garcia 1981a). Similar results were obtained by a knock-down of the salivary nitrophorins in *R. prolixus* (Araujo et al. 2009a). In a comparative analysis of the saliva of different triatomine species, successful feeding of triatomines was correlated mainly with the activity of apyrases, hydrolyzing adenosine tri- and diphosphate and the vasodilatory effect of saliva and not with inhibition of the coagulation cascade by the saliva (Ribeiro et al. 1998). In contrast, the knock-down of the Kazal-type inhibitor in the stomach of *T. brasiliensis* resulted in a reduced ingestion rate per minute as well as in a reduction of the amount of ingested blood (Araujo et al. 2007; Paim et al. 2011). The knock-down of the thrombin inhibitor in *T. brasiliensis* presumably did not affect the digestion of ingested blood since all nymphs fed again and moulted (Araujo et al. 2007).

Such investigations are complicated by host-specific effects of the blood. In *T. brasiliensis*, soluble factors of the stomach contents induce the formation of agglutinated clusters of erythrocytes in rat blood immediately after ingestion but have no effect on blood of cattle or of the common natural host *Thrichomys apereoides* (Araujo et al. 2009b). The erythrocytes are connected by flexible elastic-like fibres most probably due to the action of bug lectins and not by plasma proteins. The hemagglutination affects the feeding performance, and *T. brasiliensis* has a much higher ingestion rate when feeding on rat blood than feeding on the common natural host or cattle blood (Araujo et al. 2009b). The agglutination of erythrocytes might cause a separation of a fluid fraction of the blood with low viscosity at the top of the stomach and a semi-solid fraction of cells at the bottom of the stomach with high viscosity. The lower viscosity of the liquid fraction prevents back pressure during the pumping process when blood is arriving in the stomach, thus facilitating an unhindered blood uptake with a high rate of ingestion.

Summarizing these effects, saliva is necessary for the location of the blood vessel and perhaps an avoidance of fibrin formation at the mouthparts, and compounds in the stomach affect the amount of ingested blood (Araujo et al. 2007; Paim et al. 2011). Although the stomach contents of all triatomines inhibit thrombin and thus interfere with the fibrin formation (Campos et al. 2002; Araujo et al. 2007), the biological function of this inhibition of coagulation is not clear, since there seems to be no effect on blood digestion.

12.3 Intestine, Blood Digestion and Excretion

12.3.1 Intestinal Tract and pH Values

The intestinal tract of triatomines is a relatively simple tube and contains no diverticula like other insects. The foregut in head and thorax is modified as strong cibarial and pharyngeal pump, responsible for the ingestion of blood and the release of saliva, and ends in the oesophagus (Wenk et al. 2010). The foregut is lined by a cuticle. Thereby, the midgut is clearly separated, beginning with a short cardia, followed by a strongly distensible stomach and the small intestine. A central narrow region subdivides the small intestine into the anterior, middle and posterior small intestine. At the border of the hindgut, at a short pylorus/ileum region, the Malpighian tubules end. The major part of the cuticle-lined hindgut is the sac-like rectum.

The digestion of blood in triatomines is exceptional since they use cathepsin-like proteases active at an acid pH (Lehane 1991), and in a characterization of cathepsin D, maximum hemoglobin hydrolysis occurred at pH 2.8 (Houseman and Downe 1982). However, using universal indicator solutions, the stomach and the small intestine of *R. prolixus* possessed feeding-dependent variations between pH 5.5 and 7.4 (J. M. C. Ribeiro and E. S. Garcia, personal communication). In unfed fifth instar nymphs of *T. brasiliensis*, the pH value of the stomach and small intestine were about pH 7 and 5.5, respectively, with a sharp boundary between the different tissues (Waniek et al. 2012). Using microelectrodes, the pH value in the lumen of *T. brasiliensis* was 7.16 at 2 h after blood feeding and remained at this level for 24 h (7.02) (Barros et al. 2009). In a more comprehensive investigation, the pH values of the stomach and the three morphologically distinguishable sections of the small intestine of *Triatoma infestans* were determined separately by microelectrodes up to 20 days after feeding. In unfed fifth instar nymphs, the pH values were about 6.3 for the stomach and posterior region of the small intestine and 6.7 and 6.6 for the anterior and middle region of the small intestine, respectively (Balczun et al. 2012). Immediately after feeding, the pH value of the stomach increased up to pH 7.4, most probably due to the ingested blood, and reached its initial value 1 day after feeding. The contents of the stomach and the different regions of the small intestine were continuously acidified until 8 and 10 days after feeding, reaching minimum values of about pH 5.2. Following 15 days after feeding, the pH values of stomach and small intestine contents increased, but at 20 days after feeding, they were considerably below the values of unfed insects.

Table 12.1 Intestinal digestive enzymes of triatomines

Species	Enzymes	pH optimum	Localization
R. prolixus	β-Acetylglucosaminidase	4.5	st
R. prolixus	α-Galactosidase	4.5	st, si
R. prolixus	α-Glucosidase	4.5	st, si
R. prolixus	β-Glucosidase	4.5	si
R. prolixus	α-Mannosidase	4.5	st, si
R. prolixus	β-Mannosidase	4.5	si
R. prolixus	Alkaline phosphatase	10.4	st, si
R. prolixus	Acid phosphatase	4.5	st, si
R. prolixus	Aminopeptidase	8.0	st, si
R. prolixus	Carboxypeptidase A	n.d.	si
R. prolixus; *T. infestans*	Cathepsin D-like[a,b]	3.5	si[c]
R. prolixus; *T. infestans*	Cathepsin B-like[d]	5.5	si
R. prolixus; *T. infestans*	Cathepsin L-like[d,e]	5.5	si
R. prolixus; *T. infestans*	Lipases[f]	7.0–7.5	st, si

st stomach, *si* small intestine, *n.d.* not determined
Most enzymes are reviewed in Terra and Ferreira (1994) except for [a]Borges et al. (2006), [b]Balczun et al. (2012) [d]Kollien et al. (2004), [e]Lopez-Ordoñez et al. (2001) and [f]Rimoldi et al. (1985) and Grillo et al. (2007)
[c]Transcripts of cathepsin D-like genes are also present in stomach tissue

12.3.2 *Blood Digestion*

Immediately after ingestion, the blood is concentrated by withdrawal of ions and water (Schaub 2008). The major ingredients of the food source blood are proteins derived from the plasma and erythrocytes. Blood proteins pass the stomach of triatomines essentially undigested, while glycosidases, alkaline and acidic phosphatases, sialidases, lipases and amylases (derived from symbionts) act on their respective target molecules in the blood (Schaub et al. 2011). Only a slight aminopeptidase activity was found in the stomach being much weaker than in the small intestine (Kollien and Schaub 2000). As in the stomach, also glycosidases, alkaline and acidic phosphatases, lipases and amylases (derived from symbionts) are active in the small intestine (Table 12.1).

In all nymphal instars and adults, the blood is processed by lysis of erythrocytes (Azambuja et al. 1983). This hemolytic activity in the stomach is highest 2–4 days after feeding. Erythrocytes of different blood sources are efficiently lysed by stomach contents of *Rhodnius prolixus*, illustrating a similar lysis mechanism for a broad spectrum of hosts (Azambuja et al. 1983). Highest rate of hemolysis was induced by feeding on whole blood, while similar but considerable lower activities were observed by feeding pure hemoglobin or erythrocytes. After partial purification and biochemical characterization, the hemolytic factor of *R. prolixus* was classified as a basic peptide which is released to the stomach lumen and activated upon blood ingestion. This hemolytic activity is restricted to the lumen of the stomach, and it is not associated with the stomach wall tissue or the small intestine.

In contrast, proteolytic activity was only found in the small intestine after a transport of small portions of blood. Using azocasein as substrate, no activity was detectable in the stomach of *R. prolixus* (Azambuja et al. 1983). However, high expression rates of cathepsin-like genes—suggested to encode digestive enzymes—are also evident in the stomach tissue (Balczun et al. 2012; Ribeiro et al. unpublished). Using hemoglobin as substrate in determinations of the activity of another digestive enzyme, cathepsin D, after depletion of stomach contents, only a slight activity occurred in homogenates of the stomach wall of *T. infestans*, even at the optimal pH 3 (Gajewski and Schaub unpublished). However, the pH in the lumen of the intestine never possessed acidic conditions that are optimal for the activities of cathepsin D. A careful analysis of cathepsin D-like activity of stomach contents is necessary to elucidate if blood proteins are not processed in the stomach of triatomines, neither by cathepsin D nor by cathepsins B and L (see below).

Using homogenates of the small intestine of *R. prolixus* and *Triatoma phyllosoma pallidipennis*, activities of different exopeptidases and endopeptidases were identified (Houseman 1978; Houseman and Downe 1980, 1981a, b; Terra et al. 1988). Carboxypeptidase A and a lysosomal carboxypeptidase B were detected in homogenates of the midgut and small intestine (Garcia and Guimarães 1979; Houseman and Downe 1981b), and at least three aminopeptidases were localized between microvillar membranes or on the surface of cells of the small intestine (Ferreira et al. 1988). The aspartate protease cathepsin D and the cysteine protease cathepsin B of *R. prolixus* were identified in the luminal contents of the small intestine by their pH optima and using specific substrates (Houseman and Downe 1983a, b). For cathepsin B, D and the lysosomal carboxypeptidase B, high activities were measured 6 days after feeding, and a second peak of cathepsin B activity occurred at the time of ecdysis (Billingsley and Downe 1988; Houseman and Downe 1983b). In *R. prolixus*, cathepsin B was suggested as the major digestive protease, while cathepsin D should contribute only to a minor extent to the overall proteolytic activity (Terra et al. 1988).

However, molecular biological investigations of *R. prolixus*, *Triatoma brasiliensis* and *Triatoma infestans* identified the nucleotide sequences not only of the cysteine protease cathepsin B but also of another cysteine protease, cathepsin L (Lopez-Ordoñez et al. 2001; Kollien et al. 2004; Waniek et al. 2012). Since the previous measurements of the activities of cysteine proteases covered the activities of cathepsin B as well as cathepsin L without differentiation, the relative impact of both proteases on blood digestion is an open question. Despite the common GCDGG motif, the triatomine cysteine proteases share characteristic motifs of the respective family C1 of cysteine proteases, in particular the ERFNIN and GNFD motifs of cathepsin L and the "occluding loop" of cathepsin B (Schaub et al. 2011).

Two isoforms of cathepsin D-like protease encoding genes were characterized in *T. infestans* (Balczun et al. 2012). The expression level of *TiCatD* in the small intestine is immediately and strongly induced after feeding, while expression of *TiCatD2* is induced much later and at a considerably lower level. Both proteases were identified to be active in the lumen of the small intestine. This is remarkable since TiCatD2 possesses a motif called "proline loop" which is suggested to be a characteristic motif of lysosomal cathepsin D-like proteases (Padilha et al. 2009). This is another indication for the peculiarities of the digestion of triatomines.

Several isoforms of the cysteine and aspartate protease encoding genes of triatomines are available in GenBank. In addition, the analysis of the transcriptome of *R. prolixus* midgut reveals the existence of multiple transcript variants of the respective genes (Ribeiro et al. unpublished). Apparently, protein digestion in triatomine midguts is a complex process with a concerted interaction of endopeptidases and exopeptidases, and therefore, elucidation of putative regulatory networks of proteases in the midgut of triatomines demands further sophisticated approaches.

12.3.3 Excretion

After ingestion of such enormous volumes of blood, the nymphs can hardly move. Therefore, the most effective excretion system in the animal kingdom is already induced during blood ingestion (Maddrell 1969; Maddrell et al. 1991). Ions of the blood are transported into the hemolymph and then into the Malpighian tubules, followed by the water. The urine flows into the rectum, sweeping out the remnants of digestion. Thereby, within 24 h, the bugs excrete 76% of the imbibed fluid part of the blood (Schaub 2009).

12.4 Symbionts and Antibacterial Compounds of Triatomines

Like many blood-sucking insects, triatomines require symbionts for a normal development (summarized by Schaub 2009). These develop in the anterior midgut, cardia and stomach and are mainly lysed in the small intestine. Symbionts are obtained by coprophagy, a risk of a contamination by other bacteria. In addition, there are other occasions at which bacteria might enter the digestive tract (summarized by Schaub 2009). Therefore, the ingested sterile blood has to be protected from an uncontrolled growth of microorganisms, but the growth of the essential symbionts and their utilization has to be assured.

Saliva is the first defence mechanism against oral infections. So far, only three antibacterial proteins were described from the saliva of triatomines, trialysin, lysozyme and defensin, possessing molecular masses of 30, 13 and 5 kDa (Amino et al. 2002; Waniek et al. 2009; Klenner and Schaub unpublished). In the saliva of *T. infestans*, proteins with molecular masses of 14–17 kDa showed the highest bacteriolytic activity in zymography using Gram-positive bacteria as substrate (Klenner and Schaub unpublished). Lysozymes possess such a molecular mass, and bacteriolytic activity in extracts of the midgut of *R. prolixus* and *T. infestans* was particularly attributed to these proteins (Ribeiro and Pereira

1984). In addition, several lysis bands in higher molecular ranges are visible in such approaches. Since such zymograms do not allow an identification of the active compounds, the origin of the activities remains to be identified. Besides trialysin and lysozymes, the low-molecular-mass defensins, which are expressed in the digestive tract as well as in the salivary glands of triatomines (Waniek et al. 2009), might attribute to the antibacterial action. However, their mode of action and their putative role in immune responses require detailed investigations (Bartholomay et al. 2004).

The high number of different bacteriolytic components in the saliva indicates a more general function at the entrance of the intestine. Here, activity against a broader spectrum of bacteria is apparently necessary. The need for such a defence during oral infections, e.g. during coprophagy, is shown by the pathogenic effects occurring after feeding of blood containing faeces (Schaub unpublished). However, the saliva compounds are not effective against all bacteria indicated by the high number of different species of microorganisms isolated from triatomines (Vallejo et al. 2009).

12.5 *Trypanosoma cruzi*: Taxonomy and Developmental Cycle

The protozoan flagellate *T. cruzi* belongs to the order Kinetoplastida which have arranged concentrated extranuclear DNA including the mitochondrial DNA in a unique order in a specific region of the mitochondrium, the kinetoplast. In the family Trypanosomatidae, all members are parasites, and in the genus *Trypanosoma*, nearly all are heteroxenous, i.e. the development includes a host change between a blood-sucking invertebrate and a vertebrate. In the respective host, different forms, phenotypes, develop (Walochnik and Aspöck 2010).

According to recently established molecular biological criteria, the species *T. cruzi* is divided into six discrete typing units (DTU), TcI–VI (summarized by Schaub et al. 2011). The genetic structure of the populations is mainly clonal, but restricted genetic recombinations occur. Presumably, TcI and TcII are the most anciently divergent, TcIII and TcIV resulted from ancient hybridization events between TcI and TcII, and TcV and TcVI are hybrids from TcII and TcIII (Patterson and Guhl 2010). TcI is suggested to have developed north of Amazonia, closely associated with arboreal triatomines, e.g. *Rhodnius* sp., and arboreal marsupials, e.g. *Didelphis* sp., whereas TcII–TcVI developed in the south, associated with terrestrial triatomines, tribe Triatomini, and armadillos. At present, in some South American countries, up to four different DTUs occur, and a more geographic than host association is evident (Patterson and Guhl 2010). The strong migrations of humans introduce new strains to different regions. However, this does not generally induce a long-term establishment at the new location. Local mammals can be refractory to the respective strain, acting as "filters" (Jansen and Roque 2010).

Usually, an ingestion of parasites with the blood of an infected host initiates the development in the vector. These parasites, blood trypomastigotes, transform in the stomach to other stages, especially sphero- and epimastigotes. These stages initially multiply in the small intestine, then after passage to the rectum also there. Only in the rectum the development of the non-dividing metacyclic trypomastigotes is finished (reviewed by Kollien and Schaub 2000). All stages are present in the faeces and urine deposited by the bug during or after blood ingestion, but only this trypomastigotes stage can initiate an infection in the mammalian host. There it is phagocytized and transforms to amastigotes, which multiply and finally transform to non-dividing blood trypomastigotes. After rupturing of the exhausted host cell, they infect new cells or circulate in the blood (summarized by Schaub et al. 2011).

12.6 Effects of the Triatomines on *Trypanosoma cruzi*

In the field, the infection rates of even neighboured populations of triatomines vary strongly, indicating refractoriness mechanisms in the vector (summarized by Garcia et al. 2010). Thus, after the transport of a strain of the flagellate to another region, not only the mammalian hosts but also triatomines can act as "filters" and loose the respective strain at least during prolonged starvation (summarized by Schaub and Lösch 1989). However, susceptibilities also differ between individuals of a population, indicated by the high standard deviations in the number of intestinal flagellates after a standardized experimental infection in an optimal parasite–vector system in which both components originated from the same village (Schaub 1989b). In this system, even after eight generations of selection for differences in susceptibility, i.e. separate breeding of bugs which possessed low or high numbers of parasites in the rectum, all bugs became infected, but the "less susceptible" strain contained less parasites (Schaub unpublished). Unfortunately, these strains of *T. infestans* were killed by a failure of the incubator and its security system. Thus, the search for factors responsible for the differences in the susceptibilities could not be performed.

Of the different compounds in the saliva of triatomines, only trialysin lysed cell culture-derived trypomastigotes and epimastigotes (summarized by Schaub et al. 2011), but an effect on the initial development in the vector remains to be investigated. The initial establishment of specific strains is determined by agglutinins and hemolysins (summarized by Schaub et al. 2011). However, in these infections, epimastigotes were used, mimicking only the less important infection by coprophagy. Intestinal proteases seem to have no effects on *T. cruzi*, but knock-down studies are required.

Strong effects on *T. cruzi* are induced by starvation and feeding (summarized by Schaub 2009). During starvation, many intestinal flagellates die, and in the remaining population, the percentages of spheromastigotes and the respective intermediate forms increase from 2 to about 20%. Feeding of long-term starved bugs induces the development of giant cells, containing many nuclei, flagella and kinetoplasts. However, between 5 and 10 days after feeding, these cells disappear completely. Feeding of short-term starved bugs induces metacyclogenesis. Whereas the development of spheromastigotes and giant cells could not be correlated with specific factors in the gut, the rapid induction of metacyclogenesis could be correlated with proteins in the urine of the vector. An induction requiring a longer period of time was caused by a fragment of chicken α^D-globin and free fatty acids (summarized by Schaub et al. 2011).

12.7 Effects of *Trypanosoma cruzi* on Triatomines

T. cruzi only slightly affects the vector (summarized by Schaub 1992, 2009). Under optimal conditions, i.e. the mortality rate of control groups is <10%, the nymphal development and mortality rates are not affected by the infection. However, the competition of flagellate and vector for food components is indicated by a reduced starvation capacity (Schaub and Lösch 1989). Therefore, *T. cruzi* is classified as "subpathogenic" (Schaub 1989a, 1992). Slight effects are known for the probing behaviour, but the increased probing in infected bugs might also be caused by the competition (Schaub et al. 2011). In electron microscopy, the wax layer of the rectal cuticle is damaged in infected bugs, but the proof of an effect on the absorption processes requires precise measurements. The changes in concentrations of rectal free and protein/peptide bound amino acids (Kollien and Schaub unpublished) might be caused by cysteine proteases of the flagellate. However, in the stomach, the gene of a cysteine protease inhibitor—acting against cruzain (cruzipain) of *T. cruzi*—is expressed by *T. infestans* after an infection (Buarque et al. 2011). The activity of an aspartate protease, cathepsin D, is increased in the small intestine at 1 and 3 days after an experimental infection with epimastigotes (Borges et al. 2006). However, this protease might be involved not only in the digestion of food but also in intestinal immune responses (Balczun et al. 2012). Ingestion of trypomastigotes stimulates the immune reactions in the intestine. In *R. prolixus*, it induces the expression of the gene of the more intestinal active lysozyme RpLys-A, but not of the RpLys-B gene, which is primarily expressed in the fat body (Ursic-Bedoya et al. 2008). In *T. brasiliensis*, after infection, more mRNA encoding defensins are produced (Waniek et al. 2011). Besides these antimicrobial peptides, another antimicrobial factor, nitric oxide, is also synthesized in the intestine after an infection (Whitten et al. 2007). Since the expression rates of the genes of these compounds are also increased after blood ingestion (Araújo et al. 2006; Whitten et al. 2007), these short-term reactions to *T. cruzi* can partly be a general defence reaction against microorganisms. In long-term infected *T. infestans*, *T. cruzi* suppresses the intestinal immunity in the intestine, enabling the development of different microorganisms (summarized by Schaub et al. 2011).

12.8 Future Works

There are several aspects that need to be considered in long-term investigations to understand the co-evolution and the interactions of *T. cruzi* and triatomines. Investigations of the first topic are complicated by the human migration and the introduction of specific DTUs of *T. cruzi* in regions in which these DTUs had not been present previously. Therefore, many strains originating from mammals and vectors have to be typed to get indications for the original situation. Then detailed investigations of the development in the vector including the metacyclogenesis rates can be performed. Investigations of the interactions of *T. cruzi* and triatomines will be strongly be supported by the genome project of *R. prolixus* and large-scale applications of the RNAi technique to identify factors which are responsible for the different susceptibility of species and individuals of triatomines. These investigations with systems in which vector and parasite should originate from the same locality should be performed not only with *R. prolixus*, *T. brasiliensis* and *T. infestans* but also with triatomines engaged in the sylvatic cycle of *T. cruzi*. Thereby, factors inducing the initial killing of some strains of *T. cruzi* in the stomach and the molecular basis of the intestinal immune response and the metacyclogenesis in the rectum of the vector can be identified.

Acknowledgements We are deeply grateful for the cooperation with many colleagues—recognizable by co-authorships—and the funding of the investigations and travels by the UNDP/World Bank/WHO Special Programme for Research and Training in Tropical Diseases, the VolkswagenStiftung, the German Academic Exchange Service (DAAD), European Community programmes, the Humboldt Foundation and especially the Deutsche Forschungsgemeinschaft.

References

Abad-Franch F, Monteiro FA (2005) Molecular research and the control of Chagas disease vectors. An Acad Bras Ciênc 77:437–454

Amino R, Porto RM, Chammas R, Egami MI, Schenkman S (1998) Identification and characterization of a sialidase released by the salivary gland of the hematophagous insect *Triatoma infestans*. J Biol Chem 273:24575–24582

Amino R, Tanaka AS, Schenkman S (2001) Triapsin, an unusual activatable serine protease from the saliva of the hematophagous vector of Chagas' disease *Triatoma infestans* (Hemiptera: Reduviidae). Insect Biochem Mol Biol 31:465–472

Amino R, Martins RM, Procopio J, Hirata IY, Juliano MA, Schenkman S (2002) Trialysin, a novel pore-forming protein from saliva of hematophagous insects activated by limited proteolysis. J Biol Chem 277:6207–6213

Andrade BB, Teixeira CR, Barral A, Barral-Netto M (2005) Haematophagous arthropod saliva and host defense system: a tale of tear and blood. An Acad Bras Cienc 77:665–693

Apt W (2010) Treatment of Chagas disease. In: Telleria J, Tibayrenc M (eds) American Trypanosomiasis – Chagas disease – one hundred years of research. Elsevier, Amsterdam, pp 809–825

Araújo CAC, Waniek PJ, Stock P, Mayer C, Jansen AM, Schaub GA (2006) Sequence characterization and expression patterns of defensin and lysozyme encoding genes from the gut of the reduviid bug *Triatoma brasiliensis*. Insect Biochem Mol Biol 36:547–560

Araujo RN, Campos ITN, Tanaka AS, Santos A, Gontijo NF, Lehane MJ, Pereira MH (2007) Brasiliensin: a novel intestinal thrombin inhibitor from *Triatoma brasiliensis* (Hemiptera: Reduviidae) with an important role in blood intake. Int J Parasitol 37:1351–1358

Araujo RN, Soares AC, Paim RMM, Gontijo NF, Gontijo AF, Lehane MJ, Pereira MH (2009a) The role of salivary nitrophorins in the ingestion of blood by the triatomine bug *Rhodnius prolixus* (Reduviidae: Triatominae). Insect Biochem Mol Biol 39:83–89

Araujo RN, Pereira MH, Soares AC, Pereira IDCA, Diotaiuti L, Gontijo NF, Lehane MJ, Guarneri AA (2009b) Effect of intestinal erythrocyte agglutination on the feeding performance of *Triatoma brasiliensis* (Hemiptera: Reduviidae). J Insect Physiol 55:862–868

Assumpção TCF, Francischetti IMB, Andersen JF, Schwarz A, Santana JM, Ribeiro JMC (2008) An insight into the sialome of the blood-sucking bug *Triatoma infestans*, a vector of Chagas' disease. Insect Biochem Mol Biol 38:213–232

Assumpção TCF, Charneau S, Santiago PBM, Francischetti IMB, Meng Z, Araújo CN, Pham VM, Queiroz RML, de Castro CN, Ricart CA, Santana JM, Ribeiro JMC (2011) Insight into the salivary transcriptome and proteome of *Dipetalogaster maxima*. J Proteome Res 10:669–679

Azambuja P, Guimarães JA, Garcia ES (1983) Haemolytic factor from the crop of *Rhodnius prolixus*: evidence and partial characterization. J Insect Physiol 29:833–837

Balczun C, Siemanowski J, Pausch JK, Helling S, Marcus K, Stephan C, Meyer HE, Schneider T, Cizmowski C, Oldenburg M, Höhn S, Meiser CK, Schuhmann W, Schaub GA (2012) Intestinal aspartate proteases TiCatD and TiCatD2 of the hematophagous bug *Triatoma infestans* (Reduviidae): sequence characterisation, expression pattern and characterisation of proteolytic activity. Insect Biochem Mol Biol. doi:10.1016/j.ibmb.2011.12.006

Baptist BA (1941) The morphology and physiology of the salivary glands of Hemiptera – Heteroptera. Q J Microsc Sci 83:91–139

Barbosa SE, Soares RPP, Pires HHR, Diotaiuti L (2001) Experimental evidence for a demographic cline in *Panstrongylus megistus* populations. Mem Inst Oswaldo Cruz 96:773–775

Bargues MD, Schofield CJ, Dujardin J-P (2010) Classification and phylogeny of the Triatominae. In: Telleria J, Tibayrenc M (eds) American Trypanosomiasis – Chagas disease – one hundred years of research. Elsevier, Amsterdam, pp 117–147

Barros VC, Assumpção JG, Cadete AM, Santos VC, Cavalcante RR, Araújo RN, Pereira MH, Gontijo NF (2009) The role of salivary and intestinal complement system inhibitors in the midgut protection of triatomines and mosquitoes. PLoS One 4(6):e6047

Barth R (1954) Estudos anatômicos e histológicos sôbre a subfamília Triatominae (Heteroptera, Reduviidae). IV. parte: O complexo das glândulas salivares de Triatoma infestans. Mem Inst Oswaldo Cruz 52:517–585

Bartholomay LC, Fuchs JF, Cheng L-L, Beck ET, Vizioli J, Lowenberger C, Christensen BM (2004) Reassessing the role of defensin in the innate immune response of the mosquito, *Aedes aegypti*. Insect Mol Biol 13:125–132

Billingsley PF, Downe AER (1988) Ultrastructural localization of cathepsin B in the midgut of *Rhodnius prolixus* Stal (Hemiptera, Reduviidae) during blood digestion. Int J Insect Morphol Embryol 17:295–302

Borges EC, Machado EMM, Garcia ES, Azambuja P (2006) *Trypanosoma cruzi*: effects of infection on cathepsin D activity in the midgut of *Rhodnius prolixus*. Exp Parasitol 112:130–133

Buarque DS, Spindola LMN, Martins RM, Braz GRC, Tanaka AS (2011) Tigutcystatin, a cysteine protease inhibitor from *Triatoma infestans* midgut expressed in response to *Trypanosoma cruzi*. Biochem Biophys Res Commun 413:241–247

Campos ITN, Amino R, Sampaio CAM, Auerswald EA, Friedrich T, Lemaire HG, Schenkman S, Tanaka AS (2002) Infestin, a thrombin inhibitor present in *Triatoma infestans* midgut, a Chagas' disease vector: gene cloning, expression and characterization of the inhibitor. Insect Biochem Mol Biol 32:991–997

Campos ITN, Tanaka-Azevedo AM, Tanaka AS (2004) Identification and characterization of a novel factor XIIa inhibitor in the hematophagous insect, *Triatoma infestans* (Hemiptera: Reduviidae). FEBS Lett 577:512–516

Cavalcante RR, Pereira MH, Gontijo NF (2003) Anti-complement activity in the saliva of phlebotomine sand flies and other hematophagous insects. Parasitology 127:87–93

Chagas C (1909) Novo tripanozomiaze humana. Über eine neue Trypanosomiasis des Menschen. Mem Inst Oswaldo Cruz 1:159–281

Costa J, Almeida CE, Dotson EM, Lins A, Vinhaes M, Silveira AC, Beard CB (2003) The epidemiologic importance of *Triatoma brasiliensis* as a Chagas disease vector in Brazil: a revision of domiciliary captures during 1993–1999. Mem Inst Oswaldo Cruz 98:443–449

Costa CM, Sousa MV, Ricart CAO, Santana JM, Teixeira ARL, Roepstorff P, Charneau S (2011) 2-DE-based proteomic investigation of the saliva of the Amazonian triatomine vectors of Chagas disease: *Rhodnius brethesi* and *Rhodnius robustus*. J Proteomics 74:1652–1663

Coura JR, Viñas PA (2010) Chagas disease: a new worldwide challenge. Nature 465(Suppl.): S6–S7

Dan A, Pereira MH, Pesquero JL, Diotaiuti Lacerda Beirão PS (1999) Action of the saliva of *Triatoma infestans* (Heteroptera: Reduviidae) on sodium channels. J Med Entomol 36:875–879

Dias JCP, Machado EMM, Fernandes AL, Vinhaes MC (2000) General situation and perspectives of Chagas disease in Northeastern Region, Brazil. Cad Saúde Pública 16(Suppl. 2):S13–S34

Ferreira C, Ribeiro AF, Garcia ES, Terra WR (1988) Digestive enzymes trapped between and associated with the double plasma membranes of *Rhodnius prolixus* posterior midgut cells. Insect Biochem 18:521–530

Ferreira RA, Lazzari CR, Lorenzo MG, Pereira MH (2007) Do hematophagous bugs assess skin surface temperature to detect blood vessels? PLoS One 2(9):e932. doi:10.1371/journal.pone.0000932

Forattini OP (1980) Biogeografia, origem e distribuição da domiciliação de triatomíneos no Brasil. Rev Saúde Pública 14:265–299

Francischetti IMB, Ribeiro JMC, Champagne D, Andersen J (2000) Purification, cloning, expression, and mechanism of action of a novel platelet aggregation inhibitor from the salivary gland of the blood-sucking bug, *Rhodnius prolixus*. J Biol Chem 275:12639–12650

Friedrich T, Kröger B, Bialojan S, Lemaire HG, Höffken HW, Reuschenbach P, Otte M, Dodt J (1993) A Kazal-type inhibitor with thrombin specificity from *Rhodnius prolixus*. J Biol Chem 268:16216–16222

Galvão C, Carcavallo R, da Silva RD, Jurberg J (2003) A checklist of the current valid species of the subfamily Triatominae Jeannel, 1919 (Hemiptera, Reduviidae) and their geographical distribution, with nomenclatural and taxonomic notes. Zootaxa 202:1–36

Garcia ES, Guimarães JA (1979) Proteolytic enzymes in the *Rhodnius prolixus* midgut. Experientia 35:305–306

Garcia ES, Genta FA, de Azambuja P, Schaub GA (2010) Interactions between intestinal compounds of triatomines and *Trypanosoma cruzi*. Trends Parasitol 26:499–505

Gorla D, Noireau F (2010) Geographic distribution of Triatominae vectors in America. In: Telleria J, Tibayrenc M (eds) American Trypanosomiasis – Chagas disease – one hundred years of research. Elsevier, Amsterdam, pp 209–231

Gorla DE, Dujardin J-P, Schofield CJ (1997) Biosystematics of Old World Triatominae. Acta Trop 63:127–140

Gorla D, Ponce C, Dujardin J-P, Schofield CJ (2010) Control strategies against Triatominae. In: Telleria J, Tibayrenc M (eds) American Trypanosomiasis – Chagas disease – one hundred years of research. Elsevier, Amsterdam, pp 233–245

Grillo LAM, Majerowicz D, Gondim KC (2007) Lipid metabolism in *Rhodnius prolixus* (Hemiptera: Reduviidae): role of a midgut triacylglycerol-lipase. Insect Biochem Mol Biol 37:579–588

Haridass ET, Ananthakrishnan TN (1981) Bionomics and behaviour of the reduviid bug, *Triatoma rubrofasciata* (De Geer), the vector of *Trypanosoma* (*Megatrypanum*) *conorhini* (Donovan), in India (Insecta: Heteroptera). Proc Indian Natl Sci Acad B Biol Sci 46:884–891

Hashimoto K, Schofield CJ (2012) Elimination of *Rhodnius prolixus* in Central America. Parasit Vectors 5:45. doi:10.1186/PREACCEPT-1790853046685273

Herrera L, Pinho AP, Viegas C, Lorosa E, Xavier SCC, Emperaire L, Mangia RH, Carrasco H, Fernandes O, Jansen AM (2003) *Trypanosoma cruzi* transmission cycle in João Costa, Piauí, Brazil, an endemic area of Chagas disease. Acta Parasitol 48:294–300

Houseman J (1978) A thiol-activated digestive proteinase from adults of *Rhodnius prolixus* Stal (Hemiptera-Reduviidae). Can J Zool 56:1140–1143

Houseman JG, Downe AER (1980) Endoproteinase activity in the posterior midgut of *Rhodnius prolixus* Stal (Hemiptera, Reduviidae). Insect Biochem 10:363–366

Houseman JG, Downe AER (1981a) Exoproteinase activity in the posterior midgut of *Rhodnius prolixus* Stal (Hemiptera, Reduviidae). Insect Biochem 11:579–582

Houseman JG, Downe AER (1981b) Identification and partial characterization of digestive proteinases from *Triatoma phyllosoma pallidipennis* Stal (Hemiptera, Reduviidae). Comp Biochem Phys B 70:713–717

Houseman JG, Downe AER (1982) Characterization of an acidic proteinase from the posterior midgut of *Rhodnius prolixus* Stal (Hemiptera, Reduviidae). Insect Biochem 12:651–655

Houseman JG, Downe AER (1983a) The effects of three metabolic inhibitors on digestive proteinase production in *Rhodnius prolixus* Stal (Hemiptera, Reduviidae). J Insect Physiol 29:317–321

Houseman JG, Downe AER (1983b) Activity cycles and the control of four digestive proteinases in the posterior midgut of *Rhodnius prolixus* Stal (Hemiptera, Reduviidae). J Insect Physiol 29:141–148

Isawa H, Yuda M, Yoneda K, Chinzei Y (2000) The insect salivary protein, prolixin-S, inhibits factor IXa generation and Xase complex formation in the blood coagulation pathway. J Biol Chem 275:6636–6641

Isawa H, Orito Y, Jingushi N, Iwanaga S, Morita A, Chinzei Y, Yuda M (2007) Identification and characterization of plasma kallikrein-kinin system inhibitors from salivary glands of the blood-sucking insect *Triatoma infestans*. FEBS J 274:4271–4286

Jansen AM, Roque ALR (2010) Domestic and wild mammalian reservoirs. In: Telleria J, Tibayrenc M (eds) American Trypanosomiasis – Chagas disease – one hundred years of research. Elsevier, Amsterdam, pp 249–276

Kaneko Y, Shojo H, Yuda M, Chinzei Y (1999) Effects of recombinant nitrophorin-2 nitric oxide complex on vascular smooth muscle. Biosci Biotechnol Biochem 63:1488–1490

Kettle DS (1984) Blood-sucking Hemiptera (bugs). In: Kettle DS (ed) Medical and veterinary entomology. Wiley, New York, pp 313–328

Kollien AH, Schaub GA (2000) The development of *Trypanosoma cruzi* in Triatominae. Parasitol Today 16:381–387

Kollien AH, Waniek PJ, Nisbet AJ, Billingsley PF, Schaub GA (2004) Activity and sequence characterization of two cysteine proteases in the digestive tract of the reduviid bug *Triatoma infestans*. Insect Mol Biol 13:569–579

Lacombe D (1999) Anatomia e histologia das glândulas salivares nos Triatomíneos. Mem Inst Oswaldo Cruz 94:557–564

Lange U, Keilholz W, Schaub GA, Landmann H, Markwardt F, Nowak G (1999) Biochemical characterization of a thrombin inhibitor from the bloodsucking bug *Dipetalogaster maximus*. Haemostasis 29:204–211

Lavoipierre MMJ (1965) Feeding mechanism of blood-sucking arthropods. Nature 208:302–303

Lavoipierre MMJ, Dickerson G, Gordon RM (1959) Studies on the methods of feeding of blood-sucking arthropods. Ann Trop Med Parasitol 53:235–250

Lehane MJ (1991) Managing the blood meal. In: Lehane MJ (ed) Biology of blood-sucking insects. Harper Collins, London, pp 79–110

Lent H, Wygodzinsky P (1979) Revision of the Triatominae (Hemiptera, Reduviidae) and their significance as vectors of Chagas' disease. Bull Am Mus Nat Hist 163:123–520

Lopez-Ordoñez T, Rodriguez MH, Hernández-Hernández FD (2001) Characterization of a cDNA encoding a cathepsin L-like protein of *Rhodnius prolixus*. Insect Mol Biol 10:505–511

Lovato DV, de Campos ITN, Amino R, Tanaka AS (2006) The full-length cDNA of anticoagulant protein infestin revealed a novel releasable Kazal domain, a neutrophil elastase inhibitor lacking anticoagulant activity. Biochimie 88:673–681

Maddrell SHP (1969) Secretion by the Malpighian tubules of *Rhodnius*. The movements of ions and water. J Exp Biol 51:71–97

Maddrell SHP, Herman WS, Mooney RL, Overton JA (1991) 5-Hydroxytryptamine: a second diuretic hormone in *Rhodnius prolixus*. J Exp Biol 156:557–566

Martí MA, González Lebrero MC, Roitberg AE, Estrin DA (2008) Bond or cage effect: how nitrophorins transport and release nitric oxide. J Am Chem Soc 130:1611–1618

Meirelles RMS, Rodrigues IS, Steindel M, Soares MJ (2003) Ultrastructure of the salivary glands of *Rhodnius domesticus* Neiva & Pinto, 1923 (Hemiptera: Reduviidae). J Submicrosc Cytol Pathol 35:199–207

Meiser CK, Piechura H, Meyer HE, Warscheid B, Schaub GA, Balczun C (2010a) A salivary serine protease of the hematophagous reduviid *Panstrongylus megistus*: sequence characterization, expression pattern and characterization of proteolytic activity. Insect Mol Biol 19:409–421

Meiser CK, Piechura H, Werner T, Dittmeyer-Schäfer S, Meyer HE, Warscheid B, Schaub GA, Balczun C (2010b) Kazal-type inhibitors in the stomach of *Panstrongylus megistus* (Triatominae, Reduviidae). Insect Biochem Mol Biol 40:345–353

Mende K, Petoukhova O, Koulitchkova V, Schaub GA, Lange U, Kaufmann R, Nowak G (1999) Dipetalogastin, a potent thrombin inhibitor from the blood-sucking insect *Dipetalogaster maximus*: cDNA cloning, expression and characterization. Eur J Biochem 266:583–590

Mende K, Lange U, Nowak G (2004) Three recombinant serine proteinase inhibitors expressed from the coding region of the thrombin inhibitor dipetalogastin. Insect Biochem Mol Biol 34:971–979

Monteiro F, Marcet P, Dorn P (2010) Population genetics of triatomines. In: Telleria J, Tibayrenc M (eds) American Trypanosomiasis – Chagas disease – one hundred years of research. Elsevier, Amsterdam, pp 169–208

Morita A, Isawa H, Orito Y, Iwanaga S, Chinzei Y, Yuda M (2006) Identification and characterization of a collagen-induced platelet aggregation inhibitor, triplatin, from salivary glands of the assassin bug, *Triatoma infestans*. FEBS J 273:2955–2962

Noeske-Jungblut C, Krätzschmar J, Haendler B, Alagon A, Possani L, Verhallen P, Donner P, Schleuning W-D (1994) An inhibitor of collagen-induced platelet aggregation from the saliva of *Triatoma pallidipennis*. J Biol Chem 269:5050–5053

Noeske-Jungblut C, Haendler B, Donner P, Alagon A, Possani L, Schleuning W-D (1995) Triabin, a highly potent exosite inhibitor of thrombin. J Biol Chem 270:28629–28634

Noireau F, Dujardin J-P (2010) Biology of Triatominae. In: Telleria J, Tibayrenc M (eds) American Trypanosomiasis – Chagas disease – one hundred years of research. Elsevier, Amsterdam, pp 149–168

Padilha MHP, Pimentel AC, Ribeiro AF, Terra WR (2009) Sequence and function of lysosomal and digestive cathepsin D-like proteinases of *Musca domestica* midgut enzymes. Insect Biochem Mol Biol 39:782–791

Paim RMM, Araújo RN, Soares AC, Lemos LCD, Tanaka AS, Gontijo NF, Lehane MJ, Pereira MH (2011) Influence of the intestinal anticoagulant in the feeding performance of triatomine bugs (Hemiptera; Reduviidae). Int J Parasitol 41:765–773

Parker KR, Mant MJ (1979) Effects of tsetse (*Glossina morsitans morsitans* Westw.) (Diptera: Glossinidae) salivary gland homogenate on coagulation and fibrinolysis. Thromb Haemost 42:743–751

Patterson JS, Guhl F (2010) Geographical distribution of Chagas disease. In: Telleria J, Tibayrenc M (eds) American Trypanosomiasis – Chagas disease – one hundred years of research. Elsevier, Amsterdam, pp 83–114

Pereira MH, Souza MEL, Vargas AP, Martins MS, Penido CM, Diotaiuti L (1996) Anticoagulant activity of *Triatoma infestans* and *Panstrongylus megistus* saliva (Hemiptera/Triatominae). Acta Trop 61:255–261

Pereira MH, Gontijo NF, Guarneri AA, Sant'Anna MRV, Diotaiuti L (2006) Competitive displacement in Triatominae: the *Triatoma infestans* success. Trends Parasitol 22:516–520

Ribeiro JMC (1982) The antiserotonin and antihistamine activities of salivary secretion of *Rhodnius prolixus*. J Insect Physiol 28:69–75

Ribeiro JMC, Garcia ES (1981a) The role of the salivary glands in feeding in *Rhodnius prolixus*. J Exp Biol 94:219–230

Ribeiro JMC, Garcia ES (1981b) Platelet antiaggregating activity in the salivary secretion of the blood sucking bug *Rhodnius prolixus*. Experientia 37:384–386

Ribeiro JMC, Pereira MEA (1984) Midgut glycosidases of *Rhodnius prolixus*. Insect Biochem 14:103–108

Ribeiro JMC, Walker FA (1994) High affinity histamine-binding and antihistaminic activity of the salivary nitric oxide-carrying heme protein (nitrophorin) of *Rhodnius prolixus*. J Exp Med 180:2251–2257

Ribeiro JMC, Schneider M, Isaias T, Jurberg J, Galvão C, Guimarães JA (1998) Role of salivary antihemostatic components in blood feeding by triatomine bugs (Heteroptera). J Med Entomol 35:599–610

Rimoldi OJ, Peluffo RO, González SM, Brenner RR (1985) Lipid digestion, absorption and transport in *Triatoma infestans*. Comp Biochem Physiol B 82:187–190

Sant'Anna MRV, Diotaiuti L, Gontijo AF, Gontijo NF, Pereira MH (2001) Feeding behaviour of morphologically similar *Rhodnius* species: influence of mechanical characteristics and salivary function. J Insect Physiol 47:1459–1465

Santos A, Ribeiro JMC, Lehane MJ, Gontijo NF, Veloso AB, Sant'Anna MRV, Araujo RN, Grisard EC, Pereira MH (2007) The sialotranscriptome of the blood-sucking bug *Triatoma brasiliensis* (Hemiptera, Triatominae). Insect Biochem Mol Biol 37:702–712

Schaub GA (1989a) Does *Trypanosoma cruzi* stress its vector? Parasitol Today 5:185–188

Schaub GA (1989b) *Trypanosoma cruzi*: quantitative studies of development of two strains in small intestine and rectum of the vector *Triatoma infestans*. Exp Parasitol 6:260–273

Schaub GA (1992) The effects of trypanosomatids on insects. Adv Parasitol 31:255–319

Schaub GA (2008) Kissing bugs. In: Mehlhorn H (ed) Encyclopedia of parasitology, vol 1, 3rd edn. Springer, Heidelberg, pp 684–686

Schaub GA (2009) Interactions of trypanosomatids and triatomines. Adv Insect Physiol 37:177–242

Schaub GA, Lösch P (1989) Parasite/host-interrelationships of the trypanosomatids *Trypanosoma cruzi* and *Blastocrithidia triatomae* and the reduviid bug *Triatoma infestans*: influence of starvation of the bug. Ann Trop Med Parasitol 83:215–223

Schaub GA, Wülker W (1984) Tropische Parasitosen im Programm der Weltgesundheitsorganisation. Universitas 39:71–80

Schaub GA, Meiser CK, Balczun C (2011) Interactions of *Trypanosoma cruzi* and triatomines. In: Mehlhorn H (ed) Progress in parasitology, vol 2, Parasitology research monographs. Springer, Berlin, pp 155–178

Schofield CJ (1979) The behaviour of Triatominae (Hemiptera: Reduviidae): a review. Bull Entomol Res 69:363–379

Schofield CJ (1994) Triatominae: biology & control. Eurocommunica, West Sussex

Schofield CJ, Dias JCP (1999) The Southern Cone Initiative against Chagas disease. Adv Parasitol 42:1–27

Schofield CJ, Galvão C (2009) Classification, evolution and species groups within the Triatominae. Acta Trop 110:88–100

Schofield CJ, Diotaiuti L, Dujardin JP (1999) The process of domestication in Triatominae. Mem Inst Oswaldo Cruz 94(Suppl 1):375–378

Schofield CJ, Jannin J, Salvatella R (2006) The future of Chagas disease control. Trends Parasitol 22:583–588

Schwarz A, Helling S, Collin N, Teixeira CR, Medrano-Mercado N, Hume JCC, Assumpção TC, Marcus K, Stephan C, Meyer HE, Ribeiro JMC, Billingsley PF, Valenzuela JG, Sternberg JM, Schaub GA (2009) Immunogenic salivary proteins of *Triatoma infestans*: Development of a recombinant antigen for the detection of low-level infestation of triatomines. PLoS Negl Trop Dis 3(10):e532. doi:10.1371/journal.pntd.0000532

Soares RPP, Evangelista LG, Laranja LS, Diotaiuti L (2000) Population dynamics and feeding behavior of *Triatoma brasiliensis* and *Triatoma pseudomaculata*, main vectors of Chagas disease in Northeastern Brazil. Mem Inst Oswaldo Cruz 95:151–155

Soares AC, Carvalho-Tavares J, Gontijo NF, dos Santos VC, Teixeira MM, Pereira MH (2006) Salivation pattern of *Rhodnius prolixus* (Reduviidae; Triatominae) in mouse skin. J Insect Physiol 52:468–472

Stadler A, Meiser CK, Schaub GA (2011) "Living syringes": use of hematophagous bugs as blood samplers from small and wild animals. In: Mehlhorn H (ed) Nature helps... How plants and other organisms contribute to solve health problems, vol 1, Parasitology research monographs. Springer, Heidelberg, pp 243–271

Sun J, Yamaguchi M, Yuda M, Miura K, Takeya H, Hirai M, Matsuoka H, Ando K, Watanabe T, Suzuki K, Chinzei Y (1996) Purification, characterization and cDNA cloning of a novel anticoagulant of the intrinsic pathway, (prolixin-S) from salivary glands of the blood sucking bug, *Rhodnius prolixus*. Thromb Haemost 75:573–577

Telleria J, Tibayrenc M (2010) American Trypanosomiasis – Chagas disease – one hundred years of research. Elsevier, Amsterdam

Terra WR, Ferreira C (1994) Insect digestive enzymes – Properties, compartmentalization and function. Comp Biochem Physiol B Biochem Mol Biol 109:1–62

Terra WR, Ferreira C, Garcia ES (1988) Origin, distribution, properties and functions of the major *Rhodnius prolixus* midgut hydrolases. Insect Biochem 18:423–434

Ursic-Bedoya RJ, Nazzari H, Cooper D, Triana O, Wolff M, Lowenberger C (2008) Identification and characterization of two novel lysozymes from *Rhodnius prolixus*, a vector of Chagas disease. J Insect Physiol 54:593–603

Vallejo GA, Guhl F, Schaub GA (2009) Triatominae – *Trypanosoma cruzi/T. rangeli*: vector-parasite interactions. Acta Trop 110:137–147

Van de Locht A, Lamba D, Bauer M, Huber R, Friedrich T, Kröger B, Höffken W, Bode W (1995) Two heads are better than one: crystal structure of the insect derived double domain Kazal inhibitor rhodniin in complex with thrombin. EMBO J 14:5149–5157

Villela MM, Souza JB, Mello VP, Azeredo BVM, Dias JCP (2005) Vigilância entomológica da doença de Chagas na região centro-oeste de Minas Gerais, Brasil, entre os anos de 2000 e 2003. Cad Saúde Pública 21:878–886

Walochnik J, Aspöck H (2010) Raubwanzen, *Trypanosoma cruzi* und Morbus Chagas – die Geißel Südamerikas. In: Aspöck H (ed) Krank durch Arthropoden, Denisia, vol 30. Biologiezentrum/Oberöstereichische Landesmuseen, Linz, pp 655–672

Waniek PJ, Mendoça-Lima L, Menezes GB, Jansen AM, Araũjo CAC (2009) Recombinant expression and characterization of a lysozyme from the midgut of *Triatoma brasiliensis* (Hemiptera, Reduviidae) in comparison with intestinal muramidase activity. Physiol Entomol 34:309–317

Waniek PJ, Jansen AM, Araújo CAC (2011) *Trypanosoma cruzi* infection modulates the expression of *Triatoma brasiliensis def1* in the midgut. Vector Borne Zoonotic Dis 11:845–847

Waniek PJ, Pacheco Costa JE, Jansen AM, Costa J, Araujo CAC (2012) Cathepsin L of *Triatoma brasiliensis* (Reduviidae, Triatominae): sequence characterization, expression pattern and zymography. J Insect Physiol 58:178–187

Weichsel A, Andersen JF, Champagne DE, Roberts SA, Montford WR (2000) Nitric oxide binding to nitrophorin 4 induces complete distal pocket burial. Nat Struct Biol 7:551–554

Wenk P, Lucic S, Betz O (2010) Functional anatomy of the hypopharynx and the salivary pump in the feeding apparatus of the assassin bug *Rhodnius prolixus* (Reduviidae, Heteroptera). Zoomorphology 129:225–234

Whitten M, Sun F, Tew I, Schaub GA, Soukou C, Nappi A, Ratcliffe N (2007) Differential modulation of *Rhodnius prolixus* nitric oxide activities following challenge with *Trypanosoma rangeli*, *T. cruzi* and bacterial cell wall components. Insect Biochem Mol Biol 37:440–452

WHO (2008) Chagas disease: control and elimination. Report of the Secretariat. EB124/17:1–4

WHO (2012) Chagas disease (American trypanosomiasis) Fact sheet N°340, June 2010. http://www.who.int/mediacentre/factsheets/fs340/en/index.html. Accessed 3 Jan 2012

Wigglesworth VB (1972) The principles of insect physiology, 7th edn. Chapman and Hall, London

Wirtz H-P (1987) Eindringen der Mundwerkzeuge von Raubwanzen durch eine Membran (Hemiptera: Reduviidae). Entomol Gen 12:147–153

Zeledón R, Rabinovich JE (1981) Chagas' disease: an ecological appraisal with special emphasis on its insect vectors. Annu Rev Entomol 26:101–133

Zhang Y, Ribeiro JMC, Guimarães JA, Walsh PN (1998) Nitrophorin-2: a novel mixed-type reversible specific inhibitor of the intrinsic factor-X activating complex. Biochemistry 37:10681–10690

Chapter 13
Fleas as Underestimated Vectors of Agents of Diseases

Heinz Mehlhorn

Abstract Fleas occur worldwide and specimens of their 2,500 species suck blood daily several times at practically all warm-blooded hosts. Therefore, they may become vectors of practically all agents of diseases that are bound to the blood of this armada of varying blood donators. Since the spectacular and very controversial discovery in the year 1898 that the tropical rat flea is the vector of the plague bacterium, many other bacteria and rickettsial species were found in fleas. Due to the availability of antibiotics, the threat of flea bites had considerably decreased and the dangers of fleas were neglected even for many decades. However, since the resistance of bacteria against antibiotics increases constantly and since it was shown in transmission experiments that also viruses can easily be transmitted by fleas and their feces, the status of flea bites must be considered as being much more dangerous as before. This chapter summarizes a long list of agents of diseases that lurk in fleas, and thus it becomes clear that flea control has become an important task, especially in times of intense globalization.

Keywords Blood sucker • Ctenocephalides • Plague bacteria • Pulex • Transmission • Tunga • Vectorship • Xenopsylla

13.1 Introduction

Fleas are wingless, laterally flattened insects that have developed, during evolution, strong hind legs for jumping, which allow them to cover distances of more than 30 cm in order to reach a host (Fig. 13.1). Due to their species-dependent size of 1.5 mm up to 6 mm and the brownish color of their cuticle, they were detected even

H. Mehlhorn (✉)
Department of Parasitology, Heinrich Heine University, Universiätsstr. 1, 40225 Düsseldorf, Germany
e-mail: mehlhorn@uni-duesseldorf.de

H. Mehlhorn (ed.), *Arthropods as Vectors of Emerging Diseases*,
Parasitology Research Monographs 3, DOI 10.1007/978-3-642-28842-5_13,

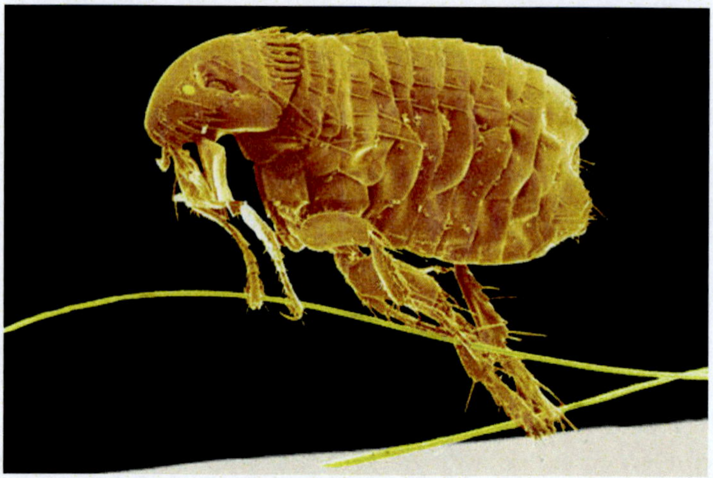

Fig. 13.1 Scanning electron micrograph (SEM) of the lateral view of a cat flea (*Ctenocephalides felis*) attached at a hair

during their "jolly jumps." Thus, they were already known and feared in the early human cultures and depicted in early human scientific textbooks (Fig. 13.2). The knowledge on fleas was permanently increased, and it was proven that they are worldwide constant companions of humans, infesting practically all warm-blooded animals (Lewis 1993; Hopkins and Rothschild 1971; Peus 1972; Eisen et al. 2006, 2008, 2009a, b; Kramer and Mencke 2001).

At first, humans had to recognize that fleas are nasty bloodsuckers and that the wounds at the biting sites reacted due to the injection of saliva with the formation of papulae and intense itching that lasts for days (Fig. 13.3a). In case of bacterial superinfections due to scratching at these biting sites, severe inflammations or even sepsis may occur (Fig. 13.3b). In addition, some flea species, e.g., the females of the sand flea *Tunga penetrans* or of the chicken flea *Echidnophaga gallinacea*, may introduce directly diseases by entering totally respectively partially the skin of their hosts (Figs. 13.4–13.6).

Since most of the specimens of more than 2,500 described species are not very host specific and may even change their hosts daily, it is not astonishing that during evolution they became vectors of different agents of diseases, which occur in the bodies of their hosts. Therefore, flea infestations of humans—especially of persons who live in camps and close neighborhoods with rats and murine species—have to be taken into serious consideration, since the outbreak of severe diseases (e.g., plague, rickettsiosis, bartonellosis, etc.) may have disastrous consequences not only for local people but also—after immigration—for inhabitants of other continents. This danger of a transport of emerging diseases from one continent to another increases daily, since the "heat" of the worldwide globalization is not yet under control due to the fact that one end of the world is just one jet-day away from the other. Thus, flea infestations of humans and their pet animals are no longer neglectable nasty events of itching but may produce health-threatening scenarios.

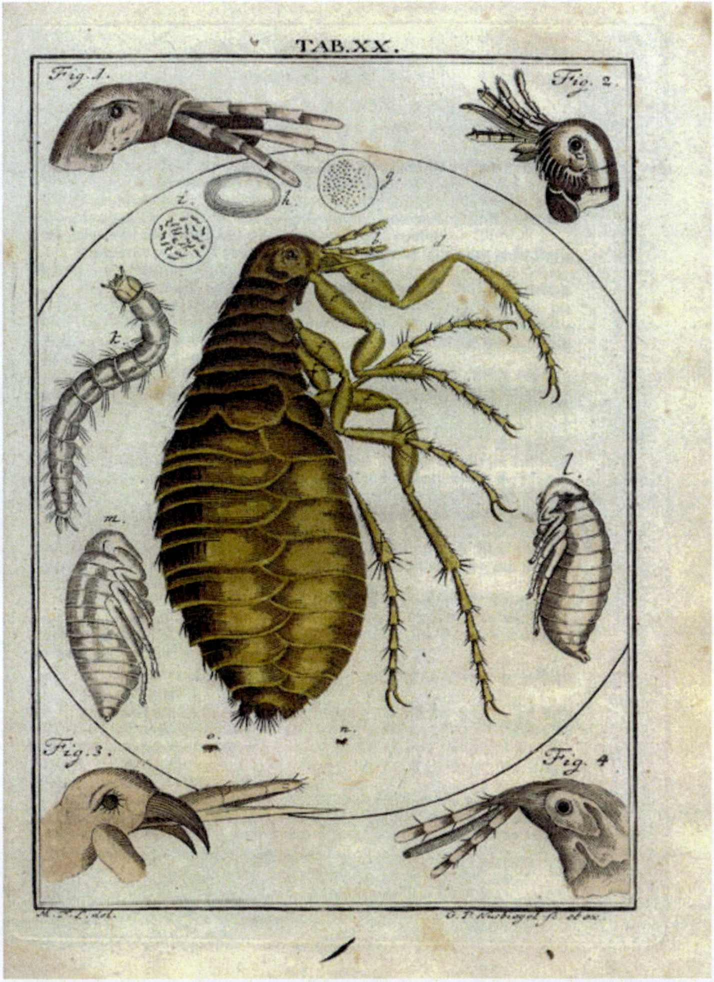

Fig. 13.2 Reproduction of a plate picturing the knowledge on fleas in the year 1760: facts and phantasy from: Martin Groben Ledermüllers: Microscopical mental and eye amusement with colors according to nature. Printer C. de Launoy at Nürnberg (Germany) (in German)

13.2 Why Are Fleas So Successful Survivors?

During evolution, fleas had to overcome many obstacles, which could threaten their survival, and spreading into constantly changing climates, since the "good old earth" exposed them to many warm and icy periods at rather short intervals during several millions of years. Therefore, the fleas developed the following skills (Figs. 13.7–13.11):

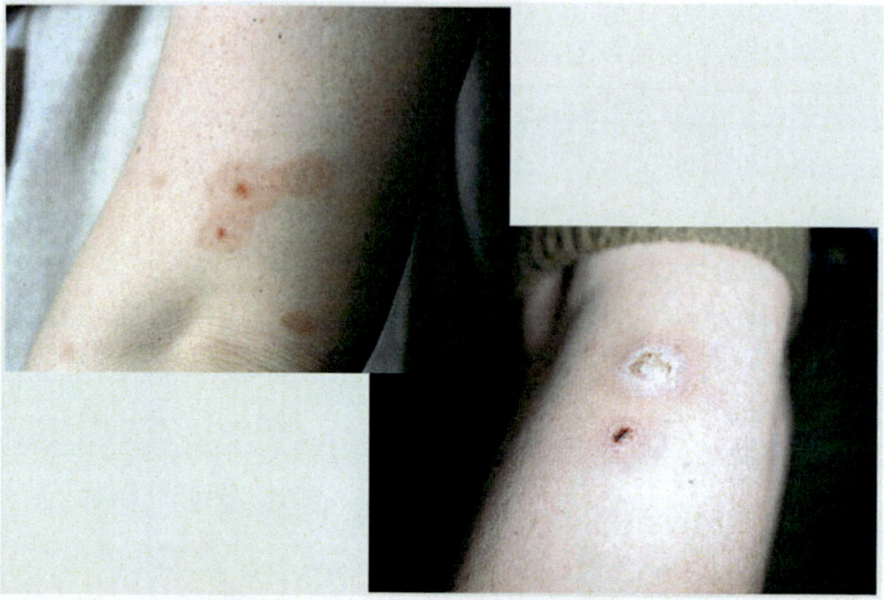

Fig. 13.3 Flea bites without (*left*) and with bacterial superinfection. Flea bites often occur in rows

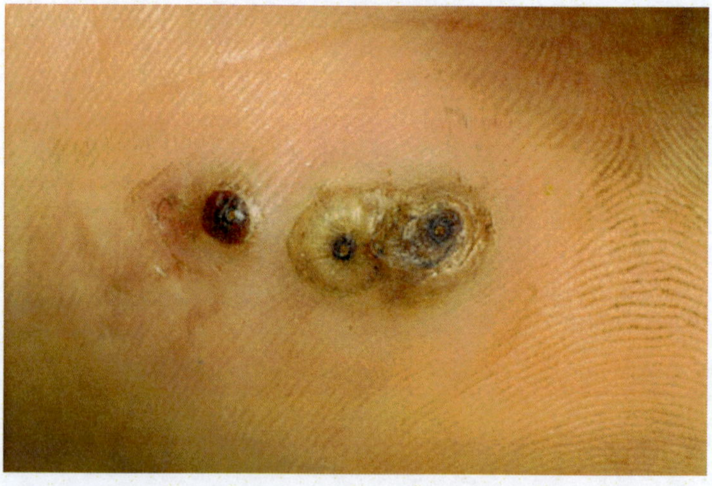

Fig. 13.4 Three females of *Tunga penetrans* just penetrated into skin

1. Flea ancestors started to suck blood at the tiny warm-blooded precursors of the mammals of our days (after the disappearance of the dinos).
2. This protein-rich food allowed a rather quick individual development so that fleas may complete their life cycle (from egg via three larvae, one pupa, and

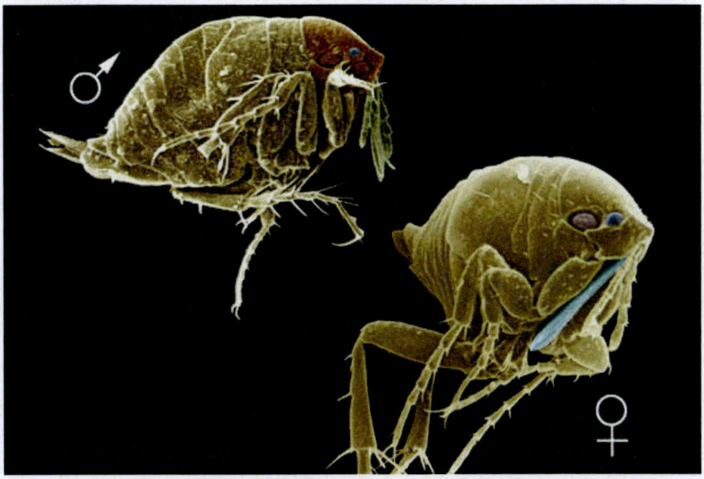

Fig. 13.5 SEM of female and male specimens of *Tunga penetrans*

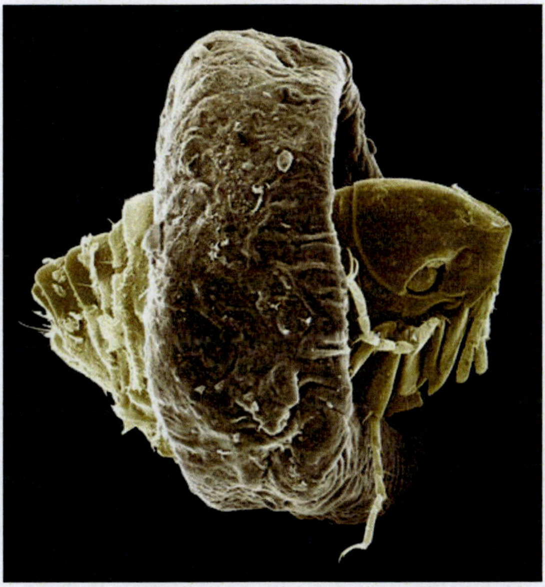

Fig. 13.6 SEM of a female *T. penetrans* from skin showing the swollen midregion of the abdomen

ending as male or female adult) under good, warm conditions (27°C and >50% relative humidity) within 3–4 weeks.

3. This admirable efficiency is supported by the capability of the females to lay (in the case of *Ctenocephalides*) up to 40 of the 0.5 mm-long whitish eggs per day while sucking every 3–4 h a total of about 14 μl blood per day representing several times of their body weight.

Fig. 13.7 Light micrograph of a cat flea egg on a dead adult inside detritus of a dwelling

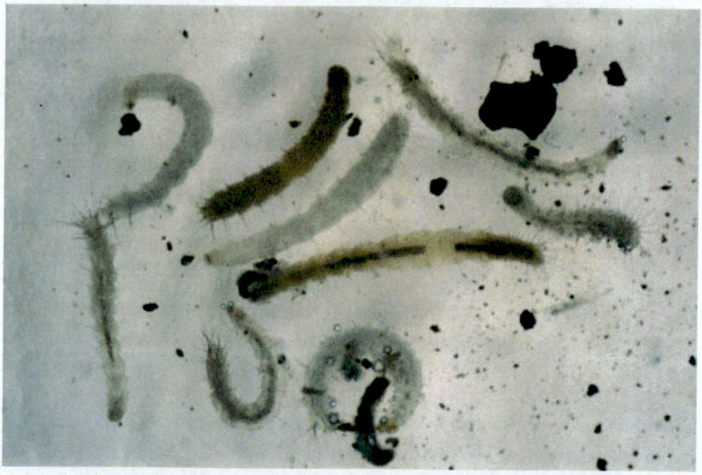

Fig. 13.8 Differently aged larvae of the cat flea, some of which had fed dried blood

4. The females and males do not suck blood to cover only their own requirements, but they excrete so much superfluous blood in their feces so that they may nourish crowds of larvae, which then appear with a dense intestine (Fig. 13.8).
5. Again, this energy-rich food allows the adults a survival period of at least 5–6 months in their pupal cocoon prior to hatch, if there is no shaking signal that announces the arrival of a potential blood donor.
6. Once hatched from the pupal cover, the male and female adults start feeding at the next warm-blooded host, since a better one might not come. Therefore, most fleas accept a very broad spectrum of hosts.

Fig. 13.9 Pupal cocoon of a cat flea covered by small stony particles

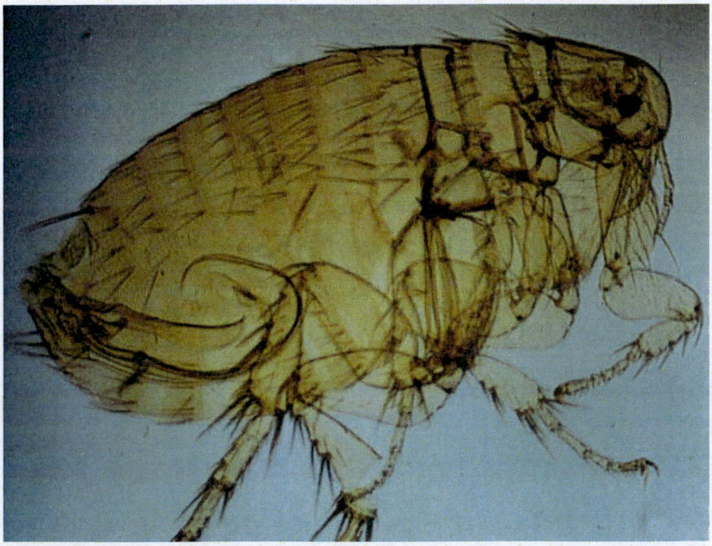

Fig. 13.10 Light micrograph of a male of the human flea *Pulex irritans*. Note that there are no bristles along the head. The protrudable copulation apparatus is visible in the abdomen of this specimen, which was made transparent

7. Furthermore, the adults have developed a laterally depressed body, which allows them to roam (crawl) easily within the hair of hosts, being supported by two strong, pointed claws at the end of the tarsus of each of the six long legs (Figs. 13.1, 13.5, and 13.10).

Fig. 13.11 SEM of *Pulex irritans*, the so-called human flea

Fig. 13.12 Light micrograph of the anterior portion of the cat flea (*C. felis*) showing the typical bristles

8. This persistence of fleas for a self-selected long period on a host also helps to distribute fleas in a given biotope, since all warm-blooded animals have to roam often for long distances in order to find appropriate food.
9. The fact that at first the eggs glue at the hair of hosts and drop down to earth after some days when the egg shell has become dry also helps that flea stages are differently distributed in a biotope and are able to find other hosts.

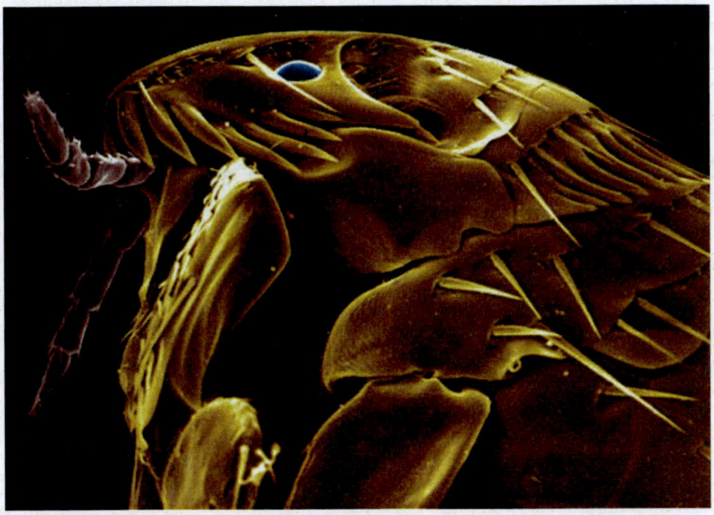

Fig. 13.13 SEM of the head of the cat flea

10. The high body temperatures along the hair—or feather—covered body surface of the hosts of fleas guarantee high survival chances.
11. If hosts of fleas leave their nests or homes even for months, these surroundings protect fleas until new inhabitants enter these "domiciles."
12. The fact that fleas are not host specific helps that they may obtain their food much easier than host-specific bloodsuckers.
13. Since the adults live hidden in the hair respectively among the feathers of their host, their number is not reduced by enemies.

Thus, nature has gifted fleas, which can be differentiated (among other criteria) by the bristles at the head and/or pronotal segment (Figs. 13.12 and 13.13) so perfectly, that under optimal conditions very often a mass production of fleas may occur, which guarantees survival of these perfect parasites—of course surely not to the pleasure and welfare of their hosts.

13.3 Why Are Fleas Successful Vectors of Agents of Diseases?

There are several crucial conditions for the successful transmission of agents of disease to humans:

1. The presence of sufficient agents of disease in the host spectrum of fleas is highly needed in a given biotope, where fleas may switch from wild animals to humans.
2. There must be a common transition of potentially infected fleas to humans, i.e., there must be a broad contact zone between flea-bearing animals and humans.

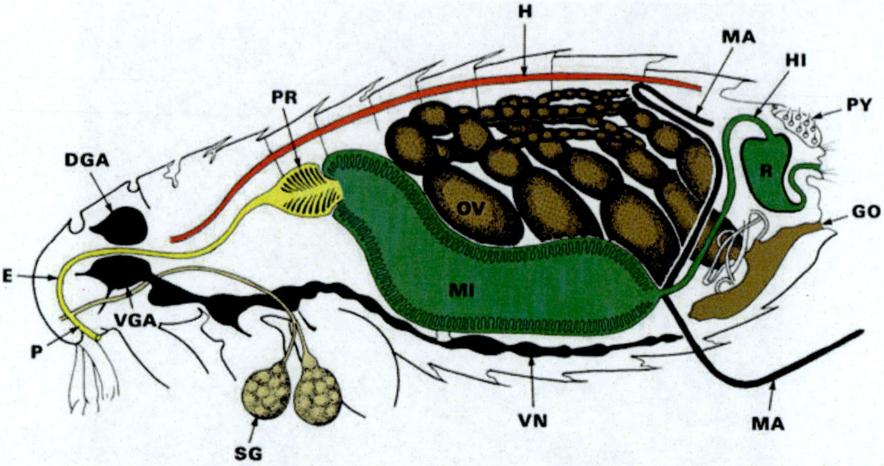

Fig. 13.14 Diagrammatic representation of the lateral view of a female flea. *DGA* dorsal ganglion (brain), *E* esophagus, *GO* genital opening, *H* heart, *HI* hind gut, *M* Malpighi channel (= excretory system), *MI* midgut, *P* pharynx, *PR* prestomach, *OV* ovary, *PY* pygidial plate = sensory bristle system, *R* rectum, *SG* salivary gland, *VGA* ventral brain ganglion, *VN* ventral nerve strand

3. The survival (and even better) reproduction of such agents of disease in the mouthparts or in the upper intestinal tract of the fleas must be possible.
4. The presence of a large number of fleas that attack different warm-blooded hosts and switch from one host to another is very helpful, since many different hosts might be infected by previously ingested pathogens.

All these conditions are fulfilled in the case of fleas:

1. They are obligatory blood feeders.
2. They suck blood several times per day (potentially at different hosts).
3. They suck large amounts of blood so that they can take up large quantities of pathogens.
4. They leave their hosts from time to time (e.g., in the nest of birds and/or rodents) and have the chance to infest other hosts.
5. They are not host specific and accept a broad spectrum of different warm-blooded hosts.
6. Fleas have developed a peculiar behavior while sucking blood, which is supported by the construction of their stomach system. The latter consists of rather stiff, not enlargeable, chitinous bristles containing prestomach and a larger posterior enlargeable portion (Fig. 13.14). Since fleas are hasty blood suckers (ingesting large amounts of blood within minutes), the transport of the masses of blood may be blocked before the prestomach, introducing a reaction called *regurgitation* (i.e., the vomiting of blood into the bite site). This regurgitated blood represents often a mixture of blood of potentially different hosts, in case a preceding meal on a host was interrupted and the flea switched to another host afterward.

7. Some agents of disease (e.g., plague bacteria) attach at the wall of the prestomach, reproduce there by repeated fissions, and thus may block—at least partially—the further inner transportation of the ingested blood. This phenomenon introduces regular common regurgitation of blood and bacteria into the wound.
8. Fleas suck hastily, a phenomenon which also may lead to regurgitation even if bacteria do not block the entrance of the stomach.

All these biological, morphological, and behavioral peculiarities of fleas listed above are favorable for the transmission of microorganisms such as typical bacteria, Rickettsiales, mycoplasms, and/or viruses. However, fleas are also known as intermediate hosts of tapeworms and/or nematodes. In these cases, the flea larvae are orally infected, when feeding worm eggs or larvae, which then are stored in the larval bodies and finally are transferred (via pupal stage) into the adults. If larvae or adults are cracked and eaten by the final hosts of these worms (e.g., cats, dogs, human kids) in the case they get into mouth contact with fleas or portions of them (in the hair of companion animals), the worm reaches maturity in these hosts.

However, although in principle each flea species might transmit practically any microorganism and/or parasite that has entered the bloodstream of any host, there are proven significant differences between different flea species with respect to their efficacy rate in transmitting a given agent of disease. This evolved transmission ability surely depended on the duration and frequency of the exposition that occurred between vectors and infected hosts. Therefore, the occurrence and recently visible or increasing successful transmission of a microorganism and/or a parasite in a given biotope depends on the following criteria:

1. *Pathogen acquisition efficacy* of the individual members of a certain flea species.
2. *Vector efficiency* of this flea species.
3. Length of the period needed until fleas may transmit an once ingested agent of disease (being called *extrinsic incubation period* (EIP)).
4. *Reproduction activity* of the agents of disease inside the flea leading to a possible blocking of the upper intestinal system and thus initiating repeated *regurgitation* with introduction of the pathogens into the new host.
5. *Host change behavior* of the flea species. High host change rates support intensely spreading of pathogens within a biotope and its inhabitants.

Considering all these conditions, Eisen and Gage (2012) showed in their excellent review that among the 30 experimentally tested flea species which had been previously confirmed as vectors of *Yersinia pestis* in North America, only a few were really good vectors. Thus, the worldwide-known vector species *Xenopsylla cheopis* (Fig. 13.15) showed in most studies a *pathogen acquisition efficacy* higher than 70%, reaching often up to 100%, a *vector efficiency rate* of around 30–70%, an *EIC-period* mostly of only 5–20 days, and in many cases a blocking of the intestinal tract in more than 70% of the cases (Douglas and Wheeler 1943; Eisen et al. 2007a, b; Eskey and Haas 1940; Wilder et al. 2008). Other flea species were much less efficient in the transmission of *Y. pestis*. As an example, the typical human flea

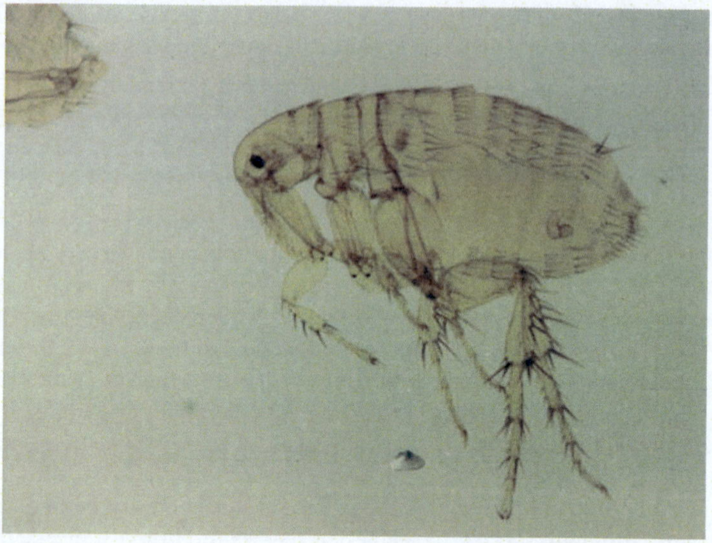

Fig. 13.15 Light micrograph of a female *Xenopsylla cheopis* flea. From the female system, only the chitinous spermatheca (where the sperms are stored) is retained after the preparation procedure

Pulex irritans (Figs. 13.10 and 13.11), that is also found on many mammals, reaches a pathogen acquisition efficiency of around 31%; however, none of 57 fleas tested was able to transmit, which surely depended on the fact that only 2% of the fleas showed a blockage of the upper intestinal system (Burroughs 1947). The same author showed that the poultry flea *Echidnophaga gallinacea* had a pathogen acquisition efficiency of 80% and a vector efficiency of 23%, apparently due to the fact that 23% of the fleas had a blocked intestinal system. The cat flea (*Ctenocephalides felis*), which is the most common flea (around 80%) on humans, dogs, and cats in Europe, showed an extreme high pathogen acquisition efficiency of 85%; however, practically no vector efficiency and blockage of the intestinal tract occurred (Wheeler and Douglas 1945a, b; Eisen et al. 2008). On the other hand, the flea species of ground squirrels (*Oropsylla montana*) reached even better values than the internationally well-known *Xenopsylla cheopis* strains, when showing in recent experiments a pathogen acquisition efficiency of 100% and low vector efficiency rates of about 10%, since there was no blocking of the stomach inside the fleas. Nevertheless, the final total transmission effect was good since there was an early-phase transmission (Eisen et al. 2008, 2009a, b).

These different capacities of the different flea species point to some facts, which have always to be considered, when evaluating a transmission risk in a certain region of the world:

1. Fleas can potentially transmit all pathogens contained in blood.
2. However, the occurrence of such pathogens in blood will not lead to an effective transmission in all cases, but on the other hand, single cases of transmission are

never excluded and often only small amounts of pathogens are sufficient to establish an infection.
3. Outbreaks of disease will only occur in cases where large numbers of hosts get into close contact with large amounts of possibly infected fleas with a high vector capacity.
4. Blockage of the anterior intestinal portions of fleas by pathogens is not needed to establish an infection of humans but is surely helpful (Fig. 13.14).
5. Thus, the best measure to avoid flea-borne transmission of pathogens is to keep away masses of flea bearers (e.g., rodents, wild dogs, etc.) from humans according to the simple rule: no mammals as pathogen carriers + no fleas = no potential transmission.
6. However, due to the crowding of humans and rodents at places (i.e., in refugee camps, at playgrounds, etc.) and as a consequence of the warming up of the climate, several emerging diseases due to flea transmission will surely occur in many places of the world.

13.4 Which Factors May Support the Increasing Vectorship of Fleas?

Recent reports (Friggens and Beier 2010) confirm what has been suggested for a long time (Patz et al. 2000; Deem et al. 2001; Daszak et al. 2001; Wilcox and Gubler 2005; Wilcox and Colwell 2005; Koontz and Daszak 2005; Crowl et al. 2008) that human influence (*anthropogenic habitat disturbances*) disrupts ecosystem processes and supports the relevant zoonotic disease dynamics. These disturbances are mainly based on the growth of human population, increase of urbanization, agricultural and forestry intensification (monocultures), and encroachment into wild (natural) areas. These factors led to severe changes and to considerable decrease of the species diversity of small mammals in a given biotope (Tikhonova et al. 2006). If one had expected that this fact would lead to a decrease of the number of fleas, it turned out to be wrong. Egoscue (1976) already showed that low host diversity often favored some common flea host species and thus led to an increase of flea species and their individual number. This was confirmed by the study of Friggens and Beier (2010), who tried to understand how fleas and flea-borne diseases might be influenced by human disturbances. They investigated the biotope flea community dynamics and flea host utilization patterns in relation of the intensity of disturbance. In their analysis of 70 small mammal communities, it turned out that human disturbance led to a decrease of host diversity but to an increase of flea infestation on small mammal hosts. They concluded that all important conditions were given—even under low disturbances—that would support an increase of transmission of flea-borne diseases, if their agents have once entered the bodies of these hosts.

Thus, *flea-borne diseases* must be added to the group of *emerging diseases*, even if they are not imported from abroad. In the latter cases, they would be described as *reemerging diseases*. These dangers are thus supported by three main factors:

1. Human-bound *change of local biotopes* as described above.
2. *Climate change*. The worldwide warming (due to human responsibility or not) brings better living conditions and higher survival rates of main hosts (small mammals) and quicker development of fleas. This gives rise to higher numbers of fleas (Githeko et al. 2000; Harvel et al. 2002; Lafferty 2009).
3. *Globalization*, which brings people and animals (including fleas) from all over the world to special habitats in new countries. Thus, a transfer of agents of disease needs only a few hours and may lead to severe outbreaks (Romi 2010).

13.5 Fleas as Vectors of Zoonotic Agents of Disease

Fleas were detected as possible vectors of agents of disease as early as mosquitoes were shown to be vectors of malaria. In both cases, the agents of the disease (e.g., plague bacteria respectively malaria parasites) had been described earlier (literature see Grüntzig and Mehlhorn 2005, 2010a, b). It was the French scientist *Paul-Louis Simond* who described the capacity of the tropical rat flea *Xenopsylla cheopis* (*Greek*: xenos = strange; psylla = flea; cheops = famous Egyptian pharaoh 2,500 B.C.) to transmit the simultaneously by the Swiss scientist André Yersin and the Japanese Shibasaburo Kitasato 1894 discovered bacterium being described today as *Yersinia pestis* (Fig. 13.16). In the last century, it was shown that these bacteria might be transmitted by many flea species. However, the vector efficacy is highest in *X. cheopis* and a special rodent flea species (e.g., *Oropsylla montana*). Studies with fleas showed that local and ubiquitous species of fleas may transport and transmit a broad spectrum of microorganisms and parasites (Table 13.1).

Most of these transmissions occur during flea's bloodsucking by regurgitation of infected blood originating from different hosts. However, transmission may also occur just mechanically by contaminated mouthparts. As was shown in literature, fleas mainly transmit bacteria and Rickettsiales, which have lost attention of physicians, since antibiotics for healthcare had been widely available. The same is true for those cases where fleas transmit, e.g., larvae of tapeworms (*Dipylidium*, *Hymenolepis* species) and nematodes of the *Dipetalonema* group (formerly *Acanthocheilonema*), which occur mainly in cats, dogs, and young children, if they swallow fleas or portions of them containing the parasitic stages (Mehlhorn 2008, 2012; Mehlhorn et al. 1995, 2012).

However, the increasing number of fleas in warmer times, the import of further microorganisms inside tourists from far away, and the invasion of small mammals and/or fleas in containers with goods from overseas make it necessary to increase the observation and intense consideration of flea-borne diseases. This is especially needed, since it was suggested (Rehacek et al. 1973; Smetana 1965) and later

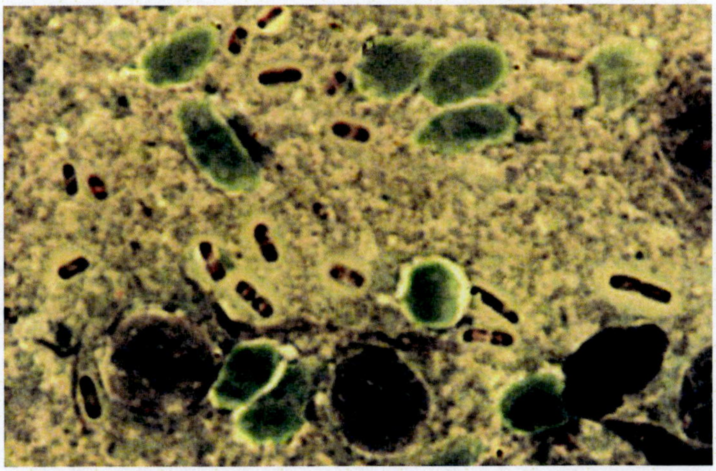

Fig. 13.16 Smear preparation of a fluid containing several plague bacteria (*Yersinia pestis*), which had been obtained from an open bubonic plague spot

proven by the experimental transmission of the myxomatosis virus (Shepherd and Edmonds 1977) that also viruses occur in fleas and thus might be transmitted during feeding and by contact to flea species.

A short glimpse on the already proven flea-borne zoonotic diseases shows how important these may become, if in the future close to large numbers of humans become infected with one of these agents of disease.

13.5.1 Proven Flea-Borne Human Diseases

13.5.1.1 Black Death, Bubonic Disease

1. *Name of disease*: *Lat.:* pestis; *Engl.:* plague; *French*: peste, bubonic maladie; *Span.:* la peste; German: Schwarzer Tod, Pestilenz; *Portuguese*: peste
2. *Epidemiology*: Today, there are focally herds of plague bacteria in rodents on all continents, from where occasionally outbreaks among humans had taken their origin (Eskey and Haas 1940; Gage and Kosoy 2005; Hinnebusch 2005; Wilder et al. 2008).
3. *Agent of disease*: *Yersinia pestis* is a 1–2 μm × 0.5 μm-sized, gram-negative, aerobe (facultatively anaerobe), flagella-less, rod-like bacterium which might also produce short chains and capsules, which may help to survive in dried blood and in flea feces. When coloring these rod-like bacteria, both of their poles are stained (Fig. 13.16).
4. *Hosts*: In nature, these bacteria propagate in rats and other rodents.

Table 13.1 Medically important flea species around the world (selection)

Species/geographical distribution	Length (mm)	Main hosts/ humans	Transmitted pathogens
Ceratophyllus gallinae/E	m 3.0 f 3.5	Chickens, turkeys, dogs, cats, humans	Mechanically many pathogens, Omsk hemorrhagic fever
Ctenocephalides canis/ worldwide	m 2.5 f 3.5	Dogs, humans	Larvae of cestodes (*Dipylidium, Hymenolepis*), *Yersinia pestis*, *Mycoplasma* (formerly, *Haemobartonella*) *haemofelis*
C. felis/worldwide	m 2.5 f 3.0	Cats, humans	Larvae of cestodes (*Dipylidium, Hymenolepis, Dipetalonema*), *Yersinia pestis*, viruses, *R. felis*, *R. typhi*
Echidnophaga gallinacea/ worldwide	m 2.0 f 2.5	Chickens, dogs, cats, pigs, humans	Bacteria, *Yersinia pestis*
Leptopsylla segnis/E, EA	m 1.6 f 1.8	Murine rodents	Mechanically many pathogens, cestodes
Nosophyllus segnis/E	m 1.8 f 2.0	Rats, humans	*Yersinia pestis*, other bacteria, erysipeloid
Pulex irritans/worldwide	m 2–2.5 f 4	Humans, domestic animals	*Yersinia pestis*, erysipeloid, larvae of cestodes
Spilopsyllus cuniculi/E	m 1.6 f 20	Rabbits, humans	Myxomatosis virus, *Francisella tularensis*
Tunga penetrans/E	m 0.7 f 5–6 (fed)	Humans, many animals	Penetrates skin, bacterial superinfections
Xenopsylla cheopis/warm countries	m 1.5 f 2.5	Rats, rodents, sheep, cats, humans (tropics)	*Yersinia* (=*Pasteurella*) *pestis*, *Rickettsia typhi*, erysipeloid, cestode larvae
Amphipsylla primaris/EA, E	m 1.6 f 2.0	Rodents, humans	*Yersinia pestis*, erysipeloid
Ctenophthalmus breviatus/E	m 2.5 f 3.0	Murine rodents, humans	*Yersinia pestis*, tularemia, pseudotuberculosis, bartonellosis
Megabothris species/E, A	m 2.7 f 3.5	Murine rodents, humans	Hemorrhagic nephrosonephritis
Archaeopsylla erinacei/E	m 2.5 f 3.1	Hedgehog, foxes, dogs, rats, humans	Bacteria, rickettsiae

A Asia, *E* Europe, *EA* East Asia, *m* male, *f* female

5. *Symptoms of disease*:

   (a) *Bubonic plague*: This form of disease occurs in 80–90% of the infections and starts 2–7 days after the infection by a flea bite with a primary blister at the infection site containing fluids and masses of the bacteria. Then follows sudden fever with chills, strong headache, and regional swellings of lymph nodes. Typically follows the formation of large, open bubonic swellings reaching infection sizes of a chicken egg (Fig. 13.16). As soon as the bacteria enter the inner organs, inner lesions occur and severe bleedings follow. Typical symptoms are cough and the black/blue color of the skin, which led to the name "black death." In untreated cases, infected persons die within 7 days.
   (b) *Septicemic plague*: In this case, the infections start with skin lesions (little wounds), which had become into contact to bacteria-containing blood of patients or infected rodents. Here occurs finally the same follow-up of symptoms as in the case of bubonic plague.
   (c) *Pneumonic plague*: The infections in these cases are acquired by inhalation of plague bacteria expectorated by patients with a severe lung affection. The progress of the disease is very quick and introduces highest lethality rates, reaching (even today) 100% of the cases within a few days. Often, 50% of treated people died.

6. *Transmission*: The bacteria are transmitted by flea bites (more than 30 species are known as efficacious vectors, Fig. 13.15), by contact of skin lesions with bacteria, and/or by inhalation of bacteria from infected people. Furthermore, infection is possible by eating infected undercooked rodents (e.g., guinea pigs in South America).
7. *Diagnosis*: Microscopy of fixed and stained bacteria taken from infected sites or from blood agar cultures at 25–28°C; ELISA.
8. *Therapy*: Streptomycin (30 mg/kg body weight), tetracyclines; all have to be given for 10 days.
9. *Legal regulations*: Patients have to be announced to health authorities, which will command strict quarantine to avoid severe outbreaks. In laboratory, the bacterium has to be kept under S3 biohazard conditions.
10. *Prophylaxis*: Ban rats and fleas from human surroundings. Use insecticidal repellents while walking in nature.

13.5.1.2 Tularemia

1. *Name*: *German*: Lemmingfieber, Hasenpest; *Engl.* deer fly fever, rabbit, hare fever, Ohara disease; *Indian*: yatobyo.
2. *Epidemiology*: Reservoir hosts are rodents, roe deers, raccoons, hares, etc.—all living on the Northern hemisphere. Thus, human cases occur exclusively in these regions.
3. *Agents of disease*: *Francisella tularensis* (occurring in four subtypes) is a short, nonmotile coccoid to rod-like, gram-negative, aerobe bacterium, which

reproduces in host macrophages but is very stable in mud and/or animal carcasses. The infection dose is very low—10–15 stages may be sufficient.

4. *Hosts*: See epidemiology.
5. *Symptoms of disease*: One-third of the infections remain without symptoms, while in the rest the following symptoms may occur within an incubation period of 3–10 days and rarely up to 21 days (Neumeister et al. 2009; Löscher and Burchard 2010):

   (a) Necrotic lesions at biting sites (if arthropod transmitted).
   (b) Chilling fevers.
   (c) Headache and general weakness.
   (d) Lymphadenopathy.
   (e) Pneumonia.
   (f) Vomiting.
   (g) Hepatosplenomegaly.
   (h) Rare meningitis cases and intraocular or oropharyngeal lesions.
   (i) Lethal cases after septic shock (in 2–3% of infections).

6. *Transmission*: Bites of bloodsucking arthropods (ticks and insects including fleas), contact to reservoir hosts (see above).
7. *Diagnosis*: Microscopy of fixed and stained bacteria, cultures of blood agar show within 3 days colonies of 2–4 mm in diameter, and serological antibody detection.
8. *Therapy*: Streptomycin for 7–10 days (20 mg/kg body weight) and gentamicin 2× 2.5 mg/kg bodyweight/day.
9. *Legal regulations*: Diagnosis and culture in laboratories to be done exclusively in personal protected systems.
10. *Prophylaxis*: Avoidance of tick and insect bites by repellents and handling of dead wild animals only with gloves.

13.5.1.3 Flea Spotted Fever

1. *Name*: Named after Ricketts Howard Taylor (1871–1910) an American scientist who died as a consequence of an infection with the agents of the Rocky Mountain Spotted Fever; *Lat.* felis = cat.
2. *Epidemiology*: The agent of disease uses as main hosts cats and rodents and is found in the Americas, but also in Europe, North Africa, and Asia (Reif and Macaluso 2009; Richter et al. 2002).
3. *Agent of disease*: *Rickettsia felis* is a small, only 0.8–2 μm-sized coccoid, rod-like bacterium, which reproduces obligatorily in cells (in cytoplasm and nucleus!).
4. *Hosts*: Cats, rodents, but also humans.
5. *Clinical symptoms*: After an incubation period of 1–2 weeks, fever, muscle pain, headache, as well as maculopapillary exanthema (in more than 50% of the cases) occur; however, lethality was found in less than 1%.

6. *Transmission*: Flea bites, however, mostly no eschar-like skin lesions occur at the biting site.
7. *Diagnosis*: Serological methods, PCR.
8. *Therapy*: Doxycycline.
9. *Legal regulations*: Nothing special.
10. *Prophylaxis*: Avoid human flea infestation and protect cat by repellents from fleas.

13.5.1.4 Murine Spotted Fever

1. *Name*: Ricketts (Am. scientist), *Lat*. typhus = intestinal diarrhea; the name spotted fever has its origin in aspects of the skin of patients that shows spots of exanthema.
2. *Epidemiology*: This agent of disease occurs worldwide in many countries: USA: in opossums; in other countries such as North Africa, Mediterranean region, South Russia, India, Thailand, Malaysia, Japan, Indonesia and Australia: rats and other rodents. Vectors are many flea species.
3. *Agent of disease*: *Rickettsia typhi* (syn. *R. mooseri*) develops exclusively in the intestinal cells of their arthropod hosts (including fleas). They are excreted within the feces of the fleas and enter the skin of the vertebrate host mostly at scratching sites. They live strictly intracellularly and are coccoid, rod-shaped bacteria with a size of 0.5–0.8 μm (Neumeister et al. 2009; Löscher and Burchard 2010).
4. *Hosts*: Rodents, cats, dogs, and humans.
5. *Clinical symptoms*: After an incubation period of 1–2 weeks, fever, muscle pain, headache, as well as maculopapillary exanthema (in more than 50% of the cases) occur; however, lethality was found in less than 1%.
6. *Transmission*: Flea bites; however, mostly no eschar-like skin lesions occur at the biting site.
7. *Diagnosis*: Serological methods, PCR.
8. *Therapy*: Doxycycline.
9. *Legal regulations*: Nothing special.
10. *Prophylaxis*: Avoid human flea infestation and protect cat by repellents from fleas.

13.5.1.5 Bartonellosis

1. *Name*: These bacteria were named in honor of the Peruvian scientist Alberto Leonardo *Barton*, who also described *B. bacilliformis* (1909: Carrion disease in South America) and *B. quintana* (1915: trench fever, Wolhynic fever) (Löscher and Burchard 2010).
2. *Epidemiology*: The flea-transmitted *Bartonella* species occur worldwide and are transmitted by flea bites and flea feces.

3. *Agents of disease*: *B. bacilliformis* (Carrion disease), *B. henselae*, *B. koehlerae*, *B. dashiae*, *B. clarridgeiae* (all cat-scratch diseases), *B. vinsonii* (bacterial fever)(Billeter et al. 2008; Breitschwerdt et al. 2005, 2007; Chomel et al. 1996; Azad et al. 1997; Rolain et al. 2003; Schmidt 1998).
4. *Hosts*: Cats, rodents, and humans.
5. *Symptoms of disease*: High fever and additional skin lesions (depending on the species). In cases of the cat-scratch disease due to *B. henselae*, papulae are developed at the biting site within 3–10 days. Furthermore, adenopathies are accompanied by fever. Only rarely severe symptoms occur.
6. *Transmission*: While *B. bacilliformis* is mainly transmitted by phlebotomid mosquitoes (*Lutzomyia* species), it also might be transmitted like other species (e.g., *B. henselae*) by fleas.
7. *Diagnosis*: PCR and microscopy of biopsies of lymph nodes.
8. *Therapy*: Azithromycin in low symptomatic cases, doxycycline (2× 100 mg/day) for 4–6 weeks in severe cases and in cases with retinitis.
9. *Legal regulations*: Nothing special.
10. *Prophylaxis*: Keep your cats free from fleas, keep away rats from houses.

13.5.1.6 Erysipelothrix rhusiopathiae

1. *Name*: *Greek*: erythros = red, pelos = mud; *German*: Rotlauf; *Engl.*: red skin (of pigs).
2. *Epidemiology*: This pathogen—identified in pigs about 100 years ago—is found in many mammals (including humans) and birds.
3. *Agent of disease*: *E. rhusiopathiae* is a facultative anaerobic, gram-positive, rod-like bacterium, which reaches enormous sizes of 0.8–60 μm in length. It may persist outside a body for a long time, e.g., in remnants of feces (Neumeister et al. 2009).
4. *Hosts*: Pigs, humans, many mammals, and birds.
5. *Symptoms of disease*: Painful inflammation of weak body regions with blue-red coloring and additional swellings of the lymph nodes. Infections of humans last mostly about 3–4 weeks.
6. *Transmission*: Mainly oral uptake of agents of disease, but also skin contact and meanwhile proven transmission by flea bites and their feces.
7. *Diagnosis*: API and VITEK products as well as PCR systems from pig control, cultures.
8. *Therapy*: Penicillin and cephalosporin.
9. *Legal regulations*: Farm workers have to stop contact with pigs while bearing symptoms.
10. *Prophylaxis*: Avoid contact with pig feces and fleas (Table 13.2).

Table 13.2 Species with a differing presence of combs (ctenidia) at their head or pronotum

Species	Size (mm)	Combs at head front	Combs at pronotum backside
Xenopsylla cheopis Tropical rat flea	2.5	–	–
Pulex irritans Human flea	4	–	–
Echidnophaga sp. Sticktight chicken flea	1.5	–	–
Nosophyllus fasciatus Northern rat flea	2	–	+
Ceratophyllus gallinae Poultry flea	3.5	–	+
Diamanus montanus Ground squirrel flea	2.5	–	+
Archaeopsyllus erinacei Hedgehog flea	3.5	+ (only a few)	+ (many)
Spilopsylla cuniculi European rabbit flea	2.5	+ (many)	+ (many)
Leptopsylla segnis European mouse flea	2	+ (many)	+ (many)
Ctenocephalides felis Cat flea	2.5–3.2	+ (many)	+ (many)
Ctenocephalides canis Dog flea	3–3.5	+ (many)	+ (many)
Tunga penetrans Sand flea = chigger	1	–	–

+, present; –, absent

Discovery of the flea as vector of the plague bacillus *Yersinia pestis*

The French scientist *Paul-Louis Simond* (1858–1947)—member of the famous Pasteur Institute at Paris—came into contact with the plague disease, which had threatened and killed as *Black Death* during centuries millions of humans, 3 years after Yersin and Kitasato had discovered 1984—both working simultaneously in Hong Kong—the plague bacillus (Yersin 1894; Kitasato 1894; Simond et al. 1998; Grüntzig and Mehlhorn 2010a, b). He was sent to Bombay (India) as successor of Yersin in order to bring an experimental serum into use that had been developed by Yersin at the Pasteur Institute. At this time, the transmission of plague was under intense discussion in the German, English, Russian, Italian, and French scientific communities and the idea of an activity of arthropods as vectors of pathogens was in its "children's shoes" or even not yet born. Simond, as an active physician, noted that at the beginning of a plague infection, small blisters occurred at the skin, containing a fluid with numerous bacilli. He concluded that this blister (phlyctène préoce) should be the infection site, which later becomes the large necrotic boubon (*French*: charbon pesteux). Simond had the idea that at the beginning this small blister could be due to an insect bite

(continued)

and that, thus, fleas as vectors would be involved in transmission. He studied cockroaches first and rather soon fleas, which were abundant there due to the occurrence of masses of rats in the poor quarters of this overcrowded Indian town. He detected the bacilli inside the fleas which he collected from dead rats. At the same time, Kitasato's colleague Ogata described in 1897 similar findings in fleas in German language in a German journal, but he missed to draw Simond's conclusions that fleas are the plague vectors and that therefore rats as hosts of fleas should be kept away from human surroundings. Simond depicted his findings and conclusions in a protocol in the year 1898, informed thereby the director Roux of the Pasteur Institute at Paris, who published the data in the Pasteur "Annales," so that this idea was spread among scientists without being officially published by Simond in a scientific journal at this early time. However, as it is usual, revolving ideas are hit by skepticism—not only by the home-proud English scene but also by French colleagues. This offense against a flea transmission of plague was kept alive, although there had been several scientific hints on the possible vectorship of arthropods before Simond's observations (e.g., Finlay 1864 discussed mosquitoes as vectors of diseases of—at this time still unknown—viruses of the yellow fever, Manson saw in 1878 relations between mosquitoes and transmission of filarial worms, as well as Bruce in 1884 described the involvement of tsetse flies in the transmission of trypanosomiasis) (Grüntzig and Mehlhorn 2010a, b). However, after many troubles and after an official publication of Simond in French in the year 1905, two English scientists confirmed finally the vectorship of fleas in the transmission of the plague bacterium that is now called *Yersinia pestis* (Bacot and Martin 1914).

13.6 Viral Transmission Experiments with Cat Fleas

Since more and more cases of flea-borne transmissions of bacteria, Rickettsiales, and parasites had been reported in the last decades, the conclusion that also viruses were involved was not very strange (Foil et al. 1983; Jarrett 1975; Jarrett et al. 1964). This idea was supported by the fact that many viruses develop a blood-based viremia, which should be included in any blood meal that is engorged by blood suckers such as fleas. The only condition for a further transmission was/is that the viruses are stable enough to survive in the engorged blood as long as it is included in the upper intestinal tract of fleas, which are known to regurgitate portions of ingested blood meals during the next sucking act occurring either on the same or on another host. Although this idea is rather convincing, literature on experiments with virus transmission by fleas was scarce when our group started in 2001 with such trials. There were the suspicion of Rehacek et al. (1973) that fleas might be vectors of the so-called Friend leukemia virus (FLV), Smetana's report (1965) of a potential transmission of the tick-borne encephalitis virus by fleas, or Shepherd's

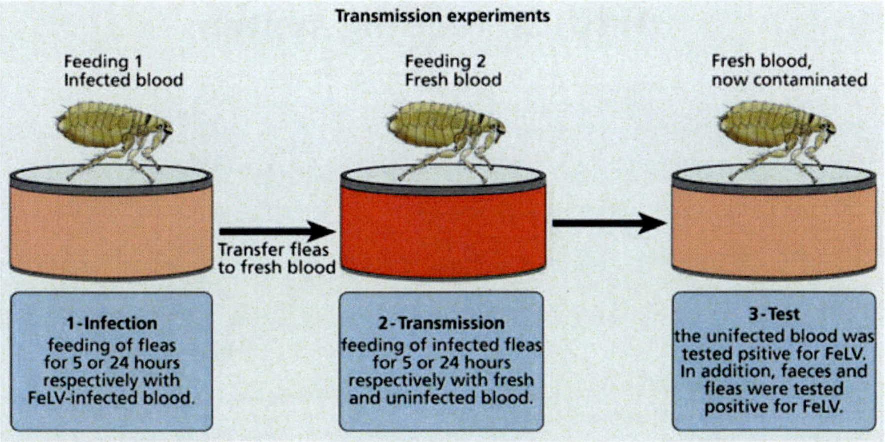

Fig. 13.17 Experimental design of the in vitro infection of cat fleas with FeLV in an artificial dog system

and Edmond's report (1977) on the myxomatosis virus in fleas; however, detailed knowledge remained scarce. Therefore, our group and those of Mencke (Bayer Company) and Truyen (University of Leipzig) started transmission experiments using laboratory-bred cat fleas (*Ctenocephalides felis*) as vectors.

13.6.1 Transmission Experiments with Feline Leukemia Virus

As described in the paper of Vobis et al. (2003), the following results were obtained in an artificial feeding system (Artificial dog from FleaData Inc., Freeville, USA), where fleas have the chance to ingest blood from small blood containers via membranes. The whole system was adjusted to keep the blood temperature at 37°C. In the first feeding, an initial population of 110 fleas was infected with feline leukemia virus (FeLV) through feeding for 24 h with 2 ml blood from a viremic cat (Fig. 13.17). To verify successful uptake of viruses, ten fleas and whole feces of this initial population were tested for FeLV RNA. The viral RNA could be detected in both samples, fleas and feces. The FeLV was incorporated by the fleas via blood feeding and was excreted with the feces. The remaining 100 infected fleas were subsequently divided in two populations of 50, where population 1 was fed for 24 h and population 2 for 5 h with 300 μl uninfected blood from a healthy cat. After the defined time, ten fleas and feces of each population were tested for FeLV. Viral RNA could be detected in fleas of both populations as well as in their feces. FeLV RNA was detectable for 5 and 24 h, respectively, after uptake by the flea and feeding with uninfected blood. To investigate whether transmission of FeLV occurred, the remaining blood from transfection 1 and 2 was examined.

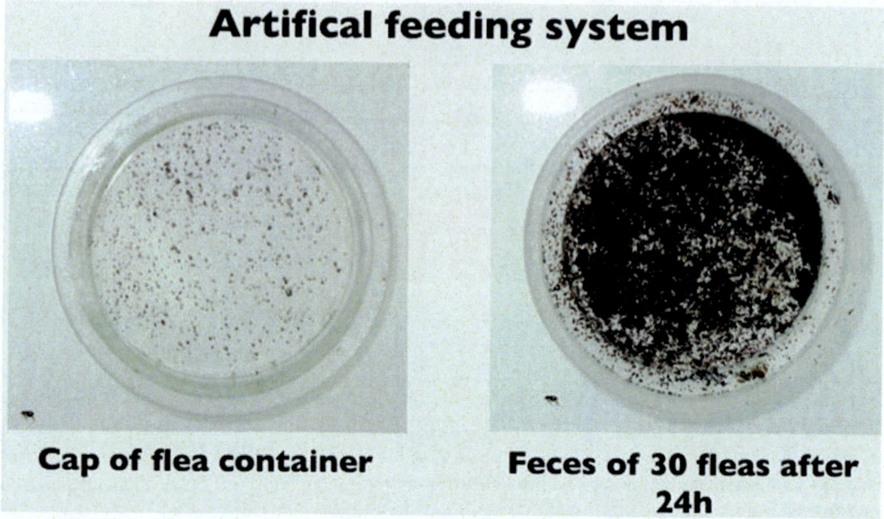

Fig. 13.18 Amount of feces produced by 30 fleas during 24 h

Viral RNA could be detected in both samples of the previously uninfected blood from a clinically healthy cat. This shows that the cat flea *C. felis* can function as a vector for feline leukemia virus RNA in this in vitro system of an artificial dog. Viral RNA was detectable in the fleas and its feces for up to 24 h. In addition, the fleas successfully transmitted FeLV RNA directly from an infected blood sample to an uninfected one. In a third feeding study, the same fleas were fed again with fresh and uninfected blood. After 5 and 24 h, respectively, fleas, feces, and the blood samples were examined for FeLV. After this third feeding, viral RNA could not be detected neither in the fleas, their feces, nor in the blood samples. FeLV RNA was detectable in the fleas up to 24 h after feeding. During this time, fleas also excreted viral RNA with their feces. In addition, the fleas were able to transmit FeLV RNA directly from one blood sample to another and functioned therefore as a vector of the virus.

In quantitation measurements (Vobis et al. 2005a, b) using the same artificial dog system as in the first series (Vobis et al. 2003), it was shown that around 70% of the ingested virus can withstand passage through the intestine with at least an intact nucleocapsid and is excreted and spread into the flea feces and thus into the surroundings (Fig. 13.18). When comparing the degradation rates of the FeLV in cell culture supernatant, the nucleocapsid seems even more stable in the feces. About 55% of the initial amount of the viruses could still be detected after 15 days at room temperature. Lower temperatures led to even higher survival rates.

These experiments showed that the viruses were transmitted by sucking (and apparent regurgitation) and that they were definitively surviving for days within the feces.

13.6.2 Transmission Experiments with Feline Calcivirus

In order to find out whether flea feces do not only contain the probably unchanged viruses, in vivo experiments were done using the feline calicivirus (FCV) (Truyen et al. 1999; Mencke et al. 2009). Fleas were inoculated with the viruses in the artificial dog system containing bovine blood free of FCV antibodies but being spiked with FCV to give a final titer of 10^6 TDC_{50}/ml. The flea feces were collected daily for 10 days and incubated at room temperature. When using Crandell-Rees feline kidney (CRFK) cells, it was shown that FCV apparently remained infectious in feces for 8 days. When exposing four pathogen-free (SPF) kittens oronasally to the virus-containing feces, all four were successfully infected (two kittens showed clinical symptoms). Four additional SPF kittens were exposed to fleas that had been fed with FCV-spiked bovine blood. One of these kittens was proven to be definitively infected in this way when testing virus isolation from pharyngeal swabs.

This series of experiments clearly shows that fleas may play an important role in the transmission of viruses under certain conditions, either during the blood meal or/and in the case that hosts get in contact with the flea feces. The only very important condition is that the virus is rather stable. However, there are many viruses known to fulfill these conditions (Löscher and Burchard 2010; Neumeister et al. 2009) *so that viruses may introduce emerging diseases under these conditions.*

13.7 Conclusions

This short review on the potential transmission activities of fleas clearly confirmed that fleas are functional vectors for many types of agents of diseases. Their role as vectors of bacteria is clear and well documented in many cases; many undetected transmissions will probably occur daily with those pathogens that introduce rather unspecific symptoms of disease. There is still a need for more examinations. However, the field of the until now apparently completely *underestimated vector capacity* of fleas for the transfer of viruses from one host to the other via blood consumption or by contact of the virus-containing flea feces lets fear for the future. There are huge dangers for outbreaks of viral emerging diseases, especially in crowded camps or under housing conditions with poor hygienic standards, occurrence of masses of potential vectors, etc. Therefore, the prophylaxis rules proposed in the year 1898 by Simond to keep away rats and their fleas from humans are still valid as was confirmed in the reminiscences of his nephew (Simond et al. 1998)!

References

Azad AF, Radulovic S, Higgins JA, Noden BH, Troyer JM (1997) Flea-borne rickettsioses: ecologic considerations. Emerg Infect Dis 3:319–327

Bacot AW, Martin CJ (1914) Observations on the mechanism of the transmission of plague by fleas. J Hyg 13(Plague suppl):423–439

Billeter SA, Levy MG, Chomel BB, Breitschwerdt EB (2008) Vector transmission of *Bartonella* species with emphasis on the potential for tick transmission. Med Vet Entomol 22:1–15

Breitschwerdt ED, Levine JF, Radulovic S, Hanby SB, Kordick DL, La Perle KMD (2005) *Bartonella henselae* and *Rickettsia* seroreactivity in a sick cat population from North Carolina. Int J Appl Res Vet 3:287–302

Breitschwerdt ED, Maggi RG, Duncan AW, Nicholson WL, Hegarty BC, Woods CW (2007) *Bartonella* species in blood from immunocompetent persons with animal and arthropod contact. Emerg Infect Dis 13:938–941

Burroughs AL (1947) Sylvatic plague studies: the vector efficiency of nine species of fleas compared with *Xenopsylla cheopis*. J Hyg 43:371–396

Chomel BB, Kasten RW, Floyd-Hawkins K, Chi B, Yamamoto K, Roberts-Wilson J, Gurfiled AN, Abbott RC, Pedersen NC, Koehler JE (1996) Experimental transmission of *Bartonella henselae* by the cat flea. J Clin Microbiol 34:1952–1956

Crowl TA, Crist TO, Parmenter RR, Lugo AE (2008) The spread of invasive species and infectious disease as drivers of ecosystem change. Front Ecol 6:238–246

Daszak P, Cunningham AA, Hyatt AD (2001) Anthropogenic environmental change and the emergence of infectious disease in wildlife. Acta Trop 78:103–116

Deem SL, Kilbourn AM, Wolfe ND, Cook RA, Karesh WB (2001) Conservation medicine. Ann N Y Acad Sci 916:370–377

Douglas JR, Wheeler CM (1943) Sylvatic plague studies. II. The fate of Pasteurella pestis in the flea. J Infect Dis 72:18–30

Egoscue HJ (1976) Flea exchange between deer mice and some associated small animals in western Utah. Great Basin Nat 36:475–480

Eisen RJ, Gage KL (2012) Transmission of flea-borne zoonotic agents. Annu Rev Entomol 57:61–82

Eisen RJ, Bearden SW, Wilder AP, Montenieri JA, Antolin MF, Gage KL (2006) Early-phase transmission of *Yersinia pestis* by unblocked fleas as a mechanism explaining rapidly spreading plague epizootics. Proc Natl Acad Sci USA 103:15380–15385

Eisen RJ, Reynolds PJ, Ettestad P, Brown T, Enscore RE et al (2007a) Residence-linked human plague in New Mexico: a habitat-suitability model. Am J Trop Med Hyg 77:121–125

Eisen RJ, Wilder AP, Bearden SW, Montenieri JA, Gage KL (2007b) Early-phase transmission of *Yersinia pestis* by unblocked *Xenopsylla cheopis* (Siphonaptera: Pulicidae) is as efficient as transmission by blocked fleas. J Med Entomol 44:678–682

Eisen RJ, Holmes JL, Schotthoefer AM, Vetter SM, Montenieri JA, Gage KL (2008) Demonstration of early-phase transmission of *Yersinia pestis* by the mouse flea, *Aetheca wagneri*, and implications for the role of deer mice as enzootic reservoirs. J Med Entomol 45:1160–1164

Eisen RJ, Eisen L, Gage KL (2009a) Studies of vector competency and efficiency of North American fleas for *Yersinia pestis*: state of the field and future research needs. J Med Entomol 46:737–744

Eisen RJ, Eisen L, Gage KL (2009b) Adaptive strategies of *Yersinia pestis* to persist during inter-epizootic and epizootic periods. Vet Res 40:1

Eskey CR, Haas VH (1940) Plague in the western part of the United States. Public Health Bull 254:1–83

Foil LD, Meek CL, Adams WV, Issel CJ (1983) Mechanical transmission of equine infectious anemia virus by deer flies (*Chrysops flavidus*) and stable flies (*Stomoxys calcitrans*). Am J Vet Res 44:155–156

Friggens MM, Beier P (2010) Anthropogenic disturbance and the risk of flea-borne disease transmission. Oecologica 164:809–820

Gage KL, Kosoy MY (2005) Natural history of plague: perspectives from more than a century of research. Annu Rev Entomol 50:505–528

Githeko AK, Lindsay SW, Confalonieri UE, Patz JA (2000) Climate change and vector-borne diseases: a regional analysis. Bull WHO 78:1136–1147

Grüntzig J, Mehlhorn H (2005) Expeditionen ins Reich der Seuchen. Spektrum Akad Verlag, Heidelberg

Grüntzig J, Mehlhorn H (2010a) Expeditions into the Empire of Plagues. Düsseldorf University Press, Düsseldorf

Grüntzig J, Mehlhorn H (2010b) Robert Koch: Seuchenjäger und Nobelpreisträger. Spektrum Akad, Heidelberg

Harvel D, Mitchell CE, Ward JR, Altizer S, Dobson AP, Ostfeld RS, Samuel MD (2002) Climate warming and disease risk for terrestrial and marine biota. Science 296:2158–2162

Hinnebusch BJ (2005) The evolution of flea-borne transmission of *Yersinia pestis*. Curr Issues Mol Biol 7:197–212

Hopkins GHE, Rothschild M (1971) An illustrated catalogue of the Rothschild collection of fleas in the British Museum (Natural History). Cambridge University Press, London

Jarrett WF (1975) Cat leukemia and its viruses. Adv Vet Sci Comp Med 19:165–193

Jarrett WF, Crawford EM, Marsin WB (1964) A virus-like particle associated with leukemia (lymphosarcoma). Nature 202:567–569

Kitasato S (1894) The bacillus of bubonic plague. Lancet 2:428–430

Koontz FW, Daszak P (2005) Sprawl and disease. In: Johnson EA, Klemens MW (eds) Nature in fragments: the legacy of sprawl. Columbia University Press, New York, pp 144–156

Kramer F, Mencke N (2001) Flea biology and control. Springer, Heidelberg

Lafferty KD (2009) The ecology of climate change and infectious diseases. Ecology 90:888–900

Lewis RE (1993) Fleas (Siphonaptera). In: Lane RP, Crosskey RW (eds) Medical insects and arachnids. Chapman and Hall, London

Löscher T, Burchard GD (2010) Tropenmedizin in Klinik und Praxis, 4th edn. Thieme, Stuttgart

Mehlhorn H (ed) (2008) Encyclopedia of Parasitology, 3rd edn. Springer, New York

Mehlhorn H (2012) Die Parasiten des Menschen und die von ihnen hervorgerufenen Erkrankungen. Spektrum Akad, Heidelberg

Mehlhorn H, Eichenlaub D, Löscher T, Peters W (1995) Diagnose und Therapie der Parasitosen des Menschen, 2nd edn. Fischer, Stuttgart

Mehlhorn B, Mehlhorn H, Walldorf V (2012) Schach den Blutsaugern und Schädlingen. Düsseldorf University Press, Düsseldorf

Mencke N, Vobis M, Mehlhorn H, D'Haese J, Rehagen M, Mangold-Gehring S, Truyen U (2009) Transmission of feline calcivirus via the cat flea (*Ctenocephalides felis*). Parasitol Res 105:185–189

Neumeister B, Geiss HK, Braun RW, Kimmig R (eds) (2009) Mikrobiologische Diagnostik, 2nd edn. Thieme, Stuttgart

Ogata M (1897) Über die Pestepidemie in Formosa. Zentralbl Bakteriol 21:769–777

Patz JA, Graczyk TK, Geller N, Vittor AY (2000) Effects of environmental change on emerging parasitic diseases. Intern J Parasitol 30:1395–1405

Peus F (1972) Zur Kenntnis der Flöhe in Deutschland. IV. Faunistik und Ökologie der Säugetierflöhe. Zool Jahrbuch 99:408–504

Rehacek J, Fischer RG, Luecke DH (1973) Friend leukemia virus (FLV) activity in certain arthropods. II. Quantitation, infectivity, determinations. Neoplasma 20:147–158

Reif KE, Macaluso KR (2009) Ecology of *Rickettsia felis*: a review. J Med Entomol 46:723–736

Richter J, Fournier PE, Petridou J, Häussinger D, Raoult D (2002) *Rickettsia felis* infection acquired in Europe and documented by polymerase chain reaction. Emerg Infect Dis 8:207–208

Rolain JM, Franc M, Davoust B, Raoult D (2003) Molecular detection of *Bartonella quintana*, *B. koehlerae*, *B. henselae*, *B. clarridgeiae*, *Rickettsia felis* and *Wolbachia pipientis* in cat fleas in France. Emerg Infect Dis 9:338–342

Romi R (2010) Arthropod borne diseases in Italy: from neglected matter to an emerging health problem. Ann Ist Super Sanita 46:436–443

Schmidt A (ed) (1998) *Bartonella* and *Afipia* species emphasizing *Bartonella henselae*. Karger, Basel

Shepherd RC, Edmonds JW (1977) Myxomatosis: the transmission of a highly virulent strain of myxoma virus by the European rabbit flea *Spilopsyllus cuniculi* (Dale) in the Mallee region of Victoria. J Hyg 79:405–409

Simond PL (1905) La question du vehicule de la peste. Rev Medico-Chirurg Bréslienne Am Lat 5:18–23

Simond M, Godley ML, Mouriquand PDE (1998) Paul-Louis Simond and his discovery of plague transmission by rat fleas: a centenary. J R Soc Med 91:101–104

Smetana A (1965) On the transmission of tick-borne encephalitis virus by fleas. Acta Virol 9:375–378

Tikhonova GN, Tikhonov IA, Pogomolov PL (2006) Impact of a small city on the structure of small mammal fauna in forests of the northeastern Moscow region. Russ J Ecol 37:278–289

Truyen U, Geissler K, Hirschberger J (1999) Tissue distribution of virus replication in cats experimentally infected with distinct felince calcivirus isolates. Berl Münchner Tierärztl Wschr 112:355–358

Vobis M, D'Haese J, Mehlhorn H, Mencke N (2003) Evidence of horizontal transmission of feline leukemia virus by the cat flea (*Ctenocephalides felis*). Parasitol Res 91:467–470

Vobis M, D'Haese J, Mehlhorn H, Mencke N (2005a) Experimental quantification of the feline leukemia virus in the cat flea (*Ctenocephalides felis*) and its feces. Parasitol Res 97:S102–S106

Vobis M, D'Haese J, Mehlhorn H, Heukelbach J, Mencke N, Feldmeier H (2005b) Molecular biological investigations of Brazilian *Tunga* sp. Isolates from man, dogs, cats, pigs and rats. Parasitol Res 96:107–112

Wheeler CM, Douglas JR (1945a) Transmission studies of sylvatic plague. Proc Soc Exp Biol Med 47:65–66

Wheeler CM, Douglas JR (1945b) Sylvatic plague studies. V. The determination of vector efficiency. J Infect Dis 77:1–12

Wilcox BA, Colwell RR (2005) Emerging and reemerging infectious diseases: biocomplexity as an interdisciplinary paradigm. EcoHealth 2:244–257

Wilcox BA, Gubler DJ (2005) Disease ecology and the global emergence of zoonotic pathogens. Environ Health Prev Med 10:263–272

Wilder AP, Eisen RJ, Bearden SW, Montenieri JA, Tripp DT et al (2008) Transmission efficiency of two flea species (*Oropsylla tuberculata cynomuris* and *Oropsylla hirsuta*) involved in plague epizootics among prairie dogs. EcoHealth 5:205–212

Yersin A (1894) La peste bubonique à Hong Kong. Ann Inst Pasteur 8:662–667

Chapter 14
Marine Crustaceans as Potential Hosts and Vectors for Metazoan Parasites

Markus Wilhelm Busch, Thomas Kuhn, Julian Münster, and Sven Klimpel

Abstract Crustaceans are highly abundant in the marine environment and play a key role as an important source of nutrition for a wide range of marine vertebrates such as fish, birds and mammals (seals, whales). In this context, marine helminth parasites have evolved complex (heteroxenous) life cycles in order to reproduce and use the trophic interactions in the marine food web to facilitate the transmission to the successive hosts. Members of the parasites taxa Digenea, Cestoda, Nematoda and Acanthocephala are common parasites in the marine environment and known to frequently include pelagic and benthic crustaceans of the subgroups Amphipoda, Cirripedia, Copepoda, Decapoda, Euphausiacea, Isopoda and Mysidacea in their life cycle. Infestation data from 52 peer-reviewed publications have been taken into consideration in order to summarize the current knowledge of crustaceans that are known to be the intermediate hosts for marine helminth parasites. This includes the discussion of life cycles, impacts of parasitism on hosts and zoonotical threats (e.g. for the nematode species of the genus *Anisakis*).

Keywords Anisakid nematodes • Crustacea • Intermediate host • Life cycle • Metazoan parasites • Vector

---

M.W. Busch (✉) • T. Kuhn • J. Münster • S. Klimpel
Medical Biodiversity and Parasitology, Biodiversity and Climate Research Centre (BiK-F), Senckenberganlage 25, 60325 Frankfurt am Main, Germany

Senckenberg Gesellschaft für Naturforschung (SGN), Senckenberganlage 25, 60325 Frankfurt am Main, Germany

Institute for Ecology, Evolution and Diversity, Goethe-University (GO), Max-von-Laue-Straße 9, 60438 Frankfurt am Main, Germany
e-mail: markus.busch@senckenberg.de; Sven.Klimpel@senckenberg.de

H. Mehlhorn (ed.), *Arthropods as Vectors of Emerging Diseases*, Parasitology Research Monographs 3, DOI 10.1007/978-3-642-28842-5_14,

14.1 Introduction

With approximately 52,000 described species worldwide, the subphylum Crustacea is one of the four most diverse groups within the Metazoa (Martin and Davis 2001). They inhabit especially aquatic (marine, limnic) ecosystems and have gone through a series of behavioural, physiological and morphological adaptations that allow them to colonize virtually any aquatic habitat. The morphological diversity is higher than in any other taxa on earth, and, not surprisingly, the current list of known species is constantly growing while there is still no consensus regarding the number of constituent crustacean classes (e.g. Spears and Abele 1997).

Oceanographers divide the ocean into regions depending on abiotic and biotic conditions of these areas. The benthic zone is the ecological region at the lowest level of a body of water in the ocean, including the sediment surface and some subsurface layers. The pelagic zone includes all open ocean regions and can be divided into further regions categorized by depth and light abundance and is based on the plankton (phytoplankton) which occupy the start of the food chain. The term plankton describes all aquatic organisms including eggs and developmental stages that are drifting passively in the water column (pelagial). Although its distribution and density varies seasonally, horizontally and vertically, it is ubiquitous in both freshwater and marine ecosystems and primarily divided into the functional groups zooplankton and phytoplankton. Crustaceans are the predominant group regarding abundance and biomass in the marine zooplankton and represent an important link between phytoplanktic organisms (primary production) and higher trophic levels (e.g. cephalopods, fish, marine mammals). Due to this key role in the aquatic food webs, the diversity of metazoan parasites utilizing crustaceans as intermediate hosts is comparatively high (Marcogliese 1995). Especially, members of the parasite taxa Digenea, Cestoda, Nematoda and Acanthocephala are known to include invertebrate organisms as intermediate or paratenic hosts to facilitate transmission in the environment even at low densities (Marcogliese 1995).

The following pages summarize the current state of knowledge about the importance of crustaceans in the marine food web and their implications on the transmission of helminth parasites. The diversity of the currently recognized crustacean species that act as intermediate or paratenic hosts for marine helminths is highlighted, and their role in the life cycle of the parasites is discussed. An overall summary of all records of helminth parasites in marine Crustacea is far beyond the scope of this chapter, but it will become apparent that the high density and abundance of marine Crustacea is also reflected in a key position within the life cycle ecology of helminth parasites.

14.2 Biodiversity and Distribution of Marine Crustacea

Copepoda with estimated 13,000 morphospecies represents the most species rich taxon of the Crustacea (Boxshall and Defaye 2008). They are highly abundant in almost all aquatic environments (limnic but predominantly marine) and occur from the Arctic to the Southern Ocean (Antarctic waters). Copepods are preyed by small predators (e.g. shrimp, small fishes) and are also on the small end of the food size spectrum of baleen whale species (Mauchline 1998). Due to their dominant part in the marine zooplankton, the calanoid Copepoda are major components of aquatic food webs (Gruner 1993).

With approximately 900 marine species, the subclass Cirripedia is far less diverse. They represent the only sessile group of Crustacea and, with the exception of parasitic forms, live attached to rocks, shells and animals (Newman and Abbott 1980; Ruppert and Barnes 1994).

The pelagic Euphausiacea is the second most common taxon within the Crustacea. They occur in large swarms as permanent members of the zooplankton, and their distribution varies from the neritic epipelagial (e.g. *Meganyctiphanes norvegica*, 100–500 m depth) to the meso- and bathypelagial (e.g. *Bentheuphausia amblyops*, 1,000–5,000 m depth) (Gruner 1993). The most famous species in the Southern Ocean is the vertical migrating Antarctic krill *Euphausia superba*, which has a circumpolar distribution and is the foundation of the Antarctic food web. Antarctic krill is the major link between primary producers and many populations of Antarctic carnivores (e.g. fishes, penguins, seals, whales) (Lascara et al. 1999). Although underlying strong variations, the biomass is estimated at more than 500 million tons (Martin and Davis 2001; Siegel et al. 1998). Investigations between the Antarctic Peninsula and South Georgia in the year 2000 show a mean *E. superba* density of 21.4 g m^{-2} (Hewitt et al. 2002).

The order Amphipoda contains approximately 7,900 species and forms the largest group of the superorder Peracarida (Wirkner and Richter 2007). Most amphipods are marine and have been diversified in pelagic and benthic habitats (Whiteley et al. 2011). Some have been found even in deep-sea trenches of 10,000 m depth (Gruner 1993). The body tends to be laterally compressed and is usually between 5 and 15 mm in length (Ruppert and Barnes 1994). Hyperiid amphipods are exclusively pelagic and distributed from the surface layers to the bathypelagial, whereas members of the Gammaridea are usually benthic, demersal or benthopelagic (Gasca et al. 2009).

The dorsoventrally flattened 4,000 Isopoda species are widely distributed invertebrates that occupy various habitats. Most specimens are between 5 and 15 mm in length and omnivore deposit feeders. They represent the largest group of truly terrestrial crustaceans, and the suborder Paraselloidea forms one of the most abundant components of the deep-sea benthic fauna (Ruppert and Barnes 1994). The probably most famous aquatic parasitic species belong to the Cymothoidae. These parasites are permanent ectoparasites of marine fish and, due to their strategy to attach to external surfaces, gills or in the buccal cavity of their hosts, commonly known as tongue biters (Hadfield et al. 2011).

The order Decapoda comprises approximately 10,000 species. Most of them are marine and colonize the oceans from the supralittoral zone to the deep sea. Specimens are on average bigger sized than all other Crustacea and have a very diverging morphology from shrimp-like elongated to crab-like. Their diet shows a wide range, but most species combine predatory feeding with scavenging (Ruppert and Barnes 1994).

Classification of the Mysidacea includes 160 genera and about 1,000 species. While there is still no consensus about monophyly, the Lophogastrida and Mysida are commonly treated as subgroups of this taxon (Meland and Willassen 2007). They are cosmopolitan and distributed throughout the water column. More than 90% of the mysid species are considered exclusively marine and have adapted to both benthic and pelagic environments (Porter et al. 2008). In addition, they have also been reported from inland fresh and brackish water habitats. Most mysidaceans reach lengths from 2 to 30 mm and live in large swarms where they form an important part of the diet of marine fish. They usually feed as omnivores or scavengers (Ruppert and Barnes 1994).

In general, the distribution of crustaceans and especially zooplankton in the marine environment is heterogenous and depends on different biotic and abiotic factors that determine the vertical and horizontal distribution patterns. Differences in the availability of nutrients as a result of reduced or increased primary production by phytoplanktic organisms often arise through the formation of frontal systems, water column stratification and the course of ocean currents. At the eastern side of the South Atlantic, for example, at the Benguela upwelling region off the coast of Namibia and Angola, zooplankton communities on the shelf across the region are characterized by low diversity and high abundance. These regions contain species which are typical for upwelling areas or cold-temperate shelf waters worldwide (Gibbons and Hutchings 1996). Along fronts primary production of organisms is, due to the permanent upwelling of nutrients, usually higher and has a direct effect on the food chain processes. In the vicinity of these fronts, concentrations of Crustacea and fish are high, and therefore, potential predators and prey organisms are accumulated. This favours the transmission of parasites (Klimpel and Rückert 2005; Munk and Nielsen 1994; Richardson et al. 1998).

14.3 Helminth Parasites of Marine Crustacea

Metazoan parasites have evolved complex life cycle strategies in order to reproduce. They can be divided into different types: monoxenous and heteroxenous life cycles. Monogenea and parasitic Crustacea are commonly monoxen. They include a single host group to complete their (direct) life cycle. On the other hand, Digenea, Cestoda, Nematoda and Acanthocephala have to switch between different hosts of various trophic levels (indirect life cycle) and can therefore be designated as heteroxen (Möller and Anders 1983; Rohde 2005). The hosts of parasites can be divided into intermediate, paratenic and final hosts, and consequently, the parasites move from hosts on lower trophic levels to those higher in the food web. In the development and distribution of heteroxen parasites, the intermediate host is a vessel for the transmission to the next host and needed for asexual reproduction (e.g. sporocysts in Digenea) and maturation. Once the parasite has reached a specific level of development or size, it is infectious for the final host in which the parasite changes from the larval stage to sexual maturity (adult stage). Paratenic or transport hosts are special forms of intermediate hosts and incorporated between the propagule stage and the definitive host. They represent an additional host and are not essential for the life cycle (facultative). The larval forms do not undergo further development but accumulate over time and facilitate a successful transmission to the next (obligatory) host (e.g. Chubb et al. 2009). In addition, they can transport and spread a parasite species in order to guarantee a wide geographical dispersal.

Digenean parasites are exclusively endoparasitic with complex indirect life cycles. The marine species commonly occur in the alimentary tract of their final hosts (Mehlhorn 2001). Eggs disperse within the host faeces and contain the miracidium stage that has to be ingested by a mollusc, the obligate first intermediate host. Once incorporated, the miracidium produces asexual sporocysts and rediae that release infectious stages known as cercariae. These will either be ingested by or penetrate actively the second intermediate host, usually a crustacean. Occasionally, the cercariae directly infect the final hosts and grow to the adult stage. If a second intermediate host is included in the life cycle, the trematode develops a metacercariae, which is in case of oral ingestion, infectious for the final host (Marcogliese 1995; Mehlhorn 2001; Möller and Anders 1983). Typical fish parasitic Digenea are represented by the families Derogenidae, Hemiuridae, Lecithasteridae and Lepocreadiidae (Klimpel et al. 2009). *Derogenes varicus* (Derogenidae) is one of the most widespread digenean species and has a distribution between the Subarctic and the Subantarctic as well as in shelf water layers and the deep sea (Klimpel et al. 2006, 2009, Køie 1979; Palm et al. 1999). The parasite utilizes gastropods (e.g. *Natica* spp.) as first intermediate hosts and demersal Decapoda as well as pelagic Copepoda as second intermediate hosts (Køie 1979, 1984). Some digenean parasites are of major medical importance (e.g. Schistosomiasis) as they infest the intestines, bile ducts, lung or blood of their final hosts and cause serious injury in these organs.

The hermaphroditic Cestoda (subclass Eucestoda, true tapeworms) are endoparasitic in their intermediate and vertebrate final hosts and encompass more than 5,000 known species (Caira and Reyda 2005). Marine cestodes are estimated to more than 1,400 species and include the orders Diphyllobothriidea, Bothriocephalidea (the latter formerly known as Pseudophyllidea), Trypanorhyncha and Tetraphyllidea (Caira and Reyda 2005; Kuchta et al. 2008). These species perform life cycles with three developmental stages. Adult cestodes live in the intestines or nearby organs of their vertebrate final hosts. The excreted egg develops a stage (known as hexacanth embryo) that remains within the egg. Among the Bothriocephalidea, Diphyllobothriidea and Trypanorhyncha, this stage is surrounded by a ciliated membrane and called coracidium (free-swimming stage). These larval stages will be ingested by an appropriate plankton organism where it reaches the so-called procercoid stage that further develops into a worm-like plerocercoid. The plerocercoids occur in the second intermediate host (plankton predator). Fishes (Elasmobranchia and Teleostei) can serve either as intermediate or as final hosts (Palm 2004). An infestation with cestodes is in many cases without serious pathogenetic consequences for the final host. Diphyllobothriasis is the most common cestode zoonosis and transmitted by the genus *Diphyllobothrium* (Diphyllobothriidae). Diarrhoea, abdominal discomfort, fatigue and anorexia are symptoms of an infection with the adult helminths (Blair 2005; Mehlhorn 2001). Copepoda are the first intermediate hosts in this life cycle, whereas fish acquire parasites by preying upon these infested crustaceans. Accumulation can occur in paratenic hosts. Birds and terrestrial mammals are common final hosts (Blair 2005). Humans can get infected by ingesting undercooked, plerocercoid-containing fish (Mehlhorn 2001). While *D. latum* has a freshwater life cycle, *D. pacificum* occurs in the marine environment (Blair 2005).

Nematodes colonize every habitat and ecosystem on earth (terrestrial, limnic, marine). With currently 2,280 known genera in 256 families and estimated 40,000 species worldwide, Nematoda represent probably one of the most species diverse phyla within the metazoan group (Anderson 2000; McClelland 2005). While most nematodes are free-living feeding on decomposing organic material, numerous species are of major significance as parasites of humans, animals and plants and cause massive losses in agriculture and livestock or cause various diseases as human pathogens. Approximately 4% of all known nematode genera are parasitic in limnic and marine fishes. The vast majority of these nematodes use intermediate and paratenic hosts, which enable them to be transferred through the food chain to the fish or mammalian host (Fig. 14.1e, f) (Anderson 1996). They perform a life cycle including four moultings and larval stages before they reach the adult stage and commonly use copepods and amphipods as well as euphausiids as crustacean intermediate hosts (Marcogliese 1995; McClelland 2005). Larvae persist in the haemocoel of their crustacean intermediate hosts and usually in the body cavity, intestines, muscles or alimentary tract of their intermediate hosts (Fig. 14.1a, b, d) (Klimpel and Palm 2011). Marine mammals (Cetacea, Pinnipedia), birds and a wide range of predatory fish species are commonly used as final hosts. As discussed below, several marine nematodes have considerable medical importance. Larvae of the family Anisakidae, for example, can cause severe gastrointestinal diseases (anisakidosis) when ingested alive, and even very small doses of the parasites antigen can be the causative agents of allergic symptoms including anaphylaxis and asthma (Audicana and Kennedy 2008).

The phylum Acanthocephala comprises approximately 1,150 species which are exclusively parasitic (Verweyen et al. 2011). The adults live in the intestine wall of their vertebrate hosts. The marine life cycle involves one or two intermediate hosts with a benthic association (Marcogliese 1995). The eggs, which contain the acanthor larvae, will be released into the water column with the faeces of the final host. Detritus-feeding zooplankton organisms (e.g. Amphipoda) acquire the parasites by oral ingestion (Fig. 14.1c). The acanthor larvae enter the body

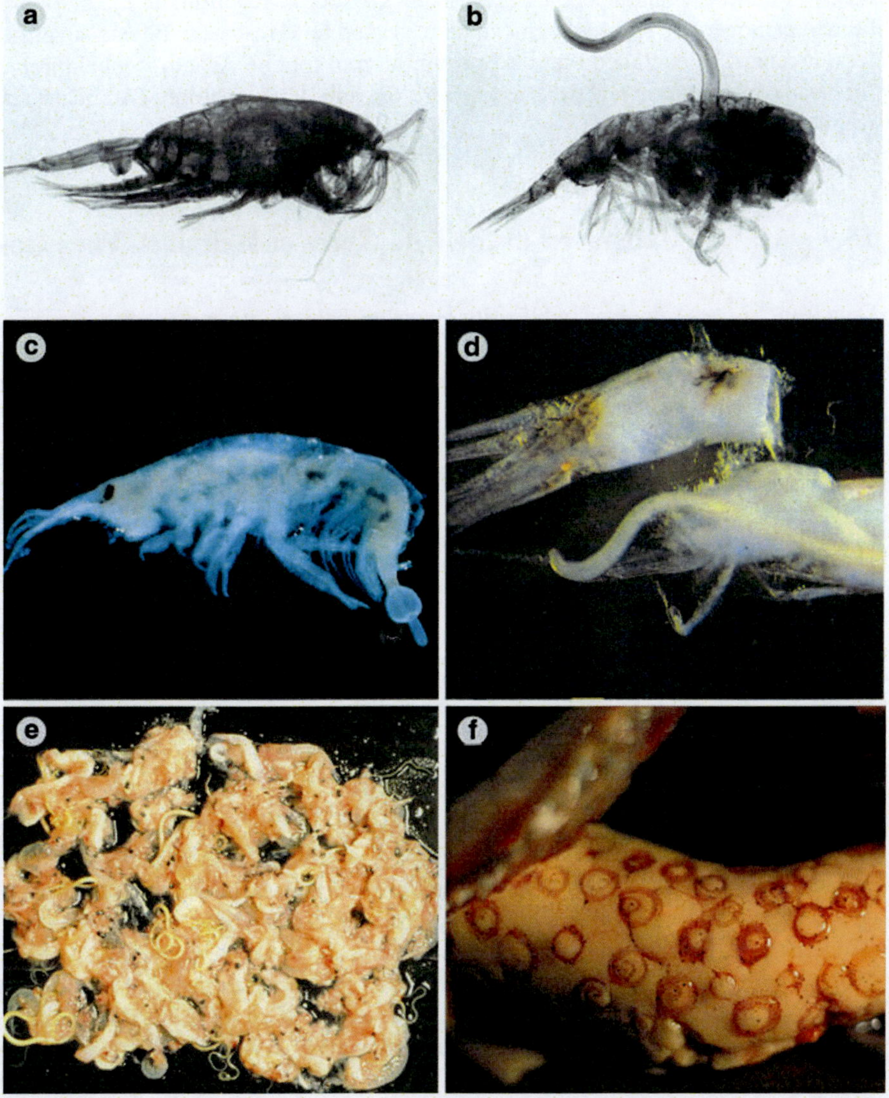

Fig. 14.1 Marine helminth parasites in intermediate hosts. **(a)** L3 larvae of *Anisakis simplex* in the haemocoel of *Paraeuchaetha norvegica* (Copepoda). **(b)** *Hysterothylacium aduncum* in a hyperiid Amphipoda. **(c)** Amphipod Crustacea hosting a *Pomphorhynchus* larva. **(d)** *Anisakis* sp. in a decapod Crustacea. **(e)** Nematodes and Crustacea in the stomach content of fish. **(f)** Fish liver with *Anisakis* sp. larvae. Photos **(a)**, **(b)**: Klimpel et al. (2004); **(d)**: Klimpel (private property); **(c)**, **(e)**, **(f)**: Möller (private property)

cavity by penetrating the intestinal wall and reach the acanthella stage. Paratenic hosts (e.g. fish) and final hosts (fish, birds, seals, whales) get infected by preying upon the intermediate hosts. In general, Amphipoda serve as main intermediate hosts for marine Acanthocephala (Taraschewski 2005). According to the literature, human pathogenetic species seems to be restricted to terrestrial habitats. Nevertheless, it cannot be excluded that the consumption of marine Acanthocephala in fish food can cause allergic reactions as known from anisakid nematodes (Audicana and Kennedy 2008).

14.4 Marine Crustacea in the Life Cycle of Helminth Parasites

Data from 52 peer-reviewed publications have been taken into consideration in order to assess the role of marine crustacean as potential hosts for helminth parasites (Tables 14.1–14.4). A total of 69 different parasites belonging to the Trematoda ($n = 28$), Cestoda ($n = 18$), Nematoda ($n = 10$) and Acanthocephala ($n = 13$) where reported among the crustacean subgroups Copepoda, Euphausiacea, Amphipoda, Decapoda, Mysidacea, Isopoda and Cirripedia (Tables 14.1–14.3). The Acanthocephala have been proven to be parasites of 38 different hosts in three subgroups Euphausiacea, Amphipoda and Decapoda, whereas the Nematoda were reported from 91 different Crustacea in all subgroups. Parasites of the Cestoda and Trematoda have been proven from a total of 27 and 32 crustacean hosts, respectively (Figs. 14.2 and 14.3).

14.4.1 Digenea of Marine Crustacea

Metacercariae of digenetic Trematoda belonging to the families Derogenidae, Hemiuridae, Lecithasteridae, Microphalidae, Opecoelidae and Syncoelidae have been documented in Copepoda, Euphausiacea, Amphipoda and Decapoda. The majority was reported from the families Hemiuridae, Microphalidae and Opecoelidae (Table 14.1). All hemiurid Digenea were detected under natural conditions in copepod Crustacea. In addition, some successful experimental infections have been proven (e.g. Køie 1979, 1989). As described above, *Derogenes varicus* is the most widespread Digenea that utilizes more than 100 teleost fishes as final hosts (Køie 1979). Køie (1979) tested the experimental infection of calanoid Copepoda and Amphipoda with metacercariae of *D. varicus* and identified six species of calanoid copepods to be capable for infection whereas not a single specimen of the Amphipoda was positive. Digenea of the families Syncoeliidae and Opecoelidae have been proven in some euphausiid intermediate hosts (Table 14.1). Adult opecoelids live in the digestive tract of marine and freshwater fishes. Metacercariae from *Opecoeloides collumbellae* have been detected in Decapoda from the species *Hippolyte inermis* (Hippolytidae) at the coast of Corsica and Le Brusc Bay (France) with prevalences between 2.5 and 4.5% (Jousson and Bartoli 2000). Furthermore, Amphipoda seem to be the typical hosts for digenetic trematodes. Most confirmations were detected from the family Microphalidae and less from Opecoelidae (Table 14.1). The common life cycle of the Microphallidae involves Gastropoda as first and intertidal crabs and Amphipoda as

Table 14.1 Reported marine crustacean intermediate hosts of Digesnea and the respective references

Host	Parasite	References
Copepoda		
Acartia clausi	*Hemiurus communis*	Marcogliese (1995)
A. tonsa	*Brachyphallus crenatus*	Køie (1992)
Acartia sp.	*Lecithaster gibbosus*[a], *Lecithaster confusus*	Hunninen and Cable (1943), Køie (1989), Marcogliese (1995)
Centropages abdominalis	*Lecithaster* sp.	Køie (1989)
C. hamatus	*Lecithaster gibbosus*[a]	Køie (1989)
Oithona similis	*Lecithocladium excisum*	Køie (1991), Marcogliese (1995), Reimer et al. (1975)
Pseudocalanus elongatus	*Lecithaster gibbosus*[a]	Køie (1989)
P. minutus	*Lecithaster* sp.	Køie (1989)
Tigriopus brevicornis	*Lecithochirium furcolabiatum*	Marcogliese (2002)
Calanoida indet.	*Derogenes varicus*, *Hemiurus luehei*, *Lecithaster gibbosus*, *Lecithocladium excisum*	Køie (1989, 1991), Marcogliese (1995), Reimer et al. (1975), Svendsen (1990)
Harpacticoida indet.	*Derogenes varicus*	Marcogliese (2002)
Euphausiacea		
Euphausia pacifica	*Paronatrema* sp.	Marcogliese (1995)
E. similis	*Neonotoporus trachuri*, *Pseudopecoelus japonicus*	Marcogliese (1995)
Nyctiphanes simplex	*Paronatrema* sp.	Gómez-Gutiérrez et al. (2010)
Euphausiacea indet.	*Syncoelium filiferum*	Marcogliese (1995)
Amphipoda		
Calliopius laeviusculus	*Microphallus papillorobustus*	Zander et al. (1994)
Gammarus duebeni	*Microphallus claviformes*, *M. papillorobustus*	Zander et al. (1994)
G. finmarchicus	*Maritrema subdolum*, *Microphallus papillorobustus*, *Podocotyle atomon*	Zander et al. (1994)
G. insensibilis	*Microphallus papillorobustus*	Brown et al. (2003)
G. locusta	*Levinseniella brachysoma*, *Maritrema subdolum*, *Microphallus papillorobustus*, *Podocotyle atomon*	Zander et al. (2002)
G. oceanicus	*Levinseniella brachysoma*, *Maritrema subdolum*, *Microphallus claviformes*, *M. papillorobustus*, *Podocotyle atomon*	Zander et al. (2002)
G. salinus	*Maritrema subdolum*, *Microphallus claviformes*, *M. papillorobustus*,	Zander et al. (2002)

(continued)

Table 14.1 (continued)

Host	Parasite	References
	Levinseniella brachysoma, *Podocotyle atomon*	
G. tigrinus	*Microphallus claviformes*, *M. papillorobustus*, *Maritrema subdolum*	Zander et al. (1994)
G. zaddachi	*Levinseniella brachysoma*, *Maritrema subdolum*, *Microphallus claviformes*, *M. papillorobustus*, *Podocotyle atomon*	Zander et al. (2002)
Paracalliope novizealandiae	*Maritrema novaezealandensis*[a]	Fredensborg and Poulin (2005), Leung and Poulin (2006)
Decapoda		
Farfantepenaeus aztecus	*Opecoeloides fimbriatus*	Feigenbaum (1975)
F. duorarum	*Microphallus* sp., *Opecoeloides fimbriatus*	Feigenbaum (1975)
Hemigrapsus crenulatus	*Maritrema novaezealandensis*, *Microphallus* sp.	Martorelli et al. (2004)
H. penicilliatus	*Levinseniella conicostoma*, *Maritrema laricola*, *M. setoenesis*, *Microphalloides japonicus*, *Probolocoryphe asadai*, *Microphallus macrorchis*	Blakeslee et al. (2009), McDermott (2011)
H. sanguineus	*Maritrema jebuensis*, *M. setoenensis*, *Microphalloides japonicus*, *Probolocoryphe asadai*, *Microphallus capellae*	Blakeslee et al. (2009), Chung et al. (2010), McDermott (2011)
H. sexdentatus	*Maritrema novaezealandensis*, *Microphallus* sp.	Koehler and Poulin (2010)
Hippolyte inermis	*Opecoeloides collumbellae*	Jousson and Bartoli (2000)
Idotea balthica	*Maritrema subdolum*, *Podocotyle atomon*	Zander et al. (2002)
I. chelipes	*Maritrema subdolum*, *Microphallus claviformes*, *M. papillorobustus*, *Levinseniella brachysoma*, *Podocotyle atomon*	Zander et al. (2002)
I. granulosa	*Maritrema subdolum*	Zander et al. (2000)
Penaeus setiferus	*Opecoeloides fimbriatus*	Feigenbaum (1975)

[a]Experimental infection

second intermediate hosts, whereas birds serve as final hosts (Leung and Poulin 2006). The digenean helminth *Maritrema novaezealandensis* induces a behavioural change of its amphipod host *Paracalliope novizealandiae* in order to be more attractive for the final host (Leung and Poulin 2006). Helluy and Thomas (2010) investigated similar consequences for the amphipod species *Gammarus insensibilis* that show an aberrant escape behaviour when infested with larvae of the Digenea *Microphallus papillorobustus* (Microphalidae).

Table 14.2 Reported marine crustacean intermediate hosts of Cestoda and the respective references

Host	Parasite	References
Copepoda		
Acartia clausi	*Bothriocephalus scorpii*	Solonchenko (1985)
Calanus finmarchicus	*Aporhynchus norvegicus*, *Scolex pleuronectis*	Klimpel et al. (2003), Marcogliese (1995)
Eurytemora affinis	*Bothriocephalus scorpii*	Marcogliese (1995)
Mesochra sp.	*Prochristianella hispida*[a]	Palm (2004)
Tigriopus californicus	*Lacistorhynchus tenuis*, *Parachristianella monomegacantha*, *Prochristianella hispida*[a]	Palm (2004)
Calanoida indet.	*Bothriocephalus barbatus*, *B. gregarius*[b], *Grillotia erinaceus*	Marcogliese (1995)
Harpacticoida indet.	*Lacistorhynchus dollfusi*	Marcogliese (1995)
Euphausiacea		
Euphausia pacifica	*Nybelinia surmenicola*	Marcogliese (1995)
E. recurva	*Pseudonybelinia odontocantha*	Marcogliese (1995)
E. similis	*Echinobothrium* sp., *Tetrarhynchobothrium* sp.	Marcogliese (1995)
Meganyctiphanes norvegica	*Aporhynchus norvegicus*	Klimpel et al. (2003)
Nyctiphanes simplex	*Echinobothrium* sp., *Tetrarhynchobothrium* sp.	Gómez-Gutiérrez et al. (2010)
Thysanoessa inermis	*Nybelinia surmenicola*	Shimazu (1975)
T. longipes	*Nybelinia surmenicola*, *Pelichnibothrium caudatum*	Marcogliese (1995), Shimazu (1975)
T. raschii	*Nybelinia surmenicola*	Shimazu (1975)
Euphausiacea indet.	*Nybelinia surmenicola*	Shimazu (1975)
Decapoda		
Callianassa sp.	*Parachristianella monomegacantha*	Palm (2004)
Callichirus islagrande	*Prochristianella hispida*	Palm (2004)
Callinectes sapidus	*Polypocephalus* sp.	Hutton (1964)
Farfantepenaeus aztecus	*Parachristianella dimegacantha*, *Prochristianella hispida*	Feigenbaum (1975)
F. brasiliensis	*Mecistobothrium penaeus*, *Parachristianella monomegacantha*, *P. heteromegacantha*, *Polypocephalus* sp., *Prochristianella hispida*	Feigenbaum (1975), Hutton (1964)
F. duorarum	*Parachristianella dimegacantha*, *P. monomegacantha*, *Polypocephalus* sp., *Prochristianella hispida*	Feigenbaum (1975), Hutton (1964)
F. setiferus	*Polypocephalus* sp.	Hutton (1964)
Penaeus setiferus	*Prochristianella hispida*	Feigenbaum (1975)
Sicyonia dorsalis	*Polypocephalus* sp.	Hutton (1964)

(continued)

Table 14.2 (continued)

Host	Parasite	References
Solenocera atlantidis	*Polypocephalus* sp.	Hutton (1964)
Squilla empusa	*Polypocephalus* sp.	Hutton (1964)
Trachypeneus constrictus	*Polypocephalus* sp.	Hutton (1964)
T. similis	*Polypocephalus* sp.	Hutton (1964)
Xiphopenaeus kroyeri	*Polypocephalus* sp.	Hutton (1964)

[a]Experimental infection
[b]Nomen nudum

14.4.2 Cestoda of Marine Crustacea

Larvae of nine cestode families have been documented in crustacean intermediate hosts from the Copepoda, Euphausiacea and Decapoda (Table 14.2). Most parasites belong to the order Trypanorhyncha, including the families Aporhynchidae (e.g. *Aporhynchus norvegicus*), Eutetrarhynchidae (e.g. *Prochristianella hispida*), Lacisthorhynchidae (e.g. *Lacistorhynchus dollfusi*), Tentaculariidae (e.g. *Nybelinia surmenicola*) and Paranybeliniidae (e.g. *Pseudonybelinia odontocantha*). A minority of the reported larvae belong to the families Bothriocephalidae, Echinobothriidae and Polypocephalidae. Trypanorhynch cestodes mature in the lumen of the spiral intestine of an elasmobranch definitive host. The eggs are dispersed within gravid segments into the seawater as part of the host faeces (Palm 2004). The coracidium stage is ingested by a Copepoda, whereupon it penetrates the intestinal wall and develops in the haemocoel to the procercoid. This stage is infective for larger invertebrates or small fish and further develops to the plerocercoid. In some cases, a second intermediate host or paratenic host, for example, larger fishes, is required (Palm 2004).

The life cycle of *Aporhynchus norvegicus* (Aporhynchidae) has been proposed by Klimpel et al. (2003). It includes three hosts with calanoid Copepoda (e.g. *Calanus finmarchicus*) as first and pelagic Euphausiacea (e.g. *Meganyctiphanes norvegica*) as second intermediate hosts. Potential final hosts are dogfish sharks such as *Etmopterus spinax* (Etmopteridae). In the Norwegian Deep, *C. finmarchicus* is the main prey of *M. norvegica*, while stomach content analyses of *E. spinax* show that *M. norvegica* is the main prey of this shark species (Klimpel et al. 2003). The life cycle of *Parachristianella monomegacantha* (Eutetrarhynchidae) includes invertebrates as obligatory second intermediate hosts. Copepoda (e.g. *Tigriopus californicus*) directly ingest the eggs. A free-swimming coracidium does not occur. Second intermediate hosts are decapod penaeid crustaceans such as the mud shrimp (*Callianassa* sp.). Elasmobranch fish like the guitarfish *Rhinobatos productus* harbour the adult helminths (Palm 2004). The cosmopolitan genus *Bothriocephalus* (Bothriocephalidae) comprises approximately 80 species that use one or two intermediate hosts in their life cycles (Blend and Dronen 2003).

Table 14.3 Reported marine crustacean intermediate hosts of Nematoda and the respective references

Host	Parasite	References
Copepoda		
Acartia bifilosa	*Hysterothylacium* sp.	Lick (1991)
A. longiremis	*Hysterothylacium* sp.[a]	Lick (1991)
A. tonsa	*Anisakis simplex*[a], *Hysterothylacium aduncum*[a]	Køie (1993, 2001)
Acartia sp.	*Contracaecum osculatum*[a]	Køie and Fagerholm (1995)
Calanus finmarchicus	*Hysterothylacium* sp.	Lick (1991), Marcogliese (1995), Svendsen (1990)
Calanus sp.	*Hysterothylacium* sp.	Lick (1991)
Centropages hamatus	*Contracaecum osculatum*[a]	Køie and Fagerholm (1995)
C. typicus	*Hysterothylacium* sp.	Svendsen (1990)
Euchoeta sp.	*Hysterothylacium* sp.	Lick (1991)
Eurytemora affinis	*Hysterothylacium* sp.	Lick (1991)
Microsetella norvegica	*Hysterothylacium* sp.[a]	Lick (1991)
Oithona similis	*Anisakis simplex*[a]	Køie (2001)
Paracalanus parvus	*Contracaecum osculatum*[a]	Køie and Fagerholm (1995)
Paraeuchaeta norvegica	*Anisakis simplex*	Klimpel et al. (2004)
Pseudocalanus elongatus	*Hysterothylacium* sp.	Lick (1991)
Pseudocalanus sp.	*Hysterothylacium* sp.	Lick (1991)
Temora longicornis	*Contracaecum osculatum*[a], *Hysterothylacium* sp.[a]	Køie and Fagerholm (1995), Lick (1991)
Calanoida indet.	*Hysterothylacium aduncum*	Hurst (1984), Køie (1993), Marcogliese (1995), Svendsen (1990)
Harpacticoida indet.	*Hysterothylacium aduncum*[a]	Klimpel et al. (2004), Køie (1993)
Cirripedia		
Balanus sp.	*Anisakis simplex*[a], *Contracaecum osculatum*[a]	Køie (2001), Køie and Fagerholm (1995)
Mysidacea		
Erythrops erythrophthalma	*Hysterothylacium* sp.	Lick (1991)
Mesopodopsis slabberi	*Anisakis* sp., *Hysterothylacium* sp., *Pseudoterranova decipiens*	Lick (1991)
Mysis gaspensis	*Paracuaria adunca*	Jackson et al. (1997)
M. mixta	*Hysterothylacium* sp.	Lick (1991)
M. stenolepis	*Paracuaria adunca*, *Pseudoterranova decipiens*	Jackson et al. (1997)
Neomysis americana	*Hysterothylacium* sp., *Paracuaria adunca*, *Pseudoterranova decipiens*	Jackson et al. (1997), Lick (1991)
N. integer	*Contracaecum osculatum*[a], *Hysterothylacium aduncum*, *Hysterothylacium* sp., *Pseudoterranova decipiens*	Klimpel and Rückert (2005), Køie and Fagerholm (1995), Lick (1991)

(continued)

Table 14.3 (continued)

Host	Parasite	References
N. intermedia	*Hysterothylacium* sp.[a]	Lick (1991)
Praunus flexuosus	*Paracuaria adunca*	Jackson et al. (1997)
Amphipoda		
Americorchestia megalophthalma	*Pseudoterranova decipiens*	Jackson et al. (1997)
Amphiporeia virginiana	*Paracuaria adunca*, *Pseudoterranova decipiens*	Jackson et al. (1997)
Calliopius laeviusculus	*Hysterothylacium* sp.	Zander et al. (1994)
Caprella septentrionalis	*Anisakis* sp., *Hysterothylacium* sp., *Pseudoterranova decipiens*	Lick (1991)
Echinogammarus obtusatus	*Paracuaria adunca*, *Pseudoterranova decipiens*, *Tetrameres* sp.	Jackson et al. (1997)
Eogammarus kygi	*Hysterothylacium* sp.	Moravec and Nagasawa (1986)
Gammarus duebeni	*Hysterothylacium* sp.	Lick (1991)
G. finmarchicus	*Hysterothylacium* sp.	Zander et al. (1994)
G. inaequicauda	*Hysterothylacium* sp.	Lick (1991)
G. lawrencianus	*Ascarophis* sp., *Hysterothylacium* sp., *Paracuaria adunca*, *Pseudoterranova decipiens*, *Tetrameres* sp.	Jackson et al. (1997), Lick (1991)
G. locusta	*Ascarophis arctica*, *Hysterothylacium* sp.	Lick (1991), Zander et al. (2002)
G. oceanicus	*Ascarophis arctica*, *Hysterothylacium* sp., *Paracuaria adunca*, *Tetrameres* sp.	Jackson et al. (1997), Lick (1991), Svendsen (1990), Zander et al. (2002)
G. salinus	*Ascarophis arctica* , *Hysterothylacium* sp.	Lick (1991), Zander et al. (2002)
G. zaddachi	*Ascarophis arctica* , *Hysterothylacium* sp.	Lick (1991), Zander et al. (2002)
Gammarus spp.	*Hysterothylacium aduncum*	Klimpel and Rückert (2005)
Hyperia galba	*Hysterothylacium aduncum*	Klimpel and Rückert (2005)
Themisto abyssorum	*Ascarophis* sp., *Hysterothylacium aduncum*	Jackson et al. (1997), Klimpel and Rückert (2005), Køie (1993), Marcogliese (1995)
T. gaudichaudii	*Hysterothylacium aduncum*	Klimpel and Rückert (2005)
Unciola irrorata	*Hysterothylacium* sp., *Pseudoterranova decipiens*	Lick (1991)
Isopoda		
Edotia triloba	*Hysterothylacium* sp.	Lick (1991)
Idotea neglecta	*Pseudoterranova decipiens*	Lick (1991)
Idotea spp.	*Hysterothylacium aduncum*	Klimpel and Rückert (2005)
Jaera albifrons	*Hysterothylacium* sp.	Zander et al. (1994)
Euphausiacea		
Euphausia krohnii	*Anisakis* sp.	Smith and Snyder (2005)
E. nana	*Anisakis simplex*, *Anisakis* sp.	Lick (1991)

(continued)

Table 14.3 (continued)

Host	Parasite	References
E. pacifica	*Anisakis simplex*, *Anisakis* sp., *Contracaecum osculatum*, *Hysterothylacium* sp.	Lick (1991), Marcogliese (1995), Smith and Snyder (2005)
E. similis	*Philometra* sp.	Marcogliese (1995)
E. vallentini	*Anisakis* sp.	Hays et al. (1998)
Meganyctiphanes norvegica	*Anisakis simplex*, *Hysterothylacium* sp.	Hays et al. (1998), Lick (1991), Smith and Snyder (2005)
Nyctiphanes australis	*Anisakis simplex*	Marcogliese (1995)
N. couchii	*Anisakis simplex*, *Hysterothylacium* sp.	Hays et al. (1998), Lick (1991)
N. simplex	*Anisakis simplex*	Gómez-Gutiérrez et al. (2010)
Thysanoessa inermis	*Anisakis simplex*, *Hysterothylacium* sp.	Lick (1991), Marcogliese (1995), Smith (1971)
T. longicaudata	*Anisakis simplex*	Smith (1971)
T. longipes	*Anisakis simplex*	Marcogliese (1995)
T. raschii	*Anisakis simplex*, *Hysterothylacium aduncum*, *Hysterothylacium* sp.	Lick (1991), Marcogliese (1995), Smith and Snyder (2005), Svendsen (1990)
Thysanoessa spp.	*Ascarophis* sp., *Hysterothylacium* sp.	Marcogliese (1995)
Euphausiacea indet.	*Anisakis simplex*, *Hysterothylacium aduncum*	Hurst (1984), Køie (1993), Marcogliese (1995), Svendsen (1990)
Decapoda		
Alpheus heterochaelis	*Pseudoterranova decipiens*	Hutton and Sogandares-Bernal (1960)
Artemesia longinaris	*Ascarophis marina*	Martorelli et al. (2000)
Carcinus maenas	*Ascarophis morrhuae*	Moravec et al. (2003)
Clibanarius vittatus	*Hysterothylacium* sp.	Lick (1991)
Emerita talpoida	*Hysterothylacium* sp.	Lick (1991)
Farfantepenaeus aztecus	*Contracaecum* sp., *Hysterothylacium* sp.	Feigenbaum (1975), Lick (1991)
F. brasiliensis	*Hysterothylacium* sp.	Lick (1991)
F. californiensis	*Hysterothylacium* sp.	Lick (1991)
F. duorarum	*Contracaecum* sp., *Hysterothylacium* sp.	Lick (1991)
Fenneropenaeus indicus	*Hysterothylacium* sp.	Lick (1991)
Hemigrapsus crenulatus	*Ascarophis* sp.	McDermott (2011)
H. oregonensis	*Ascarophis* sp.	Moravec et al. (2003)
H. sexdentatus	*Ascarophis* sp.	McDermott (2011)
Hemiplax hirtipes	Acuariidae gen. sp., *Ascarophis* sp.	Moravec et al. (2003)
Homarus americanus	*Ascarophis* sp., *Hysterothylacium* sp.	Lick (1991), Moravec et al. (2003)
Hyas araneus	*Anisakis* sp.	Lick (1991)
Litopenaeus setiferus	*Contracaecum* sp., *Hysterothylacium* sp.	Feigenbaum (1975), Lick (1991)

(continued)

Table 14.3 (continued)

Host	Parasite	References
L. stylirostris	*Contracaecum* sp., *Hysterothylacium* sp.	Lick (1991)
Litopenaeus vannamei	*Ascarophis* sp., *Hysterothylacium* sp.	Feigenbaum (1975), Lick (1991)
Marsupenaeus japonicus	*Hysterothylacium* sp.	Lick (1991)
Munida gregaria	*Hysterothylacium aduncum*	Hurst (1984), Køie (1993), Marcogliese (1995), Svendsen (1990)
Pachycheles rudis	*Ascarophis* sp.	Poinar and Kuris (1975)
Pandalus borealis	*Anisakis* sp., *Hysterothylacium* sp.	Lick (1991)
P. latirostris	*Anisakis* sp.	Lick (1991)
Peisos petrunkevitchi	*Ascarophis marina*	Martorelli et al. (2000)
Rimapenaeus constrictus	*Contracaecum* sp., *Hysterothylacium* sp.	Lick (1991)
Sclerocrangon boreas	*Pseudoterranova decipiens*	Lick (1991)
Sicyonia dorsalis	*Contracaecum* sp., *Hysterothylacium* sp.	Lick (1991)
S. typica	*Contracaecum* sp., *Hysterothylacium* sp.	Lick (1991)
Solenocera atlantidis	*Contracaecum* sp., *Hysterothylacium* sp.	Lick (1991)
Trachypeneus similis	*Contracaecum* sp., *Hysterothylacium* sp.	Hutton et al. (1962), Lick (1991)
Xiphopenaeus kroyeri	*Hysterothylacium* sp.	Lick (1991)

[a]Experimental infection

First intermediate hosts are copepod Crustacea (e.g. *Eurytemora affinis*) (Marcogliese 1995). Experimental infections succeeded to infest the Copepoda species *Acartia clausi* (Solonchenko 1985).

14.4.3 Nematoda of Marine Crustacea

Larvae of ten different nematode genera have been found in marine Crustacea distributed within all major groups, whereas Copepoda, Amphipoda, Euphausiacea and Decapoda seem to be the most important intermediate hosts (Table 14.3). Mostly parasites of the family Anisakidae (order Ascaridida) are included (*Anisakis*, *Contracaecum*, *Hysterothylacium*, *Pseudoterranova*), but members of the Cystidicolidae (*Ascarophis* sp.), Acuariidae (*Paracuaria* sp.), Tetrameridae (*Tetrameres* sp.) and Philometridae (*Philometra* sp.) (order Spirurida) have also been recognized. In addition to the naturally occurring infestations, experimental infections of the anisakids *Anisakis simplex* s.l., *Contracaecum osculatum* and

Table 14.4 Reported marine crustacean intermediate hosts of Acanthocephala and the respective references

Host	Parasite	References
Euphausiacea		
Euphausia krohnii	*Echinorhynchus* sp.	Marcogliese (1995)
Nyctiphanes simplex	*Polymorphidae* sp.	Gómez-Gutiérrez et al. (2010)
Microdeutopus gryllotalpa	*Bolbosoma caenoforme*	Marcogliese (1995)
Thysanoessa raschii	*Bolbosoma caenoforme*	Marcogliese (1995)
Thysanoessa sp.	*Bolbosoma caenoforme*	Marcogliese (1995)
Amphipoda		
Aeginina longicornis	*Echinorhynchus gadi*	Marcogliese (1994)
Ampithoe rubricata	*Echinorhynchus gadi*	Marcogliese (1994)
Calliopius laeviusculus	*Echinorhynchus gadi*	Marcogliese (1994)
C. rathkii	*Echinorhynchus gadi*[a]	Marcogliese (1994)
Caprella septentrionalis	*Echinorhynchus gadi*	Marcogliese (1994)
Cyphocaris challengeri	*Echinorhynchus gadi*	Marcogliese (1994)
Gammarellus angulosus	*Echinorhynchus gadi*	Marcogliese (1994)
Gammarus duebeni	*Echinorhynchus gadi*	Marcogliese (1994)
G. locusta	*Echinorhynchus gadi*	Zander et al. (2000)
G. oceanicus	*Echinorhynchus gadi*	Zander et al. (2002)
G. salinus	*Echinorhynchus gadi*	Zander et al. (2002)
G. tigrinus	Acanthocephala indet. (Acanthella)	Zander et al. (1994)
G. zaddachi	*Echinorhynchus gadi*	Zander et al. (2002)
Microdeutopus gryllotalpa	Acanthocephala indet. (Acanthella)	Zander et al. (1994)
Pontoporeia femorata	*Echinorhynchus gadi*	Marcogliese (1994)
Decapoda		
Brachynotus spinosus	*Polymorphus sphaerocephalus*	Pichelin et al. (1998)
Cancer irroratus	*Polymorphus major*	Schmidt and MacLean (1978)
Cyclograpsus granulosus	*Polymorphus sphaerocephalus*	Pichelin et al. (1998)
Emerita analoga	*Polymorphus kenti*, *Profilicollis altmani*	Nickol et al. (1999)
E. talpoida	*Profilicollis altmani*	Nickol et al. (2002)
Hemigrapsus crenulatus	*Profilicollis antarcticus*, *P. novaezelandensis*	Latham and Poulin (2003), McDermott (2011)
H. oregonensis	*Profilicollis botulus*	McDermott (2011)
H. sexdentatus	*Profilicollis antarcticus*, *P. novaezelandensis*	Latham and Poulin (2002, 2003), McDermott (2011)
Hemiplax hirtipes	*Profilicollis antarcticus*, *P. novaezelandensis*, *Profilicollis* spp.[a]	Fredensborg and Poulin (2005), Latham and Poulin (2002)

(continued)

Table 14.4 (continued)

Host	Parasite	References
Homarus americanus	*Polymorphus botulus*	Nickol et al. (1999)
Macrobrachium sp.	*Polymorphus formosus*	Nickol et al. (1999)
Nectocarcinus integrifrons	*Polymorphus sphaerocephalus*	Pichelin et al. (1998)
Neohelice granulata	*Polymorphus chasmagnathi*	Nickol et al. (1999)
Pagurus longicarpus	*Polymorphus* sp.	Nickol et al. (1999)
Paragrapsus gaimardii	*Polymorphus sphaerocephalus*	Pichelin et al. (1998)
P. laevis	*Polymorphus sphaerocephalus*	Pichelin et al. (1998)
P. quadridentatus	*Polymorphus sphaerocephalus*	Pichelin et al. (1998)
Uca rapax	*Arhythmorhynchus frassoni*	Nickol et al. (2002)
U. spinicarpus	*Hexaglandula corynosoma*	Nickol et al. (2002)

[a]Experimental infection

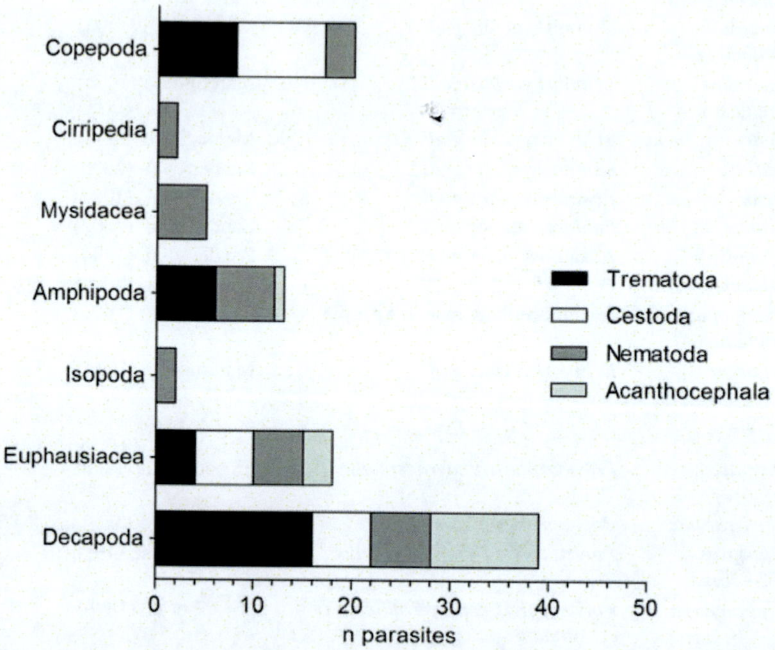

Fig. 14.2 Numbers of parasites species that have been reported in the crustacean subgroups (see Tables 14.1–14.4)

Hysterothylacium aduncum in some copepods, cirripeds and mysidacea have also been reported (Table 14.3).

The family Cystidicolidae includes some 23 genera that are considered valid (Moravec and Klimpel 2009). While they are usually found in the intestine,

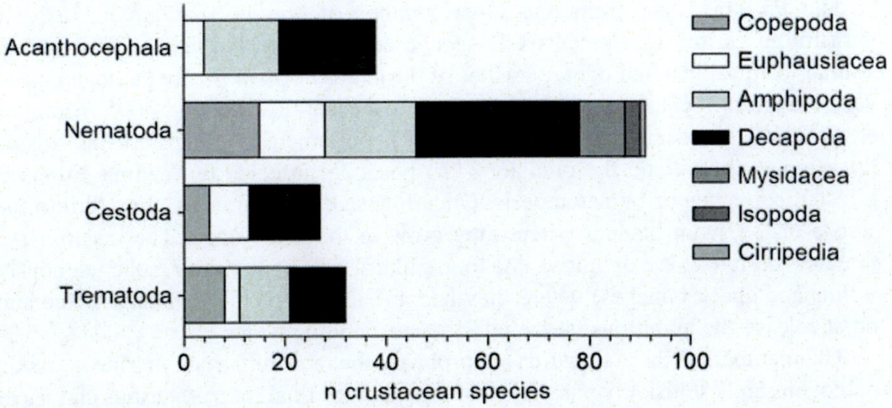

Fig. 14.3 Numbers of crustacean species that have been reported to be intermediate hosts for marine helminth parasites (see Tables 14.1–14.4)

stomach and pyloric caeca of marine and freshwater fishes, some have adapted to the swim bladder of physostomous fishes (Anderson 2000). According to Anderson (2000), the life cycle includes aquatic insects and crustaceans that serve as intermediate hosts whereas marine fish are used as final hosts. The eggs are released into the water with the faeces of the final host and are ingested by aquatic insects or crustaceans where they remain in the haemocoel or host tissue. The definitive hosts get infected by preying upon arthropods containing third stage larvae. The cosmopolitan *Ascarophis* sp. is the largest genus within this family, worldwide distributed from littoral and demersal to pelagic waters, and parasitic in the digestive tract of marine and estuarine fishes (Muñoz and George-Nascimento 2007). While *Ascarophis* has been recognized in various marine crustaceans, there is no evidence that the parasites are able to switch from one to another intermediate host (Anderson 2000).

Members of the Tetrameridae include the genera *Tetrameres* and *Microtetrameres* which are, as adults, parasites of the proventriculus of waterfowls and terrestrial birds. The larvae are not known to be very host specific regarding the intermediate hosts and include either aquatic crustaceans (aquatic species) or terrestrial insects and isopods (terrestrial species) in their life cycle, respectively. Encapsulated in the haemocoel of the arthropod host, they are infective for the definitive host within a few days (Anderson 2000).

With the exception of *Paracuaria* spp. that mainly occur in the hosts intestines, parasites of the only family of the Acuarioidea (Acuariidae) are found in the upper alimentary tract (crop, gizzard, oesophagus) of piscivorous birds (e.g. Procellariiformes, Pelecaniformes, Falconiformes). Very similar to the Tetrameridae, it is likely that two general life cycles exist, one in the terrestrial and one in the aquatic environment. The definitive avian hosts acquire their infection from ingesting larvae in the haemocoel of aquatic, arthropods or sometimes fish paratenic hosts that have preyed upon infected intermediate hosts (Anderson 2000).

The Philometridae contain a large number of species that are exclusively parasitic in various ray-finned freshwater, brackish water and marine fishes. Very similar to anisakid nematodes, parasites of the genus *Philometra* are pathogenic and very common in commercially available fish species (McClelland 2005; Moravec et al. 2010). According to Anderson (2000), philometrids utilize crustaceans as intermediate hosts, and the predatory fish hosts get infected by feeding either on infested copepods or by paratenesis. Once ingested, the parasites migrate into the serosa of the swim bladder where they grow to the adult stage. The inseminated females migrate to the definitive site (e.g. gill arteries, body cavity, subcutaneously in fins and cheek pouches) where they become gravid, break the host's tissue and burst-release the numerous larvae into the surrounding water.

The members of the Ascaridida are probably the most important marine zoonotic pathogens and exclusively parasites of the gastrointestinal tract of various classes of vertebrates, including birds and mammals. In combination with the Spirurida, they are responsible for the vast majority of nematode infections in marine fish (Rohde 2005). The family Anisakidae represents the largest group within the Ascaridoidea and can be split into the subfamilies Anisakinae (*Anisakis*, *Contracaecum*), Goeziinae and Raphidascaridinae (*Hysterothylacium*, *Pseudoterranova*) (Hartwich 1974).

The Anisakidae perform an indirect life cycle in the aquatic environment and utilize hosts on different trophic levels to be transferred through the marine food web. A wide range of marine mammals (Cetacea, Pinnipedia) and piscivorous birds are utilized as final hosts, whereas invertebrates (e.g. Crustacea) and sometimes Cephalopoda as well as a variety of fish are included as intermediate and/or paratenic hosts.

The eggs of the nematodes are released into the water column within the faeces of the respective final hosts where they are ingested by either pelagic (e.g. *Anisakis*) or benthic (e.g. *Pseudoterranova*) crustaceans. The parasites are transferred along the food chain to larger macroinvertebrates in which they reach the length and developmental stage to be infective to fish and their respective final hosts (Klimpel et al. 2004, 2008; Klimpel and Rückert 2005; Mattiucci and Nascetti 2008; McClelland 2005). These larger invertebrates as well as smaller sized schooling fish are thought to be important second intermediate hosts, whereas predatory fish serve as paratenic hosts. The capability of re-infecting larger piscivorous fish hosts without further moulting is an essential adaptation in order to accumulate enormous numbers of larvae and enhance the transmission to the successive final hosts (Klimpel and Palm 2011; Lile 1998). Due to their high prevalence in commercially available fish species, the genera *Anisakis*, *Contracaecum* and *Pseudoterranova* are probably the most important anisakid nematodes (e.g. Hochberg and Hamer 2010; Karl et al. 2010). These genera are comparatively well known and have evolved various adaptations in order to utilize a wide range of marine organisms as intermediate and final hosts. In the case of the whale worm *Anisakis* spp., marine mammals such as toothed and baleen whales of the families Delphinidae, Kogiidae, Physeteridae and Ziphiidae are considered the main definitive hosts, but members of the Balaenopteridae, Pontoporidae, Monodontidae, Phocoenidae and Neobalaenidae

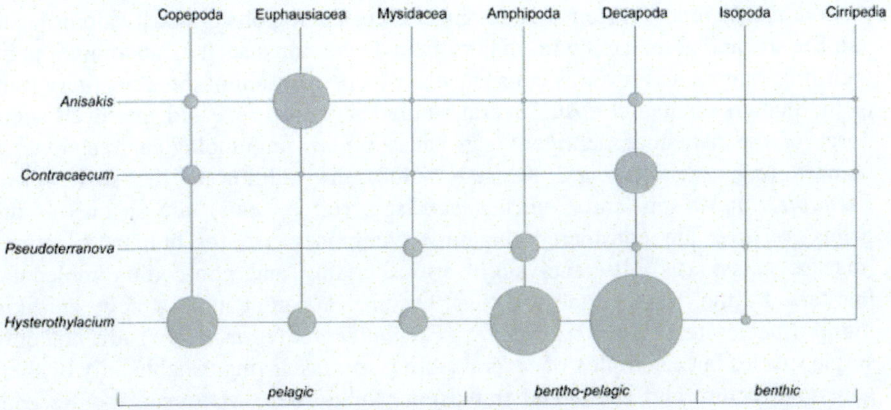

Fig. 14.4 Relative importance and pelagic, benthopelagic and benthic association, respectively, of marine crustacean intermediate hosts that have been reported for anisakid nematodes of the genera *Anisakis, Contracaecum, Pseudoterranova* and *Hysterothylacium.* Circle size reflects the number of reported hosts in the respective crustacean subgroups

have also been parasitized (e.g. Klimpel et al. 2004, 2010; Mattiucci and Nascetti 2008). The life cycle of *Anisakis* is considered to take place in the pelagic environment which is also reflected in the use of Copepoda and especially Euphausiacea as crustacean intermediate host (Fig. 14.4). The sealworm *Pseudoterranova* sp. is known to utilize a variety of ten different marine seals belonging to Otariidae and Phocidae as final hosts and is thought to perform a benthic life cycle including benthic invertebrates, teleosts and larger demersal fish species (Klimpel and Palm 2011; Mattiucci and Nascetti 2008). Very similar to *Anisakis* sp., larger piscivorous fish may serve as successive intermediate or paratenic hosts (Anderson 2000; McClelland 2005). The life cycle of the members of the genus *Contracaecum* is equally complex and includes benthic and pelagic invertebrates and fish as intermediate hosts, as well as marine mammals of the families Otariidae and Phocidae as final hosts (Fig. 14.4) (Klimpel and Palm 2011). However, in contrast to *Pseudoterranova*, *Contracaecum* is able to infect different fish-eating birds of various genera (e.g. *Larus*, *Pelecanus*) as well. The species of the raphidascarine *Hysterothylacium* are typically found as adults in the digestive tract of fishes but have although been reported from a very wide range of aquatic invertebrates including gastropods, cephalopods, chaetognaths, echinoderms and cnidarians serving as intermediate hosts (Fig. 14.4) (Anderson 2000; Klimpel and Rückert 2005; McClelland 2005). *Hysterothylacium* is therefore probably the most abundant anisakid fish parasites and extremely common throughout the marine food web.

Besides their role as important parasites of marine organisms, the anisakid nematodes have achieved a notorious dubiety as causative agents of the zoonotic disease anisakidosis when ingested alive by humans. With estimated 20,000 cases of human infections every year, this foodborne infection is of major medical importance and the result of the ingestion of larvae in raw or insufficient cooked fish products (Hochberg and Hamer 2010; Kuhn et al. 2011). The disease is predominantly reported from a wide range of coastal regions primarily along Japan and Europe and associated with major clinical symptoms such as abdominal pain, vomiting, nausea and fever as consequence of a penetration of the alimentary tract of the human accidental host. Several studies have also revealed that even small doses of the nematodes antigens can cause severe immunological reactions in humans (e.g. Audicana and Kennedy 2008). In addition to the seal worms *Pseudoterranova* spp. and *Contracaecum* spp. (birds, seals), the species of the genus *Anisakis* are considered the most common cause for human infections (disease anisakiasis). By analysing various coding and non-coding molecular markers in the last decades, it has become apparent that a few anisakid morphospecies (e.g. *Anisakis simplex*, *Pseudoterranova decipiens*) are not only single species but a complex of several sibling species distinguishable only by their genetic structure and ecological traits (e.g. Kuhn 2010; Mattiucci and Nascetti 2008). In addition, it is still unclear whether all species cause clinical symptoms in humans.

14.4.4 Acanthocephala of Marine Crustacea

Larvae of two Acanthocephala families have been documented in Euphausiacea, Amphipoda and Decapoda (Table 14.4). The acanthocephalan parasites belong to the families Echinorhynchidae (e.g. specimens of the genus *Echinorhynchus*) and Polymorphidae (e.g. specimens of the genus *Polymorphus*). All larvae that have been reported in the crustacean intermediate hosts belong to the class Palaeacanthocephala, the largest taxon within the Acanthocephala that has also been documented in limnic environments (Kennedy 2006; Klimpel et al. 2009; Moravec et al. 1999). Their general life cycle is characterized by a wide range of crustacean (first) intermediate hosts, whereas the final host specificity can be either narrow or wide (Taraschewski 2005). *Echinorhynchus gadi*, frequently occurring in North Atlantic marine hosts, uses benthic Crustacea from the families Gammaridae (e.g. specimens of the genus *Gammarus*) and Caprelidae as obligatory first intermediate hosts (Marcogliese 1994; Zander 1998) and follows a benthic life cycle strategy. Teleost fish are considered final hosts for *E. gadi* (Marcogliese 1994, 2002; Zander and Reimer 2002).

All Acanthocephala larvae documented in decapods were isolated from the family Polymorphidae (Table 14.4). Latham and Poulin (2002) investigated the impact of the Acanthocephala species *Profilicollis antarcticus* and *P. novaezelandensis* on the shore crabs of the genus *Hemigrapsus* on the coast of New Zealand. After ingestion of the parasite eggs, the acanthocephalan larvae encyst in the decapod's haemocoel and develop to the cystacanth stage. Shore-associated bird species are considered final hosts (Latham and Poulin 2002).

14.5 Impact of Parasitism on Hosts Behaviour and Viability

The infection of an organism by parasites is usually a trade-off between effective utilization of the host as a biosphere and facilitating the transmission to the next eligible host at each stage of the life cycle. Impacts on the hosts morphology, health and body condition or behavioural changes are regular phenomena during parasites infestations.

Digenean parasites have different effects on the health of their hosts. The consequences for the first intermediate host can be serious (e.g. castration) (Cribb 2005). Investigations have documented a reduced reproduction rate or an increased change of behaviour during heavy infestations of second intermediate hosts (Crustacea) (Cribb 2005). Metacercariae of the digenean family Microphalidae, for example, encyst in the protocerebrum, leading to conspicuous escape behaviour of their amphipod and decapod host, which results in an increased likelihood to be fed by a predator (Cribb 2005).

The study of the consequences of cestode infections is depending on the intensity, and research has mainly been focused on the fish host (Williams and Jones 1994). Loss of weight, anaemia, physiological stress, growth reduction and reproduction has been observed (Caira and Reyda 2005). The pathogenicity for the crustacean host is hitherto quite unknown.

Nematodes of marine Crustacea (Ascaridida, Spirurida) usually occur in the coelom of their invertebrate host either free in the haemocoel or closely associated with internal organs. Poinar and Kuris (1975) described the effects of *Ascarophis* parasitizing intertidal decapods (*Hemigrapsus oregonensis*, *Pachycheles rudis*) and indicated a decrease of growth and increase of mortality among older and larger crabs. Experiments have shown that Crustacea infected with *Hysterothylacium aduncum* demonstrate erratic behaviour and mortality (McClelland 2005). The consequences for the fish host range from mechanical compression or necrosis of the liver to lesions in the gut wall, viscera and musculature, depletion of lipids and mortality (Rohde 1984; Williams and Jones 1994). Infections with nematode larvae of the family Anisakidae have been also described to cause gastritis, ulcers, diarrhoea, dehydration and anaemia in their mammalian and avian definitive hosts (e.g. McClelland 2005).

Acanthocephalan parasites have different effects on the health of their hosts. The consequences range from change of the hosts behaviour (intermediate hosts) to serious damage of their tissue (final hosts). Studies of the effects of acanthocephalans on intermediate hosts have focused mainly on small crustaceans (Bakker et al. 1997; Hindsbo 1972). The results indicate that even an infestation with a small intensity can induce a change of behaviour or colour. Shore crabs infected with *Profilicollis* species cause alteration in the burrowing behaviour, with the consequence of elevated mortality rates due to the predation by shorebirds (Latham and Poulin 2002). The pathogenicity for the final hosts depends on the mode of attachment. As longer the presoma as higher is usually the damage of the host's intestine wall (Taraschewski 2005). Due to the short length of *Corynosoma* species and their short presoma, the infestation of seals and whales has no serious consequences. On the other hand, an infestation of sea otters and eider ducks with polymorphid Acanthocephala can have much more consequences. After perforation of the intestine wall, the adult parasites attain to the body cavity causing peritonitis and possibly mortality (Taraschewski 2005). However, it is known that acanthocephalan parasites influence host populations by affecting fitness and act as regulatory factors. The influence on the crustacean first intermediate host still needs some clarification.

14.6 Concluding Remarks

The marine environment is a large scale three-dimensional habitat, and the likelihood of finding an eligible successive intermediate or final host in a life cycle is far lower compared to the terrestrial realm. In order to ensure transmission and guarantee survival and dispersal in the aquatic environment, parasites have evolved a variety of different (highly effective) mechanisms. While distribution and dispersal of parasites is often closely related to the distribution of the respective final hosts (spread the eggs with their faeces), infection of the successive host in a life cycle is usually linked with trophic interactions within the marine food web (Klimpel et al. 2011; Kuhn et al. 2011). The parasites live in a highly diluted environment and the free-living stages are typically short-lived and have to find a suitable host within 24–48 h (Marcogliese 2005). Some parasitic larvae can actively search and penetrate the next host (e.g. cercariae of Digenea), while others have to be passively ingested (e.g. larvae of Nematoda). A successful transmission is depending on the longevity of the intermediate host, but Copepoda, for example, are short-lived organisms and their infection rates are usually very low (see Table 14.5) (Marcogliese 2005). Klimpel et al. (2004) reported *Paraeuchaetha norvegica* at the Norwegian Deep to be infected with *Anisakis simplex* s.l. third stage larvae (L3) with a prevalence of 0.26% and *Hysterothylacium aduncum* in hyperiid Amphipoda with a prevalence of 1.95%. It is very likely that even small infestation rates in crustacean intermediate hosts are sufficient to ensure high infection rates of final hosts. For example, the stomach content of a north Pacific fin whale (*Balaenoptera physalus*) contains approximately 340 kg of the Euphausiacea *Thysanoessa raschii* (Nemoto 1970). This equates to approximately 234,000 individuals of this crustacean species (Falk-Petersen 1981). With consideration of an infection rate of 0.019% (*A. simplex* s.l.), the final host would acquire at least 45 parasitic nematodes with each feeding event (Smith and Snyders 2005). This is equivalent with an weekly uptake of more than 300 infective nematode larvae (L3).

The addition of an extra intermediate or paratenic host seems to ensure the transmission and maintain parasites in a dilute environment (Marcogliese 2002). The integration of arthropod vectors is largely restricted to the terrestrial system, and only a few exceptions are reported from the marine environment

Table 14.5 Reported prevalences (min/max/mean) of parasitic helminth groups and genera in crustacean intermediate hosts

Hosts	P% min/max/mean					
	Copepoda	Mysidacea	Amphipoda	Isopoda	Euphausiacea	Decapoda
Parasite						
Digenea			0.30/100/28.05	0.20/76.00/16.68		
Cestoda					0.10/100/33.40	
Nematoda	0.26/1.95/1.10	0.02/0.59/0.25	0.10/18.70/4.51	0.90/1.00/1.03	0.001/0.49/0.12	0.09/100/20.75
Acanthocephala			0.50/16.70/4.55		3.10/3.10/3.10	1.00/91.00/25.80
Isopoda						91.10/100/95.55
Digenea						
Levinseniella			0.20/70.90/28.42			
Maritrema			2.50/63.00/31.62	1.36/76.00/32.00		
Microphallus			0.30/100/19.82	2.40/8.10/5.25		
Podocotyle			7.50/67.40/32.94	6.70/18.70/12.70		
Cestoda						
Aporhynchus	100/100/100					
Echinobothrium					0.10/0.10/0.10	
Tetrarhynchobothrium					0.10/0.10/0.10	
Nematoda						
Anisakis	0.03/0.03		0.11/0.16/0.11		0.007/0.49/0.04	0.09/0.27/0.15
Hysterothylacium	1.95/1.95/1.95	0.23/0.59/0.41	0.32/18.20/5.10	0.90/0.90/0.90		0.15/100/22.39
Contracaecum					0.001/0.001/0.001	27.27/37.50/32.38
Ascarophis			1.00/18.70/6.07			
Pseudoterranova		0.01/0.35/0.16	0.09/0.28/0.20	1.19/1.19/1.19		
Thynnascaris						3.05/3.05/3.05
Acanthocephala						
Echinorhynchus			0.50/6.60/3.54			
Polymorphus					3.10/3.10/3.10	1.00/91.00/25.80

References: Brattey et al. (1985), Brown et al. (2003), Fredensborg and Poulin (2005), Gómez-Gutiérrez et al. (2010), Hurst (1984), Hutton et al. (1962), Jackson et al. (1997), Klimpel et al. (2004), Klimpel and Rückert (2005), Lick (1991), Marcogliese (1994, 1995), Martorelli et al. (2000), Moravec and Nagasawa (1986), Nickol et al. (1999), Smith and Snyder (2005), Voigt (1986), and Zander et al. (1994, 2000, 2002) (e.g. hirudineans using isopods as vectors, McCallum et al. 2004). Complex helminth parasite life cycles develop either "upwards" by adding new hosts after the definitive host, which subsequently becomes an intermediate host (terminal addition/horizontal incorporation), or "downwards" by adding a new host, for example, Crustacea, before the definitive host (non-terminal addition/vertical incorporation) (O'Grady 1985; Parker et al. 2003). Because most helminth life cycles follow a basic scheme and often are restricted to three or a maximum of four hosts including obligate crustacean intermediate hosts, many incorporation events must have occurred long time ago. The involvement of further host species at a particular stage in the life cycle (lateral incorporation) is depending on actual niche overlap between the existing and the new host (in many helminths, via the food web) and requires generalistic parasites. Any successful mutant surviving in an extra host spreads and fixates. Though lateral incorporation does not add to helminth life cycle length, it clearly contributes to the variety of different helminth life cycles that we can observe to date. It consequently adds to the pool of helminth species available that are able to infest potential marine hosts, thus contributing to the mixture of highly specific or non-specific parasites within the various taxa and increasing metazoan parasite diversity. Palm and Klimpel (2007) suggested for marine cestodes and nematodes, lateral (involvement of alternative hosts) instead of vertical incorporation to be responsible for the recorded species biodiversity within the marine environment. So the evolution of marine fish parasites and species diversity can be driven by the parasite and its life cycle adaptation or the peculiarities of the ecosystem, leading to a better exploitation of the available ecological niches. Host–parasite interactions, however, are ubiquitous in real systems and are known to affect community structure, trophic relationships and energy flow (e.g. Lafferty 1999; Marcogliese 2002; Poulin and Morand 2000, 2004). Food web context is also thought to have been important in the evolution of many of the characteristics observed in marine parasites, including heteroxenous life cycles. Food web context has been hypothesized to have an influence on transmission rates and pathways of some parasite species in aquatic systems (Marcogliese 2002), and food webs have exerted strong selective pressures on the evolution of parasite transmission strategies, parasites are now shaping some of the ecological properties of existing food webs. Our present review shows that parasites in the marine environment are a natural component and may be viewed as an biological indicator of the relative health of the ecosystem inhabitants. The majority of metazoan parasite species present on and within invertebrates (especially Crustacea) and fish are only in some cases hazardous to human health. There are a moderate number of trematodes, cestodes and nematodes which have been reported in humans, but only a few cause serious diseases. In consideration of the biodiversity of marine Crustacea, it is not surprising that this group of invertebrates play a key role as potential hosts and vectors for metazoan parasites. The high abundance of key species in nearly all marine habitats favours the transmission potential and the maintenance of helminth life cycles. This is especially important for the distribution of zoonotic parasite species (e.g. anisakid nematodes).

Acknowledgements The present study was financially supported by the German Research Council (DFG KL 2087/1-1, 1-2).

References

Anderson RC (1996) Why do fish have so few roundworm (nematode) parasites? Environ Biol Fishes 46:1–5

Anderson RC (2000) Nematode parasites of vertebrates – their development and transmission. CAB, Wallingford, pp 650

Audicana MT, Kennedy MW (2008) *Anisakis simplex*: from obscure infectious worm to inducer of immune hypersensitivity. Clin Microbiol Rev 21:360–379

Bakker TCM, Mazzi D, Zala S (1997) Parasite-induced changes in behavior and color make *Gammarus pulex* more prone to fish predation. Ecology 78:1098–1104

Blair D (2005) Cestode and trematode infections. In: Rohde K (ed) Marine parasitology. CSIRO, Collingwood, pp 427–430

Blakeslee AMH, Keogh CL, Byers JE, Kuris AM, Lafferty KD, Torchin ME (2009) Differential escape from parasites by two competing introduced crabs. Mar Ecol Prog Ser 393:83–96

Blend CK, Dronen NO (2003) *Bothriocephalus gadellus* n. sp. (Cestoda: Bothriocephalidae) from the beardless codling *Gadella imberbis* (Vaillant) (Moridae) in the southwestern Gulf of Mexico, with a review of species of *Bothriocephalus* Rudolphi, 1808 reported from gadiform fishes. Syst Parasitol 54:33–42

Boxshall GA, Defaye D (2008) Global diversity of copepods (Crustacea: Copepoda) in freshwater. Hydrobiologia 595:195–207

Brattey J, Elner RW, Uhazy LS, Bagnall AE (1985) Metazoan parasites and commensals of five crab (Brachyura) species from eastern Canada. Can J Zool 63:2224–2229

Brown SP, De Lorgeril J, Joly C, Thoma F (2003) Field evidence for density-dependent effects in the trematode *Microphallus papillorobustus* in its manipulated host, *Gammarus insensiblis*. J Parasitol 89:668–672

Caira JN, Reyda FB (2005) Eucestoda (true tapeworms). In: Rohde K (ed) Marine parasitology. CABI, Wallingford, pp 92–104

Chubb JC, Ball MA, Parker GA (2009) Living in intermediate hosts: evolutionary adaptions in larval helminths. Trends Parasitol 26:93–102

Chung OS, Lee HJ, Sohn WM, Lee SH, Park IY, Oh SA, Chai JY, Seo M (2010) Discovery of *Maritrema jebuensis* n. sp. (Digenea: Microphallidae) from the Asian Shore Crab, *Hemigrapsus sanguineus*, in Korea. Korean J Parasitol 48:335–338

Cribb TH (2005) Digenea (endoparasitic flukes). In: Rhode K (ed) Marine parasitology. CABI, Wallingford, pp 76–87

Falk-Petersen S (1981) Ecological investigations on the zooplankton community of Balsfjorden, northern Norway: Seasonal changes in body weight and the main biochemical composition of *Thysanoessa inermis* (Krøer), *T. raschii* (M. Sars) and *Meganyctiphanes norvegica* (M. sars) in relation to environmental factors. J Exp Biol Ecol 49:103–120

Feigenbaum DL (1975) Parasites of the commercial shrimp *Penaeus vannamei* Boone and *Penaeus brasiliensis* Latreille. Bull Mar Sci 25:491–514

Fredensborg BL, Poulin R (2005) Larval helminths in intermediate hosts: does competition early in life determine the fitness of adult parasites? Int J Parasitol 35:1061–1070

Gasca R, Manzanilla H, Suárez-Morales E (2009) Distribution of hyperiid amphipods (Crustacea) of the southern Gulf of Mexico, summer and winter, 1991. J Plankton Res 31:1493–1504

Gibbons MJ, Hutchings L (1996) Zooplankton diversity and community structure around the southern Africa, with special attention to the Benguela upwelling system. S Afr J Sci 92:63–76

Gómez-Gutiérrez J, Robinson CJ, Kawaguchi S, Nicol S (2010) Parasite diversity of *Nyctiphanes simplex* and *Nematoscelis difficilis* (Crustacea: Euphausiacea) along the northwestern coast of Mexico. Dis Aquat Org 88:249–266

Gruner HE (1993) Crustacea. In: Gruner HE (ed) Lehrbuch der speziellen Zoologie. Band I: Wirbellose Tiere, 4. Teil: Arthropoda (ohne Insecta). Gustav Fischer, Jena, pp 448–1030

Hadfield KA, Bruce NL, Smit NJ (2011) *Cymothoa hermani* sp. nov. (Isopoda, Cymothoidae, Crustacea), a parasitic isopod, collected off the Zanzibar coast, Tanzania from the mouth of a parrotfish (Scaridae). Zootaxa 2876:57–68

Hartwich G (1974) Keys to genera of the Ascaridoidea. In: Anderson RC, Chabaud AG, Wilmott S (eds) CIH Keys to the nematode parasites of vertebrates. Farnham Royal, Commonwealth Agriculture Bureau, Richmond, pp 1–15

Hays R, Measures LN, Huot J (1998) Euphausiids as intermediate hosts of *Anisakis simplex* in the St. Lawrence estuary. Can J Zool 76:1226–1235

Helluy S, Thomas F (2010) Parasitic manipulation and neuroinflammation: evidence from the system *Microphallus papillorobustus* (Trematoda) – *Gammarus* (Crustacea). Parasit Vectors 3(38):1–11

Hewitt RP, Watkins JL, Naganobu M, Tshernyshkov P, Brierley AS, Demed DA, Kasatkina S, Takao Y, Goss C, Malyshk A, Brandon MA, Kawaguchi S, Siegel V, Trathan PN, Emery JH, Everson I, Milled DGM (2002) Setting a precautionary catch limit for Antarctic Krill. Oceanography 15:26–33

Hindsbo O (1972) Effects of *Polymorphus* (Acanthocephala) on colour and behaviour of *Gammarus lacustris*. Nature 238:333

Hochberg NS, Hamer DH (2010) Anisakidosis: perils of the deep. Clin Infect Dis 51:806–812

Hunninen AV, Cable RM (1943) The life history of *Lecithaster confusus* Odhner (Trematoda: Hemiuridae). J Parasitol 29:71–79

Hurst RJ (1984) Marine invertebrate hosts of New Zealand Anisakidae (Nematoda). N Z J Mar Freshwat Res 18:187–196

Hutton RF (1964) A second list of parasites from marine and coastal animals of Florida. Trans Am Microsc Soc 83:439–447

Hutton RF, Sogandares-Bernal F (1960) A list of parasites from marine and coastal animals of Florida. Trans Am Microsc Soc 79:287–292

Hutton RF, Ball T, Eldred B (1962) Immature nematodes of the genus *Contracaecum* Railliet and Henry, 1912, from shrimps. J Parasitol 48:327–332

Jackson CJ, Marcogliese DJ, Burt MDB (1997) Role of hyperbenthic crustaceans in the transmission of marine helminth parasites. Can J Aquat Sci 54:815–820

Jousson O, Bartoli P (2000) The life cycle of *Opecoeloides columbellae* (Pagenstecher, 1863) n. comb. (Digenea, Opecoelidae): evidence from molecules and morphology. Int J Parasitol 30:747–760

Karl H, Baumann F, Ostermeyer U, Kuhn T, Klimpel S (2010) *Anisakis simplex* (s.s.) larvae in wild Alaska salmon: no indication of post migration from viscera into flesh. Dis Aquat Org 94:201–209

Kennedy CR (2006) Ecology of the *Acanthocephala*. Cambridge University Press, New York, pp 240

Klimpel S, Palm HW (2011) Anisakid nematode (Ascaridoidea) life cycles and distribution: increasing zoonotic potential in the time of climate change? In: Mehlhorn H (ed) Progress in parasitology, vol 2, Parasitology research monographs. Springer, Heidelberg, pp 201–222

Klimpel S, Rückert S (2005) Life cycle strategies of *Hysterothylacium aduncum* to become the most abundant anisakid fish nematode in the North Sea. Parasitol Res 97:141–149

Klimpel S, Palm HW, Seehagen A (2003) Metazoan parasites and food composition of juvenile *Etmopterus spinax* (L., 1758) (Dalatiidae, Squaliformes) from the Norwegian Deep. Parasitol Res 89:245–251

Klimpel S, Palm HW, Rückert S, Piatkowski U (2004) The life cycle of *Anisakis simplex* in the Norwegian Deep (northern North Sea). Parasitol Res 94:1–9

Klimpel S, Palm HW, Busch MW, Kellermanns E, Rückert S (2006) Fish parasites in the Arctic deep-sea: poor diversity in pelagic fish species vs. heavy parasite load in a demersal fish. Deep Sea Res I 53:1167–1181

Klimpel S, Kellermanns E, Palm HW (2008) The role of pelagic swarm fish (Myctophidae: Teleostei) in the oceanic life cycle of *Anisakis* sibling species at the Mid-Atlantic Ridge, Central Atlantic. Parasitol Res 104:43–53

Klimpel S, Busch MW, Kellermanns E, Kleinertz S, Palm HW (2009) Metazoan deep-sea fish parasites. ACTA Biologica Benrodis, Supplementband 11. Verlag Natur & Wissenschaft, Solingen, pp 384

Klimpel S, Busch MW, Kuhn T, Rohde A, Palm HW (2010) The *Anisakis simplex* complex off the South Shetland Islands (Antarctica): endemic populations versus introduction through migratory hosts. Mar Ecol Prog Ser 403:1–11

Klimpel S, Kuhn T, Busch MW, Karl H, Palm HW (2011) Deep-water life cycle of *Anisakis paggiae* (Nematoda: Anisakidae) in the Irminger Sea indicates kogiid whale distribution in north Atlantic waters. Polar Biol 34:899–906

Koehler AV, Poulin R (2010) Host partitioning by parasites in an intertidal crustacean community. J Parasitol 96:862–868

Køie M (1979) On the morphology and life-history of *Derogenes varicus* (Müller, 1784) Loss, 1901 (Trematoda, Hemiuridae). Parasitol Res 59:67–78

Køie M (1984) Digenetic trematodes from *Gadus morhua* L. (Osteichthyes: Gadidae) from Danish and adjacent waters, with special reference to their life-histories. Ophelia 23:195–222

Køie M (1989) On the morphology and life history of *Lecithaster gibbosus* (Rudolphi, 1802) Lühe, 1901 (Digenea, Hemiuroidea). Parasitol Res 75:361–367

Køie M (1991) Aspect of the morphology and life cycle of *Lecithocladium excisum* (Digenea, Hemiuridae), a parasite of *Scomber* spp. Int J Parasitol 21:597–602

Køie M (1992) Life cycle and structure of the fish digenean *Brachyphallus crenatus* (Hemiuridae). J Parasitol 78:338–343

Køie M (1993) Aspects of the life cycle and morphology of *Hysterothylacium aduncum* (Rudolphi, 1802) (Nematoda, Ascaridoidea, Anisakidae). Can J Zool 71:1289–1296

Køie M (2001) Experimental infections of copepods and stickelbacks *Gasterosteus aculeatus* with small ensheathed and large third-stage larvae of *Anisakis simplex* (Nematoda, Ascaridoidea, Anisakidae). Parasitol Res 87:32–36

Køie M, Fagerholm HP (1995) The life cycle of *Contracaecum osculatum* (Rudolphi, 1802) sensu stricto (Nematoda, Ascaridoidea, Anisakidae) in view of experimental infection. Parasitol Res 81:481–489

Kuchta R, Scholz T, Brabec J, Bray RA (2008) Suppression of the tapeworm order Pseudophyllidea (Platyhelminthes: Eucestoda) and the proposal of two new orders, Bothriocephalidea and Diphyllobothriidea. Int J Parasitol 38:49–55

Kuhn T (2010) Molecular studies on marine ascaridoid nematodes. Diploma thesis, Heinrich Heine University, Düsseldorf, pp 100

Kuhn T, García-Màrquez J, Klimpel S (2011) Adaptive radiation within marine anisakid nematodes: a zoogeographical modeling of cosmopolitan, zoonotic parasites. PLoS One 6(12):e28642

Lafferty KD (1999) The evolution of trophic transmission. Parasitol Today 15:111–115

Lascara CM, Hofmann EE, Ross RM, Quetin LB (1999) Seasonal variability in the distribution of Antarctic Krill, *Euphausia superba*, west of the Antarctic Peninsula. Deep Sea Res I 46:951–984

Latham ADM, Poulin R (2002) Field evidence of the impact of two acanthocephalan parasites on the mortality of three species of New Zealand shore crabs (Brachyura). Mar Biol 141:1131–1139

Latham ADM, Poulin R (2003) Spatiotemporal heterogeneity in recruitment of larval parasites to shore crab intermediate hosts: the influence of shorebird definitive hosts. Can J Zool 81:1282–1291

Leung TLF, Poulin R (2006) Effects of the trematode *Maritrema novaezealandensis* on the behaviour of its amphipods host: adaptive or not? J Helminthol 80:271–275

Lick R (1991) Untersuchungen zu Lebenszyklus (Krebse-Fische-marine Säuger) und Gefrierresistenz anisakider Nematoden in Nord- und Ostsee. Ber Inst Meereskd Kiel 218:1–195

Lile NK (1998) Alimentary tract helminths of four pleuronectid flatfish in relation to host phylogeny and ecology. J Fish Biol 53:945–953

Marcogliese DJ (1994) *Aeginina longicornis* (Amphipoda: Caprellidea), new intermediate host for *Echinorhynchus gadi* (Acanthocephala: Echinorhynchidae). J Parasitol 80:1043–1045

Marcogliese DJ (1995) The role of zooplankton in the transmission of helminth parasites to fish. Rev Fish Biol Fish 5:336–371

Marcogliese DJ (2002) Food webs and the transmission of parasites to marine fish. Parasitology 763(124):83–99

Marcogliese DJ (2005) Parasites of the superorganism: Are they indicators of ecosystem health?. Int J Parasitol 35:705-716

Martin JW, Davis GE (2001) An updated classification of the recent Crustacea, Science series 39. Natural History Museum of Los Angeles County, California, pp 124

Martorelli SR, Navone GT, Ivanov V (2000) Proposed life cycle of *Ascarophis marina* (Nematoda: Cystidicolidae) in Argentine waters. J Parasitol 86:1047–1050

Martorelli SR, Fredensborg BL, Mouritsen KN, Poulin R (2004) Description and proposed life cycle of *Maritrema novaezealandensis* n. sp. (Microphallidae) parasitic in red-billed gulls, *Larus novaehollandiae scopulinus*, from Otago Harbor, South Island, New Zealand. J Parasitol 90:272–277

Mattiucci S, Nascetti G (2008) Advances and trends in the molecular systematics of anisakid nematodes, with implications for their evolutionary ecology and host–parasite co-evolutionary processes. Adv Parasitol 66:47–148

Mauchline J (1998) The biology of calanoid copepods. In: Blaxter JHS, Southward AJ, Tyler PA (eds) Advances in marine biology. Academic, San Diego, pp 710

McCallum HI, Kuris A, Harvell CD, Lafferty KD, Smith GW, Porter J (2004) Does terrestrial epidemiology apply to marine systems. Trends Ecol Evol 19:585–591

McClelland G (2005) Nematoda (roundworms). In: Rhode K (ed) Marine parasitology. CABI, Collingwood, pp 104–115

McDermott JJ (2011) Parasites of shore crabs in the genus *Hemigrapsus* (Decapoda: Brachyura: Varunidae) and their status in crabs geographically displaced: a view. J Nat Hist 45:2419–2441

Mehlhorn H (2001) Encyclopedic reference of parasitology: Diseases, treatment, therapy. Springer, Berlin, pp 676

Meland K, Willassen E (2007) The disunity of "Mysidacea" (Crustacea). Mol Phylogenet Evol 44:1083–1104

Möller H, Anders K (1983) Krankheiten und Parasiten der Meeresfische. Verlag Heino Möller, Kiel, pp 258

Moravec F, Klimpel S (2009) Two new species of cystidicolid nematodes from the digestive tract of the deep-sea fish *Coryphaenoides mediterraneus* (Gilioi) (Macrouridae) from the Mid-Atlantic Ridge. Syst Parasitol 73:37–47

Moravec F, Nagasawa K (1986) New records of amphipods as intermediate hosts for salmonid nematode parasites in Japan. Folia Parasitol 33:45–49

Moravec F, Wolter J, Körting W (1999) Some nematodes and acanthocephalans from exotic ornamental and freshwater fishes imported to Germany. Folia Parasitol 46:296–310

Moravec F, Fredensborg BL, Latham ADM, Poulin R (2003) Larval Spirurida (Nematoda) from the crab *Macrophthalmus hirtipes* in New Zealand. Folia Parasitol 50:109–114

Moravec F, Bakenhaster M, Fajer-Avila EJ (2010) New philometrids (Nematoda: Philometridae) from head tissues of two serranid fishes (*Epinephelus morio* and *Mysteroperca microleis*) off Florida, northern Gulf of Mexcio. Acta Parasitol 55:359–368

Munk P, Nielsen TG (1994) Trophodynamics of the plankton community at Dogger Bank: predatory impact by larval fish. J Plankton Res 16:1225–1245

Muñoz G, George-Nascimento M (2007) Two new species of *Ascarophis* (Nematoda: Cystidicolidae) in marine fishes from Chile. J Parasitol 93:1178–1188

Nemoto T (1970) Feeding pattern of baleen whales. In: Steele JH (ed) Marine food chains. University of California Press, Berkeley, pp 241–252

Newman WA, Abbott DP (1980) Cirripedia: the barnacles. In: Morris RH, Abbott DP, Haderlie EC (eds) Intertidal invertebrates of California. Stanford University Press, Stanford, pp 504–536

Nickol BB, Crompton DWT, Searle DW (1999) Reintroduction of *Profiliocollis* Meyer, 1931, as a genus in Acanthocephala: significance of the intermediate host. J Parasitol 85:716–718

Nickol BB, Heard RW, Smith NF (2002) Acanthocephalans from crabs in the southeastern U.S. with the first intermediate hosts known for *Arhythmorhynchus frassoni* and *Hexaglandula corynosoma*. J Parasitol 88:79–83

O'Grady RT (1985) Ontogenetic sequences and the phylogenetics of parasitic flatworm life cycles. Cladistics 1:159–170

Palm HW (2004) The Trypanorhyncha Diesing, 1863. PKSPL-IPB, Bogor, pp 710

Palm HW, Klimpel S (2007) Evolution of parasitic life in the ocean. Trends Parasitol 23:10–12

Palm HW, Klimpel S, Bucher C (1999) Checklist of metazoan fish parasites of German coastal waters. Ber Inst Meereskd Kiel 307:1–148

Parker GA, Chubb JC, Ball MA, Roberts GN (2003) Evolution of complex life cycles in helminth parasites. Nature 425:480–484

Pichelin S, Kuris AM, Gurney R (1998) Morphological and biological notes on *Polymorphus (Profilicollis) sphaerocephalus* and *Corynosoma stanleyi* (Polymorphidae: Acanthocephala). J Parasitol 84:798–801

Poinar GO, Kuris AM (1975) Juvenile *Ascarophis* (Spirurida: Nematoda) parasitizing intertidal decapod Crustacea in California: with notes on prevalence and effects on hosts growth and survival. J Invertebr Pathol 26:375–382

Porter ML, Meland K, Price W (2008) Global diversity of mysids (Crustacea-Mysida) in freshwater. Hydrobiologia 595:213–218

Poulin R, Morand S (2000) The diversity of parasites. Q Rev Biol 75:277–293

Poulin R, Morand S (2004) Parasite diversity. Smithsonian Institute Press, Washington, DC

Reimer L, Hnatiuk S, Rochner J (1975) Metacercarien in Planktontieren des mittleren Atlantik. Wiss Z Paedagog Hochsch Güstrow (Math Nat Fak) 2(75):239–358

Richardson K, Nielsen TG, Bo Pedersen F, Heilmann JP, Lokkegaard B, Kaas H (1998) Spatial heterogeneity in the structure of the planktonic food web in the North Sea. Mar Ecol Prog Ser 168:197–211

Rohde K (1984) Diseases caused by metazoans: helminths. In: Kinne O (ed) Diseases of marine animals, vol iv, part 1. Biologische Anstalt, Helgoland, pp 193–319

Rohde K (2005) Definitions, and adaptations to a parasitic way of life. In: Rohde K (ed) Marine parasitology. CABI, Wallingford, pp 1–6

Ruppert EE, Barnes RD (1994) Invertebrate zoology. Saunders College Publishing, Orlando, pp 1056

Schmidt GD, MacLean SA (1978) *Polymorphus (Profilicollis) major* Lundström 1942 juveniles in Rock Crabs, *Cancer irroratus*, from Maine. J Parasitol 64:953–954

Shimazu T (1975) Some cestode and acanthocephalan larvae from euphausiid crustaceans collected in the northern North Pacific Ocean. Bull Jpn Soc Sci Fish 41:813–821

Siegel V, Loeb V, Gröger J (1998) Krill (*Euphausia superba*) density, proportional and absolute recruitment and biomass in the Elephant Island region (Antarctic Peninsula) during the period 1977 to 1997. Polar Biol 19:393–398

Smith JW (1971) *Thysanoessa inermis* and *T. longicaudata* (Euphausiidae) as first intermediate hosts of *Anisakis* sp. (Nematoda: Ascaridata) in the northern North Sea, to the North of Scotland and at the Faroe. Nature 234:478

Smith JW, Snyder JM (2005) New locality for third-stage larvae of *Anisakis simplex* (sensu lato) (Nematoda: Ascaridoidea) in euphausiids *Euphausia pacifica* and *Thysanoessa raschii* from Prince William Sound, Alaska. Parasitol Res 97:539–542

Solonchenko AI (1985) Development of larval stages of *Bothriocephalus scorpii*. In: Hargis WJ Jr (ed) Parasitology and pathology of marine organisms of the world ocean. NOAA Technical Reprot NMFS 25, pp 83–84

Spears T, Abele LG (1997) Crustacean phylogeny inferred from 18S rDNA. In: Fortey RA, Thomas RH (eds) Arthropod relationships. Chapman & Hall, London, pp 169–187

Svendsen YS (1990) Hosts of third stage larvae of *Hysterothylacium* sp. (Nematoda, Anisakidae) in zooplankton from outer Oslofjord, Norway. Sarsia 75:161–167

Taraschewski H (2005) Acanthocephala (thorny or spiny-headed worms). In: Rohde K (ed) Marine parasitology. CABI, Collingwood, pp 116–121

Verweyen L, Klimpel S, Palm HW (2011) Molecular phylogeny of the Acanthocephala (Class Palaeacanthocaphala) with a paraphyletic assemblage of the orders Polymorphida and Echinorhynchida. PLoS One 6(12):e28285

Voigt M (1986) Gammariden (Crustacea: Amphipoda) als Zwischenwirte von Fischparasiten in Schleswig-Holsteinischen Küstengewässern. Diploma thesis, Christian-Albrechts University, Kiel, pp 63

Whiteley NM, Rastrick SPS, Lunt DH, Rock J (2011) Latidudinal variations in the physiology of marine gammarid amphipods. J Exp Mar Biol Ecol 400:70–77

Williams H, Jones A (1994) Parasitic worms of fish. Taylor & Francis, London, p 593

Wirkner CS, Richter S (2007) Comparative analysis of the circulatory system in Amphipoda (Malacostraca, Crustacea). Acta Zool 88:159–171

Zander CD (1998) Ecology of host parasite relationships in the Baltic Sea. Naturwissenschaften 85:426–436

Zander CD, Reimer LW (2002) Parasitism at the ecosystem level in the Baltic Sea. Parasitology 124:119–135

Zander CD, Groenewold S, Strohbach U (1994) Parasite transfer from crustacean to fish hosts in the Lübeck Bight, SW Baltic Sea. Helgoländer Meeresunters 48:89–105

Zander CD, Reimer LW, Barz K, Dietel G, Strohbach U (2000) Parasite communities of the Salzhaff (Northwest Mecklenburg, Baltic Sea) II. Guild communities, with special regard to snails, benthic crustaceans, and small-sized fish. Parasitol Res 86:359–372

Zander CD, Koçoglu Ö, Skroblies M (2002) Parasite populations and communities from the shallow littoral of the Orther Bight (Fehmarn, SW Baltic Sea). Parasitol Res 88:734–744

Chapter 15
Spotted Fever Rickettsiae and Rickettsioses in Germany

Gerhard Dobler and Martin Pfeffer

Abstract Rickettsiae are gram-negative vector-borne bacteria which exclusively replicate intracellularly. They cause a number of important diseases, like Rocky Mountain spotted fever, epidemic typhus, Mediterranean spotted fever or African tick bite fever. Rickettsioses are classified as emerging infections; however, they are neglected diseases in many parts of the world. Two decades ago, there was only one *Rickettsia* species known in Germany, and the available data on the occurrence of rickettsial species and their medical importance are still very limited. Nevertheless, six different *Rickettsia* species could be meanwhile detected in Germany. Five of them are transmitted by ticks, while one rickettsiosis seems to be primarily transmitted by fleas. This chapter reviews the little that is known about the distribution, vector association, abundance and medical impact of these *Rickettsia* species in Germany, thereby identifying knowledge gaps and research needs.

Keywords Germany • Imported • Rickettsia • Spotted Fever • Tick

15.1 Introduction

Among the diseases which changed the history of mankind, several vector-borne diseases are of exceptional importance. Among them are rickettsioses, whose causative agents are transmitted by fleas, lice, ticks and mites. Especially lice and fleas are well known for transmitting epidemic and endemic typhus which are

G. Dobler (✉)
Department of Virology and Rickettsiology, Bundeswehr Institute of Microbiology, Neuherbergstrasse 11, 80937 Munich, Germany
e-mail: gerharddobler@bundeswehr.org

M. Pfeffer
Institute of Animal Hygiene and Veterinary Public Health, University of Leipzig, Leipzig, Germany

H. Mehlhorn (ed.), *Arthropods as Vectors of Emerging Diseases*, Parasitology Research Monographs 3, DOI 10.1007/978-3-642-28842-5_15,

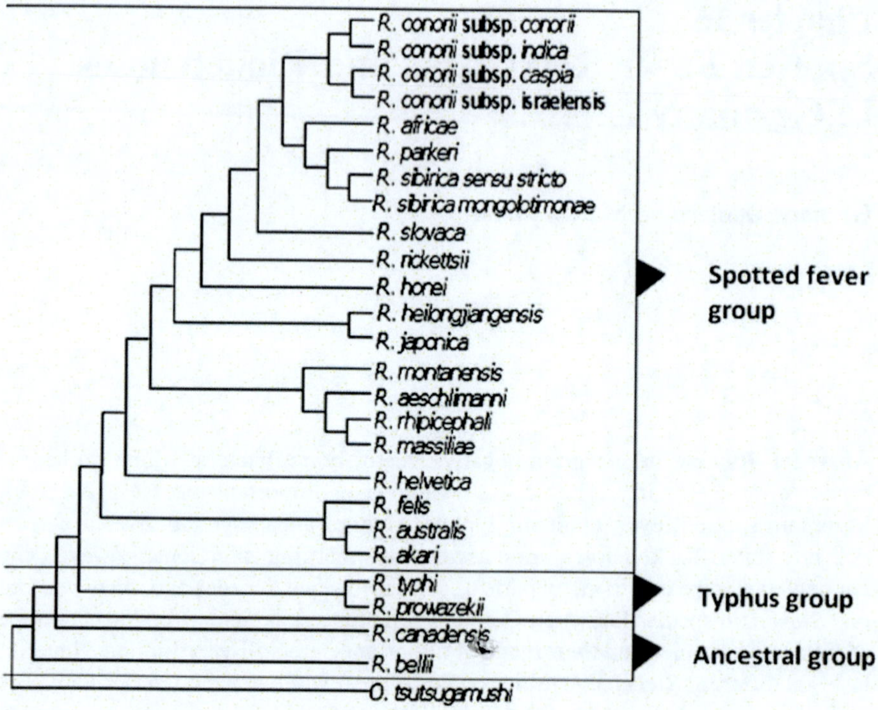

Fig. 15.1 Phylogenetic relationships between important *Rickettsia* species

infectious diseases that are little known nowadays. The epidemic typhus is of considerable historical importance and has probably helped to decide the outcome of more wars than weapons have (Stephenson 1944; Raoult et al. 2004).

Only recently, the detection of several new species of rickettsiae transmitted by ticks has shown the medical importance in different continents in the world of another group of rickettsioses, the spotted fever group. It is recognized that tick-borne rickettsiae have a rather limited geographic distribution while the louse- and flea-borne rickettsioses show a worldwide distribution.

Rickettsiae are named after Howard Ricketts, an American microbiologist who first described rickettsiae as causative agents of Rocky Mountain spotted fever in 1906 (Rickets 1906). Rickettsiae are gram-negative bacteria with an obligate intra-cellular replication. The organisms today are classified as alphaproteobacteria and together with ehrlichiae and anaplasmae form the biological order Rickettsiales (Dumler et al. 2001). Using serological and molecular techniques, rickettsiae are classified into three large groups. The group of tick-borne rickettsiae (tick-borne spotted fever group) consists of more than 30 different species. The typhus group is formed from two species (*R. prowazekii* and *R. typhi*), and the so-called ancestral group also consists of two species (*R. canadensis* and *R. bellii*) (Fig. 15.1). All rickettsial species known so far are transmitted by vectors in the natural environment.

Table 15.1 Known *Rickettsia* species in Germany

Rickettsia species	Vector (prevalence)	Possible or suggested vertebrate reservoir	Human illness
R. helvetica	*Ixodes ricinus* (3–10%)	Cattle, game animals, rodents	Aneruptive fever, endocarditis (?)
R. monacensis	*Ixodes ricinus* (0.5%)	Lizards, other reptiles	Spotted fever
R. massiliae	*Ixodes ricinus* (1.7%)	Rodents	Spotted fever
R. felis	*Ixodes ricinus* (?) *Ctenocephalides felis*	Cats	Flea-borne spotted fever
R. raoultii	*Dermacentor reticulatus* (20–50%)	Rodents	TIBOLA (?)
R. slovaca	*Dermacentor marginatus*	Sheep, goat	TIBOLA

In Germany, only limited data on the prevalence of rickettsiae in ticks have been available so far, and research on rickettsioses in ticks has been started rather recently (Hartelt et al. 2004; Dautel et al. 2006; Wölfel et al. 2006). In the available studies, mainly molecular detection methods are used to detect rickettsial DNA in ticks. All studies we have identified through a literature review together are patchwork, and only spotlights of the spatial distribution of the particular rickettsial species are possible. Neither a framework of systematic sampling of ticks in Germany nor a nationwide screening for pathogens they may harbour is existing at the moment. But based on these data, the rickettsiae of the spotted fever group transmitted by ticks seem to play a major role. A total of six *Rickettsia* species have been identified in Germany so far. *R. helvetica* and *R. raoultii* were detected in *Ixodes (I.) ricinus* and in *Dermacentor (D.) reticulatus*, respectively. A very recent study identified DNA of *R. felis* and *R. helvetica* in fleas from various animals (Gilles et al. 2008). The detection of *R. felis* is supported by the description of two human cases of flea-borne spotted fever (Richter et al. 2002). In 2002, a so far unknown rickettsial species was detected in *I. ricinus*, collected in a city park in Munich, southern Germany, and was named *R. monacensis* (Simser et al. 2002). Another rickettsial species, *R. massiliae* or a closely related species was detected in one single *Ixodes ricinus* tick from south-eastern Germany (Dobler and Wölfel 2009). Already more than 30 years ago, the isolation of a *Rickettsia* species, later to be identified as *Rickettsia slovaca* from *D. marginatus* ticks, was reported (Rehácek et al. 1977). Here, we intend to provide an overview of the current knowledge of each of the *Rickettsia* species described in Germany thus far (Table 15.1).

15.1.1 *Rickettsia helvetica*

15.1.1.1 Prevalence in Ticks

Several studies in different parts of Germany confirm the presence of *R. helvetica*. In the federal state of Baden-Württemberg in south-western Germany, adjacent to Bavaria in the West, 13% of *I. ricinus* were found positive for *R. helvetica*

(Hartelt et al. 2004), and in various public parks in Bavaria, an average of about 7% of the *I. ricinus* ticks carried *R. helvetica* (Schorn et al. 2011). In eastern Germany, 18.8% of adult *I. ricinus* were positive in Thuringia (Hildebrandt et al. 2010), in Saxony this percentage was 12.7% (Silaghi et al. 2011), while in Berlin only 14.2% of *I. ricinus* nymphs were reported positive (Pichon et al. 2006). Assuming that *R. helvetica* is the rickettsial species most frequently found in *I. ricinus*, this species would account for the majority of rickettsiae found in studies which do not distinguish between the species because a generic PCR (e.g. targeting the gltA gene, Wölfel et al. 2008) is used; the overall prevalence of *R. helvetica* is above 10%.

15.1.1.2 Seroprevalence in Animals

No detailed data on the seroprevalence rates of antibodies against *R. helvetica* in animals are available. In a study in *Myodes (M.) glareolus* and *Apodemus (A.) flavicollis*, only 1/110 transudates of *M. glareolus* and 0/98 transudates of *A. flavicollis* were found positive for *R. helvetica* antibodies, while 14/110 sera of *M. glareolus* and 11/98 transudates of *A. flavicollis* showed antibodies against *R. conorii* (Dobler and Wölfel 2009).

15.1.1.3 Natural Transmission Cycle

There is some evidence that larger game animals and larger farm animals may play a role for the transmission of *R. helvetica*. However, systematic studies are lacking. There is, however, good evidence that transovarial transmission may play a major role for the maintenance of *R. helvetica* in nature. In a study testing *I. ricinus* of different developmental stages from eastern Bavaria for *R. helvetica*, 7/136 unengorged tick larvae were found positive, implicating that the larvae had been infected via transovarial transmission.

15.1.1.4 Seroprevalence in Humans

Seroprevalence data of antibodies against *R. helvetica* in humans in Germany are missing so far. In a serosurvey in France, 9.2% of tested forest workers in the Alsace region were found positive for IgG against *R. helvetica* (Fournier et al. 2000). In a seroprevalence study in Poland, none of 129 analysed sera showed clear positive reactivity against *R. helvetica* (Podsiadly et al. 2011).

15.1.1.5 Medical and Veterinary Importance

R. helvetica seems to play a minor role in human pathogenicity. An earlier report on its role in aetiology of myocarditis has not been confirmed so far (Nilsson et al 1999). However, it seems to play a role as cause of undifferentiated flu-like disease with fever and constitutional symptoms ("aneruptive fever") during the time of activity of *I. ricinus* (Fournier et al. 2000). Rather recently, a case of meningitis was traced to *R. helvetica* in a patient in Sweden (Nilsson et al. 2010). Seroprevalence rates of 5–10% imply that infections of *R. helvetica* in humans may be more frequent than thought but that, in many cases, symptoms may be lacking or mild. There is no evidence that *R. helvetica* is causing animal disease.

15.1.2 *Rickettsia monacensis*

Rickettsia monacensis is a rickettsial species which was detected in ticks (*I. ricinus*) several years ago in a city park in Munich, southern Germany (Simser et al. 2002). Furthermore, two additional strains of *R. monacensis* could be isolated from ticks (*I. ricinus*) in the district of Amberg, north-eastern Bavaria, about 200 km north of the original location of isolation. Further characterization showed that several genes of the two new isolates showed highest homologies with *R. monacensis* and *R.* sp. *IRS4*/*IRS3* (Dobler et al. 2009).

15.1.2.1 Prevalence in Ticks

While no data on the prevalence of *R. monacensis* in ticks of the original isolation are available, additional data on the detection of this rickettsial species in ticks in north-eastern Bavaria resulted in a prevalence of 0.55% (8/1,450) for *R. monacensis* in *I. ricinus*. In Thuringia, 12 out of 64 rickettsiae-positive *I. ricinus* of which the species could be determined (147 out of 1,000 tested *I. ricinus* were positive for *Rickettsia* spp.) were *R. monacensis* (Hildebrandt et al. 2010). The detection in this region constitutes the most northern detection of this rickettsial species so far. *Rickettsia monacensis* since then has been detected in additional European countries. A recent study found *R. monacensis* in Hungary with a prevalence rate of 0.9% (Sreter-Lancz et al. 2005). The area where *R. monacensis* had been detected is close to the former Czechoslovak Republic, where *R.* sp. *IRS4*/*IRS3* was detected (Sekeyova et al. 2000). Further studies also detected *R. monacensis* or a very closely related strain in Bulgaria (Christova et al. 2003). Molecular studies comparing the two Bavarian strains of *R. monacensis* showed that *R. monacensis* is closely related or identical to *R.* sp. *IRS4*/*IRS3* (Dobler et al. 2009). These results are supported by similar results from a Hungarian group (Sreter-Lancz et al. 2005). Further studies in humans and molecular comparisons of German and Spanish strains have to show whether *R. monacensis* could also exhibit pathogenicity in humans in Germany and to determine the genetic homogeneity or plasticity of this rickettsial species over its wide geographical range.

15.1.2.2 Prevalence in Animals

There are no data available on the seroprevalence rates of antibodies against *R. monacensis* in animals. In this context, it should be stated that antibodies within the spotted fever-group and the typhus-group rickettsiae are cross reacting, but there is currently no species-specific serological test available that would allow the allocation of a particular sero-reactivity to a particular *Rickettsia* species within any of the two groups. This in turn means that any seroprevalence data can only be judged as group-specific antibodies.

15.1.2.3 Natural Transmission Cycle

There is increasing evidence that *R. monacensis* is circulating within a transmission cycle of ticks of the genus *Ixodes* and reptiles, mainly lizards (de Sousa et al. 2010). This host preference may explain the relatively low prevalence rates in ticks which have been found so far. On the other hand, a higher prevalence would be expected in climatic warmer regions, for example, the Mediterranean countries, with a year-round activity and higher abundance of reptiles in comparison to our regions north of the Alps. Further data on transovarial or transstadial transmission are not available. It is also not clear to what extent *Ixodes* ticks serve as vectors and as reservoirs for *R. monacensis*.

15.1.2.4 Seroprevalence in Humans

No data on the seroprevalence of antibodies against *R. monacensis* are available.

15.1.2.5 Medical and Veterinary Importance

More recently, *R. monacensis* was detected in the blood of two patients in Spain presenting with a non-pruritic macular rash. Both patients did not show any eschars but sought medical attention with a diagnosis of Mediterranean spotted fever (Jado et al. 2007). These two patients are the first patients with *R. monacensis* as etiologic agent and demonstrate the pathogenic potential of this rickettsial species. So far, however, no human cases with symptoms of spotted fever were described in the area where *R. monacensis* had been isolated in Germany. No data on the pathogenesis in animals are available.

15.1.3 *Rickettsia massiliae*

Rickettsia massiliae was detected in a brown dog tick (*Rhipicephalus sanguineus*) near Marseille in 1992 and finally was characterized as a distinct species (Beati and Raoult 1993). Since that time, it has been detected in southern Europe (Greece, Portugal, Spain) and central Europe (Switzerland) as well as in some African countries (Central African Republic, Mali). Recently, it was also detected on the American continent, in northern Arizona and in Argentina (Babalis et al. 1994; Bacellar et al. 1995; Beati et al. 1996; Fernandez-Soto et al. 2006; Santos-Silva et al. 2006; Bernasconi et al. 2002; Eremeeva et al. 2006).

15.1.3.1 Prevalence in Ticks

Rickettsia massiliae or a closely related (according to the *ompB* gene) rickettsial species was detected in one *I. ricinus* tick in an eastern Bavarian district (Dobler et al. 2009). To our knowledge, this constitutes the first and so far only detection of *R. massiliae* at all in Germany and also in ticks of the species *I. ricinus*. The medical importance for humans could be demonstrated by isolation of this rickettsial species from a patient with a classical spotted fever in Parma, Italy (Vitale et al. 2006). In Argentina, a patient showed a severe form of spotted fever caused by *R. massiliae* (García-García et al. 2010). Ticks of the genus *Rhipicephalus* seem to constitute the main vector and possible reservoir of *R. massiliae* in the Mediterranean, in Africa and in parts of northern and southern America. *Rh. sanguineus* is a highly abundant ixodid tick species of the warmer climate zones worldwide. If this tick species indeed serves as the main vector, this would argue for an overlapping distribution of *R. massiliae* with the corresponding geographical range of *Rh. sanguineus*. Because of high infestation rates of dogs with this particular tick species in the Mediterranean and the large number of dogs that travel from there, *Rh. sanguineus* is frequently imported into yet non-endemic countries, for example, Germany. Travelling infested dogs thus might also be carrying *R. massiliae* into different parts of the world. Due to the non-existence of *Rh. sanguineus* in Germany, the detection of this rickettsial species in *I. ricinus* is somewhat surprising. However, a former detection in ticks (*Rh. sanguineus*) in the Tessin in Switzerland seems to make a distribution north of the Alps at least possible. Furthermore, Czech scientists recently also reported on the detection of rickettsial DNA closely related or identical to *R. massiliae* in *I. ricinus* in the Slovak Republic (Derdáková et al. 2011). These results imply that *R. massiliae* or a closely related *Rickettsia* species may be distributed in east-central and maybe central Europe. Further studies in ticks will have to show to what extent *R. massiliae* is circulating in Germany and whether it could be of importance as a human pathogen in the respective area.

15.1.3.2 Prevalence in Animals

Data on prevalence of rickettsiae in animal hosts other than ticks are not available. Also no information on seroprevalence of *R. massiliae* antibodies in animals is available.

15.1.3.3 Natural Transmission Cycle

No information on the transmission cycle of *R. massiliae* is yet available. It may, however, be speculated that dogs may play a potential role as vertebrate hosts as they may constitute important vertebrate hosts for the brown dog tick, *Rh. sanguineus*, which is the main vector of *R. massiliae* in the Mediterranean and in Africa. In Slovak Republic, *R. massiliae* was detected in ear biopsies of rodents. However, the potential role of rodents for *R. massiliae* has not been elucidated so far. As with the *R. massiliae* sequence detected in *I. ricinus* from eastern Bavaria, *Rh. sanguineus* is not present in Slovak Republic. Hence, another tick species than *Rh. sanguineus*, maybe *I. ricinus*, has to serve here as a vector and potentially also as reservoir.

15.1.3.4 Seroprevalence in Humans

Few population-based seroprevalence studies on antibodies against *R. massiliae* are available. In a study on patients with fever and symptoms compatible with spotted fever, a total of 15 patients from Catalonia, Spain, exclusively reacted against *R. conorii* and *R. massiliae* with significantly higher titres against *R. massiliae* than against *R. conorii* (Cardenosa et al. 2003). In a seroprevalence study in Polish forest workers, however, 15/129 (11.6%) sera reacted against *R. massiliae* (Podsiadly et al. 2011). These results and the detection of *R. massiliae* as cause in hitherto presumably diagnosed Mediterranean spotted fever cases imply that *R. massiliae* might be responsible for a part of those cases in the Mediterranean and possibly in other areas in the world.

15.1.3.5 Medical and Veterinary Importance

Rickettsia massiliae may cause a severe form of spotted fever in humans, including fever, eschar, a maculopapular rash including palms and soles and a mild hepatomegaly. The clinical syndrome cannot be clinically differentiated from classical Mediterranean spotted fever. The incidence of infections with *R. massiliae* in humans is not known so far, but results in clinical studies show that it may be frequent at least in some areas around the Mediterranean. An important difference to other rickettsial species is the natural resistance against rifampicin

(Rolain et al. 1998). This might be of importance as rifampicin constitutes an important antibiotic in the treatment of rickettsioses in childhood.

15.1.4 *Rickettsia raoultii*

Rickettsia raoultii was detected independently several times in different parts of the world. A strain from *Dermacentor (D.) nuttallii* from Siberia was provisionally named as strain "DnS14" (Rydkina et al. 1999). Another strain, named "RpA4", was detected in Astrakhan ticks of the species *Rh. pumilio* (Shpynov et al. 2001). Molecular studies showed that both strains were genetically closely related or almost identical (Mediannikov et al. 2007). Meanwhile, the wide geographic distribution in ticks of the genus *Dermacentor* is recognized. Strains or nucleic acid sequences belonging to the species *R. raoultii* have been detected in *D. nuttallii*, *D. pumilio*, *D. reticulatus*, *D. silvarum*, *D. niveus* and *D. marginatus* through the Eurasian continent from far eastern Russia to France, Spain, Croatia and also central Europe.

15.1.4.1 Prevalence in Ticks

Rickettsia raoultii has been shown to be identical with the former *R.* sp. *RpA4* detected in Germany so far only in *D. reticulatus* (Dautel et al. 2006; Silaghi et al. 2011). Prevalence rates were found to be 32% in *Dermacentor* ticks from Lörrach (Baden-Wuerttemberg, Pluta et al. 2010a, b) to almost 60% in Leipzig (Saxony, Silaghi et al. 2011).

15.1.4.2 Prevalence in Animals

No data are available on the importance of vertebrate animals for the maintenance or as reservoir for *R. raoultii* in nature. In an ongoing study in the metropolitan area of the city of Leipzig, Saxony, rodents (bank voles and yellow-necked field mouse) were found to harbour DNA of *R. raoultii* in their skin (own unpublished data). Whether this reflects only residual material or a transient infection in an area with up to 80% *R. raoultii*-positive *D. reticulatus* ticks or provides evidence for an active role of the rodents in the transmission cycle of *R. raoultii* warrants further investigations.

15.1.4.3 Natural Transmission Cycle

So far, no detailed information is available on the natural transmission cycle of *R. raoultii*. Naturally infected *Dermacentor* ticks, mainly *D. reticulatus*, are infected lifelong and can transmit *R. raoultii* transovarially to their offspring (Samoilenko et al. 2003). Therefore, this tick species is thought to be the main vector and probably also the primary reservoir of *R. raoultii*. Whether additional vertebrate hosts are essential for the transmission cycle or whether the natural transmission cycle runs via transovarial and transstadial transmission in the ticks has not been elucidated so far.

15.1.4.4 Seroprevalence in Humans

No data on the seroprevalence of antibodies against *R. raoultii* in humans are available so far.

15.1.4.5 Medical and Veterinary Importance

Rickettsia raoultii was found in several studies to cause a clinical syndrome like TIBOLA (tick-borne lymphadenitis) (Ibarra et al. 2006; Mediannikov et al. 2007). This syndrome is, however, known to be caused by *R. slovaca*. Nevertheless, DNA of *R. raoultii* could be detected in the blood of patients with similar symptoms. There are no data available on the veterinary importance of *R. raoultii*.

15.1.5 Rickettsia slovaca

Rickettsia slovaca was isolated from the sheep tick (*D. marginatus*) in 1968 in an area of the former Czechoslovak Republic, nowadays belonging to the Slovak Republic (Rehacek 1984). Further strains could be detected in the former Armenian Socialist Soviet Republic (Makarova 1978). Subsequently, it has been detected in different European countries, among them are countries in western Europe (France, Spain, Portugal), central Europe (Germany, Switzerland), eastern Europe (Ukraine, Slovak Republic, Hungary) and Asia (Russia) (Sekeyova et al. 1998; Parola et al. 2005). But it took more than 20 years after its first detection that its role as a human pathogen was elucidated.

15.1.5.1 Prevalence in Ticks

R. slovaca was the first rickettsial species which was detected in Germany in recent times (Rehácek et al. 1977). So far, it has been detected only in ticks of the species *D. marginatus*. The known prevalence rates of *R. slovaca* in *D. marginatus* in Germany are varying depending on the study area and the method of detection reaching from 0.9% in Karlsruhe (Baden-Wuerttemberg) to 13% in Saarland to 16% in Aschaffenburg (Bavaria) (Pluta et al. 2009, 2011; Silaghi et al. 2011). In other countries, the prevalence infection rates in ticks are ranging from 1 to 17% (Parola et al. 2009).

15.1.5.2 Prevalence in Animals

There is no information available on seroprevalence rates of antibodies against *R. slovaca* in game animals, farm animals or pets.

15.1.5.3 Natural Transmission Cycle

So far, there are only limited data available on the natural transmission cycle of *R. slovaca*. *Rickettsia slovaca* is primarily transmitted by the sheep tick (*D. marginatus*). It is assumed that this tick species constitutes the primary vector and probably also the most important reservoir for the pathogen. As the sheep and other larger farm animals constitute a major host for *D. marginatus*, it may be speculated that these animals may support the natural transmission cycle of *R. slovaca* and play a role as vertebrate hosts for this rickettsial species. However, *D. marginatus* ticks are thought to require warmer and drier habitats as, for example, *D. reticulatus* or *I. ricinus*, and thus their occurrence within Germany is restricted to areas meeting these criteria. To the best of our knowledge, *D. marginatus* was not found north of the Main river, corresponding to the 50th latitude. So we assume that a natural transmission cycle of *R. slovaca* is currently limited in southern Germany.

15.1.5.4 Seroprevalence in Humans

There is no information available on the seroprevalence of antibodies in humans against *R. slovaca*.

15.1.5.5 Medical and Veterinary Importance

In 1997, *R. slovaca* was detected in the eschar of a patient in France with fever and enlarged cervical lymph nodes (Raoult et al. 1997). The syndrome was named TIBOLA/DEBONEL (*ti*ck-*bo*rne *l*ymph*a*denitis/*De*rmacentor-*bo*rne *n*ecrosis *e*rythema and *l*ymphadenopathy). However, a similar syndrome was known already earlier from France, Bulgaria and Slovak Republic. Recently, at least two human cases could be diagnosed in Germany (Pluta et al. 2009; Rieg et al. 2011). The incubation period of TIBOLA seems to be rather short with a range from 1 to 8 days. All patients show an eschar at the site of the tick bite, predominantly at the hair transition zone of the scalp to the neck. Also 100% of patients report the occurrence of enlarged lymph nodes (Raoult et al. 2002). Fever and rash are rarely reported (5–20%). The eschar may result in a long-lasting, but reversible alopecia.

Table 15.2 Published travel-imported rickettsioses

Rickettsia species	Known geographic distribution	Human disease	Frequency in travellers	Literature
R. conorii	Mediterranean, Africa, central Asia, India	Mediterranean spotted fever	>50 cases in Germany	Jelinek and Löscher (2001)
R. africae	Africa, Caribbean islands	African tick bite fever	>50 cases in Germany	Jensenius et al. (2004), own data
R. rickettsii	North America, South America	Rocky Mountain spotted fever	1 case in Switzerland	Balestra and Nüesch (2005)
R. typhi	Worldwide	Murine (endemic) typhus	>50 cases worldwide	Jensenius et al. (2004)
R. prowazekii	Africa, South America	Louse-borne (epidemic) typhus	Sporadic cases	Niang et al. (1999) and Zanetti et al. (1998)
R. sibirica	Northern Asia	Siberian tick typhus	Sporadic cases	Jensenius et al. (2004)
R. australis	Australia	Queensland spotted fever	Sporadic cases	Jensenius et al. (2004)

15.2 Imported Spotted Fever Rickettsioses

Rickettsioses occur in almost every area worldwide. Due to the increasing travel activities, an increasing number of imported rickettsioses should be expected. However, the number of reported rickettsioses in travellers is rare, in comparison with parasitic diseases (e.g. malaria) or viral diseases (e.g. dengue fever). Only few reports and even fewer reviews on rickettsioses in travellers are available. The reason that only few rickettsioses are diagnosed in travellers may be multifactorial. First of all, rickettsioses are not well known among physicians. Also their real importance in tropical countries is not well reported. Another reason may be that diagnostic tests for rickettsioses are not readily available and therefore only rarely inquired. Thus, we assume that under-diagnosing of rickettsioses is highly likely. Nevertheless, rickettsioses are among the most frequent causes of fever in tropical countries, mainly in Africa. Between 5% and 10% of all human diseases in humans are estimated to be caused by rickettsiae (Cleaveland 2011).

The most frequently reported rickettsioses in travellers from Germany are the African tick bite fever (*R. africae*) and Mediterranean spotted fever (*R. conorii*) (Jelinek and Löscher 2001). There are several other rickettsioses which have been sporadically reported in travellers, among them are the cases of epidemic typhus in health-care workers (Niang et al. 1999; Zanetti et al. 1998) or Queensland typhus or Siberian tick typhus (Jensenius et al. 2004). Also, sporadic cases of endemic typhus (murine typhus), one of the few rickettsioses occurring worldwide, are diagnosed and reported occasionally (Jensenius et al. 2004) (Table 15.2).

In a retrospective study on rickettsioses in German travellers returning with fever, 8/163 (5%) patients showed serological evidence of a confirmed (fourfold IgG titre increase) or probable (single IgG titre of >1:256) rickettsiosis of the spotted fever group. In the same study population, 2/163 (1.2%) patients showed evidence of an acute infection of endemic typhus (Dobler and Wölfel 2009). These unpublished data show that rickettsioses are frequently found in patients with fever. African tick bite fever may be among the most frequent rickettsioses acquired in African travellers. Studies show that up to 100% of *Amblyomma (A.) hebraeum* ticks in Tanzania were positive for *R. africae* (Dobler and Wölfel 2009). *A. hebraeum* is also found frequently in the national parks of various countries in southern Africa. African tick bite fever may be frequently diagnosed in travellers returning from the Kruger National Park in South Africa (Tappe et al. 2011). Mediterranean spotted fever is another rickettsiosis which may be more frequently found in travellers returning from countries around the Mediterranean Sea. The etiologic agent, *R. conorii*, is transmitted by the brown dog tick *Rh. sanguineus*. This tick species is mainly parasitizing dogs; however, it is occasionally also found on humans, especially after intense contact between humans and dogs. Exanthematous febrile illness in travellers may always induce the suspicion of a rickettsiosis and then should be confirmed or excluded.

References

Babalis T, Tselenis Y, Roux V, Psaroulaki A, Raoult D (1994) Isolation and characterization of a rickettsial strain related to *Rickettsia massiliae* in Greek ticks. Am J Trop Med Hyg 50:365–372

Bacellar F, Regnery RL, Nuncio MS, Filipe AR (1995) Genotypic evaluation of rickettsial isolates recovered of various species of ticks in Portugal. Epidemiol Infect 114:169–176

Balestra GM, Nüesch R (2005) Eine Urlaubsreise-Erinnerung: Rocky Mountain Fleckfieber. Schweiz Rundsch Med Prax 94:1869–1870

Beati L, Raoult D (1993) *Rickettsia massiliae* sp. nov., a new spotted fever group rickettsia. Int J Syst Bacteriol 43:839–840

Beati L, Roux V, Ortuño A, Castella J, Porta FS, Raoult D (1996) Phenotypic and genotypic characterization of spotted fever group rickettsiae isolated from Catalan *Rhipicephalus sanguineus* ticks. J Clin Microbiol 34:2688–2694

Bernasconi MV, Casati S, Péter O, Piffaretti JC (2002) *Rhipicephalus* ticks infected with Rickettsia and Coxiella in Southern Switzerland (Canton Ticino). Infect Genet Evol 2:111–120

Cardenosa N, Segura F, Raoult D (2003) Serosurvey among Mediterranean Spotted fever patients of a new spotted fever group rickettsial strain (Bar29). Eur J Epidemiol 18:351–356

Christova I, Van De Pol J, Yazar S, Velo E, Schouls L (2003) Identification of Borrelia burgdorferi sensu lato, Anaplasma and Ehrlichia species, and spotted fever group Rickettsiae in ticks from southeastern Europe. Eur J Clin Microbiol Infect Dis 22:535–542

Cleaveland S (2011) Invasive bacterial infections and bacterial zoonoses in sub-Saharan Africa. In: 60th Annual meeting of the American Society of Tropical Medicine and Hygiene. Philadelphia, 4–8 Dec 2011

Dautel H, Dippel C, Oehme R, Hartelt K, Schettler E (2006) Evidence of an increased geographical distribution of *Dermacentor reticulatus* in Germany and detection of *Rickettsia sp. RpA4*. Int J Med Microbiol 296(Suppl 1):149–156

de Sousa R, Lopes de Carvalho I, Bernardes C, Milhano N, Santos AS, Nuncio MS (2010) Lizards (*Teira dugesii*) as a potential reservoir of *Rickettsia monacensis* in Madeira Island, Portugal. In: 24th Annual meeting of the American Society for Rickettsiology, Stephenson, 31 July–3 Aug 2010; Abstract #77

Derdáková M, Tarageľová V, Selyemová D, Špitalská E, Štefančíková A, Stanko M, Peťko B (2011) Circulation and genetic variability of Anaplasma spp. and other tick-borne intracellular microorganisms in the natural foci of Slovakia. In: XI International Jena Symposium on Tick-borne diseases, Weimar, 24–26 Mar 2011

Dobler G, Wölfel R (2009) Typhus and other rickettsioses: emerging infections in Germany. Dtsch Ärztebl Int 106:348–354

Dobler G, Essbauer S, Wölfel R (2009) Isolation and preliminary characterization of "*Rickettsia monacensis*" in South-Eastern Germany. Clin Microbiol Infect 15(S2):263–264

Dumler JS, Barbet AF, Bekker CP, Dasch GA, Palmer GH, Ray SC, Rikihisa Y, Rurangirwa FR (2001) Reorganization of genera in the families *Rickettsiaceae* and *Anaplasmataceae* in the order *Rickettsiales*: unification of some species of *Ehrlichia* with *Anaplasma*, *Cowdria* with *Ehrlichia* and *Ehrlichia* with *Neorickettsia*, descriptions of six new species combinations and designation of *Ehrlichia equi* and "HGE agent" as subjective synonyms of *Ehrlichia phagocytophila*. Int J Syst Evol Microbiol 51:2145–2165

Eremeeva ME, Bosserman EA, Demma LJ, Zambrano ML, Blau DM, Dasch GA (2006) Isolation and identification of *Rickettsia massiliae* from *Rhipicephalus sanguineus* ticks collected in Arizona. Appl Environ Microbiol 72:5569–5577

Fernandez-Soto P, Perez-Sanchez R, Diaz Martin V, Encinas-Grandes A, Alamo Sanz R (2006) *Rickettsia massiliae* in ticks removed from humans in Castilla y Leon, Spain. Eur J Clin Microbiol Infect Dis 25:811–813

Fournier PE, Grunnenberger F, Jaulhac B, Gastinger G, Raoult D (2000) Evidence of *Rickettsia helvetica* infection in humans, eastern France. Emerg Infect Dis 6:389–392

García-García JC, Portillo A, Núñez MJ, Santibáñez S, Castro B, Oteo JA (2010) A patient from Argentina infected with *Rickettsia massiliae*. Am J Trop Med Hyg 82:691–692

Gilles J, Just FT, Silaghi C, Pradel I, Passos LMF, Lengauer H, Hellmann K, Pfister K (2008) *Rickettsia felis* in fleas, Germany. Emerg Infect Dis 14:1294–1296

Hartelt K, Oehme R, Hernning F, Brockmann SO, Hassler D, Kimmig P (2004) Pathogens and symbionts in ticks: prevalence of *Anaplasma phagocytophilum* (*Ehrlichia* sp.), *Wolbachia sp.*, *Rickettsia sp.* and *Babesia spp.* in Southern Germany. Int J Med Microbiol 293(suppl 37):86–92

Hildebrandt A, Krämer A, Sachse S, Straube E (2010) Detection of *Rickettsia spp.* and *Anaplasma phagocytophilum* in *Ixodes ricinus* in a region of middle Germany (Thuringia). Ticks Tick Borne Dis 1:52–56

Hildebrandt A, Straube E, Neubauer H, Schmoock G (2011) *Coxiella burnetii* and coinfections in *Ixodes ricinus* ticks in central Germany. Vector Borne Zoonotic Dis 11:1205–1207

Ibarra.V, Oteo J.A, Portillo .A, Santibáňez. S, Blanco J.R, Metola. L, Eiros J.M, Perez-Martínez . L, Sanz. M (2006) *Rickettsia slovaca* Infection: DEBONEL/TIBOLA. Ann N Y Acad Sci 1078:206–214

Jado I, Oteo JA, Aldamiz M, Gil H, Escudero R, Ibara V, Portu J, Portillo A, Lezaun MJ, Garcia-Amil C, Rodrigo-Moreno I, Anda P (2007) *Rickettsia monacensis* and human disease, Spain. Emerg Infect Dis 13:1405–1407

Jelinek T, Löscher T (2001) Clinical features and epidemiology of tick typhus in travelers. J Travel Med 8:57–59

Jensenius M, Fournier P-E, Raoult D (2004) Rickettsioses in travellers. Clin Infect Dis 39:1493–1499

Makarova VA (1978) Antigenic structure of spotted fever group rickettsiae isolated in Armenia SSR. Zh Microbiol Epidemiol Immunol 4:112–116

Mediannikov O, Matsumoto K, Samoylenko I, Drancourt M, Roux V, Rydkina E, Davoust B, Tarasevich I, Brouqui P, Fournier P (2007) *Rickettsia raoultii* sp. nov., a new pathogenic spotted fever group rickettsia associated with tick-borne lymphadenitis. Int J Syst Evol Microbiol 58:1635–1639

Niang M, Brouqui P, Raoult D (1999) Epidemic typhus imported from Algeria. Emerg Infect Dis 5:716–718

Nilsson K, Lindquist O, Pahlson C (1999) Association of *Rickettsia helvetica* with chronic perimyocarditis in sudden cardiac death. Lancet 354:1169–1173

Nilsson K, Elfvig K, Pahlson C (2010) *Rickettsia helvetica* in patient with meningitis, Sweden. Emerg Infect Dis 16:490–492

Parola P, Paddock CD, Raoult D (2005) Tick-borne rickettsioses around the world: emerging diseases challenging old concepts. Clin Microbiol Rev 18:719–756

Parola P, Rovery C, Rolain JM, Brouqui P, Davoust B, Raoult D (2009) *Rickettsia slovaca* and *R. raoultii* in tick-borne rickettsioses. Emerg Infect Dis 15:1105–1108

Pichon B, Kahl O, Hammer B, Gray JS (2006) Pathogens and host DNA in *Ixodes ricinus* nymphal ticks from a German forest. Vector Borne Zoonotic Dis 6:382–387

Pluta S, Tewald F, Hartelt K, Oehme R, Kimmig P, Mackenstedt U (2009) *Rickettsia slovaca* in *Dermacentor marginatus* ticks, Germany. Emerg Infect Dis 15:2077–2078

Pluta S, Hartelt K, Oehme R, Mackenstedt U, Kimmig P (2010a) Prevalence of *Coxiella burnetii* and *Rickettsia spp.* in ticks and rodents in southern Germany. Ticks Tick Borne Dis 1:145–147

Pluta S, Hartelt K, Oehme R, Kimmig P, Mackenstedt U (2010b) What else besides TBE and Borreliosis? Tick-transmitted pathogens in Germany and beyond. Parasitol Res Monogr 2:297–323

Podsiadly E, Chmielewski T, Karbowiak G, Kedra E, Tylewska-Wierzbanowska S (2011) The occurrence of spotted fever rickettsioses and other tick-borne infections in forest workers in Poland. Emerg Infect Dis 11:985–989

Raoult D, Berbis P, Roux V, Xu W, Maurin M (1997) A new tick-transmitted disease due to *Rickettsia slovaca*. Lancet 350:112–113

Raoult D, Lakos A, Fenollar F, Beytout J, Brouqui P, Fournier PE (2002) Spotless rickettsiosis caused by *Rickettsia slovaca* and associated with *Dermacentor* ticks. Clin Infect Dis 34:1331–1336

Raoult D, Woodward T, Dumler JS (2004) The history of epidemic typhus. Infect Dis Clin North Am 18:127–140

Rehacek J (1984) *Rickettsia slovaca*, the organism and its ecology. Acta Sci Nat Brno 18:1–50

Rehácek J, Liebisch A, Urvölgyi J, Kovácová E (1977) Rickettsiae of the spotted fever isolated from *Dermacentor marginatus* ticks in South Germany. Zentralbl Bakteriol (Orig A) 239:275–281

Richter J, Fournier PE, Petridou J, Häussinger D, Raoult D (2002) *Rickettsia felis* infection acquired in Europe and documented by polymerase chain reaction. Emerg Infect Dis 8:207–208

Rickets HT (1906) The transmission of Rocky Mountain spotted fever by the bite of the wood tick (Dermacentor occidentalis). J Am Med Assoc 47:358

Rieg S, Schmoldt S, Theilacker C, de With K, Wölfel S, Kern WV, Dobler G (2011) Tick-borne lymphadenopathy (TIBOLA) acquired in Southwestern Germany. BMC Infect Dis 11:167

Rolain JM, Maurin M, Vestris G, Raoult D (1998) In vitro susceptibilities of 27 rickettsiae to 13 antimicrobials. Antimicrob Agents Chemother 42:1537–1541

Rydkina E, Roux V, Fetisova N, Rudakov N, Gafarova M, Tarasevich I, Raoult D (1999) New rickettsiae in ticks collected in territories of the former Soviet Union. Emerg Infect Dis 5:811–814

Samoilenko IE, Rudakov N, Shpynov SN, Tankibaev MA, Yakimenko VV, Kumpan LV (2003) Study of biological characteristics of spotted fever group rickettsial genotypes RpA4, DnS14 and DnS28. Ann N Y Acad Sci 990:612–626

Santos-Silva MM, Sousa R, Santos AS, Melo P, Encarnacao V, Bacellar F (2006) Ticks parasitizing wild birds in Portugal: detection of *Rickettsia aeschlimannii*, *R. helvetica* and *R. massiliae*. Exp Appl Acarol 39:331–338

Schorn S, Pfister K, Reulen H, Mahling M, Silaghi C (2011) Occurrence of *Babesia spp.*, *Rickettsia spp.* and *Bartonella spp.* in *Ixodes ricinus* in Bavarian public parks, Germany. Parasit Vectors 4:135

Sekeyova Z, Roux V, Xu W, Rehácek J, Raoult D (1998) *Rickettsia slovaca* sp. nov., a member of the spotted fever group rickettsiae. Int J Syst Bacteriol 48:1455–1468

Sekeyova Z, Fournier PE, Rehacek J, Raoult D (2000) Characterization of a new spotted fever group rickettsia detected in *Ixodes ricinus* (Acari: Ixodidae) collected in Slovakia. J Med Entomol 37:707–713

Shpynov S, Parola P, Rudakov N, Samoilenko I, Tankibaev M, Tarasevich I (2001) Detection and identification of spotted fever group rickettsiae in *Dermacentor* ticks from Russia and central Kazakhstan. Eur J Clin Microbiol Infect Dis 20:903–905

Silaghi C, Hamel D, Thiel C, Pfister K, Pfeffer M (2011) Spotted-fever group rickettsia in ticks, Germany. Emerg Infect Dis 17:890–892

Simser JA, Palmer AT, Fingerle V, Wilske B, Kurtti TJ, Munderloh UG (2002) *Rickettsia monacensis sp. nov.*, a spotted fever group *Rickettsia*, from ticks (*Ixodes ricinus*) collected in a European city park. Appl Environ Microbiol 68:4559–4566

Sreter-Lancz Z, Sreter T, Szell Z, Egyed L (2005) Molecular evidence of *Rickettsia helvetica* and *Rickettsia monacensis* infections in *Ixodes ricinus* in Hungary. Ann Trop Med Parasitol 99:325–330

Stephenson CS (1944) Epidemic typhus fever and other rickettsial diseases of military importance. N Engl J Med 231:407–413

Tappe A, Dobler G, Stich A (2011) Tache noire in African tick bite fever. Am J Trop Med Hyg 81:733–734

Vitale G, Mansueto S, Rolain JM, Raoult D (2006) *Rickettsia massiliae* human isolation. Emerg Infect Dis 12:174–175

Wölfel R, Terzioglu R, Kiessling J, Wilhelm S, Essbauer S, Pfeffer M, Dobler G (2006) *Rickettsia spp.* in *Ixodes ricinus* ticks in Bavaria, Germany. Ann N Y Acad Sci 1078:509–511

Wölfel R, Essbauer S, Dobler G (2008) Diagnostics of tick-borne rickettsioses in Germany: a modern concepts for a neglected disease. Int J Med Microbiol 298(S1):368–374

Zanetti G, Francioli P, Tagan D, Paddock CD, Zaki SR (1998) Imported epidemic typhus. Lancet 352:1709

Index

H. Mehlhorn (ed.), *Arthropods as Vectors of Emerging Diseases*,
Parasitology Research Monographs 3, DOI 10.1007/978-3-642-28842-5,

D

E

U

V

W

X

Y

Z